AF606523

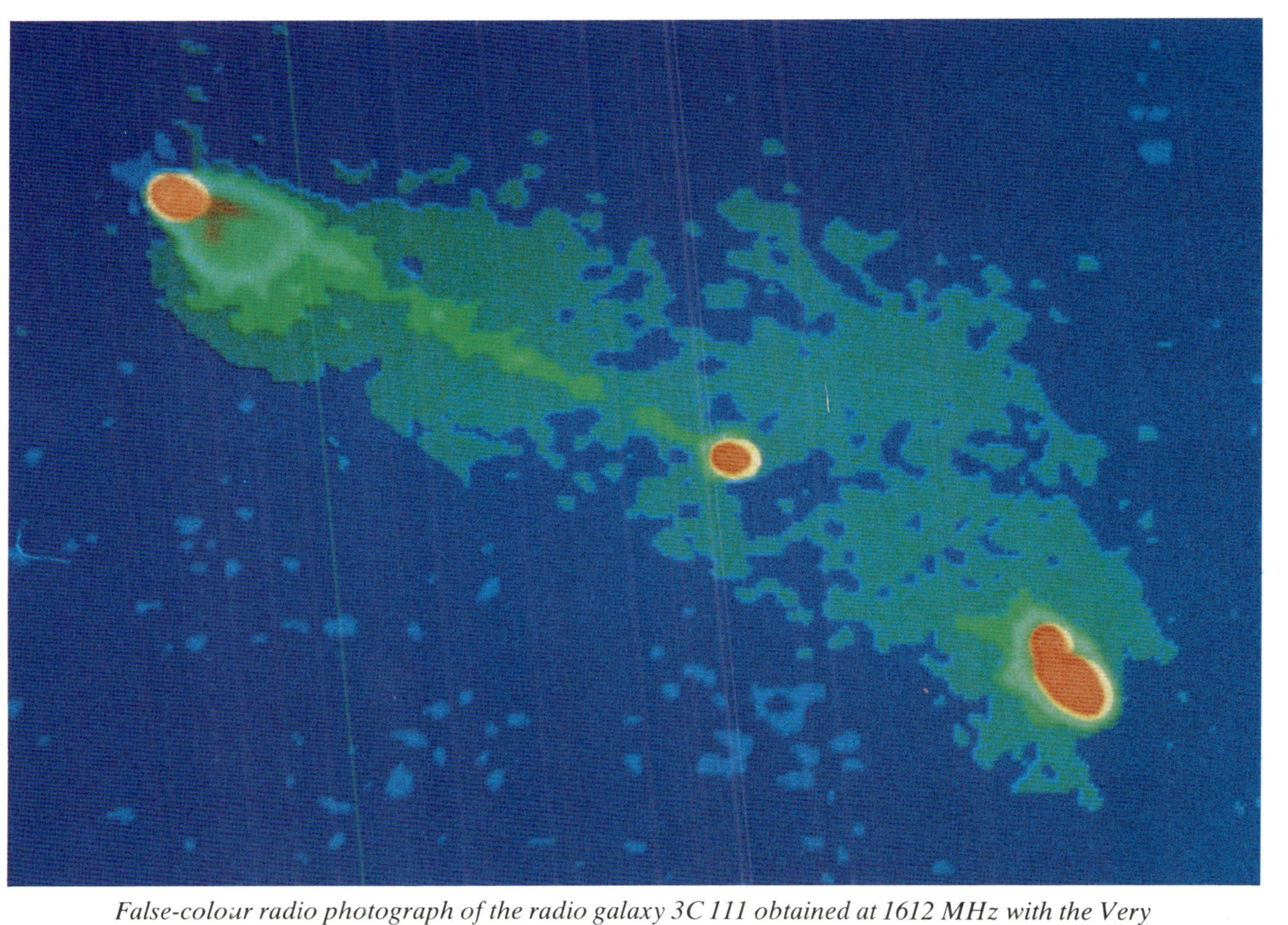

False-colour radio photograph of the radio galaxy 3C 111 obtained at 1612 MHz with the Very Large Array of the National Radio Astronomy Observatory at Socorro, New Mexico, USA.

THE MICROWAVE ENGINEERING HANDBOOK

VOLUME 3

Microwave Technology Series

The *Microwave Technology Series* publishes authoritative works for professional engineers, researchers and advanced students across the entire range of microwave devices, sub-systems, systems and applications. The series aims to meet the reader's needs for relevant information useful in practical applications. Engineers involved in microwave devices and circuits, antennas, broadcasting communications, radar, infra-red and avionics will find the series an invaluable source of design and reference information.

Series editors:
Michel-Henri Carpentier
Professor in 'Grandes Écoles', France,
Fellow of the IEEE, and President of the French SEE

Bradford L. Smith
International Patents Consultant and Engineer
with the Alcatel group in Paris, France,
and a Senior Member of the IEEE and French SEE

Titles available

1. **The Microwave Engineering Handbook Volume 1**
   Microwave components
   Edited by Bradford L. Smith and Michel-Henri Carpentier

2. **The Microwave Engineering Handbook Volume 2**
   Microwave circuits, antennas and propagation
   Edited by Bradford L. Smith and Michel-Henri Carpentier

3. **The Microwave Engineering Handbook Volume 3**
   Microwave systems and applications
   Edited by Bradford L. Smith and Michel-Henri Carpentier

4. **Solid-state Microwave Generation**
   J. Anastassiades, D. Kaminsky, E. Perea and A. Poezevara

5. **Infrared Thermography**
   C. Gaussorgues
   Translated by D. Häusermann and S. Chomet

THE MICROWAVE ENGINEERING HANDBOOK

VOLUME 3

Microwave systems and applications

Edited by

Bradford L. Smith

International Patents Consultant and Engineer
with the Alcatel group in Paris, France, and
Senior Member of the IEEE and the French SEE

and

Michel-Henri Carpentier

Professor in the 'Grandes Ecoles', France,
Fellow of the IEEE and President of the French SEE

CHAPMAN & HALL
London · Glasgow · New York · Tokyo · Melbourne · Madras

Published by Chapman & Hall, 2–6 Boundary Row, London SE1 8HN

Chapman & Hall, 2–6 Boundary Row, London SE1 8HN, UK

Blackie Academic & Professional, Wester Cleddens Road, Bishopbriggs, Glasgow G64 2NZ, UK

Van Nostrand Reinhold Inc., 115 5th Avenue, New York NY10003, USA

Chapman & Hall Japan, Thomson Publishing Japan, Hirakawacho Nemoto Building, 6F, 1-7-11 Hirakawa-cho, Chiyoda-ku, Tokyo 102, Japan

Chapman & Hall Australia, Thomas Nelson Australia, 102 Dodds Street, South Melbourne, Victoria 3205, Australia

Chapman & Hall India, R. Seshadri, 32 Second Main Road, CIT East, Madras 600035, India

---

First edition 1993

Typeset in 10/12 pt Times by Thomson Press (India) Ltd, New Delhi
Printed in Great Britain by The University Press, Cambridge

ISBN 0 412 45680 X   0 442 31657 7 (USA)

A catalogue record for this book is available from the British Library

Library of Congress Cataloging-in-Publication data available

Contents

Contributors **xi**

1 Point-to-point transmission, terrestrial line-of-sight links, terrestrial troposcatter links **1**
Philippe Magne, Jean Iltis, Claude Brémenson, Philippe Legendre

1.1 General 1
1.2 Principles of radio relays 2
1.2.1 Repeater functions 2
1.2.2 Carrier frequency utilization 4
1.2.3 Different types of repeaters 6
1.2.4 Different types of terminals 7
1.2.5 Special features of antennas for radio relay links 7
1.2.6 Feeders 10
1.3 Analogue microwave links 10
1.3.1 General 10
1.3.2 Characteristics of the signals transmitted 11
1.3.3 Analogue modulation 12
1.3.4 Technological aspects 16
1.3.5 Operating aid facilities 26
1.3.6 Frequency modulation distortions 27
1.3.7 Performances of analogue microwave links 28
1.3.8 Quality improvement 40
1.3.9 Transmission quality 45
1.4 Digital microwave links 49
1.4.1 Characteristics of the signals transmitted 49
1.4.2 Digital modulation 61
1.4.3 Technological aspects 75
1.4.4 Other specific features of digital microwave links 80
1.4.5 Performances of digital microwave links 85
1.4.6 Techniques used for improving quality 98
1.4.7 Predicting outages due to propagation 106

1.5 Specific nature of over-the-horizon microwave links 109
1.5.1 Properties of the propagation medium 109
1.5.2 Equipment characteristics 112
1.5.3 Transmission quality 122
References 122

2 Satellite links **125**
Jean Salomon
2.1 General 125
2.1.1 Introduction 125
2.1.2 Communication satellite systems 125
2.1.3 Utilization of the radio frequency spectrum 126
2.1.4 Specific characteristics 127
2.1.5 Main applications and techniques 132
2.1.6 Historical overview 133
2.1.7 Existing satellite systems 135
2.2 Basic principles 137
2.2.1 The basic satellite communication link 137
2.2.2 Definitions and formulae 139
2.2.3 Other topics 146
2.2.4 The link budget 151
2.2.5 Link quality 158
2.3 Communication satellite technology 167
2.3.1 Communication satellite construction 167
2.3.2 Communication satellite payload 172
2.4 Earth stations 191
2.4.1 General 191
2.4.2 The antenna system 198
2.4.3 The low noise amplifier (LNA) 212
2.4.4 Measurements of noise temperatures and antenna G/T 215
2.4.5 The high power amplifier (HPA) 217
2.4.6 The up- and down-converters (U/C, D/C) 222
2.5 Conclusions and prospects 224
References 227

3 Low and medium power translators and transmitters **229**
Claude Cluniat
3.1 Optimization of the input design of a television translator 229
3.1.1 Brief review: what are the design possibilities? 229
3.1.2 Optimization of antinomic couple noise factor/input stage linearity 231
3.2 Development of television transmitter modulation stages 240
3.2.1 Vision and sound IF modulation 240
3.2.2 IF vision corrector 242

3.2.3 Non-linearity correction of the vision channel 243
3.2.4 Output wideband converter 246
3.3 Optimization of the output design of a television transmitter or translator and enhancing the level of transistorization 246
3.3.1 Analysis of the distortions generated in a power amplifier 246
3.3.2 Non-linearity correctors 249
3.3.3 Amplifier assemblies 251
3.3.4 Improvement of reliability 252
3.3.5 Overall characteristics of transmitters 253
3.4 Conclusion 253
Appendix 3.A The frequency spectrum and broadcasting channels 257

4 Radar systems **267**
Michel-Henri Carpentier
4.1 The history of radar 267
4.1.1 Before 1935 267
4.1.2 Since 1935; the pulse radar 268
4.1.3 The angular measurement 269
4.1.4 Pulse compression and coded radars 270
4.1.5 Doppler filtering 271
4.1.6 Electronic scanning 272
4.2 General description of radar systems 273
4.2.1 Basic principles derived from the theory of radar systems 273
4.2.2 About the parasitic noise 280
4.2.3 Radar block diagram 282
4.2.4 About antennas 286
4.2.5 About transmitters 287
4.2.6 About receivers 287
4.2.7 Choice of wavelengths—basic examples of radar parameters 292
4.2.8 Radar cross-section—target fluctuation—stealth targets 294
4.2.9 Problems: analysis of a multifunction radar 296
4.2.10 Pulse compression 304
4.2.11 About digital processing 313
4.2.12 Action against clutter 315
4.2.13 Pulse-Doppler radars 315
4.3 Main applications of radar systems 323
4.3.1 Surveillance radars 323
4.3.2 Fire control radar systems 325
4.3.3 Radar systems on board aircraft 325
4.3.4 Instrumentation radars 326

4.3.5 Other applications 326
4.4 Expected evolution of radar systems 326
4.4.1 Multifunction and multimode in radar systems 326
4.4.2 Present and future implementation of ancient ideas—active antennas 327
4.4.3 High resolution in distance 329
4.4.4 New wavelengths 330
References 330

5 Electronic confrontation 331
François Naville
5.1 Introduction 331
5.2 Electronic support measures (ESM) 333
5.2.1 General 333
5.2.2 Reception techniques 334
5.2.3 Direction-finding techniques 338
5.2.4 Location measurements 341
5.2.5 Evolution of the systems 342
5.3 Electronic countermeasures (ECM) 342
5.3.1 Introduction 342
5.3.2 Main operational uses of jammers 342
5.3.3 Jamming techniques 343
5.3.4 Main effects of jammers 346
5.3.5 Evolution of jamming facilities 347
5.4 ECCM applied to radio frequency links 350
5.4.1 General 350
5.4.2 Jamming protection techniques 350
5.4.3 Signal interception protection techniques 354
5.4.4 Conclusion 355
5.5 ECCM applied to radars 355
5.5.1 General 355
5.5.2 Radar range in the presence of jamming 356
5.5.3 General principles used against jamming 357
5.5.4 Main ECCM techniques 357
5.6 System design methodology 371
5.6.1 New tools to be incorporated into operational systems 371
5.6.2 Strategic intelligence: an in-depth analysis 371
5.6.3 Tactical intelligence: quick analysis of situations and priorities 372
5.6.4 Self-protection of weapon systems: a highly complex function 372
5.6.5 Air strike: active support at several levels 373
5.6.6 Counter-mobility: neutralization by jamming 373
5.6.7 Radioelectric superiority: preventing the use of the spectrum by the enemy 374

5.6.8 Elimination of the enemy anti-aircraft defence: extensive use of ESM and ECM 374

6 Infrared 377
Jean Dansac, Yves Cojan and Jean Louis Meyzonnette
6.1 Introduction 377
6.1.1 General definition 377
6.1.2 Spectral bands 378
6.1.3 Infrared system classification 380
6.2 Short historical background 382
6.3 Theory notes 386
6.3.1 Optical quantities and relationships 386
6.3.2 Photometry and radiometry 392
6.3.3 Atmosphere 395
6.3.4 Sources 395
6.3.5 Optical materials 408
6.3.6 Detectors 408
6.4 Infrared techniques 415
6.4.1 Instrument design considerations for passive IR detection 415
6.4.2 Performances of passive infrared optronic systems 422
6.4.3 Laser detection techniques 428
6.4.4 IR laser system performance 433
6.5 Military applications of infrared 444
6.5.1 Military applications of passive infrared 445
6.5.2 Military applications of active infrared systems 453
6.5.3 Military applications of semi-active infrared 462
6.5.4 Military applications of point-to-point links 465
6.6 Developments and trends in the infrared field 466
6.6.1 Optical windows and IR domes 466
6.6.2 Stabilization and scanning 466
6.6.3 Optical systems 467
6.6.4 Detectors 467
6.6.5 Cooling devices 468
6.6.6 Laser emitters 468
6.6.7 Processing devices 468
6.6.8 Display 469

7 Industrial, scientific and medical (ISM) applications of microwaves present and prospective 471
Bernard Epsztein, Yves Leroy, J. Vindevoghel and Eugene Constant
7.1 Introduction 471
7.2 High power applications 474
7.2.1 Microwave heating 475
7.2.2 High energy scientific applications 481

7.2.3 The problems of leakage: the personnel exposure standards 493
7.2.4 Conclusion 495
7.3 Active sensors and systems 495
7.3.1 Radar type sensors and miscellaneous 495
7.3.2 Non-destructive control 498
7.3.3 Microwave active imaging 500
7.3.4 Conclusion 501
7.4 Passive sensors and systems 501
7.4.1 Principles 501
7.4.2 Radiometric receivers 504
7.4.3 Applications of radiometry 505
7.4.4 Conclusion 507
7.5 Conclusion 507
References 507

8 Radioastronomy **511**
Nguyen-Quang Rieu
8.1 Introduction 511
8.2 Radio telescopes 512
8.2.1 Single dish 512
8.2.2 Interferometry and aperture synthesis 514
8.3 Cosmic radio emission 520
8.3.1 Continuum emission 521
8.3.2 Line emission 522
8.4 Continuum radio sources 524
8.4.1 The Galaxy 524
8.4.2 Extragalactic radio sources 528
8.5 The 21 cm hydrogen line 530
8.6 Interstellar molecules 531
8.6.1 The discovery 531
8.6.2 Astrochemistry 534
8.6.3 Cosmic maser amplification 536
8.6.4 The concept of a two-level maser 538
8.7 Conclusion and prospects 540
Appendix 8.A Units and constants in astronomy 541
References 542
Problems 542

Index **545**

Contributors

Claude Brémenson
Senior Engineer (retired)
Alcatel-Thomson
Courbevoie
France

Michel-Henri Carpentier
Vice-President, Scientific and Technical Director (retired)
Thomson-CSF
Paris
France
presently Professor in 'Grandes Écoles', Fellow of the IEEE and President of the French SEE

Claude Cluniat
Scientific Director
Thomson-CSF/LGT
Conflans
France

Yves Cojan
Senior Engineer
Thomson TRT Défense
Guyancourt
France

Eugene Constant
Professor
Université de Lille
France

Jean Dansac
Scientific Director (retired)
Thomson TRT Défense
Guyancourt
France

Bernard Epsztein
Scientific Director
Microwave Tubes Division
Thomson-CSF
Velizy
France

Jean Iltis
Senior Engineer (retired)
Alcatel-Thomson
Courbevoie
France

Philippe Legendre
Senior Engineer and Patents Officer
Alcatel Telspace
Nanterre
France

Yves Leroy
Professor
Université de Lille
France

Philippe Magne
Scientific Director (retired)
Alcatel-Telspace
Courbevoie
France

Jean-Louis Meyzonette
Professor
Institut d'Optique
Orsay
France

François Naville
Director of the Test Flight Base and Engineer of the Armed Forces
Délégation Générale de l'Armement
Istres
France

Nguyen-Quang Rieu
Researcher
Observatoire de Paris
Meudon
France

Jean Salomon
Telspace Consultant and French Delegate, CCIR
Revil-Malmaison
France

J. Vindevoghel
Professor
Université de Lille
France

1

Point-to-point transmissions, terrestrial line-of-sight links, terrestrial troposcatter links

Philippe Magne, Jean Iltis, Claude Brémenson and Philippe Legendre

1.1 GENERAL

This chapter will deal with microwave links (radio relay links), the technology of which is founded on the possibilities offered by the very high frequencies, 1 GHz $< f <$ 100 GHz, to which correspond certain very short wavelengths called microwaves, 30 cm $> \lambda >$ 3 mm. These two aspects make it possible to obtain both wide passbands and a very directive radiation, particularly suitable for transmitting large quantities of signals or data from point to point.

Three types of propagation are used.

1. Line-of-sight: to avoid the problem of the earth's roundness, repeaters (receivers and transmitters) are installed at approximately every 50 km. A series of radio hops is called a radio link (Fig. 1.1(a)). As seen in Volume 2, the propagation is fairly stable.
2. Troposcatter: in order to reach a radio hop extremity which would be beyond the optical horizon (Fig. 1.1(b)). The reception is weak and fluctuating (Volume 2).
3. Line-of-sight with an artificial satellite: placed in a geostationary orbit at an altitude of 36 000 km (Fig. 1.1(c)) and which contains an active repeater and antennas oriented toward the earth (see chapter 2).

Three types of access are necessary. The distinction between the types of access will appear further on in this chapter.

1. Accesses in radio frequency (RF) are, in particular, those of the antennas and of the link feeders with the transmitters and the receivers.
2. Accesses in intermediate frequency (IF) are those of the low level amplication and the modulator and demodulators for which the signals are injected and extracted at the use points.

3. Accesses in baseband (BB), defined as the frequency band occupied by all the signals transmitted and which, in a modulation process, is used in the first place, to modulate a carrier. Consequently, this is also the band injected or introduced at the access of the modulation–demodulation.

Two types of signal are transmitted by microwaves: analogue signals, and digital signals.

A microwave link is called analogue when the signals carried are of analogue type. These signals are used with multichannel telephony by frequency division multiplex (FDM), and colour television transmission. In an analogue microwave link, the noises from the various hops accumulate power. As the signal goes through a number of repeaters (approximately five) without being demodulated, the distortions brought in by the various hops are summed, according to their type, in power or sometimes in voltage, and the transmission quality must be defined for a modulation–demodulation section.

A microwave link is called digital when the signals transmitted are digital. These are signals from the time division multiplex, carrying telephone, television and other data signals. In a digital microwave link, we almost always regenerate the signal at each hop. It is the error rates of the various hops which are summed, and the transmission quality can be defined hop by hop.

Although the details of the performance are expressed in very different ways (noise and error rates for example), the performance seen from the users' point of view (link availability, intelligibility of a telephone conversation) are the same.

The capacity per RF channel goes up to 3600 voice channels or 560 Mbit/s for colour television programs (pictures: 525 or 625 lines plus four audio channels). Several RF channels (up to eight) can follow the same route and be radiated by the same antenna.

1.2 PRINCIPLES OF RADIO RELAYS

1.2.1 Repeater functions

Line-of-sight propagation requires that the message be repeated by a receiver associated with a transmitter (Fig. 1.1(a)). (It should be noted that this is the same

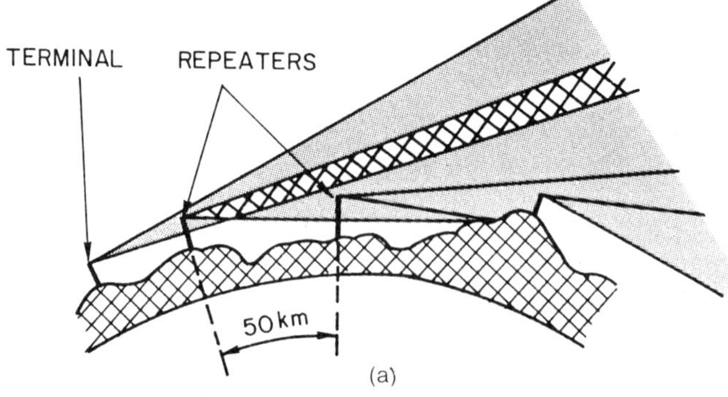

(a)

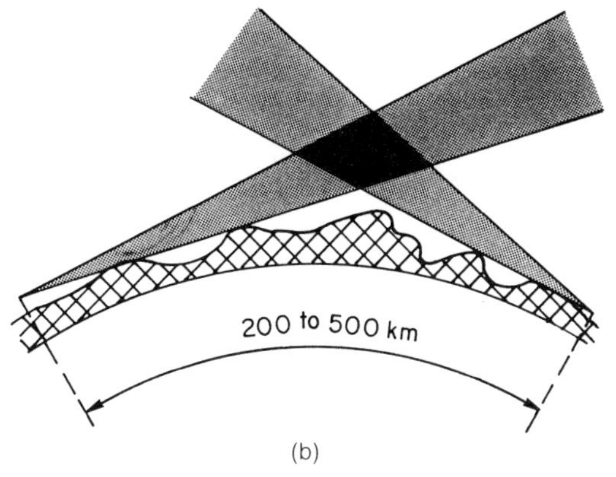

(b)

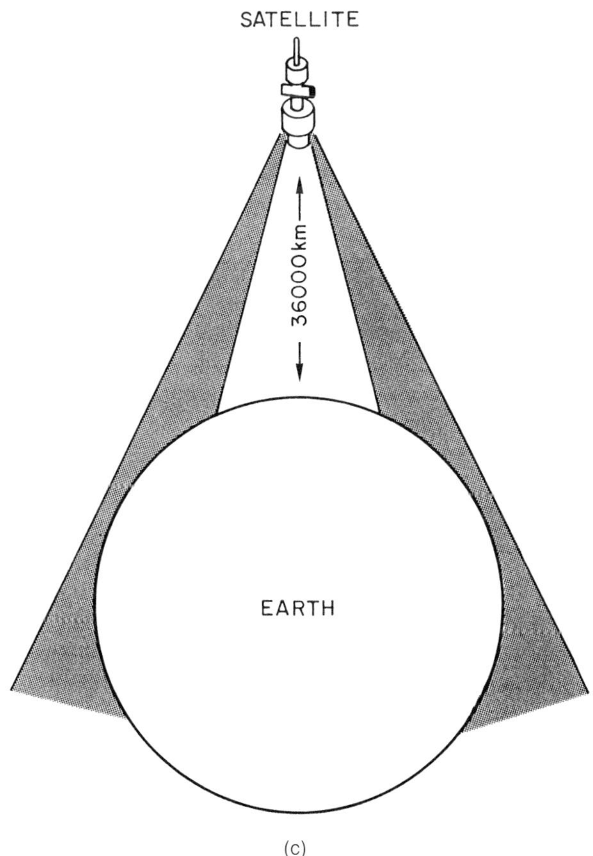

(c)

Fig. 1.1 Types of propagation: (a) line-of-sight; (b) forward scatter; (c) geostationary satellite.

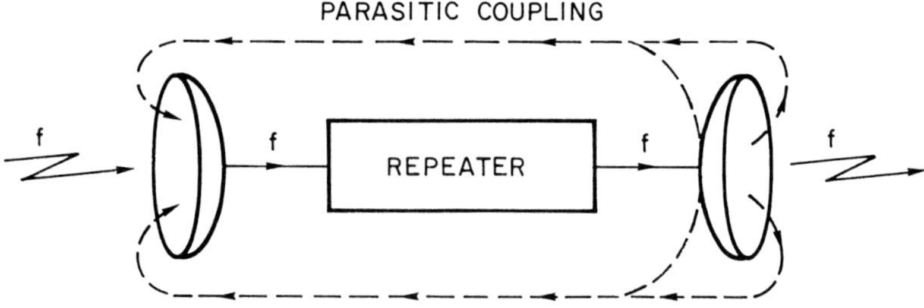

Fig. 1.2 Parasitic coupling between transmit and receive channels.

for a transmission beyond the horizon). So the first function of a repeater is to amplify the weak signal received in order to restore the transmitting level, in other words, to compensate at each instant for the propagation attenuation. One difficulty arises because of this, and that is the necessity of continually sending and receiving at a single location.

Without any special precautions, the receiver input would be powered by a small part of the transmitter (Fig. 1.2) and the complete repeater (receiver and transmitter) would oscillate all by itself in spite of the absence of the input signal. This difficulty is avoided by a translation of the carrier frequency. In this case, the receiver input is protected by a bandpass filter.

So, the second function of a repeater is to convert the carrier frequency. The transmitted modulation spectrum is at another centre frequency, close to that of the received spectrum, but sufficiently offset to eliminate the disturbance due to the transmitter by filtering.

Finally, an important point to take into account is the congestion of the radio spectrum due to the continually increasing number of microwave users. For this reason, planning must anticipate from the very beginning, an intensive reuse of the varied frequencies in a network of radio links.

1.2.2 Carrier frequency utilization

To intensively reutilize the carrier frequencies, it is necessary to take advantage of the decouplings which may exist between the various paths at a single wavelength. There are several types of decoupling.

1. By angular discrimination due to the directivity of the antennas (see Volume 2). Typically, an attenuation equal to or greater than 60 dB results from an angular shift equal to or greater than 30°. The front-to-back decoupling of the antennas is 40 to 50 dB at the lowest carrier frequencies of the radio links (2 GHz) and can reach 70 dB at the highest carrier frequencies (> 6 GHz). These data particularly concern the nodal point (Fig. 1.3(a)).
2. By discrimination of the polarization of the radiated electromagnetic field (generally it is linear in the case of radio links), due to the properties of the

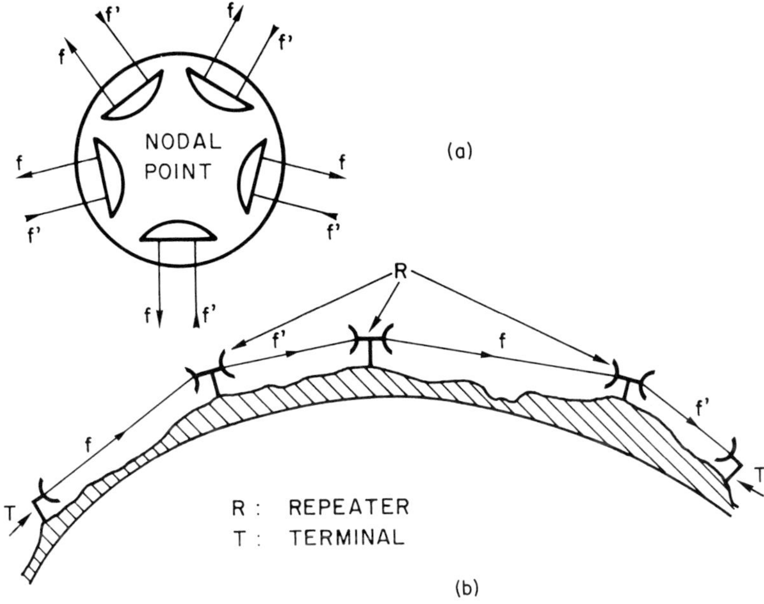

Fig. 1.3 Geometrical considerations allow frequency re-use.

microwave antennas (see Volume 2). 30 to 45 dB can be obtained between vertical and horizontal polarization.

3. By a screen effect due to the curvature of the earth (Fig. 1.3(b)). One hop out of two is at the same frequency, often in cross-polarization. In every case, locally, in a single station, transmission or reception takes place at the same frequencies but never transmission and reception at the same frequencies (Fig. 1.3(a)). Normally, two frequencies are enough for a bilateral transmission, used alternately. But if the front-to-back decoupling is not enough, then four frequencies are used.

Figure 1.3 concerns the case of two frequencies only (f and f'). On this figure, a single transmitting direction is shown. The opposite direction would also use f' and f, but if f is used in one direction then f' is used in the other.

RF channel arrangement To facilitate the development and the growth of radio link networks, international and federal committees specify frequency plans comprising several channels in the bands alloted to microwave links, i.e. the International Radio Consultative Committee (CCIR), the Federal Communications Commission (FCC), the International Frequency Registration Board (IFRB), etc. The specific frequency plans used with the analogue and digital signals are given in publications from these committees and are updated periodically.

1.2.3 Different types of repeaters

The different types of repeaters result from the know-how and a technology–cost compromise. Most of the amplification is generally obtained in intermediate frequency (IF) after conversion of the RF signal to an IF signal by a down-converter, the receiver being a superheterodyne type. For the transmission, various diagrams can be used.

Remodulating repeater

In this type of repeater, the baseband signal passes through the connection between the receiver and the transmitter (Fig. 1.4). The frequency conversions are RF–IF, IF–BB, BB–RF. The amplifier gain in intermediate frequency (IF) is automatically controlled so that the output level in IF remains constant whatever the variations of the propagation attenuation due to fading.

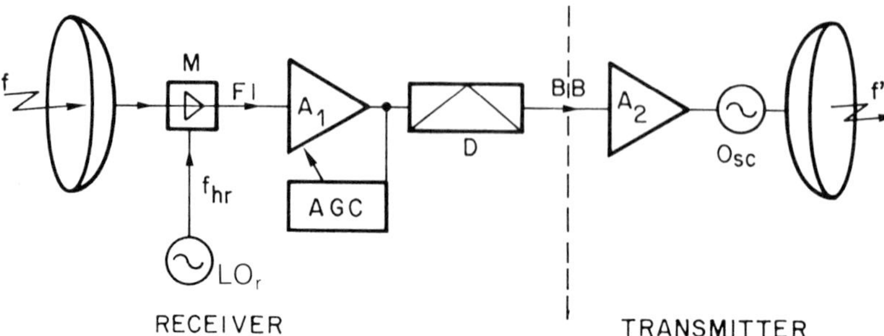

Fig. 1.4 Remodulating repeater. A_1, intermediate frequency amplifier (IFA); A_2, base band amplifier; AGC, automatic gain control; D, demodulator; f_{hr}, Rx heterodyne frequency; M, mixer (down converter); O_{sc}, frequency-modulated oscillator; LO_r, Rx local oscillator.

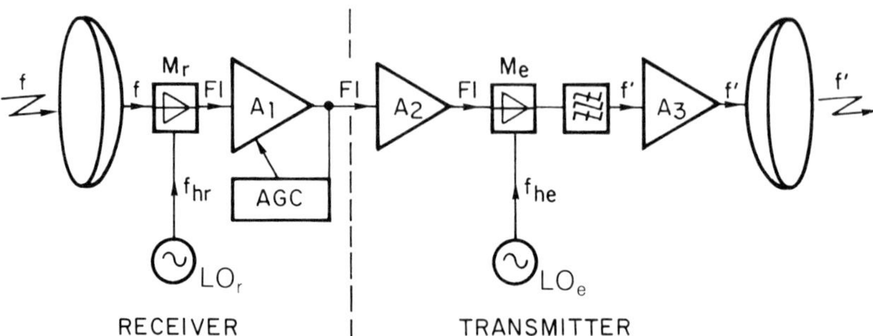

Fig. 1.5 Heterodyne repeater. A_1, A_2, intermediate frequency amplifiers; A_3, RF amplifier; AGC, automatic gain control; f_{he}, Tx heterodyne frequency; f_{hr}, Rx heterodyne frequency; M_e, Tx mixer (up converter); M_r, Rx mixer (down converter); LO_e, Tx local oscillator; LO_r, Rx local oscillator.

Heterodyne repeater

This differs from the preceding one by the introduction of an IF–RF conversion at high level. The heterodyne frequencies are chosen in order to set up a translation from f to f', and in this way, we have, for example Fig. 1.5:

$$f - f_{\mathrm{hr}} = \mathrm{IF} \qquad \text{and} \qquad f_{\mathrm{he}} + \mathrm{IF} = f'.$$

In this type of repeater, the IF signal passes through the connection between the receiver and the transmitter. The frequency conversions are: RF–IF and IF–RF.

Direct RF amplification

The amplification is obtained directly in RF (Fig. 1.6). In this case, there will be only one frequency conversion in order that $f' - f$ or $f - f' = f_t$ may be approximately 200 to 500 MHz. The RF gain is automatically monitored so that the RF output level will remain constant within the range planned for the propagation attenuation.

1.2.4 Different types of terminals

These can be inferred from the diagrams of the repeaters. The main types of modulations are FM with a large frequency deviation and phase shift keying (PSK).

1.2.5 Special features of antennas for radio relay links

The antennas intended for use with radio relay links are optimized to best satisfy the needs of line-of-sight microwave links (MWs) which result from the organization concerning the optimal use of carrier frequencies described in section 1.2.2.

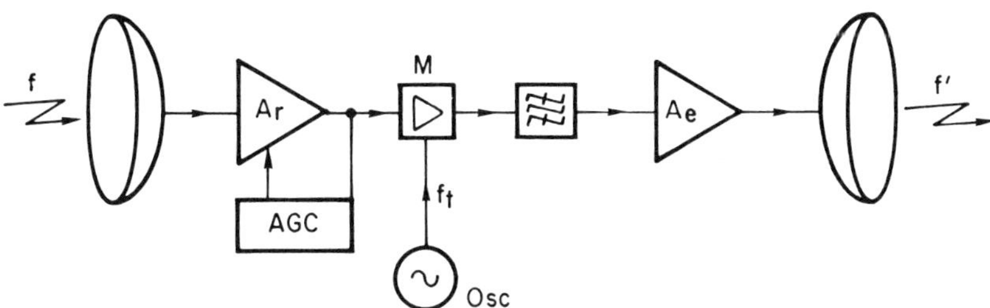

Fig. 1.6 Direct RF amplification. A_e, Tx RF amplifier; A_r, Rx RF amplifier; AGC, automatic gain control; f_t, translation frequency; M, translation mixer; Osc, oscillator.

Important parameters

1. Gain G (in the direction of the main radiation) is limited at the top by the narrowness of the main lobe, which must be compatible with the unwanted twisting of the antenna support or, with the tolerable wind effect (restriction concerning the diameter D).
2. Side lobe attenuation: the ratio of the maximum radiation of the side lobe to that of the main lobe.
3. Front-to-back ratio.
4. Standing wave ratio (SWR) at the accesses.
5. Local decoupling between counter-polarized accesses and the decoupling between the radiated or received polarizations XPDO. Figure 1.7 defines these parameters, considering the energy exchanges between the four accesses 1,2,1′,2′.

We have:

$$\mathrm{XPDO} = 10\log\left(\frac{P'_1}{P'_2}\right) = 10\log\left(\frac{t_{11'}}{t_{12'}}\right)$$

where $t_{11'}$ and $t_{12'}$ are the transfer coefficients of 1 to 1′ (copolarized) and from 1 to 2′ (counter-polarized). The local decoupling of the counter-polarized accesses

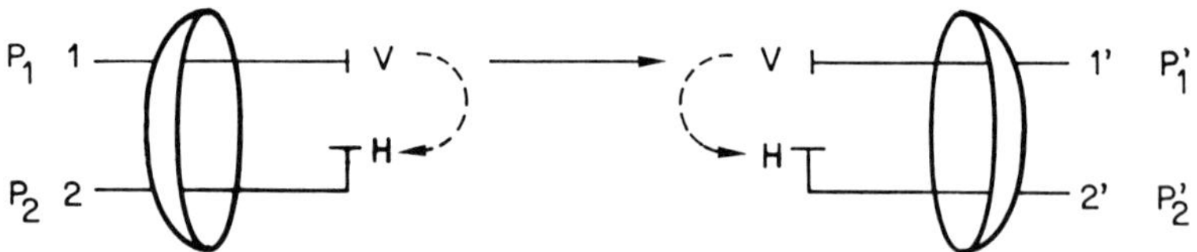

Fig. 1.7 Orthogonal polarization amplifier. H, horizontal polarization; V, vertical polarization.

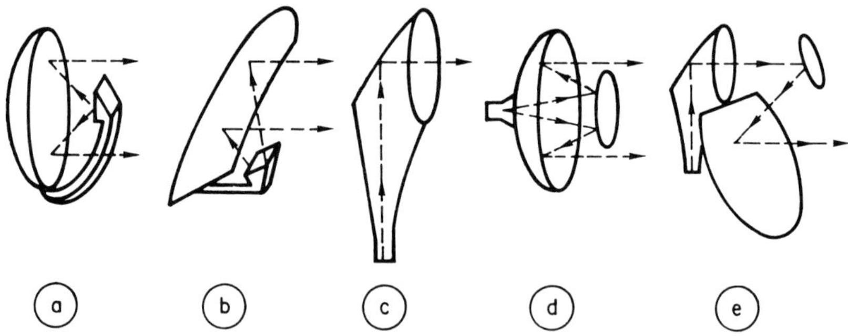

Fig. 1.8 Microwave optics: (a) focus-feed horn; (b) offset-feed horn; (c) horn parabolic reflector; (d) Cassegrain reflector system; (e) folded-horn parabolic reflector.

1 to 2 or 1′ to 2′ and vice versa, is:

$$10 \log t_{12} \text{ or } 10 \log t_{21} \qquad \text{and} \qquad 10 \log t_{1'2'} \text{ or } 10 \log t_{2'1'}.$$

The order of magnitude is 30 to 40 dB.

Main optics

These optics are chosen based upon technical and cost compromises and are shown in Fig. 1.8. The orders of magnitude are given in Table 1.1.

Utilization of the antennas

By employing polarization multiplexers and demultiplexers, the natural aperiodicity of the paraboloids makes it possible to group several frequency bands in order to radiate them together. The advantage of this approach is to diminish the cost of the 'antenna support' infrastructure by reducing by a factor of 2, 3 or 4 the number of antennas according to whether they are used in double-band, triple-band or quadruple-band. For example, it is possible in bands 3.6 to 4.2 GHz, 5.9 to 6.4 GHz, 6.4 GHz, 6.4 to 7.1 GHz and 10.7 to 11.7 GHz to obtain a capacity of 30 000 to 40 000 voice channels per antenna.

Use of passive mirrors

In certain cases, instead of installing active repeaters in the line-of-sight locations, it is preferable to install passive mirrors, to take advantage of the quasi-optical

Table 1.1 Some typical parameters of antennas for microwave links

f (GHz)	2	4	6.175	6.77	7.575	8.35	11.2	13	18.7	22.4	31.15	38
λ	15 cm	7.5 cm	4.85 cm	4.43 cm	3.96 cm	3.59 cm	2.67 cm	2.3 cm	16 mm	13.4 mm	9.63 mm	7.9 mm
Diameter (m)	1.8 to 3.7	1.8 to 3.7	1.8 to 3.7	1.8 to 3.7	1.2 to 3	1.8 to 3.7	2.4 to 3.7	1.2 to 3	1.2	0.45	0.25	0.27
Gain in the access (dB)	29.4 to 35.4	37.3 to 41	38.8 to 44.8	39.8 to 45.6	37 to 44.7	41 to 47	46.4 to 49.8	41.5 to 48.8	45.2	37	35	37
Directivity front to back (dB)	36 to 42	42 to 68	46 66 75	47 66 75	44 to 70	48 to 71	80	49 to 71	66			
XPDO (dB)	30	30	30	30	30	25	25	25	25			
Local decoupling, VH access	30	35	40	40	35	35	40	35	40			
Standing wave ratio	1.08	1.06	1.04 to 1.06	1.04 to 1.06	1.04 to 1.06	1.04 to 1.06	1.04	1.1	1.15			

behaviour of the microwaves. In ITT (1968) we find the formulae for calculating the gain due to the parabola and reflector association.

1.2.6 Feeders

To connect the equipment to the antenna, we mainly use the following.

1. Rectangular waveguides in mode TE_{10} for lengths of a few dozen metres. An order of magnitude of losses is 0.05 dB/m at 4 GHz. The losses increase with the frequency.
2. Semi-flexible elliptical waveguides in TE_{11} mode, easy to install, with lengths which can reach 100 to 200 m per radio hop. The order of magnitude for the losses is 2.8×10^{-2} dB/m to 4 GHz.
3. Circular waveguides, in mode TE_{11} which can simultaneously carry two polarizations but which are difficult to install. The lengths may be a few hundred metres per hop. An order of magnitude for the losses is 1.2×10^{-2} dB/m at 4 GHz, they go down to 0.9×10^{-2} dB/m at 6 GHz with a waveguide whose diameter is 71.5 mm.
4. 'Semi-airspaced' coaxial cables are used for frequencies less than 3 GHz for lengths of a few dozen metres. An order of magnitude for the losses is from 0.1 dB/m (S.M. 6.3/13.6) to 4.8×10^{-2} dB/m (S.M. 17.2/39.7) at 2.5 GHz.

Non-linear phenomena may occur in the passive microwave circuits because of imperfect contacts (waveguide flanges, tuning screws). They are very weak but are troublesome when the waveguide is carrying both transmission and reception signals, with regard to the very large difference in levels, and it is easy to imagine that certain intermodulation products can jam the receivers.

1.3 ANALOGUE MICROWAVE LINKS

1.3.1 General

The first analogue microwave links were installed at the end of the Second World War. Their technology has evolved considerably since then (transistorizing of circuits in baseband and in intermediate frequency, then microwave circuits, reduction of power consumption. etc.), and today they will constitute a considerable proportion of transmission networks. The new installations for telephone service now use digital technology, but analogue technology still predominates in television signal transmission.

The state of the art makes is possible to transmit up to 3600 subchannels per radio carrier and to group up to 8 bilateral channels (hence 16 carriers) on the same antennas, and in this way to establish lines with 20 000 telephone subchannels. One RF channel of 1800 voice channels can be replaced by one television channel with colour picture and sound. By associating several groups of this type in various frequency bands, links with much higher capacities can be set up.

1.3.2 Characteristics of the signals transmitted

There are two main types of signals transmitted: telephony and television. A general definition of the baseband is: the baseband is the band occupied by the signals to be transmitted, which, in a modulation process, are the signals used in the first place to modulate a carrier.

Telephony The limits of the baseband are 8 to 12 MHz for 1800 and 2700 voice channels. The pattern of the voltage developed by the multiplex signal is shown in Fig. 1.9.

The average power, which corresponds to the root-mean-square voltage V_{rms} is given at a point of zero relative level, according to the following expressions, depending upon the capacity of telephone channels N.

$$P_{mo} = \begin{cases} -1 + 4\log N & \text{for } N < 240 \\ -15 + 10\log N & \text{for } N \geqslant 240. \end{cases}$$

Probability p of exceeding $\pm V_c$ (peak voltage) is, respectively, $p = 10^{-3}$ and 10^{-2}, for $20\log(V_c/V_{eff}) = 10.3\,\text{dB}$ and $8.2\,\text{dB}$.

Television Composite picture signals: the signal consists of the luminance and the chrominance as well as the synchronization (Fig. 1.10). The video band is 5 to 6 MHz. Subcarriers, frequency modulated by the sound (30 to 10 or 15 kHz): the sound subcarriers are placed above the maximum frequency of the video band (7 to 9 MHz).

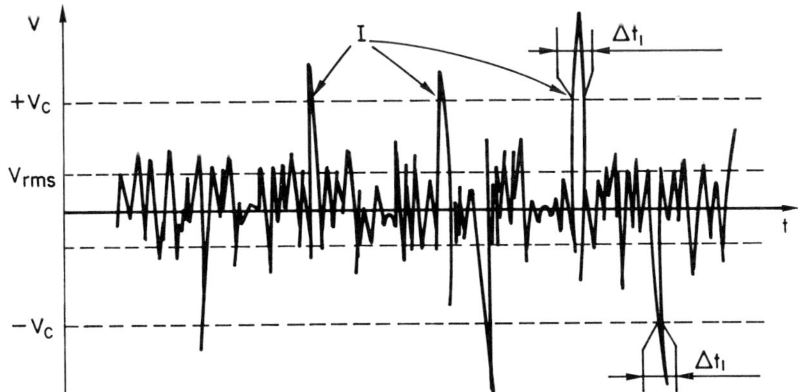

Fig. 1.9 Multiplex signal. I = instants when V exceeds V_c; Δt_I = duration of time for which V exceeds V_c.

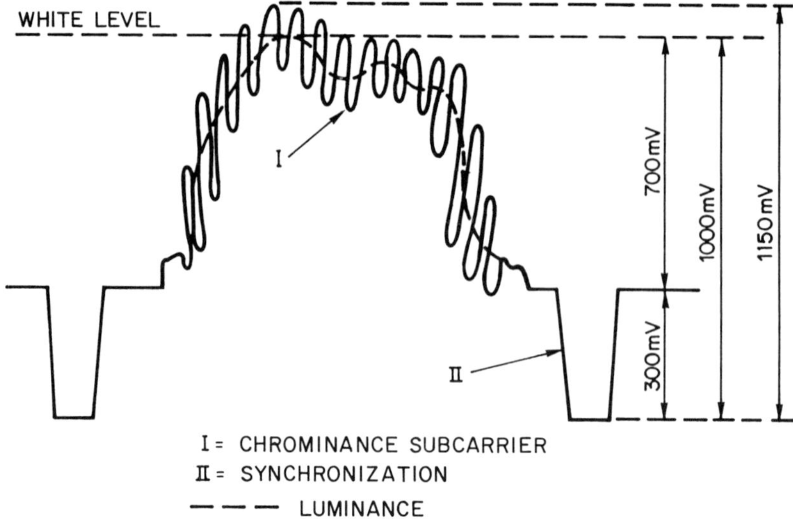

Fig. 1.10 Colour image signal (voltage at 75 Ω). I, chrominance subcarrier; II, synchronization; ---, luminance.

1.3.3 Analogue modulation

Basic purpose

The baseband signals which are mentioned above are to be transmitted while respecting severe linearity requirements, i.e. discrepancy not to exceed 10^{-3} over the range swept by the instantaneous voltage of these signals.

The choice of frequency modulation

The above considerations partially justify this choice with regard to the known properties of the frequency modulation. To better explain another advantage not yet mentioned, it should be noted that the frequency demodulation by discrimination, preceded by a limiter, has the property of converting the frequency swing into baseband voltage with a sensitivity which is substantially independent of the demodulator input level, which leads us to affirm that the equivalent of transmission (overall loss in baseband) on frequency modulation is independent, as an initial approximation, from the fluctuations of the propagation attenuation.

Finally, frequency modulation is constant envelope modulation, a feature which solves the difficulties which might arise with regard to the amplitude–amplitude non-linearity or from the amplitude–phase conversion of the power amplifiers operating in microwave frequency.

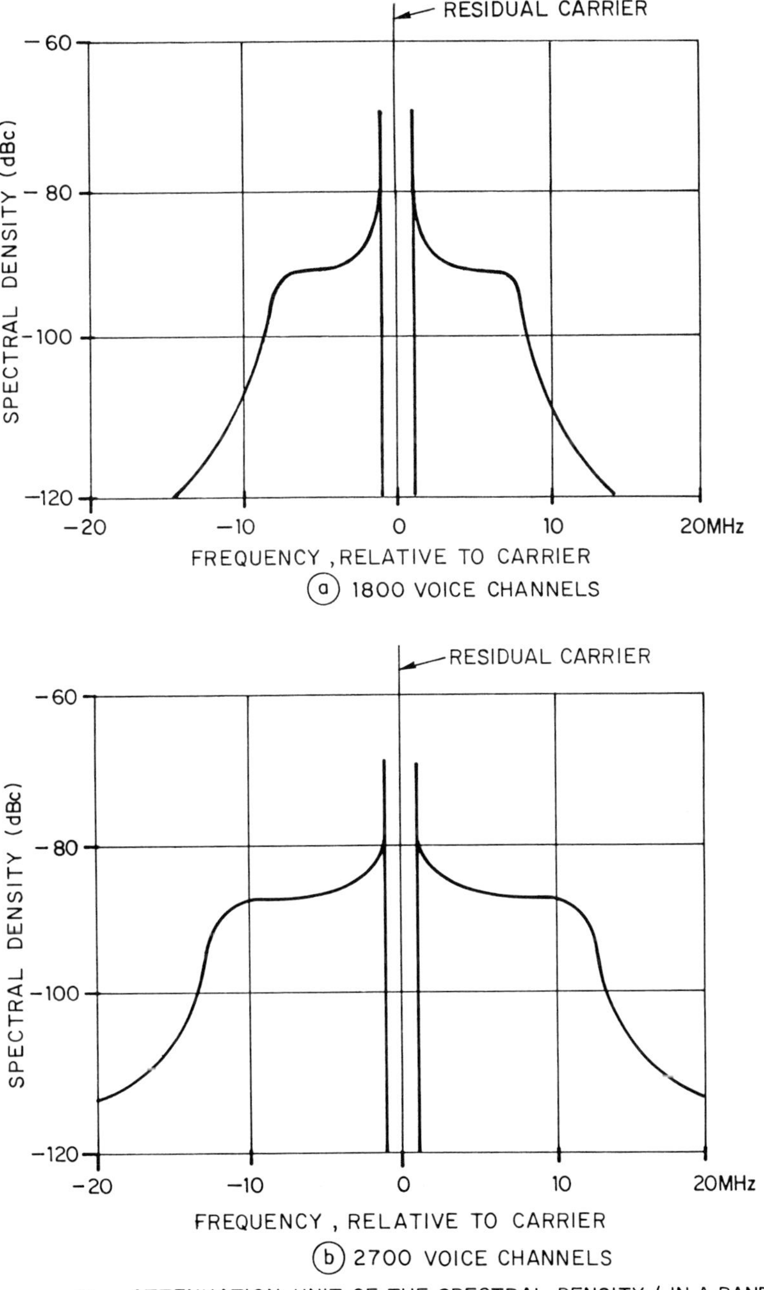

Fig. 1.11 Frequency modulation spectrum for different carrier capacities.

Frequency modulation spectra

Telephony The modulating signal appears as noise whose continuous spectrum occupies the whole multiplex band which only begins at 60 or 300 kHz (no very low frequency component). The characteristics of the modulation correspond to an optimization based on technological know-how: the frequency deviation at the neutral point of the pre-emphasis, caused by sending the signal of 1 mW at 800 Hz at a zero relative level point, and the pre-emphasis. For example for capacities 2700 telephone channels, the frequency deviations are ± 4228 kHz.

Figure 1.11 shows the frequency modulation spectrum obtained at the analyser for two capacities 1800 and 2700 subchannels. We find a very substantial line at the carrier frequency and a continuous spectrum whose spectral density comes in dBc, that is:

$$10\log\left(\frac{\text{power in 1 Hz}}{\text{power of the unmodulated carrier}}\right)$$

It should be remembered that dBc is a spectral density attenuation unit (in a band of 1 Hz) with relation to the unmodulated carrier (c = carrier).

Television The modulating signal appears highly unbalanced and, from the spectral point of view (in the baseband), very rich in components at very low frequencies. The modulation characteristics correspond to optimization which takes into account the technological know-how and above all, the specific nature of the television signal. They are the peak-to-peak frequency deviation, i.e. 8 MHz caused by the video signal; and the pre-emphasis—this has a beneficial effect by making the signal more symmetrical and the modulation spectrum more narrow, which results in a better transmission of the video signal with less differential gain and differential phase due to distortions of the radio channels.

Figure 1.12 shows the effect of the pre-emphasis in the simple case of a black and white picture; the luminance, in this particular example, is at white. We can compare these spectra without pre-emphasis (Fig. 1.12(a)) and with pre-emphasis (Fig. 1.12(b)). In the simple case, the modulation spectrum is a spectrum of lines and we see the reduction in the width which results from the pre-emphasis.

In the case of a more complex and coloured picture (test pattern), and with a sound modulation subcarrier, the spectrum is wider and we see a fairly large energy dispersion as shown in Fig. 1.13. The spectral density is given with regard to the unmodulated carrier as in the case of telephony and in a band of 1 Hz.

Linearization of the modulation and of the demodulation

By designating the input of a nearly linear system as x and its output as y, we generally express a weak distortion by a polynomial development in the following

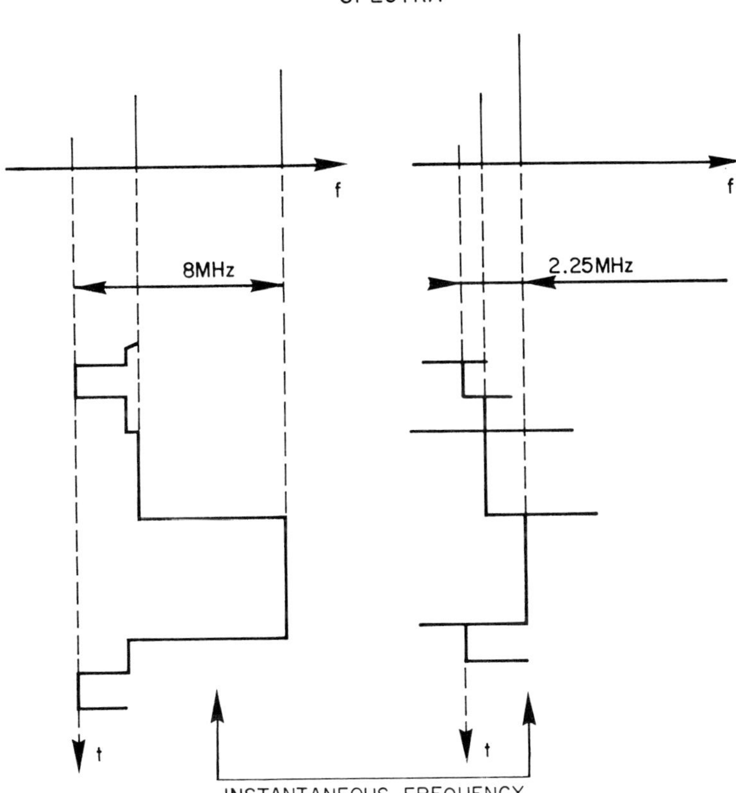

Fig. 1.12 White image frequency modulation spectra: (left) without pre-emphasis; (right) with pre-emphasis.

form:

$$y = a_1 x + a_2 x^2 + a_3 x^3 + \cdots.$$

This development makes it possible to indicate the harmonic distortion of the intermodulation lines, for example when x takes the form $X\ \cos\omega t$ or when it takes the form $X(\cos\omega_1 t + \cos\omega_2 t)$. The goal is naturally to make a_2 and a_3 as weak as possible.

In practice, the linearity is obtained by compensations, localized in the modulator itself and in the demodulator; they are optimized by adjusting the settings. After having displayed the non-linearity with a frequency sweep test, we seek to display an inflection and a symmetry which will make it possible to cancel the even terms (a_2, a_4, etc.).

We should note that single sideband modulation and vestigal amplitude modulation are considered elsewhere for the specific applications.

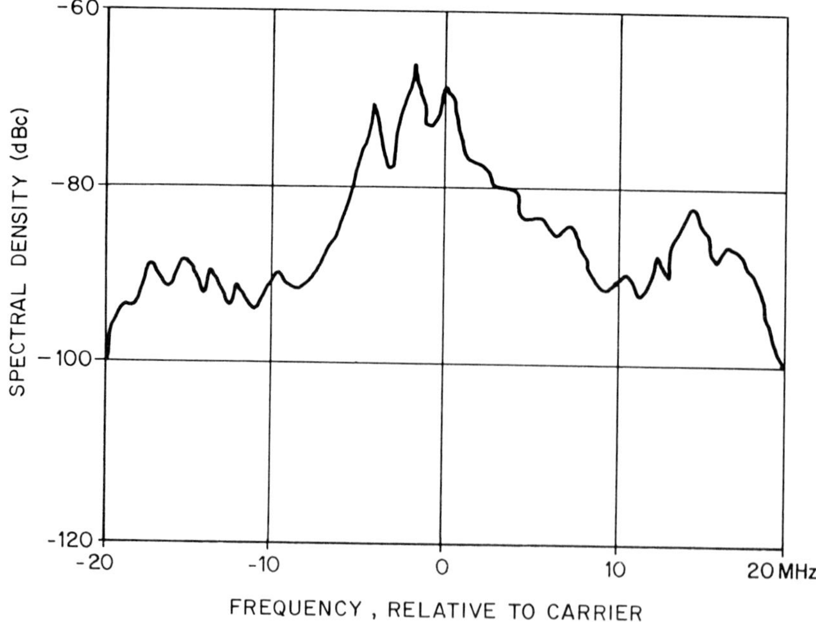

Fig. 1.13 Frequency modulation spectrum of normal image TV plus four sounds.

1.3.4 Technological aspects

This paragraph describes subunits used in a microwave link (MW) which have nearly the same characteristics for analogue MWs as for digital MWs. So their description will not be repeated in sections 1.4 and 1.5.

Amplifiers

For line-of-sight microwave links, there will be an attenuation between antenna accesses varying from 50 to 110 dB. The amplification of the radio repeater must therefore compensate these attenuations at each instant, in spite of their permanent variation, which implies an automatic gain control (AGC). It follows that the amplifier to be set up is to be presented under two aspects: a low-level amplification, preceded by a low-noise amplification at reception, and a high-level amplification ending with a final power stage at transmission.

In line-of-sight microwave links, the noise factor is 3 to 10 dB and the sending power is from 50 mW to 20 W.

Low level amplifiers There are two types: intermediate frequency amplifiers and low noise amplifiers.

Intermediate frequency (IF) amplification

1. Standard stage: this amplification is set up almost aperiodically, based on transistorized two transistors assemblies which can be used either at 70 MHz plus or minus 20 MHz, or at 140 MHz plus or minus 20 MHz. In order to facilitate placing such stages in cascade, the input impedance and the internal impedance of the source as seen from the outside are both 75 Ω.
2. Automatic gain control (AGC): this is set up by p-i-n diodes which can be controlled by the automatic gain control voltage.
3. Intermediate frequency amplifiers: these consist of an association of amplification stages and variable attenuators.

RF amplifiers In the case of tubes, the main RF tubes used for the line-of-sight microwave links have been triodes (now abandoned), and travelling-wave tubes (TWT). Table 1.2 summarizes the most important characteristics of these tubes. In general, the power supply is built with semiconductors such as diodes, transistors, thyristors, etc. We often use switching type power supplies, with regulation by pulse width modulation (PWM).

In the case of transistors, the power versus frequency curve of the transistors is such that we use them for power levels going from 100 mW to 5 to 10 W and by placing them in parallel from 10 to 100 W.

A distinction is made between: bipolar silicon transistors, which can operate at least up to 4 GHz, in class A at low level and in class C at high level; and field-effect gallium arsenide transistors, today used from 1 up to 30 to 40 GHz in class A only, even at high level. Table 1.3 gives the main characteristics of the transistors and of the types of power amplifiers built with these transistors.

Problems which transistorized microwave frequency amplifiers pose to the designer essentially concern the building of the interstage connections; we need to transform, by a network of reactors, the impedance at the input of a transistor into an optimal impedance as seen from the preceding transistor. Generally speaking, these are the measured parameters S_{11} and S_{22} with which we begin to optimize the matching reactors in order to obtain the gain, the passband and

Table 1.2 Characteristics of microwave tubes (SHF)

Tube	*Power* (W)	*Efficiency* *Tube alone* (%)	*Efficiency* *With power supply* (%)	*Gain* (dB)	*Frequency band* (GHz)	*Noise factor* (dB)	*Amplitude/phase conversion* (°/dB)	*Application*
TWT 2 GHz to 18 GHz	10 20	40	30	40 to 50	1	27	3	Line-of-sight microwave links

Table 1.3 Characteristics of the SHF power transistors

Transistor		*Maximum output power at 1 dB compression* (W)	*Efficiency* (%)	*Gain (linear)* (dB)	*Frequency* (GHz)	*Passband* (GHz)	*Amplitude/phase conversion* (°/dB)
Bipolar silicon	Class A ($V_1 = 20$ V)[1] [a]	5	25	7	1.9	1.7–2.1	1
	Class C ($V_a = 24$ V)	5	30	6	4	3.8–4.2	
		25	45	7	1.9	1.7–2.1	
	Standard amplifier ($V_a = 28$ V)	50	20	30	2.1	2 –2.3	
Field-effect type gallium arsenide	Class A ($V_a = 10$ V)	4	25	5	14.2	1.4–14.5	1
		8	25	6	8.1	7.7–8.5	
	Standard amplifier ($V_a = 10$ V)	2	10	50	11.2	10.7–11.7	1

[1] V_a power supply voltage

the output power. The input impedance comprises a very weak resistant term, from 1 to 10 Ω, and a high reactance.

Low-noise transistorized amplifiers require optimized components, that is, components whose main causes of noise have been minimized. The main causes are the Schottky effect, and unwanted resistance. To reduce the Schottky effect, we choose a weak collector current in order to obtain the desired result. However, when this becomes weak, the gain diminishes and we must not forget that the noise factor depends mainly upon the first and second stages of an amplifier, therefore, indirectly on the gain of the first one, according to the Friis formula.

$$F = F_1 + \frac{F_2 - 1}{G_1}$$

where F_1 is the noise factor of the first stage of the amplifier, F_2 is the noise factor of the second stage of the amplifier, and G_1 is the gain of the first stage of the amplifier.

In general, the optimization leads to a current from 2 to 5 to 10 mA for the first stage. However, it will be noted that the behaviour of the bipolar transistors and that of the field-effect transistors is a little different. Furthermore, the noise factor also depends on the signal source impedance as seen from the input of the first transistor. The optimization generally leads to a mismatch as seen from the signal source. To avoid this difficulty, we often place a circulator, which functions as an isolator, at the input of the low-noise preamplifiers (Fig. 1.14). Table 1.4 summarizes the present possibilities for low noise amplifiers.

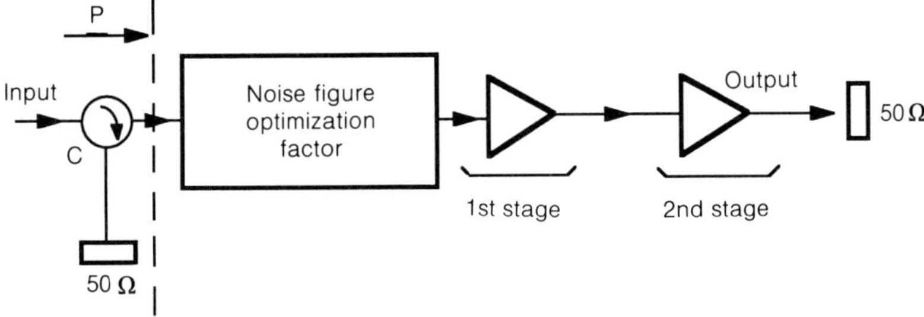

Fig. 1.14 Low-noise preamplifier: C, circulator; P, mismatched plane, on the amplifier side.

Table 1.4 Characteristics of the low noise SHF amplifiers

Frequency (GHz)	*Noise factor* (dB)	*Gain* (dB)	*Number of stages*	*1 dB compression point at the output* (dBm)	*Types of transistors*	*Circulator at input*
0.61 to 0.96 1.35 to 1.85 1.7 to 2.4	2	20	2	+7	Bipolar	0
3.8 to 4.2 6.4 to 7.1 7.7 to 8.4 10.7 to 11.7	2	20	2	+7	Field effect (GaAs)	1

Mixers

There are two main categories of needs which correspond to modulation spectrum translations along the access of the frequencies: frequency change by lowering the frequency RF to IF (down-converter), as needed for reception; and changing the frequency by raising the frequency IF to RF (up-converter), necessary for transmission. To implement a frequency change, three devices are necessary: a mixer, an oscillator, and in certain cases, a filter to eliminate the image frequency.

Conversion loss The conversion loss of a mixer in microwave frequency is normally approximately 6 dB. It is possible to reduce this conversion loss to approximately 3 dB by 'energy recovery' of the image frequency, by loading the reception mixer input or the transmission mixer output with a suitable reactive impedance. This technique was used when solid-state, low-noise or power microwave amplifiers were not yet available. This is no longer the case today when it is preferable to load the mixer resistively throughout the whole range of working frequencies, which makes the mixing more aperiodic. The conversion loss of 6 dB is compensated by an IF or RF amplification, at medium level.

Linearity A mixer is said to be linear when the envelope of the output signal is proportional to that of the input signal. So the conversion loss is independent of the level. For the symmetrical mixers which do not produce even harmonics, the linearity is expressed quantitatively by the interception point level.

For frequency modulation analogue microwave links, the modulated signal level is constant, and it is not necessary for the mixers to be linear. A good linearity is necessary for the single sideband analogue microwave links and for the QAM (quadrature amplitude modulation) digital microwave links (see section 1.4.2).

A mixer is all the more linear as the local oscillator signal power increases in relation to the power of the input and output signals. Naturally the transmission mixers are the most critical ones.

Image frequency It is possible to reduce the sensitivity of a reception mixer to the image frequency, or to reduce the output level of the signal at the image frequency of a transmission mixer, using a microwave filter, placed respectively at the input or at the output of the mixer.

At reception, the filter should be placed at the input of the mixer and not at the input of the low-noise amplifier which precedes it (Fig. 1.15), otherwise the noise factor will be increased by approximately 3 dB because of the noise generated at the image frequency by the amplifier. It is also possible to use an 'image rejection' mixer (not to be confused with the 'image frequency energy recovery' mentioned earlier), which attenuates the image frequency by approximately 30 dB. Its diagram is shown on Fig. 1.16 in the case of a transmission mixer. We would obtain the diagram of a reception mixer by reversing the direction of the hori-

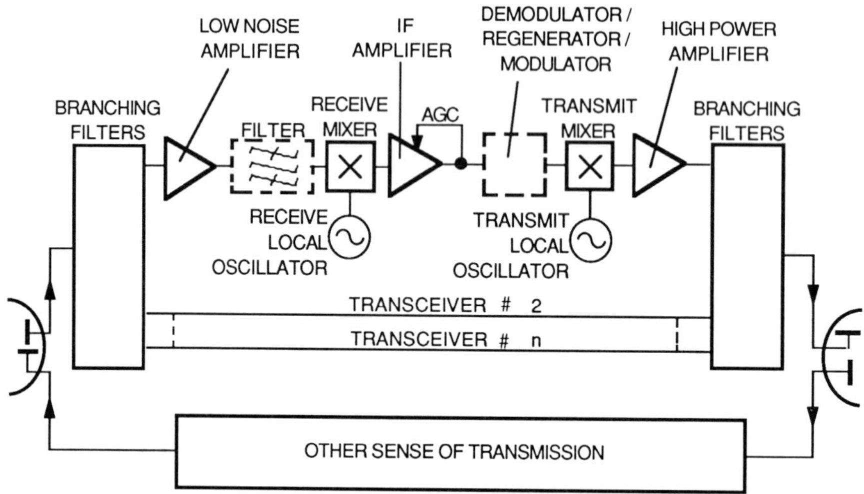

Fig. 1.15 Block diagram of a radio-link transceiver.

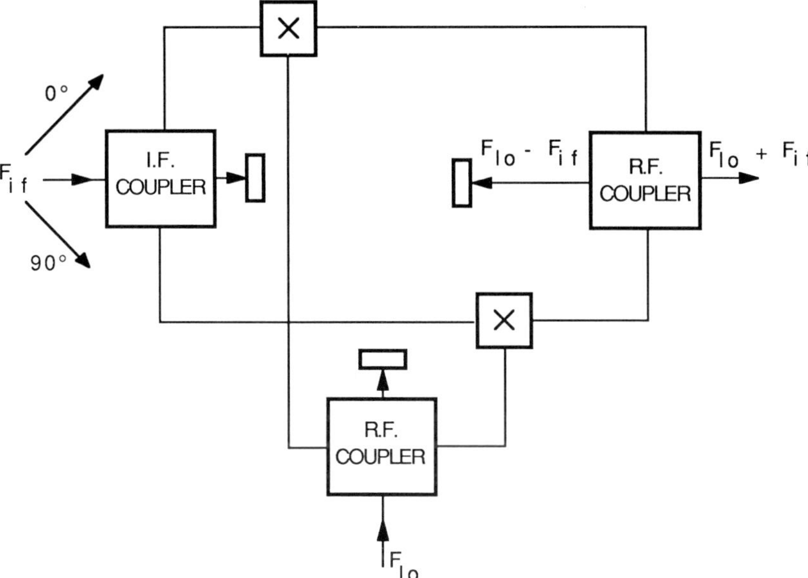

Fig. 1.16 Image rejection transmit mixer.

zontal arrows. This requires the use of at least two diodes and three couplers. The intermediate frequency coupler and one of the two microwave frequency couplers must be of the type 0°/90°.

Local oscillators

The local oscillators, at heterodyne frequency f_{LO} supply the RF power necessary to perform the RF–IF or IF–RF conversions. The wave that they supply is a pure sine wave, characterized by an accuracy and a frequency stability of approximately 10^{-5} to 10^{-6}; and for the analogue microwaves, a spectral purity, in particular a very weak phase or frequency scintillation, whose tolerable magnitude results from the balance of the contributions to the equipment noise. A general noise magnitude is 5 pW_{op}^* in telephony, and 7 μV (weighted value) in television at a point where the luminance signal develops 700 mV.

Two main types of local oscillators are used in modern microwave links. A first type is based on the use of a Gunn oscillator or a transistor oscillator followed, in certain cases, by a frequency multiplier by two or three. The oscillator is voltage controlled by a varactor, and its frequency is slaved to the frequency of a quartz crystal, after a possible frequency change and a frequency division. When we wish to obtain a frequency 'agility', that is, the possibility of changing

* 1 pW_{o} = 1 pW (picowatt), noise mean power measured at a zero relative level point, and 1 pW_{op} = 1 pW (picowatt), weighted, noise mean power measured at a zero relative level point.

the carrier frequency during operation, as in the case of mobile microwave links for television, the frequency divider is programmable and the local oscillator becomes a frequency synthesizer.

A second type consists of stabilizing the frequency of a Gunn oscillator or of a transistorized oscillator by a dielectric resonator. We obtain a very low manufacturing cost and a frequency stability which approaches that of a quartz crystal.

Filters

The purpose of the filters is to protect the channels from neighbouring frequency interference, or to considerably attenuate the non-essential frequencies created by the frequency changes (residual heterodyne, image frequencies, etc.) so as not to jam the adjacent channels of any other microwave users.

The filtering is obtained by the use of filters in microwave frequency, in intermediate frequency, in baseband. These are bandpass, bandstop, and low-pass filters, respectively.

In this section, we consider only the RF filters because the methods used are more particularly specific to microwave links. Furthermore, RF filtering, while contributing to the elimination of interference, also enables the branching of several channels at the same access.

RF filtering idealization Figure 1.17 shows an ideal microwave frequency filter whose transfer function H is such that $H = 1$ in a frequency band B centred on f_o (which is the carrier frequency of a radio channel), and $H = 0$ outside that band. The approximate magnitude of B is from 50 to 80 MHz for carrier frequencies of approximately 6 GHz.

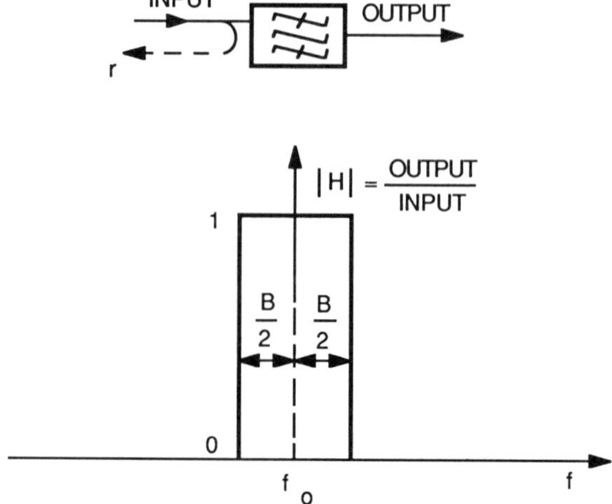

Fig. 1.17 SHF filter idealization.

If the losses are low or negligible, the result is that the reflection coefficient r at the filter input is such that $r = 0$ in band B and $r = 1$ outside that band. It is often said that a filter presents a short-circuit plane outside its passband; when seen through a quarter-wave, the short circuit is thus transformed into infinite impedance, if we so desire, allowing to associate several filters. The transmission and reflection properties of the bandpass filters enable setting up branchings on the common accesses as we see further on.

Branching of RF channels Generally speaking, the branching of two or more different frequency generators on a single access involves energy losses if bandpass filters are not used, because each generator output sees the access in parallel with the internal impedances of the other generators. This phenomenon can be avoided by using the directive properties of microwave frequency couplers (3 dB coupler) as shown in Fig. 1.18. However, there is a resulting loss of 3 dB, consumed in a dissipative load.

Another solution is to use the propagation of two orthogonal polarizations (Fig. 1.19), using a polarization duplexer and a feeder which can propagate these two polarizations (circular waveguide, for example). This solution entails no energy loss. We can even couple two generators f_o, at the same nominal frequency in this way, but only two. The decoupling between the two signals is approximately 20 to 30 dB.

The principles which have just been mentioned also apply in reception by reversing the propagation direction of the signals. The common access then behaves like a multisignal source.

Use of bandpass filters and circulators The introduction of bandpass filters makes it possible to eliminate the inherent loss of 3 dB, as shown in Fig. 1.18, and to set up branchings of more than two channels.

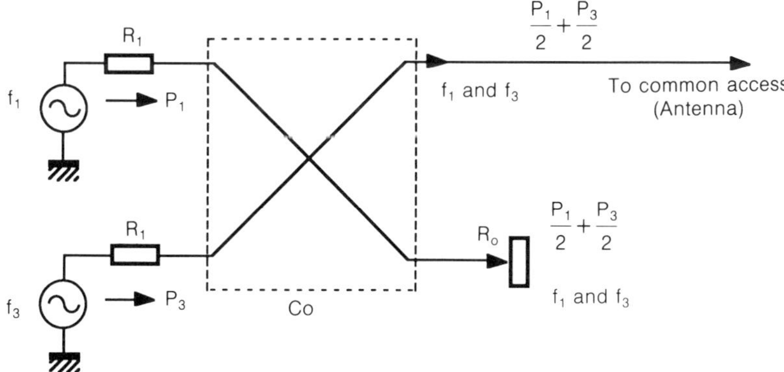

Fig. 1.18 Branching of two frequency generators f_1 and f_3 on a common access by a directive 3 dB coupler: Co, SHF 3 dB coupler; R_0, matched load; R_1, internal resistance.

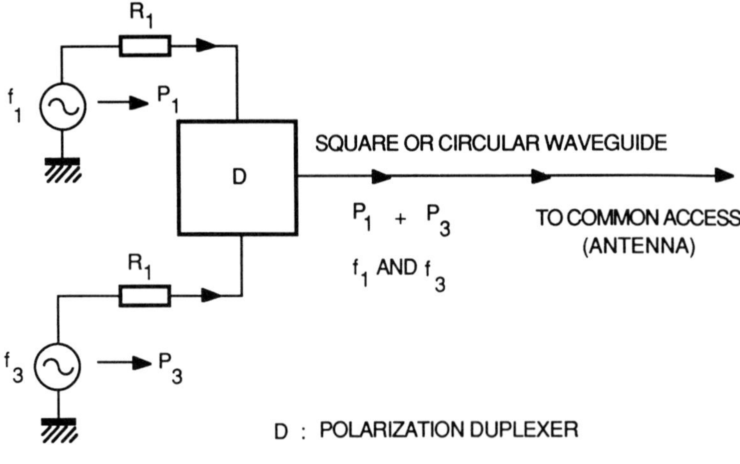

Fig. 1.19 Branching of two frequency generators f_1 and f_3 on a common access by a polarization duplexer: D, polarization duplexer; R_1, internal resistance.

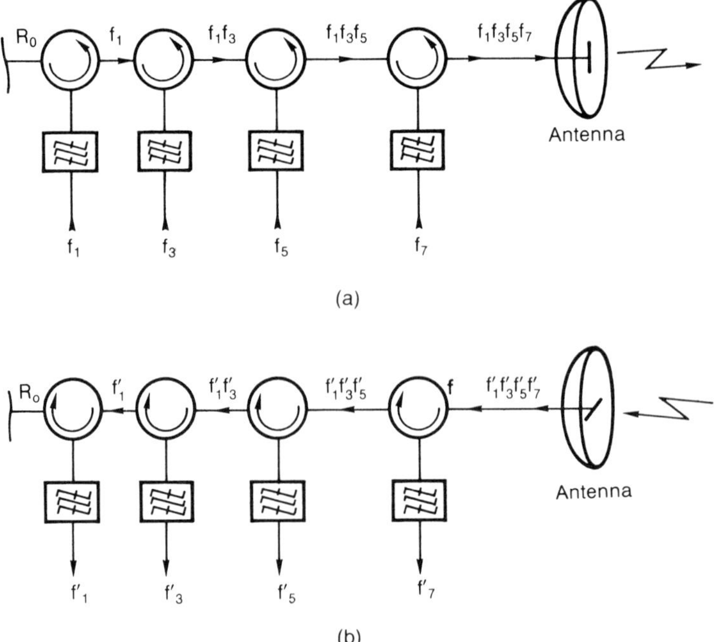

Fig. 1.20 Branching of channels by associating filters and circulators: (a) transmission; (b) reception.

There are a number of ways of associating bandpass filters, for example, by an appropriate positioning of the short-circuit planes, each with relation to the others, so that, in the passbands, the bandstop filters will contribute an infinite impedance via a suitable line length.

Another widespread method is to associate bandpass filters with the aid of circulators. Figure 1.20 shows such an arrangement.

Branching of transmissions and receptions on the same antennas Figure 1.21 shows two ways of branching the transmissions and receptions on a single antenna. The solution in Fig. 1.21(a) is preferable when the power of the transmitters is approximately 5 to 20 W; and the solution in Fig. 1.21(b) is preferable when it is only 0.1 to 1 W.

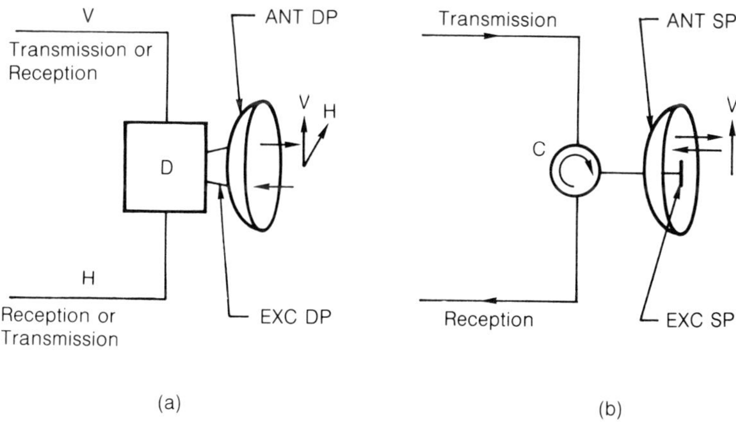

Fig. 1.21 Branching of transmissions and receptions on a single antenna: ANT DP, double polarization antenna; ANT SP, single polarization antenna; C, circulator; D, polarization duplexer; EXC DP, double polarization excitation; EXC SP, single polarization excitation; H, horizontal polarization; V, vertical polarization.

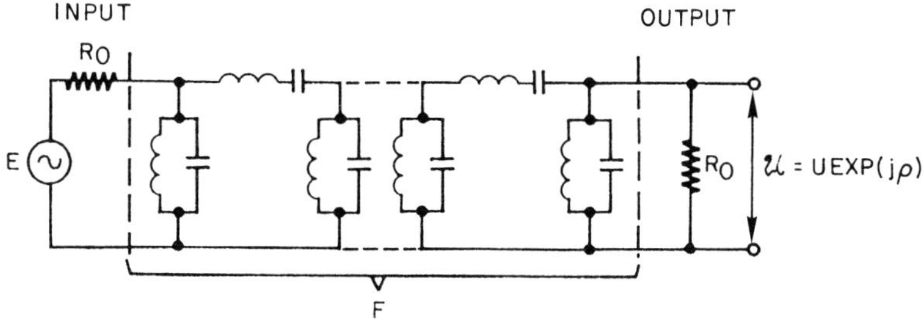

Fig. 1.22 Butterworth filter: F, bandpass filter between characteristic impedances R_0.

Principle of the filtering function – Butterworth functions The ideal function, shown in Fig. 1.17, cannot be obtained; we content ourselves with an approximation, given by the Butterworth functions (see, for instance, ITT, 1968) obtained with filters such as represented in Fig. 1.22.

1.3.5 Operating aid facilities

These facilities have three main functions.

Protection switching

The protection switching of the radio channels helps avoid link downtime due to equipment failures and, to a lesser degree, propagation faults. Switching takes place at the extremities of a switching section which generally includes several hops (about five hops). The organization of a switching section is given in Fig. 1.23 for a type $N+1$ system in which N radio channels are backed up by a single standby channel.

When a propagation fading or equipment failure appears in one transmission direction, on channel 1 for example, the quality assessor, placed downstream from the section, detects the fault and informs a control logic of it. This logic, via an engineering order wire, requests the logic located upstream to feed signal number 1 at the input of the standby channel. Signal number 1 is allowed to be propagated at the same time, on channel number 1 and on the standby channel during a time slightly longer than the link propagation time, then a switching is performed downstream on the output of signal number 1 on the standby channel.

By performing these operations and by using special switches at the downstream extremity, we perform a hitless switching of the data in case the signal quality degradation is slow (increase of noise due to propagation fading) or null (switching for maintenance purposes).

Monitoring

For monitoring the analogue microwave link signal quality, the quality is measured by feeding, at the switching section input, a sine wave signal (pilot) at a frequency slightly above the highest frequency of the multiplex. At the downstream extremity of the section, the quality is assessed by measuring the level of the pilot and the noise near it. These signals are used for the automatic switching, and are also sent to a master station for any recordings and statistics, along with other signals such as a switching state, the alarm indications from the equipment and power stations, etc.

Engineering order wires

These are used, between the various terminal or relay stations, to transmit the telephone signals and the remote display and remote control signals necessary

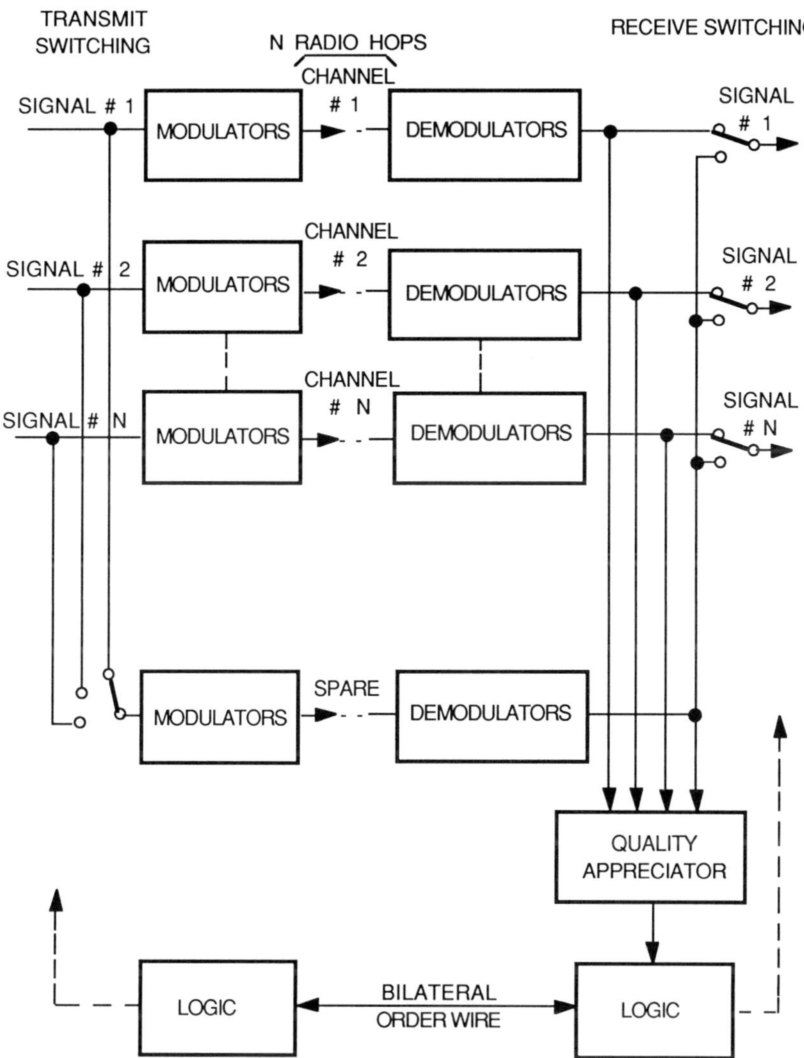

Fig. 1.23 Organization of a switching section.

for maintenance. These engineering order wires are generally transmitted below the baseband of one of the signals to be transmitted. In this way we take advantage of the protection provided by the automatic channel switching.

1.3.6 Frequency modulation distortions

There are three distinct categories of distortions:

1. those created by the weak non-linearity of the modulation and of the demodulation in the terminals;

2. those generated by the imperfections in the transparency of the repeaters in the band occupied by the frequency modulation spectrum;
3. those produced by the echoes, generally speaking, either because of the imperfection in the matching of the feeders or, in the multiple propagation paths.

These distortions are well known in classical analogue systems, so we shall not give other details.

1.3.7 Performances of analogue microwave links

These performances are analysed based on reference circuits and the so-called 'quality and availability objectives'.

(CCIR systems) hypothetical reference circuit – 'quality objectives'

This circuit is defined for frequency division multiplexing microwave links with capacities greater than 60 telephone voice channels per RF channel; the length of the circuit is 2500 km. It is represented in Fig. 1.24. The 2500 km is divided into radio hops. (In practice, the hops average 45 km, so approximately 55 repeaters are placed in cascade.) This circuit is organized in sections of 280 km, at the extremities of which the signal is demodulated to the baseband.

In television, the 2500 km is divided into three sections of 833 km at the extremities of which the signal is demodulated to the video band.

The quality objectives essentially concern the signal-to-noise ratio and the tolerable distortions in the baseband. They take into account the specificity of the signals transmitted: telephone, television, sound modulation channels.

Availability objectives

They are defined as the probability that a device will fulfill its required function under given conditions and in a given time. Consequently, the quality goals are specified during the availability period. The hypothetical reference analogue circuit is considered as unavailable, in at least one transmitting direction, when one or both of the following conditions is fulfilled during at least 10 consecutive seconds.

1. The baseband level is brought to 10 dB below the reference level.
2. The unweighted noise power, with an integration constant of 5 ms, in any telephone channel, is greater than $10^6\ \mathrm{pW_o}$.

After long discussions, the CCIR has set for the hypothetical reference circuit an availability objective of 99.7% of the time over a period equal to or greater than one year (99.7 is a compromise between 99.5 desired by some experts, and 99.9 requested by others). This objective can be used on a kilometric basis for real circuits with lengths $\geqslant 280$ km.

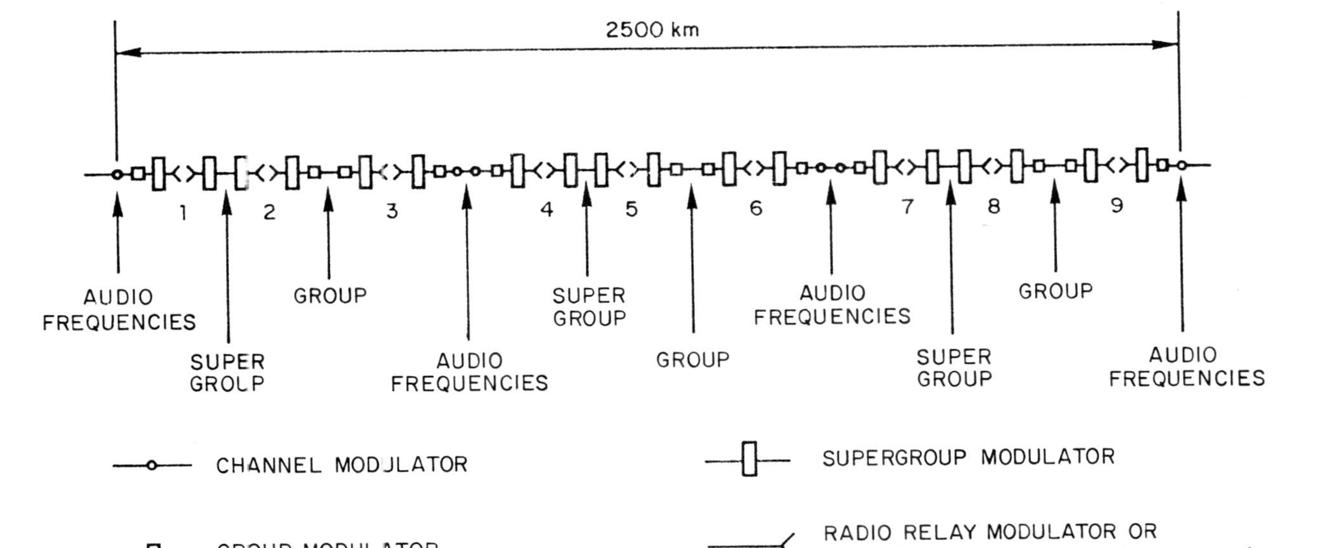

Fig. 1.24 Hypothetical reference circuit.

Main contributors to circuit noise

They are as follows.

1. Thermal noise which depends upon the propagation attenuation. This is an additive noise. It occurs even if the baseband signals are not fed to the modulation input.
2. Basic equipment noise which is independent of the propagation attenuation. It is considered an additive type.
3. Intermodulation noise which is created following the non-linearities described in section 1.3.6. In the case of telephony, it depends upon the telephone load (number of voice channels simultaneously active), that is, upon the average power P of the multiplex signal which modulates the microwave link. This noise is generally considered as multiplier type and shows up when the multiplex signal is fed to the modulation input.

In television, the non-linear distortion of the first and second types, via sound modulation subcarrier intermodulation, creates spurious lines in the part of the baseband occupied by the video. Generally, we can consider these lines as periodic noises. In fact, in operation, these subcarriers are frequency modulated. The result is an energy dispersion effect which reduces this noise by about 20 dB. Furthermore, the distortions due to the presence of the video signal produce noise components around subcarrier frequencies, components which are due to video spectrum harmonics.

Influence of the propagation

Figure 1.25 shows reference plans for transmitting power P_e at the access of antenna A_1 with gain G, and receiving P_r at the access of antenna A_2 with G_2. So we have:

$$10 \log\left(\frac{P_e}{P_r}\right) = 92.45 + 20 \log f + 20 \log d - G_1 - G_2$$

where d is the distance between the antennas (km), f is the frequency of the carrier (GHz) and L_0 is the loss in free space $= 92.45 + 20 \log f + 20 \log d$. In the

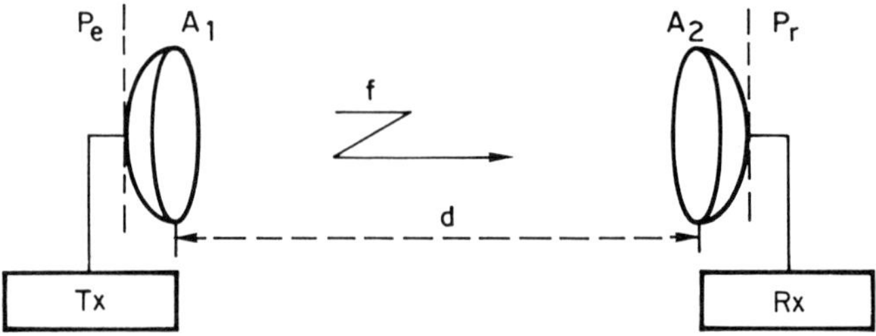

Fig. 1.25 Antenna accesses.

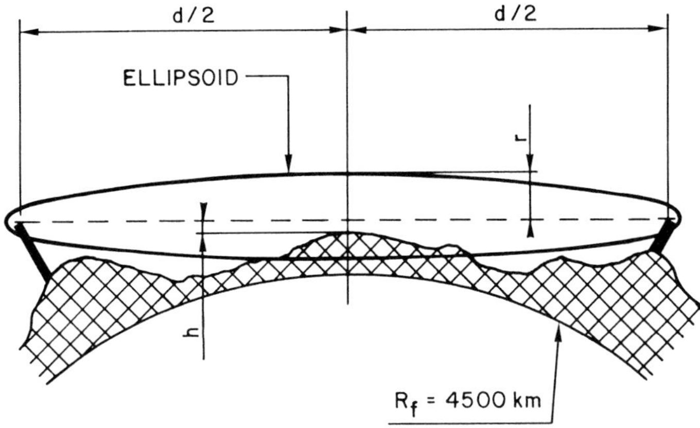

Fig. 1.26 Path profile and Fresnel zone.

propagation loss calculation, it is necessary to add to L_o, the losses due to the atmospheric refraction and to the obstacle effect on the Fresnel ellipsoid.

A variation of the refraction index n with altitude h is the cause of a curve in the microwaves. In the microwave link performance study, we position ourselves in the unfavourable case where the effective earth radius is $R_f = 4500\,\text{km}$ ($6500\,\text{km} \times k$, with $k = 0.7$).

The clearance of the RF path is studied based on the Fresnel ellipsoid. Figure 1.26 shows an earth section and a clearance rule:

$$h = 0.2r \qquad \text{with } r = \frac{\sqrt{\lambda d}}{2}$$

where r and $d/2$ are the Fresnel ellipsoidal radii.

Multiple path propagation It must be attributed to reflections on the ground or to atmospheric irregularities which result from stratification into horizontal layers; the weather conditions assume an absence of wind and a high index of the air (Fig. 1.27). Multiple paths cause fading.

When the amplitudes of the secondary waves (reflected or refracted) are weak in comparison with that of the main wave, reception is affected only by scintillations of about 1 dB or a few decibels. On the other hand, if the amplitudes of the secondary and main waves are similar, reception is affected by very strong subtractive fluctuations from 20 to 40 dB. In certain cases, additive fluctuations of about 10 dB have been observed. They can sometimes saturate receivers.

In the case of line-of-sight microwave links, there is a fairly low probability that such phenomena will appear. This can be determined based on measurements in the field. Figure 1.28 shows a recording of the level received during a very troubled period.

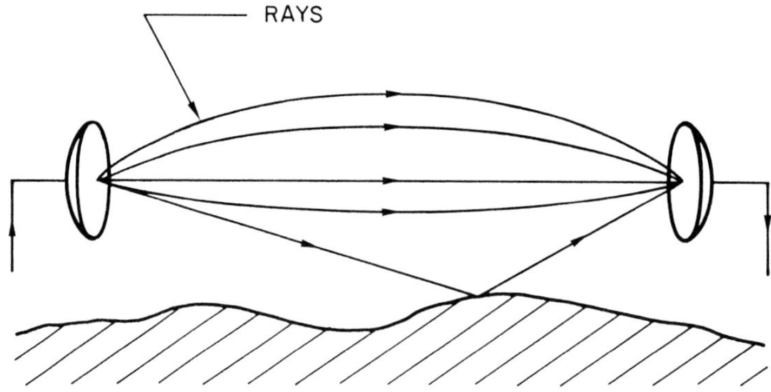

Fig. 1.27 Multipath propagation.

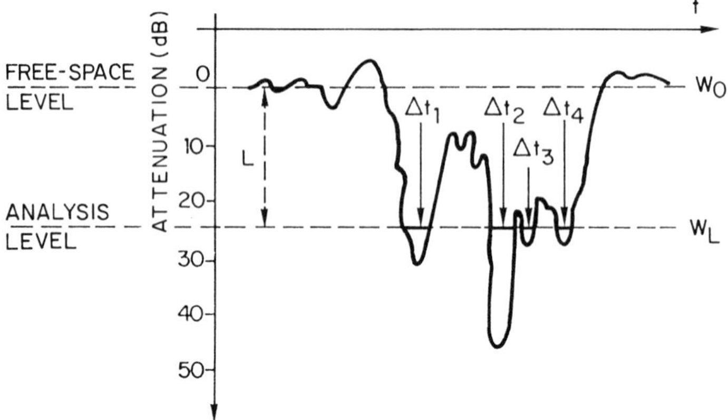

Fig. 1.28 Received level recording.

So we have shown that for $L > 15\,\text{dB}$, the probability p that the received power W will be equal to or less than the level analysed W_L, is given by the formula

$$p(W \leqslant W_L) = KQf^B d^c \frac{W_L}{W_o}$$

where d is the radio hop length (km), K is the climate coefficient, Q is the coefficient dependent upon the land, W_o is the received power in free space and B and C are coefficients dependent on the profile of the link and of the climate.

The average duration of the fading corresponds to the time interval during which the received power remains below L (dB) with relation to the free space which depends upon the geographical position. For the USA it is given by:

$$\Delta t = 56.6 \times \frac{1}{l} \times \sqrt{\frac{d}{f}}$$

where d is in km, f is in GHz, Δt is in s and L is the depth of the fading in dB ($10 \log l = L$).

Influence of hydrometeors and gases Their influence is expressed by an increase in the propagation attenuation.

For the hydrometeors, this attenuation is negligible below 8 GHz. From 8 to 20 GHz, it increases rapidly with the frequency (dB/km $= kf^2$). Furthermore, the presence of this phenomenon can cause a depolarization effect on the propagation trajectories. When precipitation is intense, propagation by multiple paths is very weak in general, one has only to add the time percentages corresponding to these two causes of fading.

The attenuation due to absorption by atmospheric gases is taken into consideration for frequencies greater than 12 GHz. These phenomena are analysed in Volume 2.

Influence of the frequency and of the length of the radio hops If the radiating surface of the antennas S is constant, which is equivalent to taking the wind effect at the top of the pylons as the only strain, we indicate that the maximum attenuation l between transmitter and receiver takes the following form:

$$l = K \frac{d^{5.5}}{S^2 f}$$

This relationship leads to the following conclusions.

1. The average length d of radio hops must remain moderate. An order of magnitude of 45 km is desirable.
2. The frequency f must also be as high as possible as long as hydrometeors and gases do not come into play (up to 11 or 12 GHz).
3. For the short links (5 to 15 km), the use of frequencies higher than 12 GHz is desirable.

Frequency arrangements

General The following rules can be defined.

1. For an unilateral radio channel, at least two frequencies are needed per repeater, one for sending and one for receiving.
2. In the case of telephony, which requires bilateral channels, two or four frequencies are needed per channel. A choice between two and four frequencies depends upon the front-to-back ratio which can be taken from the antennas. This decoupling increases with the frequency and can reach 65 to 70 dB once $f \geqslant 4$ GHz.
3. For technical and economic reasons, it is necessary to connect several channels on a single antenna (see section 1.2.2.).

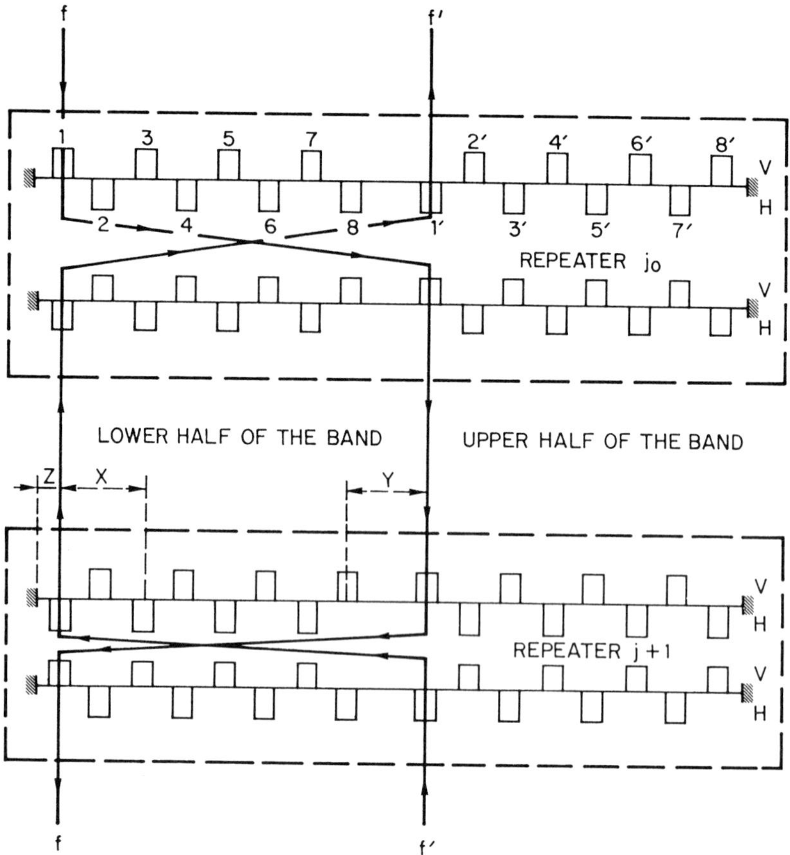

Fig. 1.29 Frequency assignment pattern: H, horizontal polarization; V, vertical polarization.

Organization of frequency arrangements CCIR defines the fundamental criteria for determining the frequency plans used with analogue systems. In analogue systems, we generally use the frequency arrangements with alternating polarizations (vertical and horizontal) of the RF signals for the adjacent channels. Figure 1.29 shows such an arrangement.

Some examples of frequency arrangements, defined by the CCIR, are shown in Table 1.5. (f_0 is the central frequency of the frequency band; Δf is the spacing between the central frequencies of adjacent channels; and ΔF is the spacing between the central frequencies of adjacent transmitting and receiving channels). Other frequency bands are used from 13 to 40 GHz. Their use is different from one country to another.

Certain authorities propose rules accompanying the application of frequency plans. Among them, the Federal Communication Commission (FCC) in the

Table 1.5

Frequency band (GHz)	*Band limits* (GHz)	*Number of channels*	f_o (MHz)	Δf (MHz)	ΔF (MHz)	*RF channel capacity in voice channel*	CCIR Rec
2	1700–1900	6	1808				
	1900–2100	6	2000				
	2100–2300	6	2203	14[b]	49	60–120–300	283
	2500–2700	6	2586				
	1700–2100[a]	6	1903	29[b]	68	600–1800	382
4	3700–4200[a]	6	4003.5	29[b]	68	600–1800	382
	3700–4200	6	—	40	—	1260	382
6	5925–6425[a]	8	6175	29.65[b]	44.5	1800	383
	6430–7110	8	6770	40	60	2700	384
7	7425–7725	20	7575	7–14	28	60–120–300	385
8	8200–8500[a]	6	8350	11.662	70	960	386
11	10700–11700	12	11120	40[b]	90	1800	387
13	12750–13250[a]	8	12996	28	70	960	497
	12750–13250	—	12996	14	—	300	497

f_0 is the central frequency of the frequency band; Δf is the spacing between the central frequencies of adjacent channels; ΔF is the spacing between the central frequencies of adjacent transmitting and receiving channels.
[a] The RF channels of these frequency bands can be used for transmitting television.
[b] Possibility of creating interleaved channels, offset by $\Delta f/2$ from the main channels.

United States, proposes restrictive rules. They are often adopted as models by other authorities. The main recommendations are about:

the frequency bands reserved for point-to-point communication;
the maximum bandwidth used to perform a given service;
the maximum average power of a transmitter;
the repartition of the transmitted power versus frequency;
antenna characteristics (nature, gain, side lobe level);
the frequency stability of transmitters.

Interference

Channel arrangements with two frequencies are the ones which are utilized the most with microwave links and the same pair of transmitted–received frequencies is used at all the hops.

Under these conditions, the interference in a system using microwave links can be broken down into two parts:

1. the interference caused by signals transmitted on single radio hop;
2. the interference produced by signals coming from different directions in a nodal station.

In these two possible cases, the interference degrades the reception threshold of equipment for a certain quality value, thus reducing the available margin against fading. The result is an increase in the accumulated duration of service interruptions. From then on, it is necessary that the level of the source of disturbance is less than the threshold level of the receivers, to prevent the receiving and sending of false signals when the normal link is cut (transmitter failure).

Interference on a single radio hop

Figure 1.30 shows the disturbance paths. On each radio hop, the useful and disturbing signals undergo the same fading. There are three types of interference:

1. the interference between channels using the same frequency and the same polarization;
2. the interference between co-polarized adjacent channels;
3. the interference between cross-polarized adjacent channels.

The magnitude of the interference of the first two types depends upon radio equipment characteristics (type of modulation, distribution of the filtering, etc.) and upon the performances of the antennas (radiation patterns, XPDO, front-to-back ratio, etc.).

The influence of interference between cross-polarized channels depends upon the discrimination factor (XPD) and on its degradation during troubled propagation periods. This is the transpolarization phenomenon. It has two origins.

1. The depolarization of a signal due to turbulence, reflection and scatter in the atmosphere. In this case, the radiation diagram in cross-polarization of the antenna has no effect on this phenomenon.

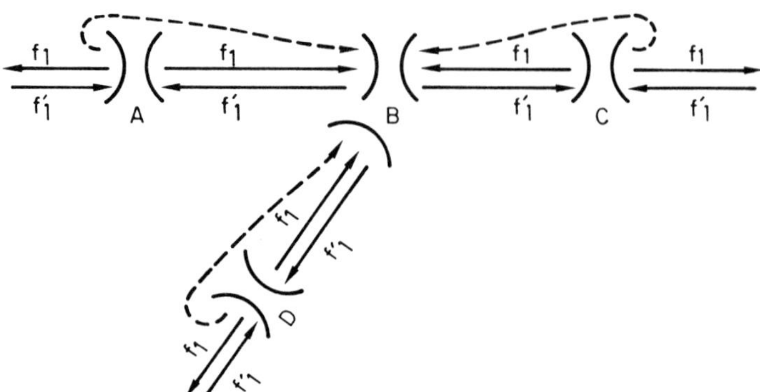

Fig. 1.30 Interferences on the same radioelectric hop.

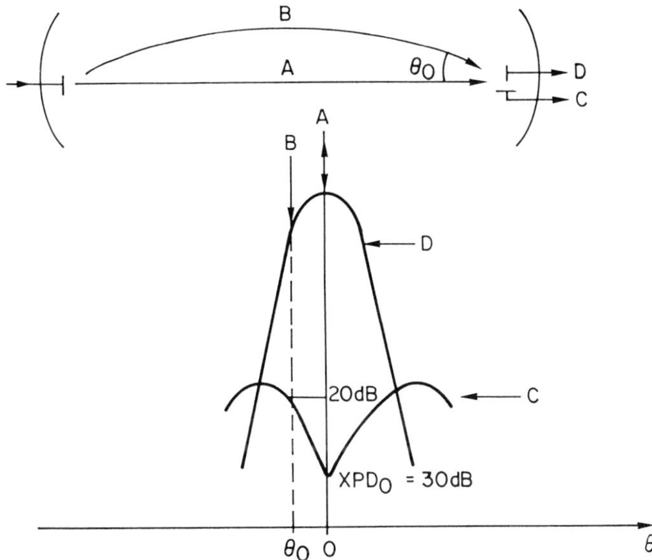

Fig. 1.31 Additional effects of the antenna cross-polarized pattern.
D: $A = 1$, $B = 0.9$, $A - B = 0.1 \rightarrow 20$ dB.
C: $A_c = \text{XPD}_0$, $B_c = 0.9 \times \text{XPD}_0$, $A_c - B_c = 0.03 - 0.9 \times 1 = 0.06 \rightarrow$ 24.5 dB.
XPD = 4.5 dB for 20 dB.

2. The additional effects of the cross-polarized patterns of the antennas. The refraction index variations cause a modification in the incoming angle of radio waves with regard to the maximum decoupling direction and the XPD drops. Figure 1.31 shows this phenomenon.

The reductions in XPD are associated with multipath fading and rain attenuation of the co-polarized signal (CCIR proposes empirical formulae to take that into account).

Interference in a nodal station Figure 1.32 shows the disturbance paths. The useful and disturbing signals do not undergo the same fading. The influence of the disturbance sources is maximal when the useful signal is being subjected to a fading. The influence of the interference depends upon the following parameters:

1. difference between the attenuations of the useful parts and of the disturbing parts;
2. characteristics of the radio equipment (sending power, type of modulation, distribution of the filtering, etc.);
3. performances of the antennas (radiation diagrams, XPDO, etc.).

Interference effect In the baseband, it is due to the interaction of the spectral

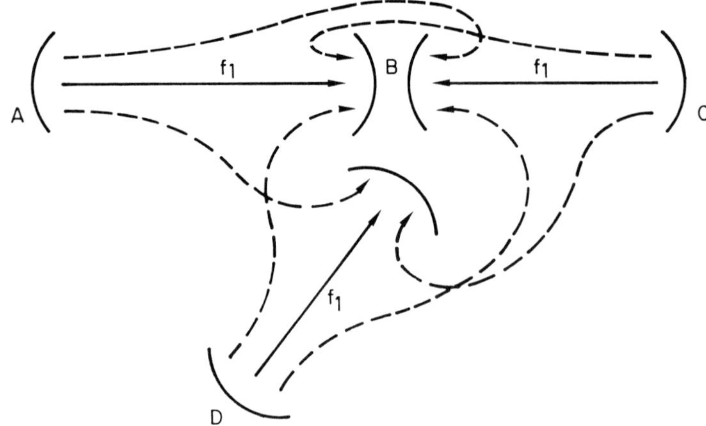

Fig. 1.32 Interferences in the nodal station.

densities of the useful and spurious signals. In the case of voice channel transmission, the following must be considered:

1. interference on a pure frequency;
2. interference between channels which are adjacent or which are working on the same frequency, and modulated by multiplex signals;
3. sine frequencies.

Additional information can be found in CCIR publications. The effects of interference in a TV channel of frequency modulation microwave link systems are also examined in CCIR publications.

Linear distortion in the baseband

Telephony The amplitude–frequency distortion in the baseband is due to the amplitude–phase conversion and to the spurious amplitude modulation due to the residual slope of the group delay as a function of the frequency in each non-demodulating repeater (see CCIR recommendations).

The propagation can be the cause of the distortion of the amplitude–frequency response in baseband by disturbing the modulation spectrum because of the selectivity of the fading. The consequence of the phenomenon is an increase in the transmission at the upper extremity of the baseband (selective attenuation of the carrier, increasing the modulation index). This very transient problem (a few milliseconds) is substantial (3 to 10 dB).

Television The main distortions can be summarized as follows.

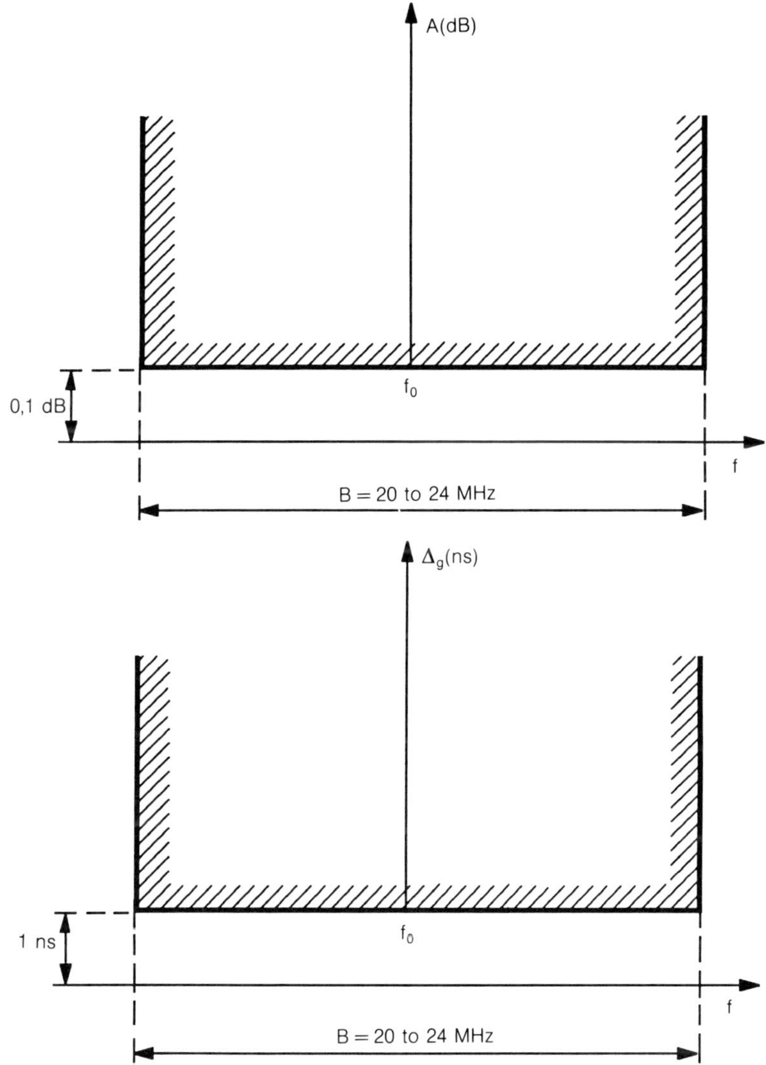

Fig. 1.33 Residual distortions. A, attenuation; τ_g, group delay.

LINEAR DISTORTION – DIFFERENTIAL GAIN (DG) AND DIFFERENTIAL PHASE (DP)

For each repeater, the CCIR recommendations are satisfied by keeping DG $< 1\%$ at ± 12 MHz and for the DP, by obtaining the group delay response, fitting the envelope, Fig. 1.33.

SYNCHRONIZATION (SYNC) SIGNAL NON-LINEARITY DISTORTION

The main causes of this distortion are located in the transmission and reception video amplifiers and in the non-linearity of the modulation and demodulation.

However, the characteristics of modems, after the application of the linearization method, can be used to minimize the non-linearity of this origin.

LINEAR DISTORTIONS OBSERVABLE BETWEEN BASEBAND ACCESSES

A very good response in very low frequency of video amplifiers is fundamental to satisfy the recommendations concerning the signals with a duration equal to one frame or to longer durations. The structure of the coupling and band limitation filters in baseband for the video associated with the sound modulation subchannels is substantial for the transients (rise time, overswing, etc.). This remark is valid for the analysis of differences between the luminance and the chrominance.

1.3.8 Quality improvement

Section 1.3.7 defines the main contributors to the noise of the circuit. It is possible to reduce the magnitude of certain contributors by using appropriate means. In particular, the intermodulation noise can be reduced by using amplitude correctors and group delay equalizers. The general idea is to create a distortion with the sign opposite the initial distortion in order to cancel it.

Amplitude correctors The losses increase the attenuation distortion in passband B at the filters as shown in Fig. 1.34. It is shown as symmetrical with relation to f_o. This is only true if we have taken certain precautions for the choice of the structure of the filters, mainly those made in intermediate frequency (IF). The correction involves introducing a circuit whose amplitude–frequency response compensates for the distortion. Figure 1.35 shows the correction circuit in IF, designed to work between characteristic impedances (source and load $R_o = 75\ \Omega$).

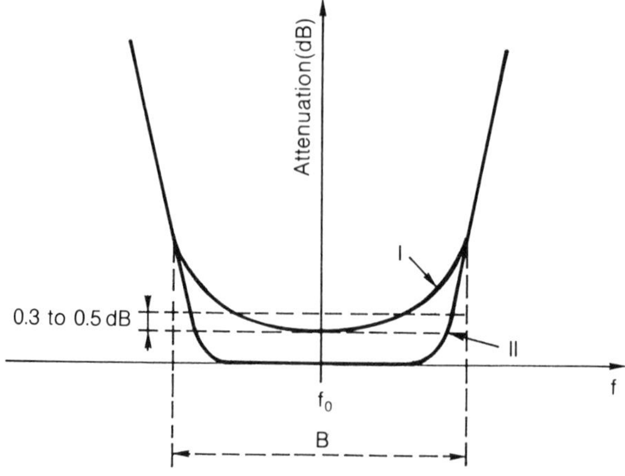

Fig. 1.34 Attenuation in the pass band: I, with losses; II, without losses.

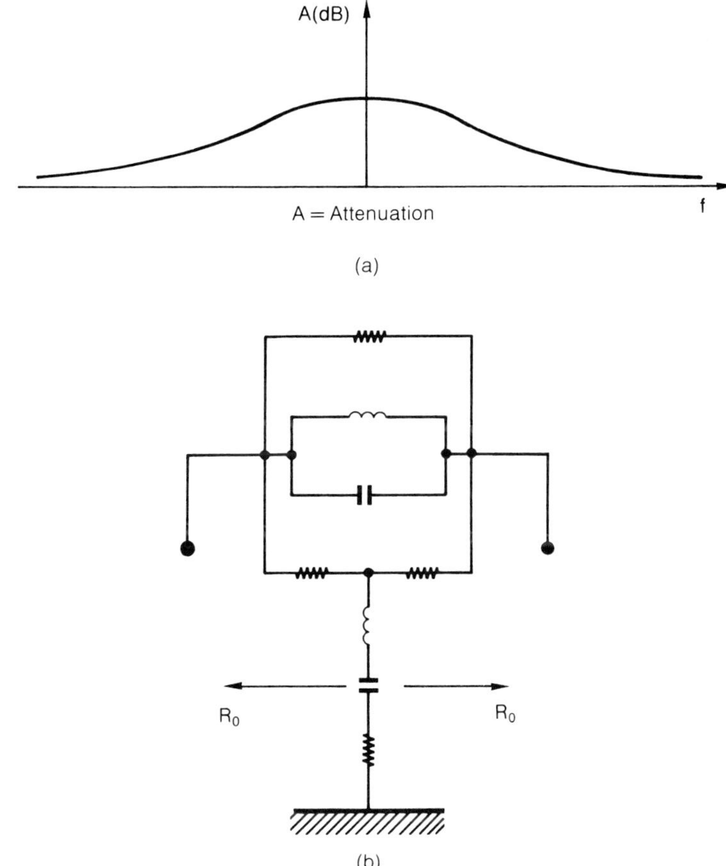

Fig. 1.35 Amplitude equalization: (a) equalization; (b) equalizer with constant characteristic impedance R_0.

Group delay equalizers Expressed in another way, what is involved is correcting the phase distortion, defined as the non-linearity of the phase as a function of the frequency. Since the group delay is a derivative of the phase with relation to the frequency to within the factor 2π, the correction involves making the group delay as constant as possible as a function of the frequency.

Figure 1.36 shows the typical group delay response of an RF repeater. Using a filter, partly compensating that curve is possible and Fig. 1.37 shows the typical correction which we can obtain, using all-pass networks with staggered tuned circuits, operating in IF. It is also possible to build RF group delay equalizers, by associating a cavity resonator with a circulator, as shown in Fig. 1.38.

In the present state of the art, residual distortion can satisfy the CCIR recommendations for television.

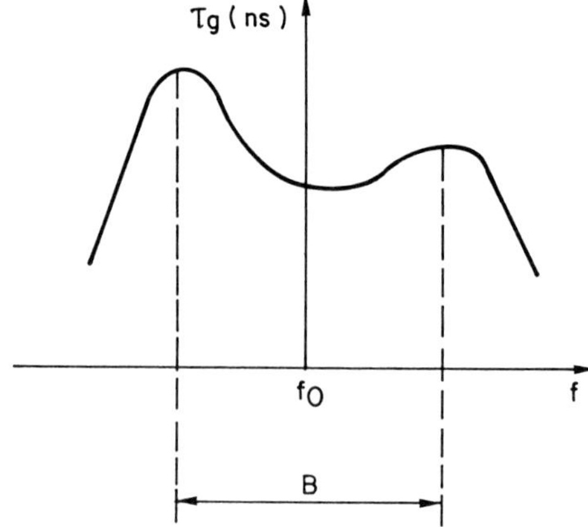

Fig. 1.36 Group delay distortion for a repeater.

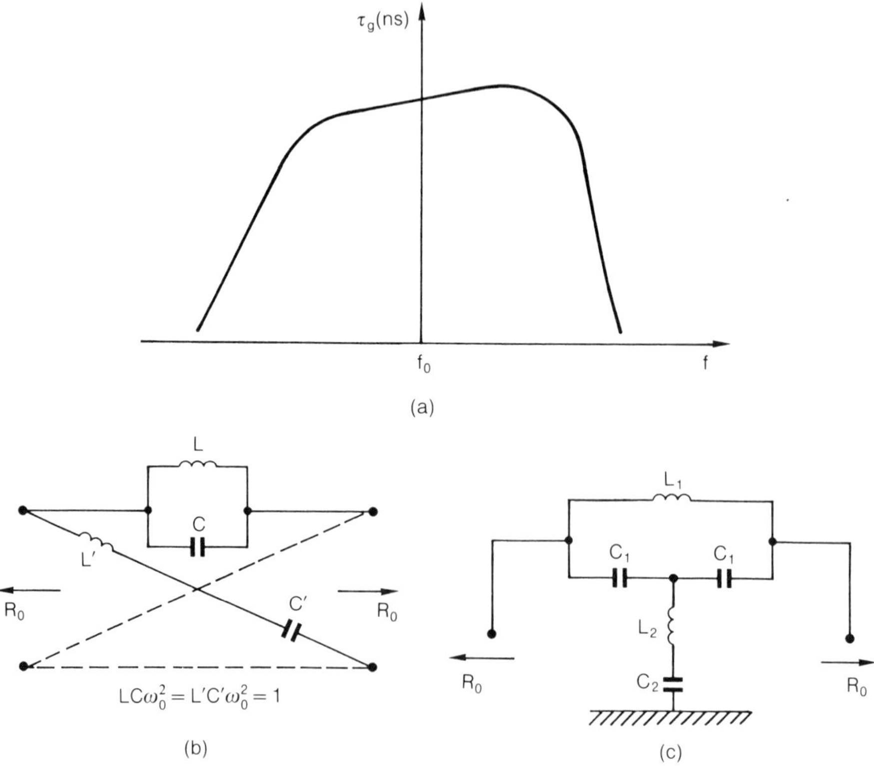

Fig. 1.37 Group delay equalization: (a) equalization; (b) equalizer; (c) equivalent bridged T network.

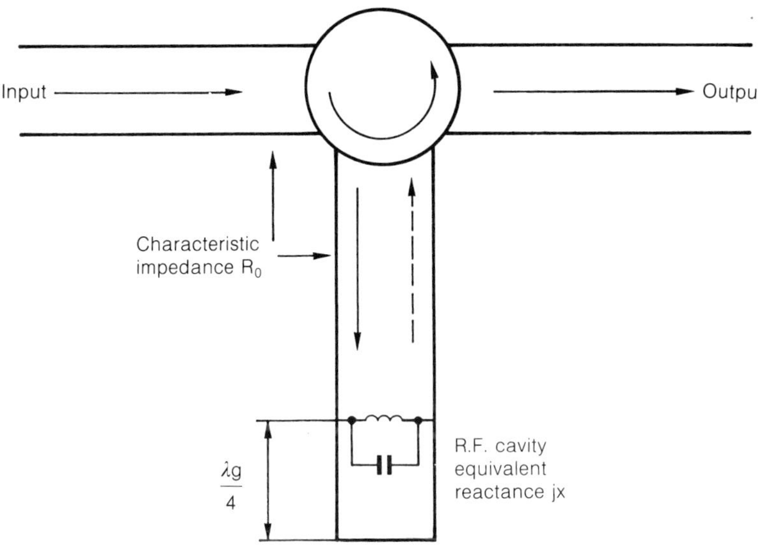

Fig. 1.38 Group delay equalization.

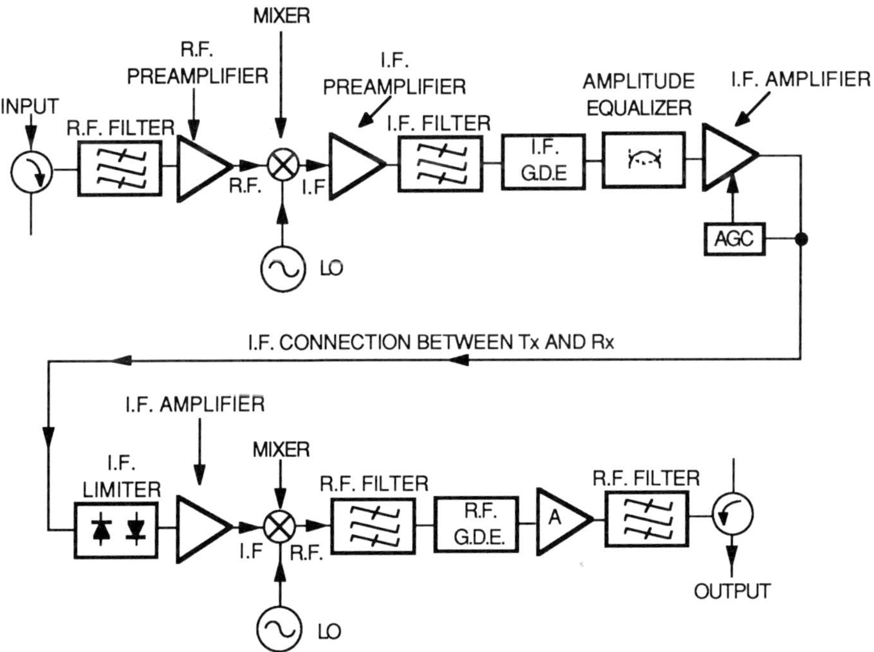

Fig. 1.39 Optimal position for the correctors: AGC, automatic gain control; GDE, group delay equalizer; LO, local oscillator.

Optimal position for the correctors Although we have obtained residual distortions as mentioned in the preceding paragraph, after demodulation we sometimes find different distortions which are incompatible with expectations, if insufficient attention has been given to the amplitude–phase conversion and to the consequences of the saturations which mask the selectivity of the passband filters and circuits in an amplification system. To avoid this anomaly, the arrangement in Fig. 1.39 should be adopted.

Means for countering propagation effects

For a microwave link, the consequence of a fading is a brief increase in the noise in the voice channel or on the television picture, following the reduction in the

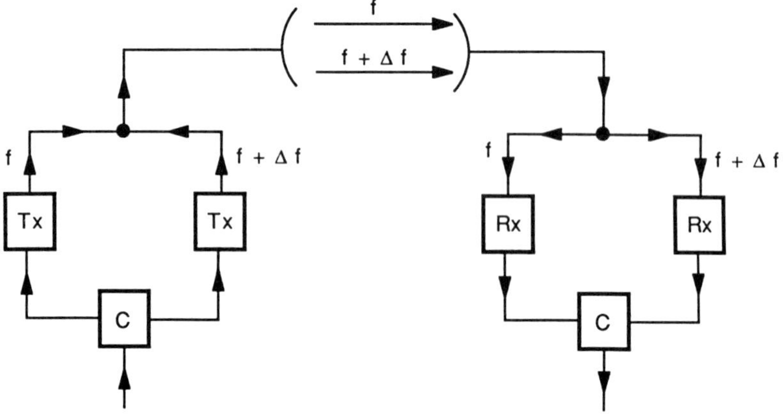

Fig. 1.40 Frequency diversity: C, combiner; Tx, transmitter; Rx, receiver.

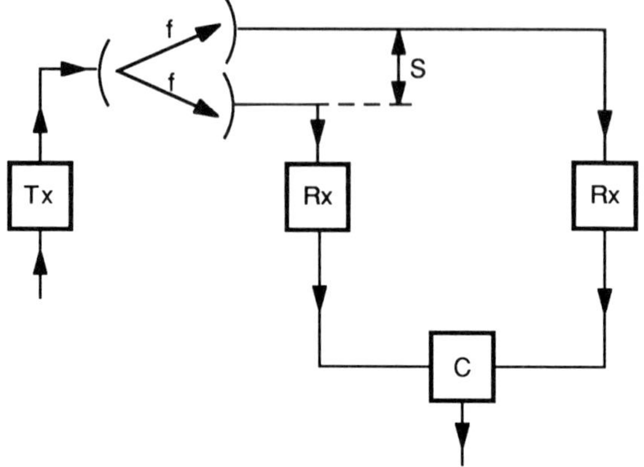

Fig. 1.41 Space diversity: C, combiner; Tx, transmitter; Rx, receiver; S, vertical separation of received space diversity antennas.

receiving power. A first way of alleviating this phenomenon is to increase the transmitting power to compensate the brief extra attenuation (approximately 40 dB). This is not always possible due to the transmitting level choice and certain restrictions (choice of itineraries, length of hops, etc.).

Another way is seeking two or more means to carry the same data. They can be characterized by the carrier frequency (frequency diversity, Fig. 1.40), or by the path (space diversity, Fig. 1.41). The main idea is that we seek paths so that there will be no correlation between the fadings.

1.3.9 Transmission quality

This quality essentially concerns the signal-to-noise ratio and the tolerable distortions. The quality calculation method takes into account the specific nature of the transmitted signals, i.e. telephony, television, sound modulation subchannels. The calculation is performed for a radio hop, and then for n hops in cascade.

Telephony

This involves calculating the signal-to-noise ratio (S/N) in any voice channel which is up- or down-converted around a frequency F of the baseband whose limits are F_a and F_b.

Thermal noise The S/N ratio is computed in a conventional manner (see Volume 2). Assuming that CCIR recommendations are followed (regarding modulation, pre-emphasis gain, etc.) the signal-to-noise ratio is essentially given by

$$\frac{\mathrm{S}}{\mathrm{N}}(\mathrm{dB}) = P'_{\mathrm{e}}(\mathrm{dBm}) - A(\mathrm{dB}) - \mathrm{NF}(\mathrm{dB}) + K(\mathrm{dBm})$$

where P'_{e} is the transmission power, NF the noise factor, $A(\mathrm{dB}) = 10 \log(P'_{\mathrm{e}}/P'_{\mathrm{r}})$, P'_{r} is the reception power and K is given by Table 1.6.

Basic equipment noise By planning the S/N curve as a function of P'_{r} (Fig. 1.42), we bring out the basic equipment noise which is independent of RF input level,

Table 1.6 Thermal noise figuring in S/N

RF channel capacity in voice channels N	960 (3886 kHz)[a]	1800 (7600 kHz)[a]	2700 (11 700 kHz)[a]
K (dBm)	119.2	110.27	106.65

[a] Baseband frequencies to perform the measurements (CCIR recommendation).

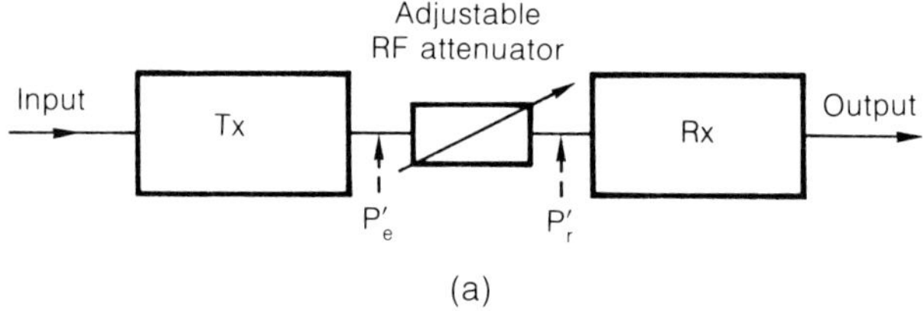

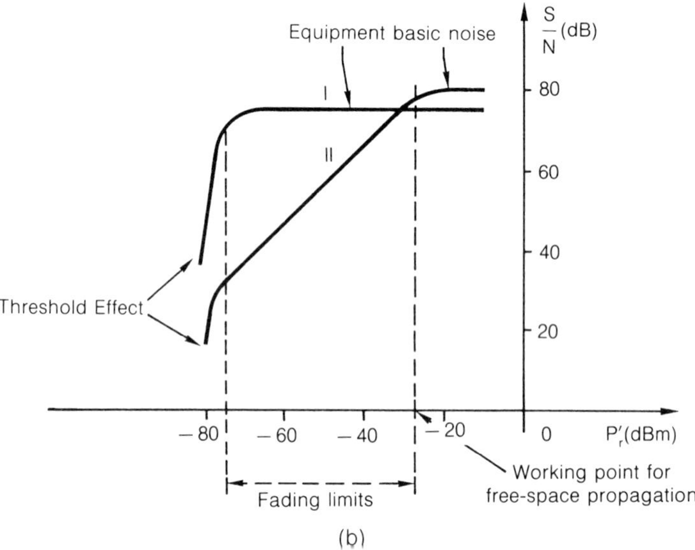

Fig. 1.42 Basic noise power of the equipment and frequency modulation threshold: (a) measurement conditions; (b) typical curve. I = low channel (frequency, F_a); II = high channel (frequency, F_b).

and the frequency modulation threshold power P_{rth}, characterized by the following relationship:

$$P_{rth} = \frac{\text{RF carrier with power } P'_r}{\text{Noise power in the reception band}} \simeq 10\,\text{dB}.$$

Intermodulation noise This noise is created following non-linearities (section 1.3.5). Table 1.7 gives an example of intermodulation noise (I_{22}) due to the amplitude–phase conversion.

In the case of weakly mismatched waveguides, the intermodulation signal-to-noise ratio is given by:

$$\frac{S}{I} = \text{echo attenuation (dB)} + 22\,\text{dB}$$

Table 1.7 Intermodulation noise due to AM/PM

Voice channel capacity N	l_{22} *with* n *repeaters without demodulation* (pW)	
	$n = 5$	$n = 10$
1800	70.7	1132
2700	239	3821

Evaluation of noise contributions

It is generally admitted that the various noise contributions are added together in power. To illustrate the method, we shall refer to Fig. 1.43 which models a cascade of heterodyne repeaters on a 280 km section and we obtain:

$$\left(\frac{\mathrm{S}}{\mathrm{N}}\right)_{\mathrm{t}} = \frac{S_{\mathrm{o}}}{N_1 + N_2 + N_3 + N_4 + N_5}.$$

By adopting for S_{o} a power of 1 mW and for the different values of N as a conventional unit the picowatt.

Table 1.8 gives an initial approximation of the noise contributions in telephony in the case proposed in Fig. 1.43 (1800 voice channels). But for the intermodulations, it is necessary to take into account the fact that the relevant noise does not always vary like the number n of hops but a part of it varies like n^2 or like n^4. In our present example, the total noise in pW due to the intermodulation is found to be given by the following relationship:

$$30n + 6n^2 + 0.02n^4.$$

So, with $n = 5$ the column to the right of the table should be replaced by:

$$30 + 30 \times 5 + 6 \times 25 + 0.02 \times 625 = 342.5\,\mathrm{pW}.$$

This is instead of 180 pW. The total noise is found to be equal to 442.5 pW. This leads to a signal-to-noise ratio in free space of:

$$\frac{\mathrm{S}}{\mathrm{N}} = 63.54\,\mathrm{dB}.$$

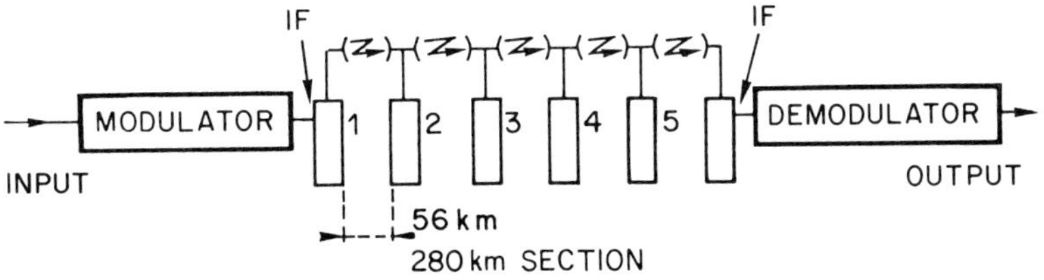

Fig. 1.43 Cascade of heterodyne repeaters.

Table 1.8 Noise approximations for cascaded heterodyne repeaters

Noise contribution origins	*Dependent on the propagation attenuation* (pw)	*Independent of the propagation attenuation* (pw)
Modulator–demodulator		
frequency modulation noise + first kind intermodulation		30
First repeater		
thermal receiver noise	20	
intermodulation of RF repeater + waveguide		30
Second repeater		
thermal receiver noise	20	
intermodulation of RF repeater + waveguide		30
Third repeater		
thermal receiver noise	20	
intermodulation of RF repeater + waveguide		30
Fourth repeater		
thermal receiver noise	20	
intermodulation of RF repeater + waveguide		30
Terminals		
thermal receiver noise	20	
intermodulation of RF repeater + waveguide		30
	100	180
Total	280	

Television

The evaluation of the continuous random noise is calculated based on the fundamental modulation parameters. The other noises depend much more on the particular technological design of the equipment. For frequency modulation microwave links, if the recommendations are respected for the luminance signal-to-noise ratio, the chrominance is generally good enough.

Thermal noise In television, the signal-to-noise ratio is expressed by the ratio, in decibels, of the nominal amplitude of the luminance at the rms voltage of the

Table 1.9 Thermal noise for video bandwidths

Video bandwidth (MHz)	6	5
K_{TV} (dBm)	124.77	126

noise after limitation of the band and weighting. It is shown that, with the modulation and weighting parameters given by the CCIR, we obtain, as in the case of telephony (NF = noise factor)

$$\frac{S}{N}(\text{dB}) = P'_e(\text{dBm}) - \text{NF}(\text{dB}) + K_{TV}(\text{dBm}) - \text{A}(\text{dB}).$$

Table 1.9 gives the values of K_{TV}.

For example, for a video band ($F_c = 6\,\text{MHz}$), a transmission power ($P'_e = 30\,\text{dBm}$), a noise factor (NF = 3 dB) and free space conditions ($A = 64.5\,\text{dB}$) (link of 56 km at 6.175 GHz), we obtain: S/N = 87.27 dB. With a fading of 40 dB, the signal-to-noise ratio becomes 47 dB.

Based on these results, we can calculate the rms voltage of noise U_N, which is added to the video signal, a point where the luminance develops a maximum voltage of 700 V (the peak-to-peak voltage of the complete video signal with the synchronization at 1 V):

$$20 \log \frac{700}{U_N} = \frac{S}{N}(\text{dB})$$

$U_N = 0.03\,\text{V}$ without fading and $U_N = 3\,\text{V}$ with fading.

1.4 DIGITAL MICROWAVE LINKS

1.4.1 Characteristics of the signals transmitted

Introduction

The digital signals contain data at discreet and generally repetitive instants. The value (voltage) of the signal is quantified at a finite number of levels; so the fidelity when restoring the signal is less critical. However, it is necessary to regenerate this signal, usually at each repeater station, whereas the analogue microwave repeater stations limit themselves to amplifying the signal and to changing its frequency.

Primary multiplex

To create a time division multiplex (TDM), the signal of each voice channel is limited by filtering the 300 to 3400 Hz band, and is sampled at 8000 Hz, that is, one sampling every 125 μs. (That is valid for commercial equipment; military equipment is different and could request less data rate per voice channel.)

The samples are quantified at $2^8 = 256$ levels, symmetrical with relation to zero voltage. These levels are not regularly spaced, the spacing (A-law or μ-law companding: CCITT Recommendation G.711) increasing for the high voltages, to make the quantification signal-to-noise ratio less dependent upon the signal level. Each sample is then represented by eight bits, which corresponds to a digital rate of 8 bits 8000 times per second, that is 64 kbit/s. The quantification and the encoding are performed either by an encoder shared by the channels, directly delivering the TDM signal, or by channel specific encoders, working at 64 kbit/s.

In Europe, the 125 μs frame comprises 32 equal time slots. The first 30 intervals (IT_1 to IT_{15} and IT_{17} to IT_{31}) each contain 8 bits (1 octet), the sign bit and other high order bits being transmitted first. The A companding law is used.

Time slot IT_0 is reserved for synchronizing the frame and time slot IT_{16} is used for sequentially transmitting the signalling data of the 30 channels. So the rate is 32×64 kbit/s, or 2048 kbit/s. At each frame, we transmit in the IT_{16}, the signalling data of two of the channels, 4 bits for each. So we must set up a 'superframe' of 15 frames plus 1 frame for its synchronization, with a duration of 16×125 μs, or 2 ms.

In the USA, Canada and Japan, the organization of the primary multiplex is different; the frame only has 24 channel slots, and the bit rate of the primary multiplex (D2 channel bank), 1544 kbit/s, is not a multiple of 64 kbit/s. The quantification is only performed in eight bits 5/6 of the time, and during the remaining proportion of the time, 1/6, the eight bit is used for signalling. Furthermore, to avoid the appearance of a long series of zeros which might disturb the operation of some transmission systems, certain quantification levels are not taken into account (we say that there is no 'transparency' of bits at level of 64 kbit/s, which is troublesome for data transmission). The logarithmic companding law (CCITT law μ) is different, etc. Therefore, in an international link, it is sometimes necessary to perform a code translation at the voice channel level. This code translation is the responsibility of the country which uses law μ.

Higher order multiplexing

To build a higher order multiplexing, we interleave, bit by bit, the bits of several lower order incoming multiplexes, very often four of these incoming multiplexes. A difficulty then appears which is specific to time-division multiplexing, due to the non-synchronism of the incoming streams. In general, these have not been generated in the same centre. So their clocks come from oscillators with different quartz crystals, and their digital bit rate is the same, but only to within the accuracy of the quartz crystal oscillators, that is $\pm 5 \times 10^{-5}$. They are called plesiochronous. To solve this difficulty, the bits of each incoming stream are written in memory, at the rate of the clock of the incoming stream. The frame of the higher order multiplex is obtained by reading these memories sequentially and by periodically inserting a sync signal so that the distant demultiplexer can distinguish the streams.

The read speed must be greater than the sum of the maximal bit rates of the incoming streams, to prevent the fastest incoming stream memories from filling up and therefore losing data bits. There are then necessarily instants when certain memories, those of the slowest incoming streams, will be empty at the moment of the read. We then insert, in the flow of output bits, a meaningless bit, called the 'justification bit' or 'stuffing bit'; to indicate to the distant demultiplexer that it must take the stuffing bit into account or leave it aside, we also insert 'justification indication bits'. The demultiplexer distinguishes the data bits from the justification bits and the justification indication bits by the positions (which are well defined in advance) which these last bits occupy in the multiplex frame. So plesiochronous multiplexing and demultiplexing operations do not introduce errors.

The bit rates Only the 64 kbit/s level is common to the whole world, although the characteristics of the signal are not always the same. For higher levels, two bit rates coexist, based respectively on a primary multiplex at 2.048 Mbit/s in Europe and 1.544 Mbit/s in North America and in Japan.

The levels in Table 1.10 are standardized for Europe. Furthermore, certain transmission systems operate at 560 Mbit/s (4 × TN4).

In the USA and in Canada, the bit rate levels are as shown in Table 1.11. Furthermore, certain systems operate at a fourth level of 274.176 Mbit/s, which is not standardized.

In Japan, the upper levels are different, see Table 1.12. Furthermore, certain systems operate at a fourth level 396.2 Mbit/s, which is not standardized.

Synchronizing networks Certain countries, such as France, have partially synchronized their network to eliminate the errors created by time-division

Table 1.10 Bit rates in Europe

Designation	*Number of channels*	*Binary rate* (Mbit/s)	*Duration of the bit* (μs)
TN1	30	2.048	0.488
TN2 = 4 TN1	120	8.448	0.118
TN3 = 4 TN2	480	34.368	0.029
TN4 = 4 TN3	1920	139.264	0.007

Table 1.11 American bit rates

Level	*Number of subchannels*	*Binary rate* (Mbit/s)	*Duration of the bit* (μs)
1	24	1.544	0.648
2	96	6.312	0.158
3	672	44.736	0.022

Table 1.12 Bit rates in Japan

Level	Number of subchannels	Binary rate (Mbit/s)	Duration of the bit (μs)
1	24	1.544	0.648
2	96	6.312	0.158
3	480	32.064	0.031
4	1440	97.728	0.010

switching and to prepare the ISDN of the future, for which it is planned to have, inside a national network, all the clocks synchronized from a cesium master clock, with long term stability better than 10^{-11}. The communications (switching and transmission) will then be error-free within a country, and with only one error octet every two months in an international communication.

Another consequence of network synchronization, in the future, will be to considerably simplify the multiplexing operations because the incoming streams will be synchronized. At the CCITT, the standardizing of a new range of synchronous multiplexing levels is now being prepared.

Properties of the digital signals in baseband

When the demodulation is coherent as is the case for high capacity microwave links, the modulation and demodulation operations lend themselves easily to this reasoning because the correspondence between the modulated signal and the baseband signals is simple. At sending, the digital signal to be transmitted almost always comes in the form of a binary signal whose voltage takes only two discreet values, constituted by a series of regularly spaced bits. The integrated circuits generally supply signals with the same polarity. At the modulator input, it is necessary to make these voltages symmetrical, by using a network of resistors for example (an asymmetry in the modulated signals would introduce, into the spectrum of the modulated signal, a line at the carrier frequency which would drain energy uselessly).

The symbols What we call a symbol is an element of the digital signal, ready to be transmitted. We often feel the need to simultaneously transmit the data contained in several consecutive bits, whose number n is, in practice, equal to 1, 2, 3 or 4. So it is a symbol with $N = 2^n$ levels which is fed to the modulator input (Table 1.13). At the output of the distant demodulator, the recognition of one of these special levels is used to determine, in parallel (that is simultaneously, on n wires), and without ambiguity, the value of n bits. A simple logic operation is then used to put these bits back into serial form (sequential appearance, on a single wire). So the duration of the symbol is n times greater than that of the bit.

Table 1.13 Values of the levels

Number of bits n *transmitted in serial form*	*Number of levels* 2^n	*Levels (balanced voltages)*							
1	2	−1							1
2	4	−1		−1/3			1/3		1
3	8	−1	−5/7	−3/7	−1/7	1/7	3/7	5/7	1
4	16	−1	−13/15............................					13/15	1

The passband necessary for the transmission depends upon the timing (clock) of the state changing of the symbols, a timing which is n times lower than the timing of the bits, which does not depend upon the number of levels. We therefore economize the passband by transmitting the bits in parallel. With a given passband, the data capacity (number of bits transmitted per unit of time) is proportional to n, therefore to the base two logarithm of the number of levels N.

We show that the minimal passband (maximal frequency of the baseband) which is necessary to the transmission of symbols of duration T is equal to $1/2T$ (Nyquist band). In this band, it is possible to transmit a digital rate of n/T bits per second.

Effect of received thermal noise The increase in the number of levels of the symbols does not only have advantages. At reception (Fig. 1.44), at the sampling instants and assuming that the channel filtering function is taking place without intersymbol distortion (see below), at the demodulator output we find the same signals as those which were sent, plus the thermal noise generated in the receiver input states.

Let us standardize at ± 1 values corresponding to the peak powers received and sent. For a given antenna diameter, a given transmitting power and a given receiver noise factor, this range of ± 1 is affected by the same thermal noise amplitude, whatever the number of levels N. The uncertainty concerning the determination of the levels due to noise increases as N increases. This uncertainty is expressed by a probability of error (or error rate). There is an error in the determination of the level of a symbol if the noise, at a sampling instant, brings the signal from one decision zone to another. For a given error rate, it is necessary to size the peak power of the transmitter according to the square of $N-1$. The above illustrates the general principle of the 'band–power trade-off', also valid for analogue modulations.

The error rate calculation is based upon probability p that a Gaussian noise of zero average and rms voltage σ will exceed the threshold voltage algebraically (in one polarity):

$$p = 0.5\,\mathrm{erfc}\,(a/\sigma\sqrt{2})$$

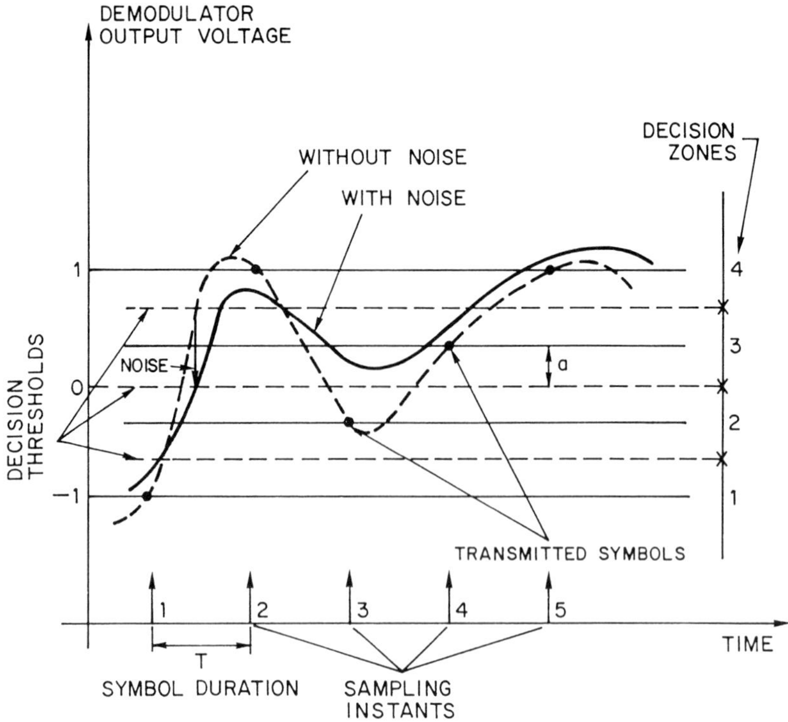

Fig. 1.44 Errors due to thermal noise, for the case of a four-levels signal. Only symbol no. 3 will be erroneous, after regeneration.

erfc(x) being the complementary error function:

$$\operatorname{erfc}(x) = 1 - (2/\sqrt{\pi}) \int_0^x \exp(-u^2)\,du.$$

Let us suppose that the N levels have the same probability, which is the general case for microwave links.

Case of a signal with two levels The probability of error on the bits is equal to the value p calculated above with the threshold voltage a being equal to the peak amplitude of one pulse ($a = 1$ in Fig. 1.44).

Case of a signal with N levels If we ignore the probability of the appearance of a noise peak so strong that it would bring the signal into the decision zone of a non-adjacent symbol, the probability that a symbol will be erroneous is p for the extreme levels, and $2p$ for the others whose recognition is confronted both by negative noise and by positive noise.

If the symbols have equal probability, the error probability on the symbols is the average of these probabilities, that is:

$$p_s = 2p(N-1)/N.$$

For each error concerning a symbol, in general there is only one error bit along the $\log_2 N$ bits which define it (this assumes, that we are using Gray coding, so that two adjacent symbols only differ by a single bit).

So the probability of error on the bits is equal to:

$$p_b = \frac{2p(N-1)}{N \log_2 N}$$

But, in the formula giving p, threshold voltage a is $N-1$ times lower than in the case of two levels, that is $1/(N-1)$ for Fig. 1.44. So the error probability on the bits is:

$$p_b = \frac{N-1}{N \log_2 N} \operatorname{erfc}\left(\frac{1}{(N-1)\sigma\sqrt{2}}\right)$$

The error rate varies very quickly as a function of the noise level; for approximately 4 dB noise level variation, the error rate goes from 10^{-3}, the value above which the quality of a telephone channel is considered unacceptable, to 10^{-6}, which corresponds to an inaudible deterioration.

Intersymbol distortion There is a second cause of error on the determination of the sampled value of a particular symbol which is the 'intersymbol distortion', due to symbols transmitted before and after the symbol considered.

When we apply a single narrow pulse (Dirac pulse) at the input of a four

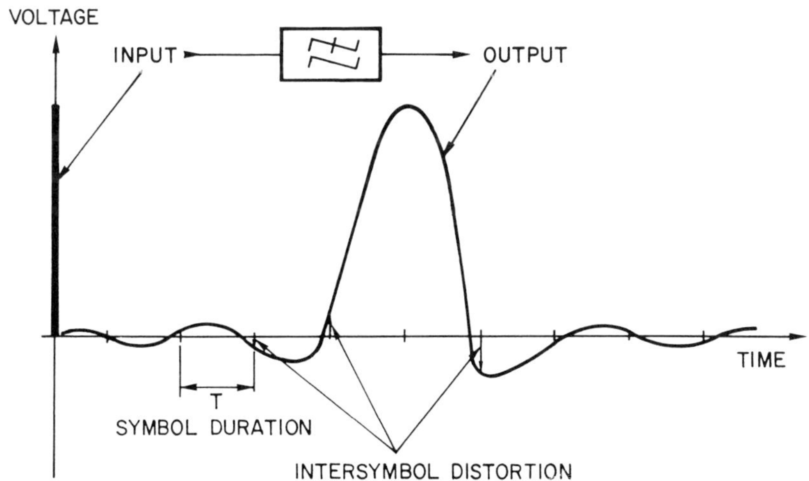

Fig. 1.45 Percussional response of a low-pass filter.

terminals network, constituted, for example, by the equipment of a microwave link, falling between the modulator input and the demodulator output, the signal observed at the output is delayed and widened by the limitation of the passband, and generally presents ripples, one of which is greater than the others and corresponds to the input pulse, in sign and in amplitude (Fig. 1.45).

In reality, the modulator input signal consists of a series of pulses with different signs and, for a number of levels greater than 2, with different amplitudes. If the network is linear, the demodulator output signal is equal to the sum of the output signals corresponding to the individual pulses. So the contribution of the other pulses, before and after, are algebraically added to the value obtained at the sampling instant for a particular pulse.

There generally exists a particular phase of the clock signal which controls the sampling for which the sum of these contributions is minimal. This optimal sampling instant is definitely visible on the 'eye pattern' (Fig. 1.46), obtained by observing the demodulator output signal on the oscilloscope, and by synchronizing the horizontal time base of the oscilloscope in order to superpose all the pulses.

Families of filters exist for which this minimum is zero. Their impulse response is reduced to zero periodically, at separated repeating instants of T, except at a

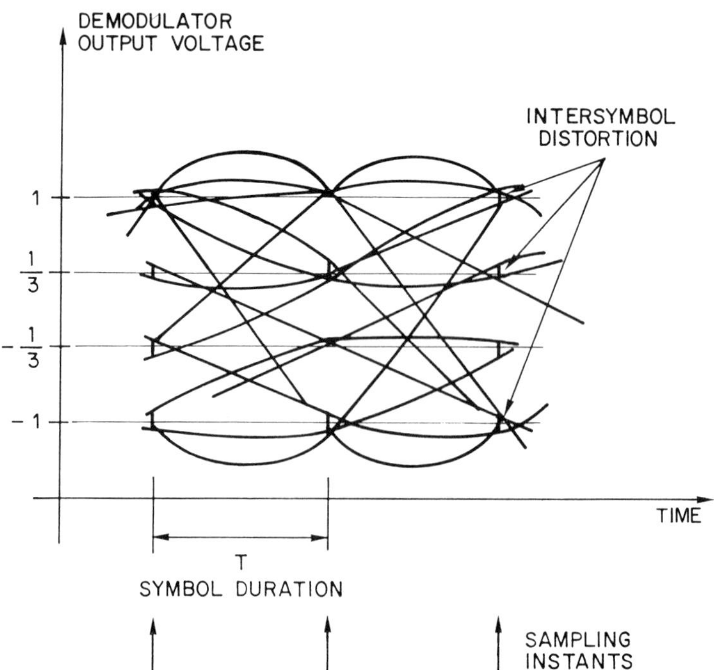

Fig. 1.46 Eye diagram.

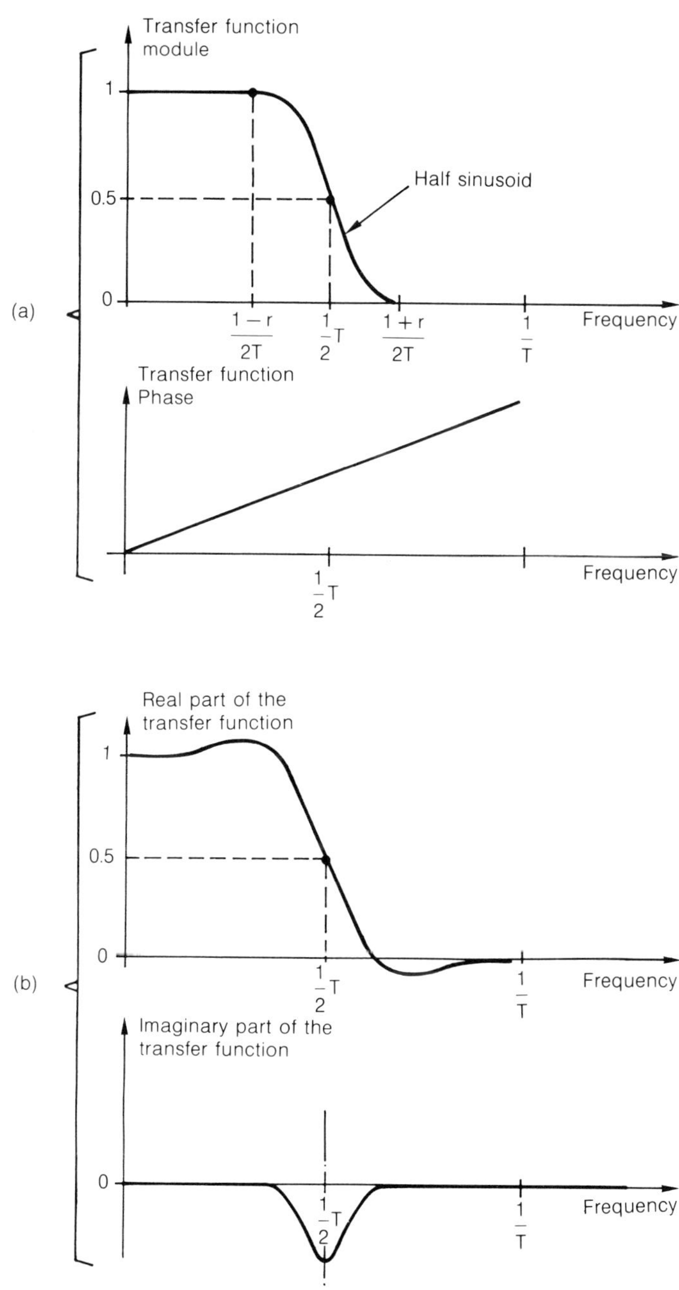

Fig. 1.47 Intersymbol free filters: (a) particular case of raised cosine filters; (b) general case.

single instant which is representative of the level of the transmitted pulse. The best known is the one for the so-called 'raised cosine' filters (Fig. 1.47(a)) characterized by the 'roll-off' factor r $(0 < r < 1)$.

1. The modulus of their transfer function is, as a function of the frequency, equal to 1 up to $(1 - r)/2T$, zero beyond $(1 + r)/2T$ and describes a half cosine between these two values.
2. Their group propagation time delay is constant.

More generally, we can be sure that the transfer function of a low-pass filter is free of intersymbol distortion as follows:

1. From the transfer function group delay, subtract its value at zero frequency, by placing in series a negative length dummy delay line, then express the resulting transfer function in real part and in imaginary part, then normalize to 1 the real part at 0 frequency. The absence of intersymbol distortion will then occur at time $t = 0$.
2. The real part must be symmetrical to the point with the abscissa $1/2T$ and ordinate $1/2$, and must reduce to zero beyond $1/T$.

The intersymbol distortion should be smaller, and therefore the filters should be all the more accurate as the number of levels increase. The same is true of the linearity of the circuits.

Consequence of the logic circuit memory The above assumes that the modulating signals are carried by narrow pulses (Dirac pulses), present at characteristic instants of the digital signal. In reality, the modulators are driven by signals obtained from 'off-the-shelf' logic circuits, which store in memory the data present at a characteristic instant (for example, clock pulse rise) until the following characteristic instant. An isolated pulse has a rectangular form with duration T, and therefore has a spectrum at $(\sin x)/x$ with $x = 2\pi FT$, instead of the white spectrum which is characteristic of a Dirac pulse.

It is understandable that by inserting into the transmitting channel, generally when sending, a linear phase filter whose transfer function modulus is at $x/\sin x$ (Fig. 1.48), 'whitening' the spectrum in this way, everything occurs as if the modulation pulses were narrow. This filter cannot be set beyond $F > 1/T$ but, very fortunately, since the channel transfer function must be cancelled beyond $(1 + r)/2T$, the whitening filter transfer function can be anything beyond that limit.

Review concerning the adapted filter It has been shown that the probability of error in detecting a pulse in the presence of a Gaussian white noise is minimal if the receiver transfer function is the imaginary conjugate of the Fourier transform of the pulse. For white spectrum Dirac pulses and linear phase raised cosine filters, it follows that the filtering function must be shared equally by transmission and reception. The square of the modulus of the receiver transfer function therefore takes the form shown in Fig. 1.47(a); from this we deduce that the

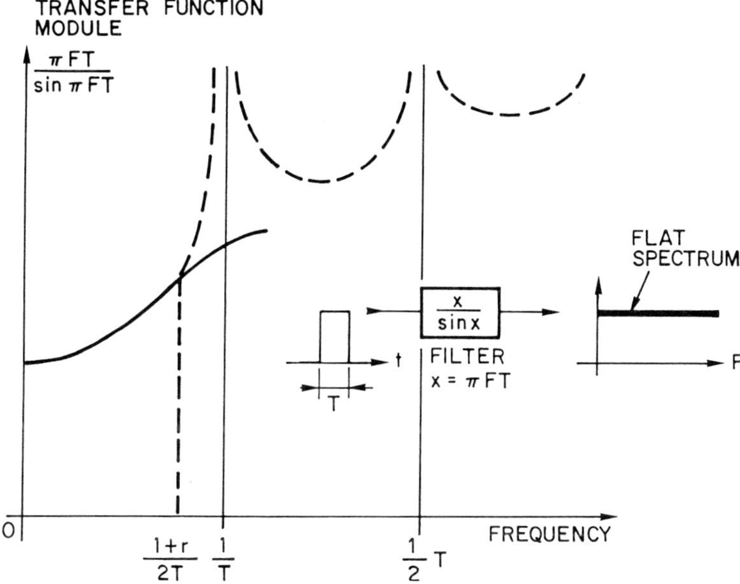

Fig. 1.48 Spectrum whitening filter.

equivalent receiver noise band is equal to the Nyquist band $1/2T$ ($1/T$ after modulation).

Regeneration Because of the deliberate limitation of the passband of the RF channels (we always try to use the least band possible for a given digital bit rate), the demodulator output signal is the continuous type; a unit called a regenerator, specific to digital transmissions, performs the following functions:

1. based on the demodulated signal, it reconstitutes the clock signal optimal frequency and phase;
2. it uses this clock signal to sample the demodulator output signal and to determine the most probable quantification level (the so-called 'decision' operation).

The clock used to sample the signal at the demodulator output must be restored based on the signal itself. To do this, we generally use its zero crossings.

The spectrum of the signal does not include a line at frequency $1/T$ because it definitely goes to zero, on the average, one symbol out of two, near regularly spaced instants of a multiple of T, but as often in one direction as in the other. The most currently used method to create a line at the timing frequency is to generate a pulse of the same polarity each time it goes to zero, whatever its direction.

The line obtained in this way is assigned a phase noise (jitter) because the times when it goes to zero are not separated exactly by a multiple of T. So it is necessary to filter it, using a bandpass filter with circuits L and C, or using a phase-lock loop if we wish to obtain a small passband. The band must become weaker as the number of levels becomes higher because, when this number increases the jitter increases, and the eye opening width is narrower. The order of magnitude of the passband is a few kilohertz for a four level signal at 70 Mbit/s.

To avoid having to use an excessively narrow passband, which would increase the microphonics and the acquisition time of the loop, we sometimes seek to establish, at the timing regeneration circuit input, a high-pass type transfer function, without intersymbol distortion each time it goes to zero, whose pulse response goes through zero at an infinite series of recurrent instants separated by T, without exception (Fig. 1.49).

Such a transfer function must have the following characteristics, which are the opposite of those which cancel the distortion at the sampling instants. For a certain delay added to the filter's group delay, the impulse response is reduced to zero at every instant $t = T/2 \pm kT$, if:

1. the real part of the transfer function is symmetrical to abscissa straight line $1/2T$, and zero beyond $1/T$;

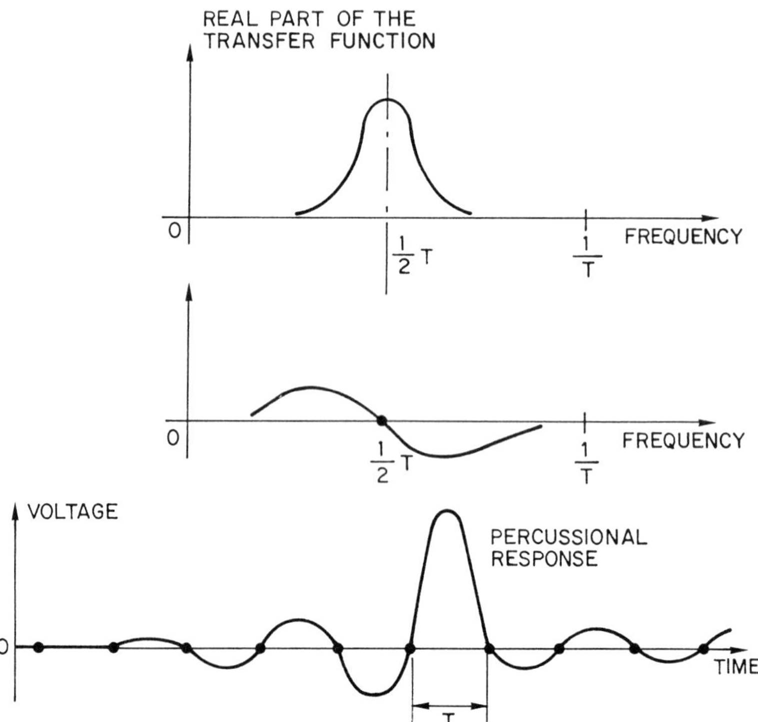

Fig. 1.49 Filters without jitter at zero crossings.

2. the imaginary part is symmetrical to abscissa point $1/2T$ at ordinate zero, and zero beyond $1/T$.

1.4.2 Digital modulation

Introduction

As for the analogue signals, modulation makes it possible to transmit by radio waves the data contained in a baseband signal. The spectral efficiency, which takes into account the proper utilization of the microwave frequency band, can be expressed in number of voice channels per unit of band, but is also given in a more general way in bits per second per hertz of band, setting up the ratio of the digital bit rate transmitted in one propagation direction to the passband of the corresponding microwave channel.

In the case of a 'co-channel' arrangement, for which the same nominal central frequency is used on two orthogonal polarizations for the transmission of two different signals, it is naturally necessary to double the value obtained for the spectral efficiency.

Figure 1.50 gives the organization of the subunits of a microwave link relative to the process of modulation–demodulation.

We are going to begin by describing the actual modulation and demodulation operation.

Amplitude modulation

The digital modulation of a carrier (Fig. 1.51) takes place either directly, or via an intermediate frequency. Let ω_0 be the pulsation of the carrier. The modulation of carrier $\cos \omega_0 t$ by digital signal $A(t)$ is done by means of a mixer, a non-linear device, in order to optimally simulate the mathematical operation multiplication $A(t) \cos \omega_0 t$ which supplies a modulated signal whose spectrum is centred on ω_0. This is an amplitude modulation, with the carrier suppressed since the modulating signal has a zero average.

After transmission via the microwave link, the demodulator reconstructs a local carrier and performs a new multiplication:

$$A(t) \cos \omega_0 t \times 2 \cos \omega_0 t = A(t) + A(t) \cos 2\omega_0 t$$

which makes it possible to come back to the modulating signal in baseband, after filtering the term in $\cos 2\omega_0 t$. We say that what is involved here is a 'coherent' demodulation.

In practice, the mixer also creates harmonics $A^m(t) \cos m\omega_0 t$. The harmonics characterized by $m = 1$ are the only troublesome ones because their spectrum is centred on ω_0. Everything then occurs as if we were transmitting:

$$A(t) + a_2 A^2(t) + a_3 A^3(t) + \cdots$$

which has two disadvantages.

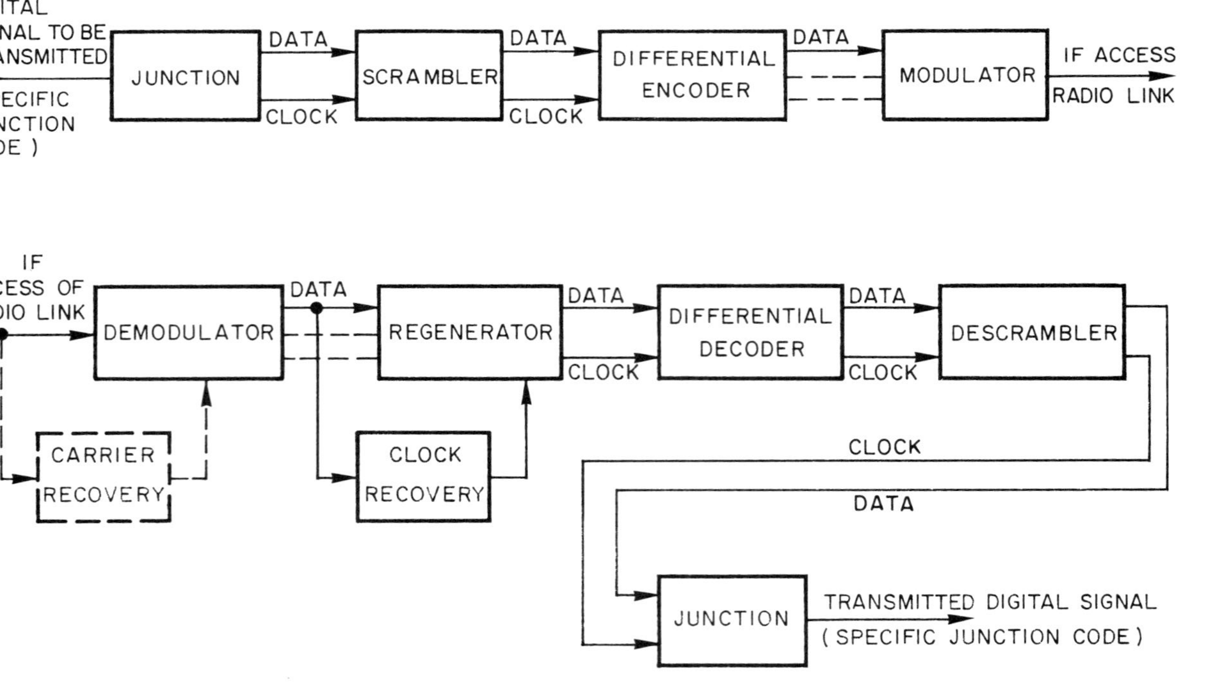

Fig. 1.50 Modem-associated functions.

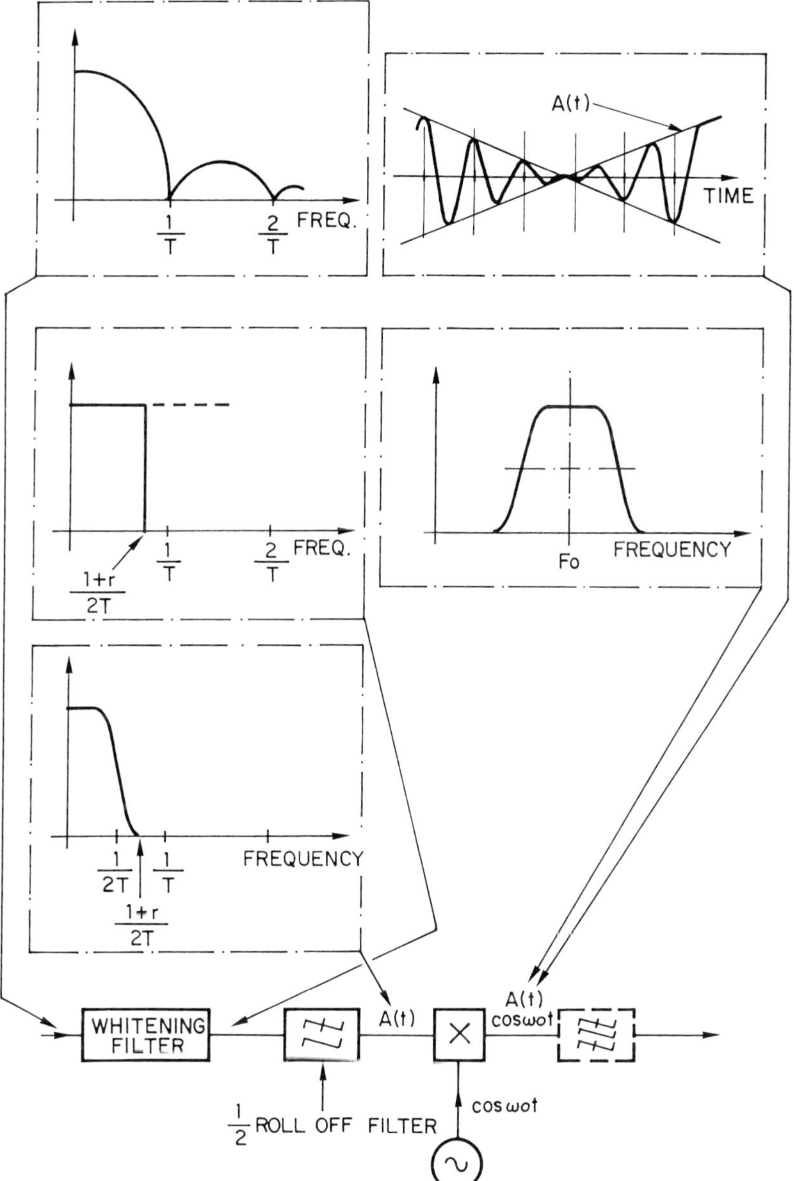

Fig. 1.51 Modulation of one carrier axis.

1. In the case of a number of levels greater than 2, the non-linearity destroys the regular spacing of the levels, generally by bringing the extreme levels closer together, which makes the demodulation more sensitive to noise.
2. The terms at $A^m(t)$ have a wider spectrum than for $A(t)$ which can overlap that of the neighbouring channels, and therefore jam them.

It is possible to reduce the level of these harmonics as much as we wish by reducing the level of the modulating signal $A(t)$ at the input of the mixer. On the other hand, there is no disadvantage in the local carrier level's being high. We then say the modulation is linear, for the modulating signal. When the modulation is linear, a low-pass filtering upstream from the modulator is equivalent to a bandpass filtering downstream.

Calculating the error rate The value of the error rate on the bits for a signal in baseband with N equiprobable levels was calculated earlier:

$$p_b = \frac{2p(N-1)}{N\log_2 N}$$

with $p = 0.5\,\mathrm{erfc}\,(a/\sigma\sqrt{2})$. In the case of a modulated signal, what is usually done is to express the error rate as a function of the ratio of the peak power of signal C to the noise power N_oB.

B is the measurement band, for example, the Nyquist band $B = 1/T$ (in the general case where the receiver is adapted and where the filtering function is in raised cosine, the equivalent noise band of the receiver is equal to the Nyquist band). N_o is the spectral density of the thermal noise, product of the equivalent noise temperature of the receiver times the Boltzmann constant.

The receiver input voltage is written as follows:

$$V(t)\cos\omega_o t + n_c(t)\cos\omega_o t + n_s(t)\sin\omega_o t$$

where $V(t)\cos\omega_o t$ represents the modulated signal on one of the axes of the carrier; $n_c(t)\cos\omega_o t$ and $n_s(t)\sin\omega_o t$ represent respectively, the noise in phase and in phase quadrature with the signal.

At the output of the coherent demodulator, the term at $n_s(t)\sin\omega_o t$ is eliminated and we obtain, at a characteristic instant t_o, (K being a proportionality coefficient):

$$\frac{(N-1)a}{\sigma} = KV(t_o) = K\sqrt{2R_o C}$$

for the signal, when the reception power reaches its peak value $C = V^2(t_o)/2R_o$. $Kn_c(t_o)$ for the noise, with an rms value $\sigma = K\sqrt{N_o B R_o}$. So we have $(N-1)a/\sigma = \sqrt{2C/N_o B}$ and the error rate is written as follows:

$$p_b = \frac{N-1}{N\log_2 N}\,\mathrm{erfc}\sqrt{\frac{C}{(N-1)^2 N_o B}}$$

and, if D represents the digital bit rate, $D = B\log_2 N$, we obtain:

$$p_b = \frac{N-1}{N\log_2 N}\operatorname{erfc}\sqrt{\frac{C\log_2 N}{(N-1)^2 N_o D}}.$$

The representation of the modulated signal in the Fresnel plan, at the characteristic instants, is constituted by aligned points. The modulated signal envelope is cancelled at each sign change of the modulated signal, that is, an average of one symbol out of two. This type of modulation is not used as such except for two phase modulation, where the number of levels is reduced to two.

Quadrature amplitude modulation

Let us establish two digital signals $A(t)$ and $B(t)$, each modulating a carrier with the same angular frequency, but whose phases differ by 90° (Fig. 1.52). We therefore obtain $A(t)\cos\omega_o t$ and $A(t)\sin\omega_o t$. If we sum these two so-called orthogonal signals in a coupler or in a resistor network, and if we transmit these signals via a microwave link, the distant demodulator may retrieve $A(t)$ and $B(t)$ respectively, by means of a multiplication by $\cos\omega_o t$ and by $\sin\omega_o t$. This is because:

$$2[A(t)\cos\omega_o t + B(t)\sin\omega_o t]\cos\omega_o t = A(t) + A(t)\cos 2\omega_o t + B(t)\sin 2\omega_o t$$
$$2[A(t)\cos\omega_o t + B(t)\sin\omega_o t]\sin\omega_o t = B(t) - B(t)\cos 2\omega_o t + A(t)\sin 2\omega_o t$$

since the pulsation terms $2\omega_o$ are easily eliminated by filtering.

The demodulator has a structure which is analogous to that of the modulator. The above described operation is advantageous in comparison with the amplitude

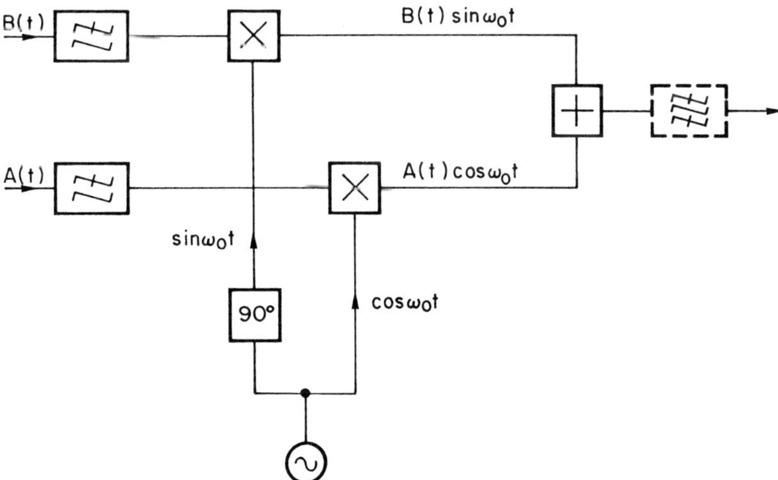

Fig. 1.52 Modulation of both carrier axes.

modulation of a single axis of the carrier: if $B(t)$ has the same digital bit rate at $A(t)$, we double the data flow transmitted in a given bandwidth and this only costs us an increase of 3 dB in the peak power of the transmitter. Nevertheless, it has a number of disadvantages.

1. If the filters in intermediate frequency and in microwave frequency are not perfectly symmetrical with relation to the frequency of the carrier, an additional intersymbol distortion appears between signals $A(t)$ and $B(t)$; a part of the modulation of the channel in sine appears on the channel in cosine, and vice versa.
2. The non-linearity of the transmitter also creates an intersymbol distortion; on the same axis by non-linearity of amplitude (non-linear variation of the level of the output signal envelope as a function of the level of the modulated input signal envelope) and from one axis to the other by non-linearity of phase (variation of the phase of the output signal as a function of the level of the modulated input signal envelope).
3. It is sometimes necessary to employ a system for removing the ambiguity between signals $A(t)$ and $B(t)$ at demodulation.
4. The noise of the regenerated carrier phase is much greater.

Table 1.14 Representation in the Fresnel plane of modulation states

Fresnel plane	Denomination	Number of levels/ numbers of states	Spectral efficiency (Bit/S/Hz)	Theoretical peak power for a same bit rate and BER $= 10^{-6}$ (dB)
• •	2 PSK	2/	1	0
• • / • •	4 PSK	2/4	2	0
 4 × 4 dots	16 QAM	4/16	4	+ 6.5
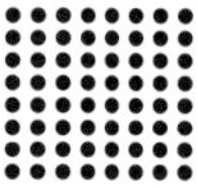 8 × 8 dots	64 QAM	8/64	6	+ 11.9
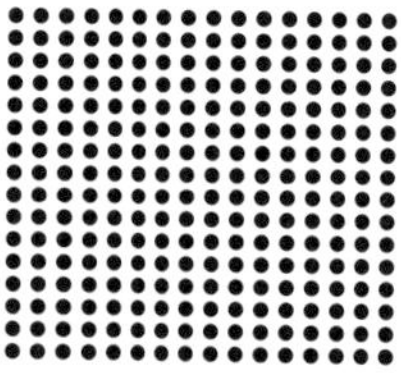 16 × 16 dots	256 QAM	16/256	8	+ 17.3

The two modulating signals may only be plesiochronous. In this case, it is necessary to use two different clock recovery circuits for the regeneration. However, in most cases, these two signals are synchronous, and furthermore, their transitions before filtering take place simultaneously. They come, for example, from the odd and even bits of a single digital signal. We then speak of QAM modulation (quadrature amplitude modulation). It is these modulations which are most used for high capacity microwave links.

Calculating the error rate The error rate on the bits corresponding to the modulation of one of the two axes was calculated previously:

$$p' = \frac{N-1}{N\log_2 N}\operatorname{erfc}\sqrt{\frac{C\log_2 N}{(N-1)^2 N_o D}}.$$

The representation of the modulated signal in the Fresnel plane, at the characteristic instants which are the same for the two axes, is constituted by horizontally and vertically aligned points. The number of points or 'states' is equal to the square of the number of levels of each axis (Table 1.14).

The envelope of the modulated signal varies considerably, as in the case of a modulation of a single access of the carrier, but it is almost never reduced to zero because this would require that the two modulating signals in baseband be cancelled out simultaneously, which is very improbable.

Carrier recovery

As mentioned earlier, the coherent demodulation requires having a sine wave which represents the unmodulated carrier. Since the demodulator is several dozen kilometres from the transmitter where the carrier was generated, it is necessary to reconstruct (to recover) this carrier, based on the modulated signal.

Residual carrier transmission The most simple method for indicating the direction of the carrier in the Fresnel plan is to create a dissymmetry in its direction or in a known direction relative to it (Fig. 1.53). This can be done simply by using slightly unbalanced signals at one or two inputs in the modulator baseband. In this way, we create a spectral line in direct current in the baseband, which the modulation translates at the carrier frequency. One has only to filter this line at reception to retrieve the carrier. Given the weak filtering band necessary to sufficiently reduce the phase noise due to the modulation, and the frequency drifts caused by the local RF oscillators at sending and at receiving, we use a filtering by phase-lock loop.

A compromise needs to be found between the substantial dissymmetry which consumes an excessively large proportion of the available transmitted peak power, and an excessively weak dissymmetry which, at a given phase noise, requires an excessively narrow filtering band. This method is only rarely used in spite of its

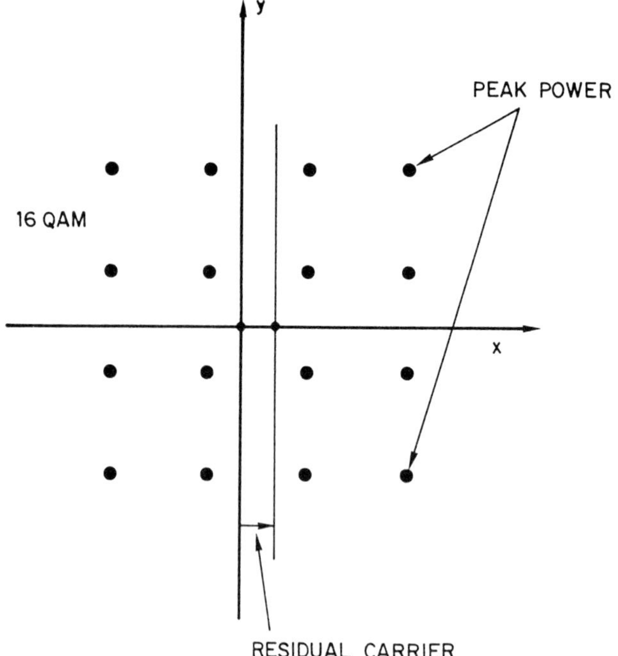

Fig. 1.53 Modulation of a residual carrier.

simplicity. It operates in open loop like the methods described in the following paragraph.

Carrier recovery in open loop. When a single axis is modulated, the extremity of the representative vector of the modulated signal in the Fresnel plane describes a straight line. The phase of the modulated signal therefore only varies suddenly in jumps of 180°, since the signal envelope is reduced to zero at each phase reversal. A rectification of the two half-cycles, followed by a voltage limiter, therefore gives square wave signals at a frequency twice that of the carrier, with no phase noise due to the modulation lines. In fact, every non-linear operation causes a line to appear at a frequency twice that of the carrier, without neighbouring modulation lines. By dividing the frequency by two (Fig. 1.54(a)), we obtain the carrier, or the carrier with its phase shifted by 180° (ambiguity of 180°).

When the two axes of the carrier are modulated, the extremity of the vector sweeps a surface. However, this surface extends more in the direction of the diagonals than in the direction of the axes, and we show that a non-linear operation on the modulated signal causes a line to appear at a frequency four times that of the carrier. Unlike the previous case, this line is accompanied by unwanted modulation lines. After narrow band filtering in a phase-lock loop and

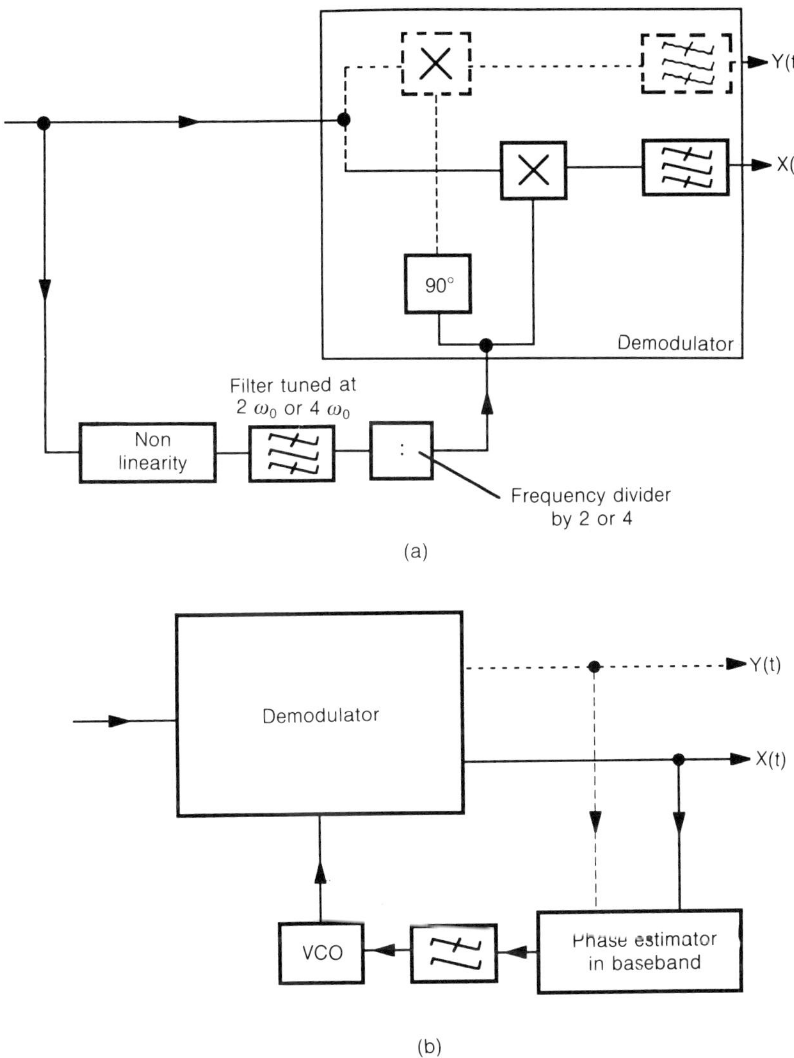

Fig. 1.54 Carrier recovery circuits: (a) open-loop arrangement; (b) closed-loop arrangement.

after dividing the frequency by four, we obtain a regenerated carrier with an ambiguity of 90°.

The carrier generation methods which we have just described are in open loop. It is inevitable that phase shifts take place in the recovered carrier at the various circuit points, due to the variations of the temperature and of the delay time caused by the semiconductors, by ageing of the components, etc. These variations which have a harmful influence on the operation of the demodulator, particularly when the two axes of the carrier are modulated (intersymbol distortion between signals A and B), are not corrected by the loop.

Carrier recovery in closed loop The recovered carrier is created by a voltage controlled oscillator driven, after low-pass filtering, by a signal obtained on the demodulator outputs by means of a baseband processing circuit called the phase estimator. So we have a closed loop, called the Costas loop; any phase shift of the carrier for one of the previous causes is corrected by the variation of the demodulator output signals which results from it.

Various algorithms can be used to obtain the oscillator control signal based on analogue signals $X(t)$ and $Y(t)$, present at the demodulator output. The most used are those which only utilize signs, and can therefore be employed by phase estimators which only use logic circuits (Fig. 1.54(b)).

Modulation of the two axes (QAM) Let $x(t)$ and $y(t)$ be the signals which we would obtain at the demodulator output if the carrier were in its nominal position, as at transmission, and let $r(t)$ and $\alpha(t)$ be the corresponding polar coordinates.

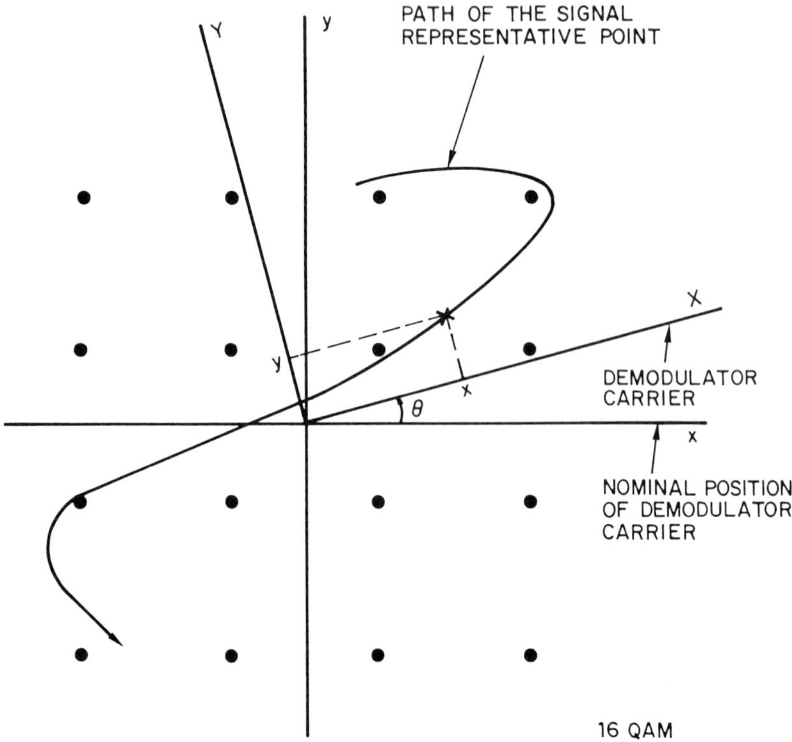

Fig. 1.55 Effect of a phase shift of the demodulator carrier.

Then, let $X(t)$ and $Y(t)$ be the same signals when the carrier has turned by θ (Fig. 1.55). So we have:

$$X = x\cos\theta + y\sin\theta = r\cos(\alpha - \theta)$$

and

$$Y = y\cos\theta - x\sin\theta = r\sin(\alpha - \theta).$$

Using 'exclusive OR' logic circuits, the estimator performs the following mathematical operation:

$$\text{sgn}(X)\,\text{sgn}(Y)\,\text{sgn}(X+Y)\,\text{sgn}(X-Y)$$

whose result is equal to the sign of the expression

$$\begin{aligned} XY(X+Y)(X-Y) & \\ &= r\cos(\alpha-\theta)r\sin(\alpha-\theta)[r\cos(\alpha-\theta)+r\sin(\alpha-\theta)][r\cos(\alpha-\theta)-r\sin(\alpha-\theta)] \\ &= r\cos(\alpha-\theta)r\sin(\alpha-\theta)[r\sqrt{2}\cos(90^\circ-\alpha+\theta)][r\sqrt{2}\sin(90^\circ-\alpha+\theta)] \\ &= r^4\sin(4\alpha-4\theta). \end{aligned}$$

So at each instant, the estimator supplies $\text{sgn}[\sin(4\alpha-\theta)]$, and the low-pass filter which follows the estimator calculates the temporal mean of this expression. Let us examine a long series of symbols during which θ is constant and $\alpha(t)$ varies as a function of the modulation. It suffices to calculate the mean only for the modulation states of the first quadrant because the modulation states, having the same modulus and phases α spaced by a multiple of 90°, give the same result.

Let us suppose that the band limitation being very broad, the signal is delayed during the duration of each symbol on the position which it must take at the characteristic instant and jumps rapidly from one position to another. In this simple case, the output voltage of the estimator therefore goes, before filtering, from $+1$ to -1 at instants spaced exactly by one or several symbols.

Let us consider two states which are symmetrical with relation to the first bisector. Their contributions to the mean are cancelled out, because:

$$\sin(4\alpha-4\theta)+\sin[4(90^\circ-\alpha)-4\theta] = 2\sin(-4\theta)\cos(90^\circ) = 0.$$

So the only useful contributions are those of the states located on the first bisector, for which $4\alpha = 180^\circ$, equal to $\text{sgn}[\sin(4\theta)]$. We therefore obtain, after low-pass filtering, $K\text{sgn}[\sin(4\theta)]$. K is the ratio of the number of states located on the two bisectors to the total number of states ($K = 2/N$, that is, 1 in 4 QAM, 1/2 in 16 QAM, 1/4 in 64 QAM, etc.; Table 1.15).

The characteristic curve of the phase detector, obtained in this way, is periodic and has a 90° period, in the form of square wave signals, amplitude $\pm K$ (Fig. 1.56).

In reality, the filtering of the digital signal and of the additive thermal noise modifies the instants when the estimator output voltage changes sign; the characteristic curve becomes closer to a sine wave than to a square wave signal; its peak amplitude is less than K, but it keeps the periodicity of 90° which ensures the loop lock.

Table 1.15 QAM modulations

Number of levels	*Number of states*	*Type of QAM*	*Capacity*
2	4	4 QAM (4PSK)	small ($\leqslant$ 34 Mbit/s)
4	16	16 QAM	high ($\geqslant$ 34 Mbit/s)
8	64	64 QAM	high ($\geqslant$ 34 Mbit/s)
16	256	256 QAM (draft)	high ($\geqslant$ 34 Mbit/s)

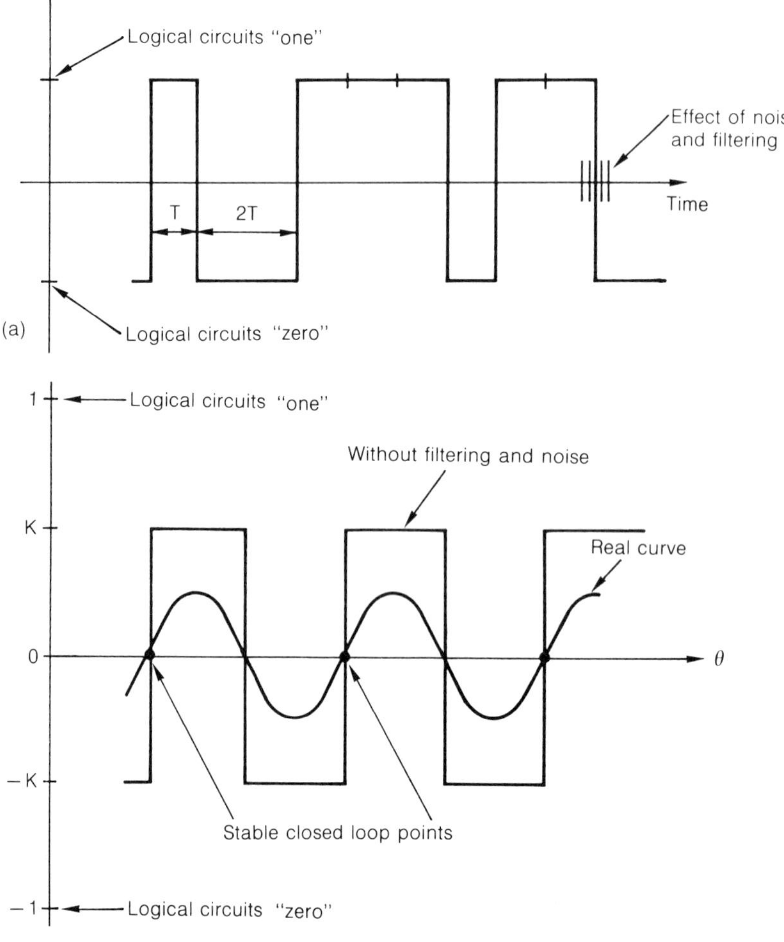

Fig. 1.56 Phase estimator output signals: (a) before low-pass filtering; (b) after low-pass filtering (phase estimator characteristic curve).

Modulation of a single axis (AM) The relationships connecting X and Y to x and y are the same as above, but are considerably simplified because y is zero:

$$X = x\cos(\theta) \qquad \text{and} \qquad Y = -x\sin(\theta).$$

For example, a suitable algorithm is $\operatorname{sgn}(XY) = -\operatorname{sgn}[\sin(2\theta)]$. This time we see that the modulation does not come into play since $\alpha(t)$ does not appear in the formula. Only the thermal noise whatever the filtering of the signal, modifies the pattern in square wave signals of the characteristic curve of the phase estimator.

Sampled loops We should just mention the possibility of only feeding the frequency multiplier or the phase estimator a time-domain fraction of the signals close to the characteristic instants, which makes it possible to considerably reduce, if not reduced to zero, the regenerated carrier phase noise. This supposes, after an accidental cut-off of the link, that we have the timing before being able to recover the carrier.

Phase ambiguity suppression All the carrier regeneration methods, except for the one which entails transmitting a residual carrier, create an ambiguity of 180° (single axis modulation) or of 90° (two axis modulation). The respective results, in the absence of an adequate system, are the reversal of a signal or an interchanging of data signals A and B.

Generally speaking, the ambiguity is removed by associating the data, not at the relative position of the vector representing the signal with relation to a carrier which is only known ambiguously, but by associating it with the variation of this vector between one symbol and the following symbol. This process which we call differential encoding, is conducted in baseband, by means of logic circuits, on sets of n bits of one symbol, at the modulator input and at the output of the demodulator regenerator unit. At this point, the bits are obtained from a wrong reference (carrier), a reference which we rightly hope will maintain its false nature until the following symbol.

So a given symbol is processed twice successively. If it has an error, two successive bit groups are erroneous and therefore include, a minimum of one erroneous bit per group. The differential encoding should therefore double the error rate. In fact, for the QAMs with a great number of states, certain encoding algorithms, for which we only differentially encode the two bits of the symbol which define the quadrant to which it belongs, lead to an increase in the error rate less than two.

The description of the diagrams of the encoders and decoders, whose complexity increases rapidly as a function of n, are not covered in this chapter. Figure 1.57 gives the extremely simple diagrams used in the two phase modulation.

Other types of modulation–demodulation

Phase shift keying modulations (PSK): 8 to 16 phases For these modulations, everything occurs as if we suddenly varied the signal phase, and filtered the

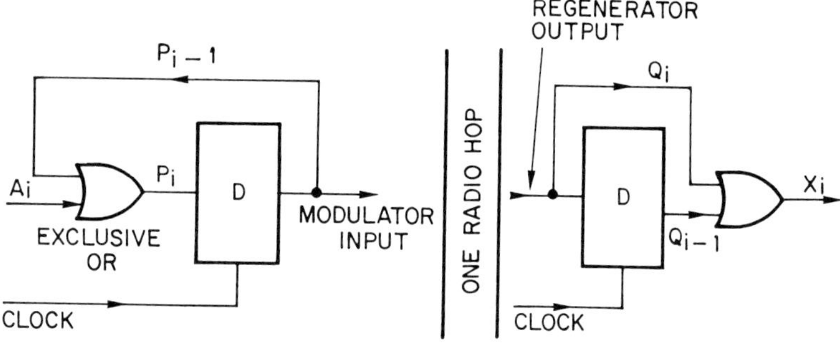

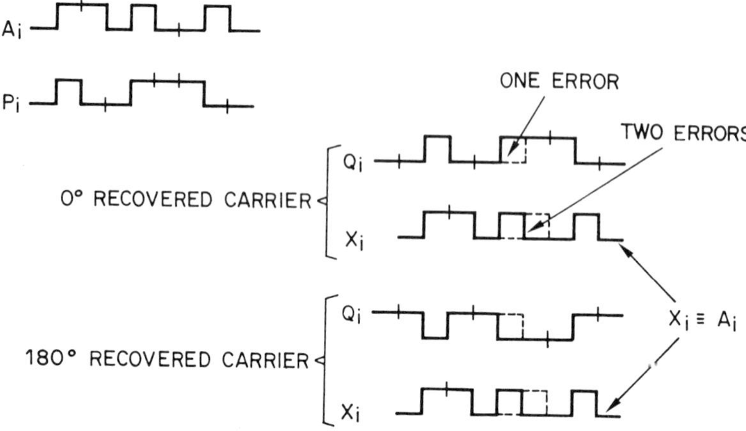

Fig. 1.57 Differential coding and decoding in 2 PSK. A_i, digital signal to be transmitted; P_i, modulator input; Q_i, regenerator output; X_i, received signal. The purpose of the D flip-flop is to keep the memory of the preceding bit.

modulated signal downstream. The envelope is only constant at the sampling instants, if filtering is intersymbol-free. It is highly variable at the other instants and it is necessary to utilize linear amplifiers.

Differential demodulation A PSK signal can also be demodulated differentially by comparing the phase of each symbol with that of the previous symbol, placed in memory in a delay line. The flow diagram of the demodulator becomes simplified by eliminating the carrier regeneration circuits.

Frequency shift modulation (FSK) When the spectral efficiency requirements are not very high, as in the case of low capacity regional microwave links, this type of modulation enables a simple structuring of the receiver and particularly of the transmitter.

1.4.3 Technological aspects

In the following section, we are going to examine the different variants used by microwave link designers to optimize cost and performance, which involve establishing operations of modulation, demodulation and filtering in baseband, in intermediate frequency or in microwave frequency.

Filtering technology

The five filtering functions in a digital microwave link are as follows:

1. connecting several channels to a single antenna (branching filters);
2. eliminating the image frequency of a mixer;
3. selectivity, mainly with relation to adjacent channels;
4. noise equivalent band limitation;
5. shaping pulses, to cancel the intersymbol distortion.

These functions are performed by several separate filters of a transceiver, and a single filter can perform several functions. The first two functions are necessarily performed in microwave frequency (section 1.3), at least for the diagrams with a single frequency change. In the following texts, only the last three functions will be described; these require the use of more selective filters and can be performed in baseband, in intermediate frequency or in microwave frequency.

Filtering the baseband This is the type of filtering which has been implicitly considered until here. It is necessary that the modulator and demodulator be linearized by a strong carrier signal power so that the effect of the filtering in baseband will exactly fit the modulated signal. This is the type of filtering which enables the high working precision necessary for the modulations with a large number of states.

Filtering in intermediate frequency At the demodulator output or at the modulator input, a single filter in IF replaces the two baseband filters placed at each modulation or demodulation access but, given equal selectivity, an IF filter contains twice as many components as in a baseband filter. An IF filter is more sensitive to the accuracy of the components and to their ohmic losses than a baseband filter, in a ratio close to the loaded Q factor.

If it is not perfectly symmetrical to the central frequency, or if at reception, the frequency drifts of the microwave frequency local oscillators are too great, an intersymbol distortion appears between the modulating signals of the two axes. On the other hand, a filter in IF can be placed far upstream from the amplification circuit, which prevents the downstream appearance of intermodulation products due to energy from thermal noise or from the neighbouring channels.

In case of a modulation at only two levels, a filtering in IF makes it possible to use high level logic signals because it is no longer necessary for the modulator to be linear. This makes it possible to tolerate a poor isolation of the ring modulator in the direction carrier access to output access, and therefore to reduce the residual level of the carrier frequency line of the modulation spectrum and also to limit the gain of the amplification circuit which follows. However, two faults appear if no baseband filtering is there to limit the spectrum in $\sin(x)/x$ of the modulating signal.

1. At the modulator output access, appearance of spectral components of the modulating signal close to the IF, within the isolation limit of the ring modulator in the direction modulation access to output access.
2. At the output access, appearance of spectral components of the neighbouring modulating signal close to the double of the intermediate frequency, by 'spectral folding' or 'aliasing' (Fig. 1.58).

It should be noted that surface wave IF filters exist which combine accuracy and physical stability, strong equivalent unloaded Q factor and linearity in the phase-frequency characteristic.

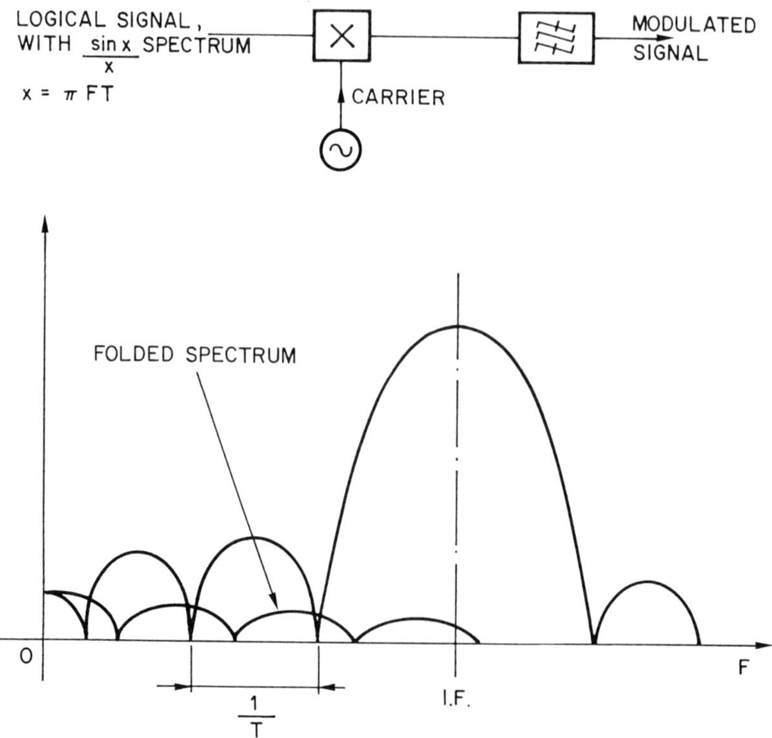

Fig. 1.58 Spectrum folding in a modulator.

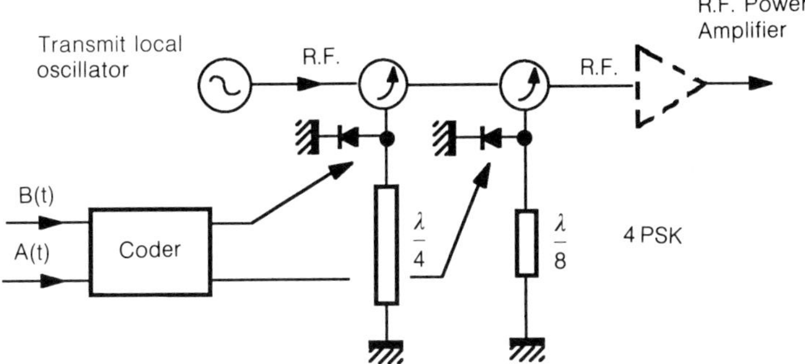

Fig. 1.59 Transmitter block diagram: modulation at RF, with path length variation.

Filtering in microwave frequency The shaping of signals in microwave frequency is complex because the faults due to inaccuracy of the mechanical structure and to ohmic losses are amplified by the strong value of the loaded Q factor. This type of shaping filtering is the only one which can be used in the case of modulation in microwave frequency with variation of the length of the path (Fig. 1.59).

Modulation technology

Modulation in intermediate frequency The basic working component of a modulator in IF is the so-called 'double balanced' ring modulator, shown on Fig. 1.60. It has four diodes installed between two ferrite core transformers. The modulating signals must be fed to the access which lets the d.c. component pass. The three accesses are insulated; for a modulator operation, we can count on an isolation (attenuation) of 30 dB from the carrier access and from the modulation access to the output access. The power fed at the carrier access ensures the linearity of the modulation. It is approximately 7 dBm, sometimes 17 dBm if we

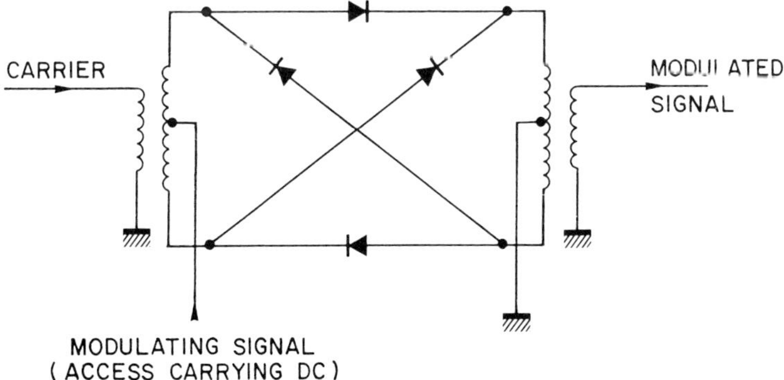

Fig. 1.60 Ring modulator.

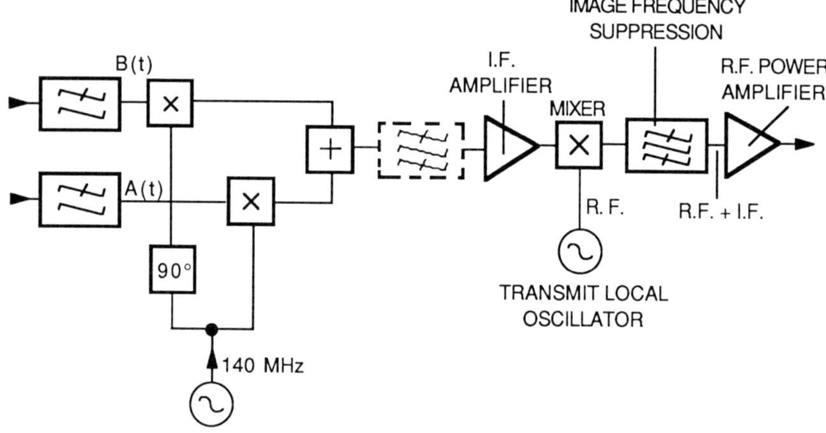

Fig. 1.61 Transmitter block diagram: modulation at IF.

design for very good linearity. Figure 1.61 shows the general diagram of an IF modulation digital microwave transmitter.

Modulation in microwave frequency As indicated in Fig. 1.62, the same modulation principle, applied in microwave frequency, can be used to simplify the flow diagram of the transmitter. We are generally satisfied with two diode balanced microwave mixers. However, the mixer performance is inferior in microwave frequency than in intermediate frequency with regard to linearity, isolation, etc.

Figure 1.59 shows a different type of modulation which can be used in practice for the modulations with two, four and eight phases, and which can operate at

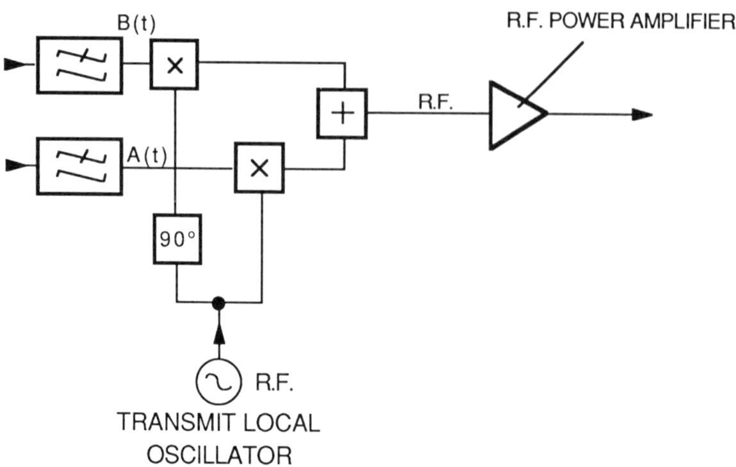

Fig. 1.62 Transmitter block diagram: linear modulation at RF.

high level. P-i-n diodes placed at the input of the lines in quarter-wave, eighth wave, etc. and controlled by the modulation logic signals, vary the path of a microwave which is reflected either at the line output (non-conducting diode) or at its input (conducting diode).

Demodulation technology

The demodulation generally takes place in intermediate frequency, with shaping filters shaping in baseband or in IF, as indicated in Fig. 1.63. Figure 1.64 shows

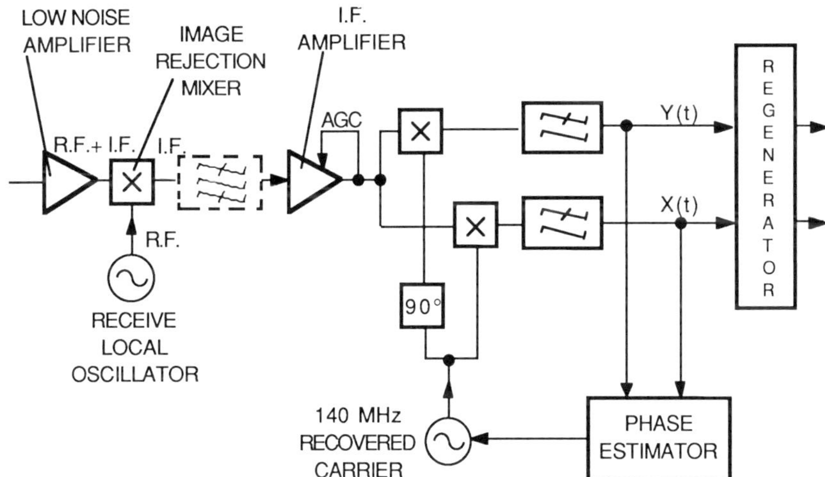

Fig. 1.63 Receiver block diagram: demodulation at IF.

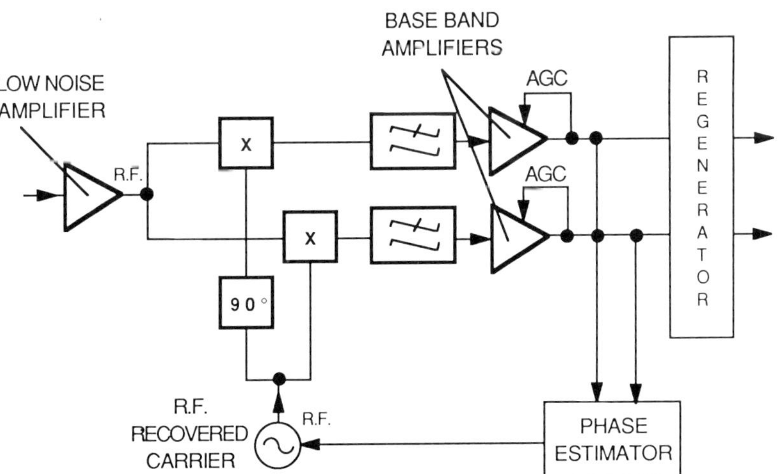

Fig. 1.64 Receiver block diagram: demodulation at RF.

a direct demodulation in microwave frequency, which enables a considerable simplification of the flow diagram. Almost all of the amplification is with automatic gain control, and the shaping filtering takes place in the baseband. The phase detector controls the frequency of the microwave-frequency-receiving local oscillator.

1.4.4 Other specific features of digital microwave links

In the following sections, we are going to briefly describe certain digital microwave functions which are not directly linked to modulation, i.e. automatic switching of radio channels, quality monitoring, transmission of engineering order wires and signalling data, etc.

Transmission of auxiliary signals

The method which is most generally used for transmitting specific microwave link signals is the use of a 'radio frame' (Fig. 1.65), especially created for the needs of the microwave link and deleted at its output access (junction). To do so, the useful signal is divided into successive blocks of about 100 bits, and it is framed by sync words which synchronize the frame, and by the specific service bits of

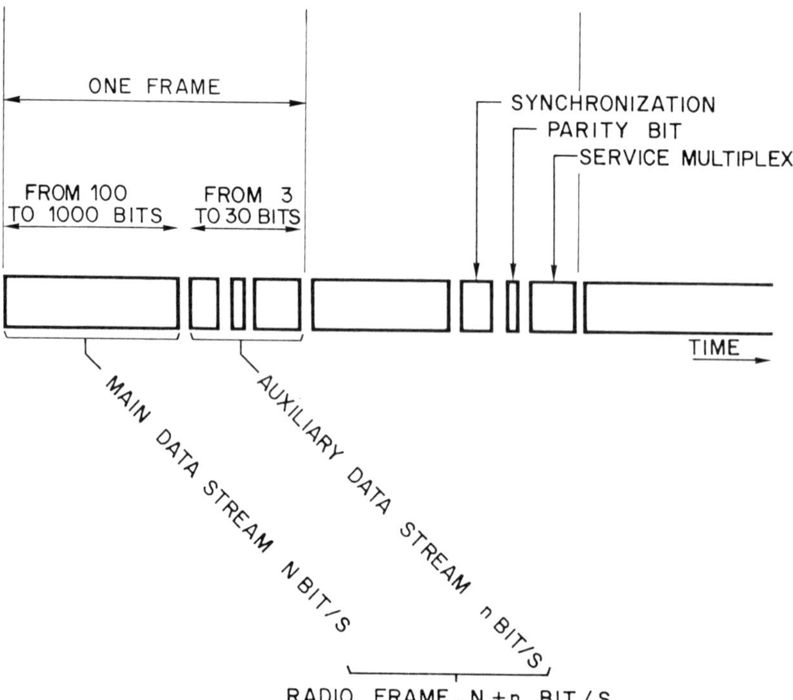

Fig. 1.65 Principle of a radio frame.

the microwave link (engineering order wires in audio frequency, switching control signals, telesignalling, quality monitoring, etc.). The resulting digital bit rate increase is approximately 3%.

Other methods have been and are still used for low capacity microwave links, consisting, for example, of performing a low index frequency modulation superposed on the digital modulation, or of frequency modulating the clock of the modulator. These methods avoid the forming of a frame, but are generally limited in transmission capacity and must, in principle, be implemented hop by hop.

Quality monitoring

The transmission quality of a digital signal is mainly characterized by its error rate. We obtain a good approximation of the error rate by the so-called 'parity bit' method. A service bit in the frame is used to transmit this parity; at the input of a MW section which may include several hops, if we find that the number of 1s in the 100 data bits of a frame is even, this bit is set to 1, and if it is odd, this bit is set to 0. At the end of the section, we measure the parity of the block of 100 bits again and we compare it with the one indicated by the parity bit set with the block. If there is a discrepancy, we conclude that one of the bits of the block and one only is erroneous. Naturally, this method is only valid when the error rate to be assessed is clearly less than 1/100, which is generally the case. Furthermore, special arrangements must be made to account for the multiplication of errors brought in by differential encoding.

Channel switching

As for the analogue MWs (see section 1.3.5), the reliability of the components of a long microwave link is generally insufficient for warranting the availability goal. So it is necessary to keep a radio channel on standby, ready to replace a deficient channel. The switching criteria generally constituted by the error rate, is assessed by means of the parity bit. If the error rate measurement device, placed downstream from a channel detects an anomaly it actuates the switching of the signal to be transmitted onto the standby channel locally, on the reception side, and also, at a distance, on the transmission side, by a switching command, transmitted via a service link. In case of a sudden failure, it is not possible to avoid a brief outage of approximately a fraction of a second, at the moment of switching.

Hitless switching However, it is desirable to avoid any signal hits during deliberate switching, for maintenance requirements, or during switchings performed to benefit the frequency diversity effect between channels. This is because the few nanoseconds of a single bit lost or added at the moment of a switching, because of the difference of propagation time delay of the switched channels (difference in the physical length of connecting cables, in the microwave branching filters, etc.), would bring about the loss of synchronization of the primary multiplexers which would extend the outage time by several milliseconds.

For those switchings which we can predict (in case of switching for frequency diversity, the speed with which the propagation fading is established is very slow in comparison with the digital bit rate), we begin by transmitting the signal simultaneously on the two channels on the transmission side and, on the reception side, we compare, bit by bit, the signals received in a logic device which delays one or the other in order to bring them into phase. The switching is performed downstream, with no addition or subtraction of bits, and we may disconnect the channel to be isolated upstream for maintenance purposes. We can also use equipment-by-equipment automatic switching (so-called 'hot standby'), instead of link-by-link switching.

Scramblers and descramblers

The digital signal, present at the input of an MW, may have a periodic structure or may have a long series of ones or zeros. This can disturb the proper operation of the carrier and timing recovery circuits, and create spectral lines which jam other transmissions. To avoid this, we insert a scrambler and a descrambler, at, respectively, the input and at the output of the link. These logic devices are built around pseudo-random sequence generators.

Generation of a pseudo-random sequence The diagram of a pseudo-random sequence generator is given in Fig. 1.66. It comprises N flip-flops which, at the leading edge of the clock, transfer to their output, the signal present at their input. The input of the first flip-flop is the result of an 'exclusive OR' between the output of the last flip-flop and the output of a special intermediate flip-flop. For certain values of N, it is necessary to have the exclusive OR between the outputs of more than two fip-flops. We show that:

1. At the output of one of the flip-flops there will appear a periodic series of period $2^N - 1$ in which all the arrangements of N successive bits are present, with the exception of the series consisting of N zeros.
2. If we take off one bit from this series every $2q$ bits, q being an integer, we obtain an identical series.
3. The spectrum of the balanced signal (zeros replaced by minus ones) is constituted by lines spaced by $1/(2^N - 1)T$. After going through a whitening filter, all the spectral lines have the same power, except for the d.c. line whose power is 2^N times weaker.

Synchronous scrambler A synchronous scrambler (Fig. 1.67(a)) performs the modulo 2 addition of the signal to be transmitted and a pseudo-random sequence. To retrieve the signal, the descrambler performs a new modulo 2 addition with

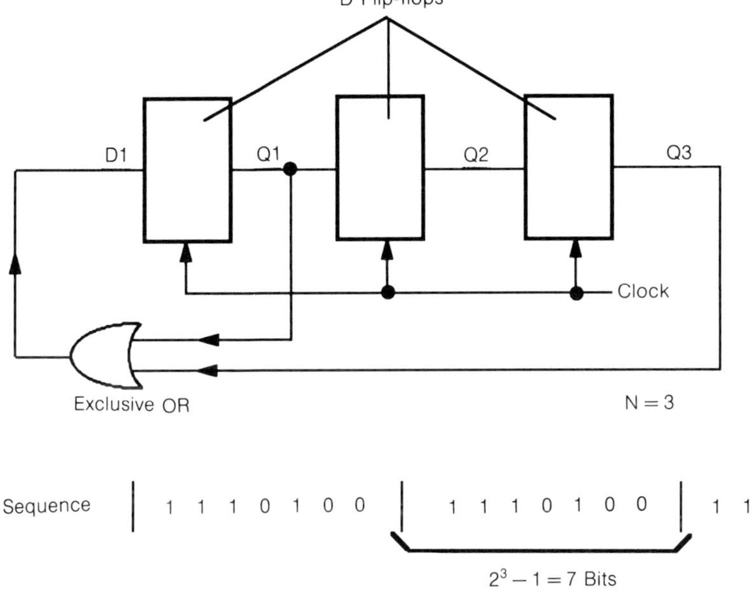

Exclusive or truth table

Input #1	Input #2	Output
0	0	0
1	1	0
0	1	1
1	0	1

N	N° of Intermediate connection		
3	1		
4	1		
5	2		
6	1		
7	1		
8	2	3	4
9	4		
10	3		
11	2		
12	1	4	6
13	1	3	4
14	1	3	5
15	1		
16	2	3	5
17	3		
18	1	2	5
19	1	2	5
20	3		
21	2		
22	1		
23	5		
24	1	3	4

Fig. 1.66 Pseudo-random sequence generator.

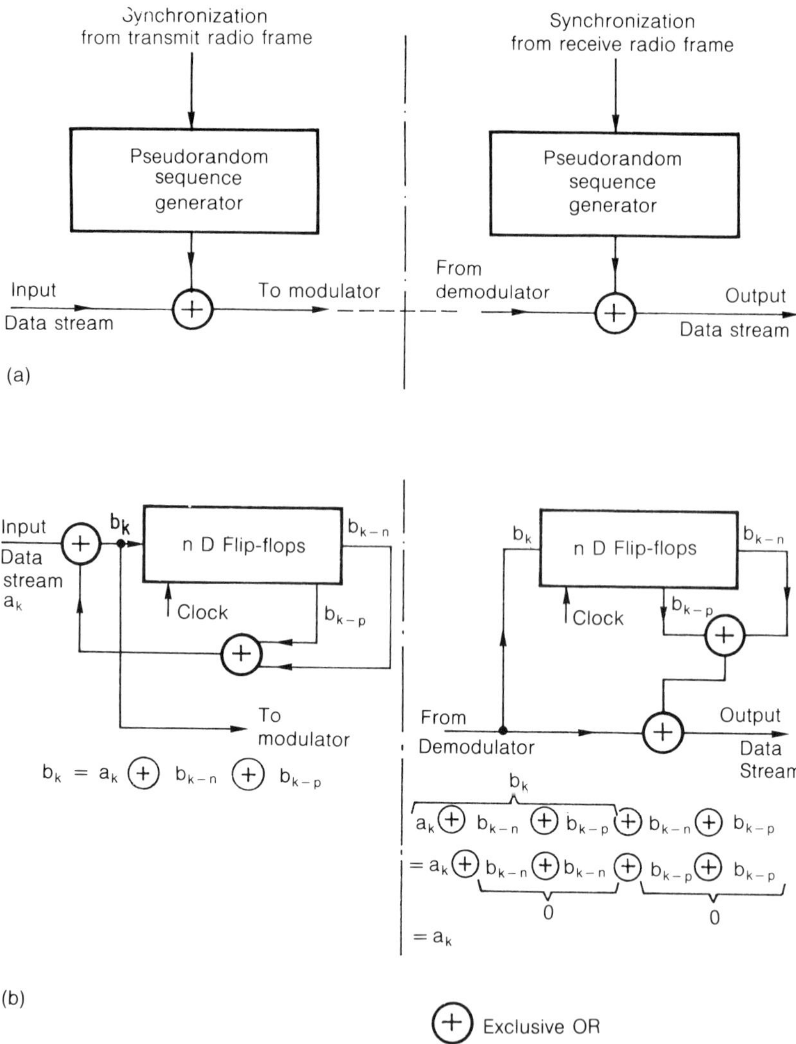

Fig. 1.67 Scramblers: (a) synchronous scrambler; (b) autosynchronizing scrambler.

the same pseudo-random sequence. The error rate is not increased, but since the descrambler must synchronize the locally generated pseudo-random series, this type of scrambling is reserved for the case where we have a radio frame.

Self-synchronizing scrambling Self-sychronizing scramblers–descramblers also exist (Fig. 1.67(b)). The diagrams of the scramblers and descramblers are similar to those of a pseudo-random sequence generator. Their disadvantage is that they increase the error rate by a factor of 3.

Note that in the MW technique, the pseudo-random sequence generators are also used to generate a test sequence, for example, to measure the error rate. Like a purely random sequence, the test sequence generated by the transmitter must, to the degree possible, include all the sequential arrangements of zero and of one, in order, for example, to obtain the worst case of an intersymbol distortion. It must also be known by the receiver so that we can compare it to the signal received; therefore the receiver uses a generator which is identical to the one at the transmitter, and must synchronize it so that the bits to be compared will be simultaneous.

Junctions

The junctions constitute the interface of the MWs and of the other systems of transmission among them, or with multiplexing equipment. For each hierarchic level, the CCITT specifies the junction code, the amplitude and the form of an isolated pulse, the connection line type (two wire pair or coaxial cable), the impedance and the return loss at the accesses. It is at the junctions where the performances are measured (error rate, jitter, etc.).

The signal code (AMI, CMI; HDB3, B3ZS, B6ZS, etc.) must lend itself to a transmission on coaxial cable (low spectral density at low frequencies) and must include enough transition elements so that the clock can easily be extracted.

We should be able to use a certain length of coaxial cable to connect two systems. It is at the input of the systems that a special purpose device must be installed which, in general, automatically matches the cable length. The attenuation to be compensated is 6 or 12 dB at the Nyquist frequency, and varies as $\sqrt{f}$ at the other frequencies.

1.4.5 Performances of digital microwave links

These performances are analysed based on the hypothetical reference digital path and the quality and availability objectives.

CCIR systems: hypothetical reference digital path

This reference path is defined for digital microwave links having capacities greater than 8 Mbit/s. The length of the path is 2500 km. It is represented with the various modulation and demodulation sections on Fig. 1.68.

The digital microwave link quality objectives are expressed in the form of five parameters: two error ratio values, two error-free second criteria, and one error burst criterion.

Low bit error ratio value (BER). This guarantees the operating quality of the link during most of the time. The error rate for the hypothetical reference path

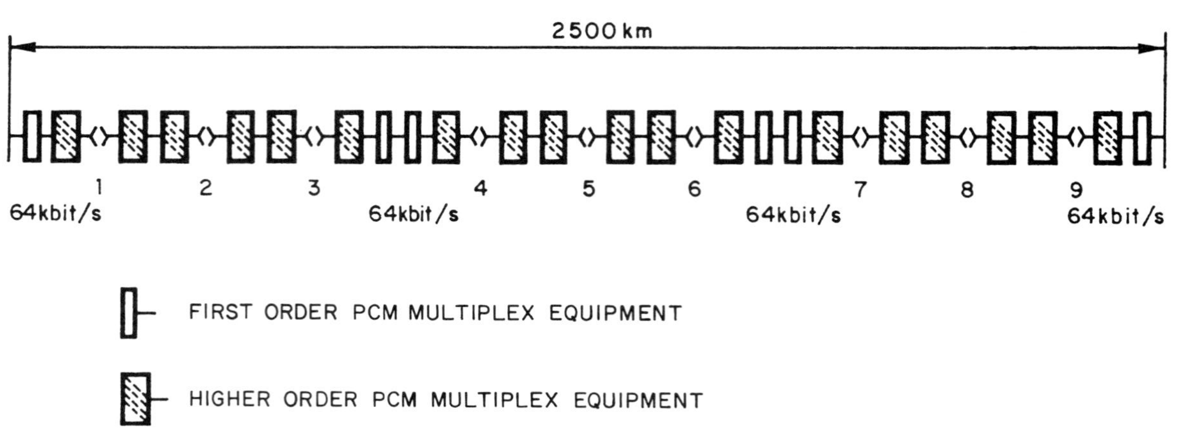

Fig. 1.68 Hypothetical reference digital path.

at 64 kbit/s, must not exceed 1×10^{-6} during more than 0.4% of the duration of any one month, with an integration time of one minute.

High bit error ratio value (BER) This is one of the most important parameters. More often it determines the spacing between repeaters. The error rate for the hypothetical reference path must not exceed 1×10^{-3} during more than 0.054% of the duration of any one month (integration time of 1 s).

Long term error-free seconds (EFS) This is a way of estimating the quality of a data transmission circuit. The criterion of 99.68% of error-free seconds is defined on any one month for the hypothetical reference path.

Short term error-free seconds (EFS) Its purpose would be to guarantee a satisfactory operating quality during an acceptable percentage of data communications. This parameter is under study.

Error bursts The CCIR and the CCITT are designing a limit for error bursts. This is because the arrival of error bursts, in particular on high capacity digital microwave links, causes undesirable effects on telecommunication networks. The limit will closely depend upon the qualities sought for an acceptable operation of the demultiplexing and signalling equipment. For real circuits, the quality objectives are generally fulfilled by a residual error rate from 10^{-8} to 10^{-11} for the hypothetical reference path.

US and Canadian systems

Two digital reference paths are defined: a long haul system whose circuit length L can reach 6500 km and a short haul system where $L \leqslant 400$ km. For these two systems, the binary error rate must be less than 10^{-3} during 99.98% of the time. This goal is generally divided in equal parts between the equipment and the propagation effects. For the long haul system, the long term objectives in error-free seconds in 99.5% and the residual error rate must be less than 10^{-10}.

Availability objectives

Availability has been defined in section 1.3.7 for analogue transmission; unavailability is characterized by long-term quality impairment, for at least 10 consecutive seconds (signal outage, high bit error rate (BER $< 10^{-3}$), etc.). CCITT specifies that the availability objective appropriate to a 2500 km hypothetical reference digital path, should be 99.7% of the time, the percentage being considered over a long period of time (more than one year). For the long and short haul systems in the USA and Canada, the availability must be 99.98% of the year, for both directions of transmission.

Main causes of error

The origins of the bit errors in digital microwave systems are the following.

1. Thermal noise: its effects are analysed in section 1.4.2. This noise is determined from propagation conditions (free space attenuation, flat fading, hydrometeors and gas absorption);
2. Multipath propagation distortion: it is due to interference of several radioelectric rays.
3. Equipment imperfections:
   (a) intersymbol distortion;
   (b) amplitude and phase error on modems;
   (c) amplitude and group delay distortions: filter bandwidth limitations (RF and IF);
   (d) AM–AM and AM–PM distortions in power amplifiers;
   (e) feeder and echo distortions;
4. Jitter;
5. Temperature;
6. Ageing;
7. Interferences: these are due to equipment imperfections (antenna cross-polarization factor) and depolarization phenomena produced by the multipath effects and hydrometeors.

Propagation influence

In the design of digital radio systems, two problems linked to propagation conditions, are to be taken into consideration: multiple path effects and the influence of the atmosphere on the propagation in free space.

Multiple path effects Multipath propagation, causes amplitude and phase distortions, in the signal transmission band, and a depolarization of the waves received. These are causes of errors in digital microwave links. The first effect is called selective fading.

Various models have been proposed to analyse this phenomenon. In particular, the three ray model has been proposed. Figure 1.69 illustrates this model. Rays P_1 and P_2 have generally a smaller power and follow a longer trajectory than a direct ray, N, and require a time supplement $t_1 = 0.1$ to 0.4 ns and $t_2 = 4$ to 6 ns to reach the receiving antenna.

After a simplification of this model, the propagation medium transfer function is written as follows:

$$H(\omega) = a[1 - b\exp(\pm j(\omega - \omega_o)t)]$$

where a is the sum of amplitudes of rays P_1 and N, b is the delayed P_2 ray relative amplitude, with $ab = a_2$, t is ray P_2 relative delay with $t = t_2$, ω is $2\pi f$, radio channel radian frequency and ω_o is the difference between the central radian

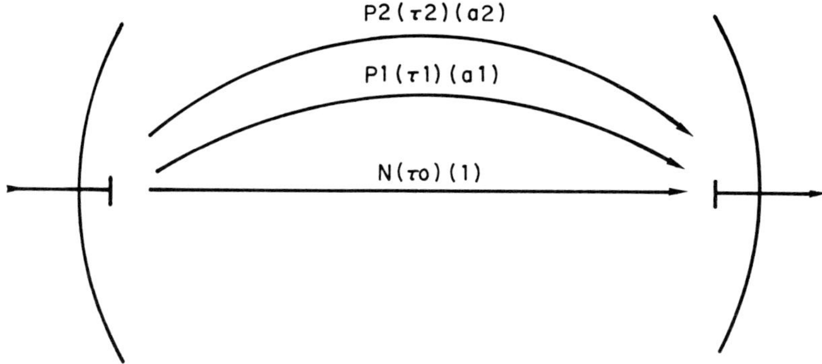

Fig. 1.69 Three-ray model.

frequency of the channel and the corresponding radian frequency at the lowest point of the fading.

The modulus of this function presents a minimum each time

$$\omega t_o = (2k + 1)\pi.$$

These minima, spaced $1/t_o$, characterize the notches of the fading process.

The group delay (GD), differential of the phase of $H(j\omega)$ versus ω, presents maxima or minima at the same frequencies when the fading notches are observed. The sign of the GD disturbance is negative if $b < 1$ (minimum phase fading). It is positive if $b > 1$ (non-minimum phase fading).

Flat fading is characterized by the amplitude variations of sum vector $P_1 + N$ and selective fading is defined by the interference between vectors $P_1 + N$ and P_2.

By convention, we call the selectivity of a fading S the ratio expressed in decibels between the field which would result from the strongest beam (taken as reference 0 dB) and the field at the most attenuated frequency.

$$S = 20\log(1 - b).$$

Figure 1.70 shows the pattern of the selectivity and of GD.

Concept of 'equipment signatures'. The concept is used to compare the sensitivity of digital systems to the selective fading effects and, in certain cases, for calculating the interruptions. It is defined by a laboratory test, using a two ray propagation model. Figure 1.71 represents the measurement block diagram.

The following parameters used:

1. bit rate;
2. modulation type;
3. proportionality factor between direct and delayed rays: b;
4. delay time of the delayed ray, in nanoseconds;
5. the bit error rate;
6. difference between the central frequency of the channel and the notch frequency.

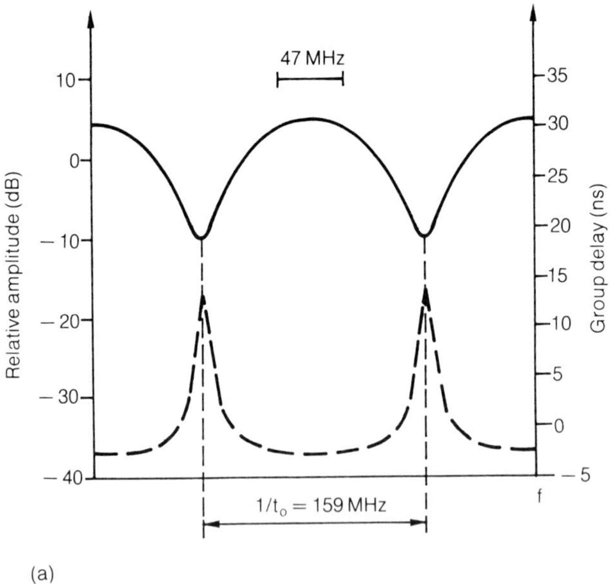

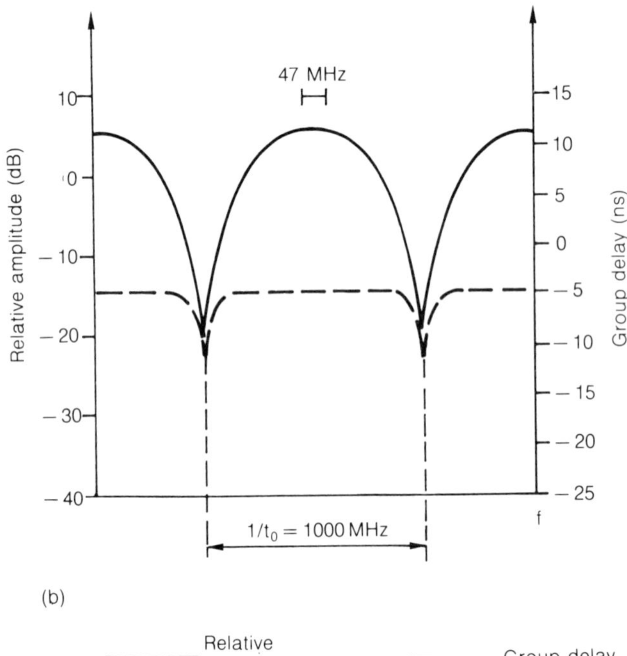

Fig. 1.70 (a) Delay, 6.3 ns; fading selectivity, 10 dB; ray ratio, 3.3 dB. (b) Delay, 1 ns; fading selectivity, 20 dB; ray ratio, 0.9 dB.

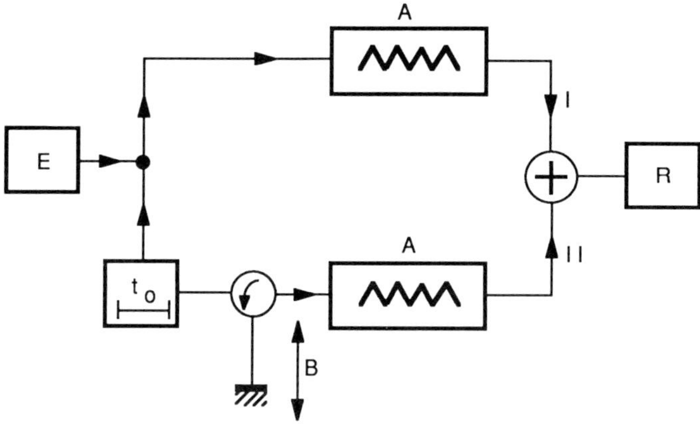

Fig. 1.71 Signature measurement: A, attenuators; B, notch position adjustment; E, Tx; R, Rx; I, main ray; II, delayed ray.

Curve $S = 20\log(1 - b) = -20\log\lambda$ versus f_1 with a given BER represents the system signature. The critical selectivity S_c is generally obtained for $t_o = 6.3$ ns, and BER $= 10^{-3}$ corresponding to an interruption of the transmission, or outage on digital microwave lengths. For different values of τ, the following can be adopted:

1. the width of a signature is independent of t_o;
2. the height is proportional to t_o.

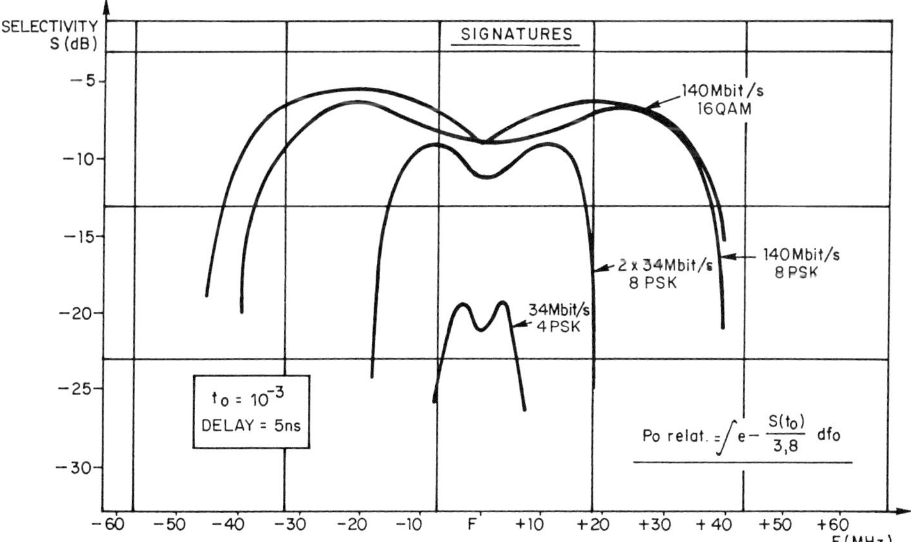

Fig. 1.72 Equipment signature.

Figure 1.72 shows several signatures with the following parameters: indirect signal delay in nanoseconds, BER $= 10^{-3}$, and digital bit rate and modulation type.

An approximation of the outage relative to selective fading can be given based on the signature $S_c = -20 \log \lambda_c$ by the following relationship:

$$P_{\mathrm{ere1}} = K \int \lambda_c^{2.28}(f_1)\,\mathrm{d}f_1$$

Consequently, by raising λ to the power 2.28, the area contained under this curve is proportional to the outage. This property is used to evaluate the selective fading outages.

Interference

The analysis of interference in a digital transmission system using microwave links is similar to the one described in section 1.3.7 for analogue systems. The interferences degrade the level of the reception threshold of the equipment for a given BER and thus reduces the margin of the flat fading.

Frequency arrangements

Spectral improvement The modulation processes (multistate, reduced band, etc.) and the filtering in the Nyquist band have appeared as ways to increase the number of bits transmitted per second in the radio band unit (hertz). The orthogonal crossing of the polarization of the propagated waves can also be used to increase the quantity of data to be transmitted if we have a sufficient XPD

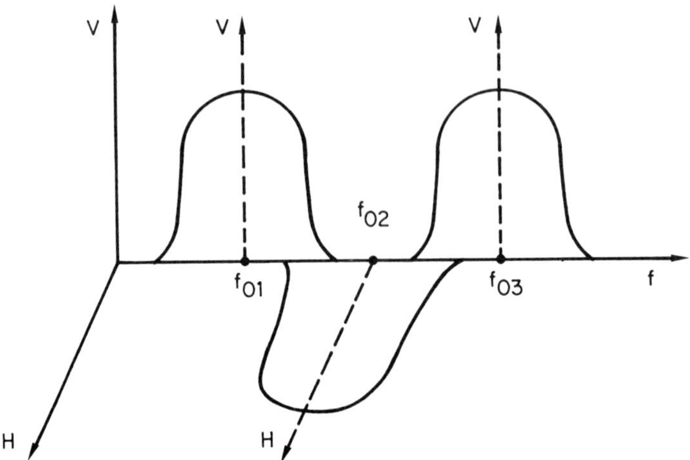

Fig. 1.73 Interleaved channels.

(25 to 40 dB). The alternating polarization between adjacent channels of microwave lengths facilitates the filtering and the branching of these channels, and in this way enables the modulation spectral components to be interleaved (Fig. 1.73).

By extrapolating, it is also possible to reuse the same frequency twice in both polarizations as long as we obtain a sufficient XPD during the worst periods of

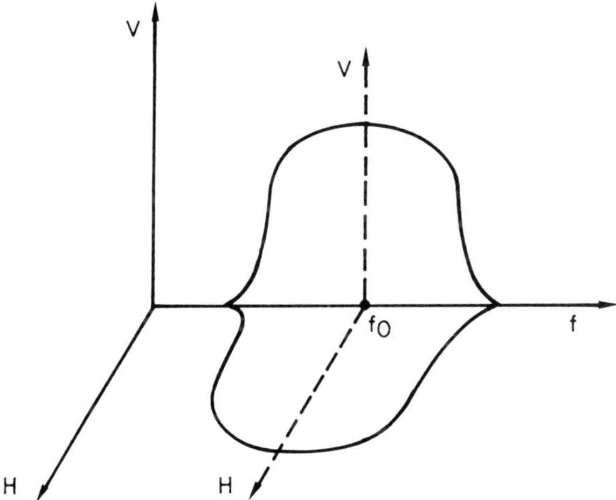

Fig. 1.74 Co-channels.

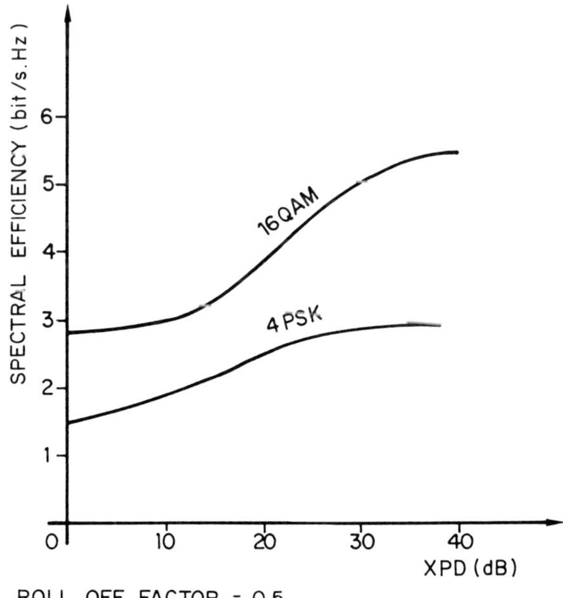

Fig. 1.75 Spectral efficiency.

propagation (Fig. 1.74). So we define the spectral efficiency as the ratio of the total number of bits transmitted to the total band occupied by the frequency plan chosen. Figure 1.75 shows the spectral efficiency which we should obtain as a function of a value taken by XPD during the worst propagation periods.

Figure 1.76 shows that, by associating the following orthogonal patterns: two orthogonal carrier modulation and crossing of the polarizations (vertical and horizontal), we obtain up to four accesses per carrier frequency.

Organization of the frequency arrangements The CCIR determination of the frequency arrangements depends upon:

1. XS: spacing between the central frequencies of adjacent channel;
2. YS: spacing between the central frequencies of adjacent channels for transmission and reception;
3. ZS: difference between the central frequency of the outside channels and the edges of the frequency band.

In digital systems, we generally use two types of frequency plans, alternated channel type or co-channel type. They are shown in Fig. 1.77. In an alternated channel plan, an extra decoupling may be necessary between adjacent channels by using a complementary filter, called the net filter discrimination (NFD) which depends upon XS to satisfy the minimal value of carrier–interference $(C/I)_{min}$. Some examples of frequency plans defined by the CCIR are presented in Table 1.16.

Other frequency bands are also considered at 23, 26, 31 and 39 GHz. They are generally used for digital systems designed for local distribution. These frequency bands have different capacities which are variable from one country to another. Other frequency plan organizations are proposed by the US Federal Communication Commission (FCC), the Department of Communication of Canada (DOC) or various telecommunication administrations.

The FCC recommendations are generally the most restrictive and they are often used as a model for other administrations. The most important rules are about:

1. the frequency bands reserved for digital microwave links (Table 1.16);
2. the maximum bandwidth used by each RF channel (Table 1.17);
3. the digital rate in bits per second (equal to or greater than the band width in hertz);
4. the repartition of the average power of transmitters in a band of 4 kHz contained within diagram of Fig. 1.78;
5. antenna characteristics;
6. stability of the carrier frequencies.

Compatibility of digital and analogue channels The gradual digitizing of communication networks poses the problem of the coexistence on a single route

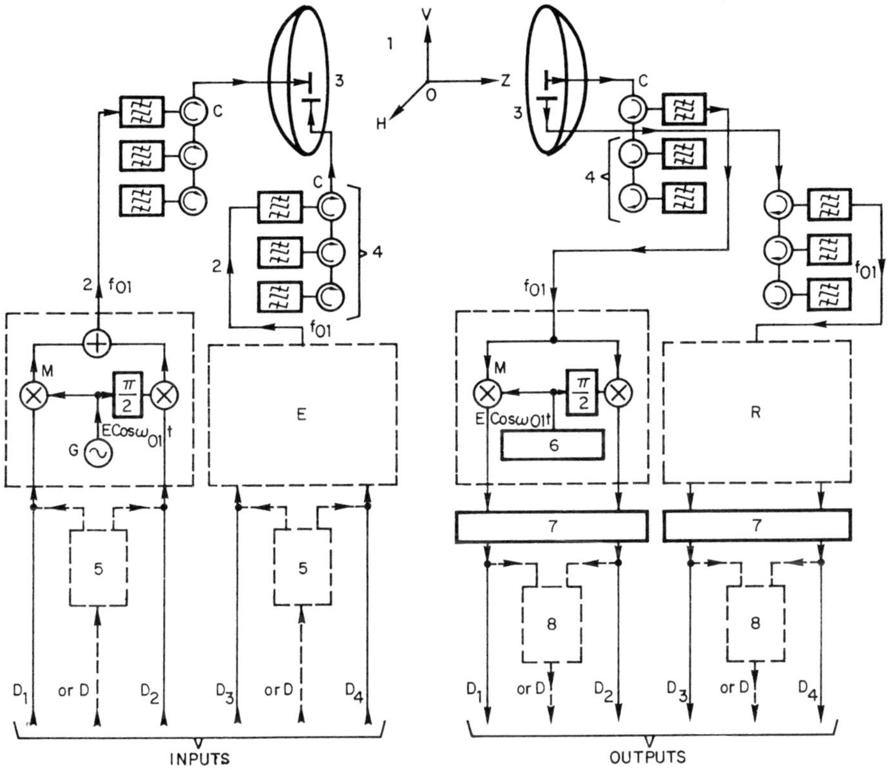

Fig. 1.76 Four accesses per carrier frequency. C, circulators; E, phase shift keying Tx; G, carrier generator; H, horizontal polarization; M, modulators; R, coherent detection receiver; V, vertical polarization; 1, propagation—two orthogonal polarizations; 2, two carriers orthogonal with f_{01}; 3, f_{01} reuse (four times); 4, other channels; 5, serial/parallel converter and differential coding; 6, carrier recovery; 7, phase ambiguity suppression; 8, serial/parallel decoding.

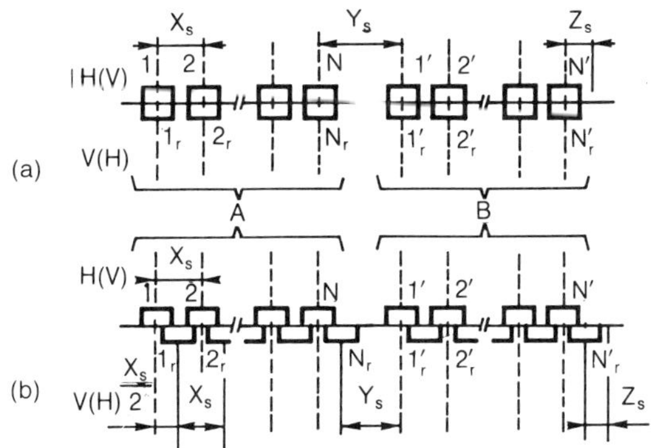

Fig. 1.77 Frequency plans: (a) frequency reuse; (b) interleaved channels.

Table 1.16 CCIR digital microwave link frequencies and capacities

Frequency band (GHz)	*Band limits* (GHz)	*Number of channels*	*XS/2* (MHz)	*YS* (MHz)	*ZS* (MHz)	*Capacity* (Mbit/s)	*Efficiency* (bit/s/Hz)	*CCIR Reference*
2	1.9–2.3	6	29	68	20	2 × 34	2.1	Rapp. 934
4	3.8–4.2	6	29	68	17.5	2 × 34	2.1	Rapp. 934
4	3.8–4.2	6	29	68	17.5	140		Rcc. AA9
6	5.925–6.425	8	29.65	44.5	20.2	140	4.72	Rapp. 934
6	5.925–6.425	8	29.65	44.5	20.2	2 × 34	2.24	Rapp. 934
6	6.430–7.110	8	40	60	30	140	3.29	Rcc. 384
7	7.425–7.725	5	28	42	17	34	1.13	Rapp. AD9
8	8.275–8.500	6	14	49	18	34	1.81	Rapp. AD9
8	8.275–8.500	12	14	49	11	2 × 8	1.70	Rapp. AD9
11	10.7–11.7	12	40	90	15	140	3.36	Rcc. 387
13	12.75–13.25	8	28	70	15	2 × 34	2.17	Rcc. 497
15	14.4–15.35	16	28	70	17	2 × 34	2.29	Rcc. AB9
19	17.7–19.7	4[a]	220	460	110	2 × 140	2.24	Rcc. 595
19	17.7–19.7	8[a]	110	240	110	140	2.24	Rcc. 595
19	17.7–19.7	35[a]	27.5	75	27.5	34	2.38	Rcc. 595

[a] In this frequency band, the number of channels is doubled by a frequency reuse plan.

Table 1.17 FCC digital microwave link frequencies and bandwidths

Frequency band (GHz)	*Band limits* (GHz)	*Number of channels*	*Band width* (MHz)	*Capacity* (Mbit/s)	*Efficiency* (bit/s/Hz)	*Minimal distance* (km)	*Minimal capacity* (Mbit/s)
2	2.11–2.13		3.5	6.3	1.8	5	
2	2.16–2.18		3.5	6.3	1.8	5	
4	3.7–4.2	6	20	2 × 45	4.5	17	10
6	5.975–6.425	8	30	2 × 45	3.0	17	10
11	10.7–11.7	12	40	2 × 45	2.25	5	5–10

of FDM–FM analogue systems and digital systems (PSK or QAM). The criteria leading to a compatible operation can be defined as follows.

Limitation of the spectra transmitted (Fig. 1.79) is by an efficient filtering: In particular, it appears necessary to keep the digital spectra within the FCC mask to enable digital systems to operate without disturbance in the adjacent channel. For an operation of adjacent digital and analogue channels, it is necessary to apply strict standards to the spectrum of the digital signal. Consequently, the tones (residual carrier $f_c = 70$ MHz, the harmonics of the clock signal f_s and the frequencies $f_c \pm f_s$) must be at least 76 dB less than the total digital power, or

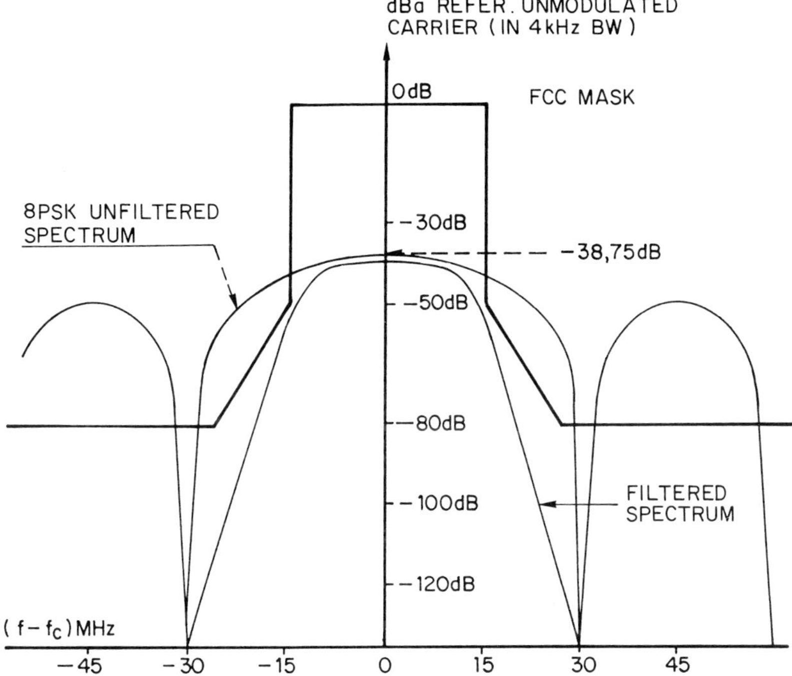

Fig. 1.78 FCC diagram.

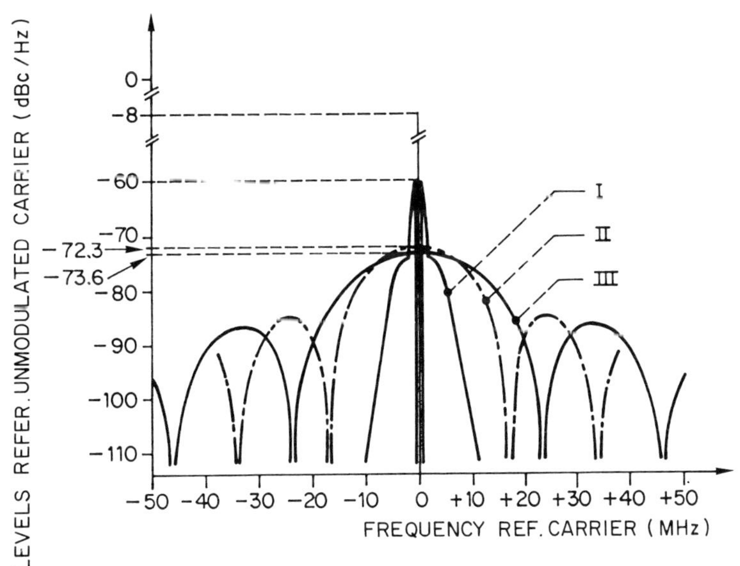

Fig. 1.79 Analogue and digital spectra: I, 960 FDM channels; II, 34 Mbit/s 4PSK; III, 2 × 34 Mbit/s 8PSK.

in compliance with the FCC standard as long as the jamming caused at the FDM–FM system remains acceptable.

Definition of the filtering is necessary, taking into account the frequency difference XS between channels of the same polarization, or XS/2 between cross polarized channels: The filtering depends to a great degree on the expected XPD during the propagation troubles. In practice, we can find radio hops whose propagation causes the XPD to drop to a few dB or even zero dB during a small percentage of the time.

Reduction of the transmitting power of digital systems in comparison with analogue systems (from a few dB to 10 dB, is also possible).

The jamming in a single channel or in an adjacent channel to FDM–FM analogue systems caused by digital systems must be limited to 2.5 pWop for each type of interference. The jamming caused to digital systems in the same channel or in an adjacent channel shall not disturb radio transmission during fading periods. In general, it is admitted that there will be a reduction of 1 dB in the uniform attenuation margin.

1.4.6 Techniques used for improving quality

Behaviour of digital microwave links

Paragraph 2 on p. 100 defines the origin of errors, and the experimental results both in the laboratory and in the field show that appropriate means must be implemented to reach the error rate specified in the next section.

During good propagation periods, the errors find their origins in the imperfections of the equipment and mainly in the amplitude–amplitude distortion (AM–AM) and in the amplitude–phase distortion (AM–PM) of the output stages of the transmitter. The faults may be corrected by using a linearizer. In addition to the effects of these imperfections, we find a very weak error rate, designated as 'residual' whose causes are the distortions remaining in the equipment after 'best possible' equalizing of the amplitude–amplitude linearity, of the amplitude–phase non-linearity, of the imperfections of modems and of the filtering of the modulation spectrum, etc. These errors turn up in an isolated and erratic way. They may be eliminated by forward error correcting code (FEC).

As in the case of analogue microwave links, protection systems, using the principle of redundancy and maintenance aids are used. They are described in section 1.3.5. They make it possible to ensure the availability sought for the radio system and to ensure the maintenance operations, in particular in unstaffed stations.

During the periods of propagation troubled by multiple paths, we find fairly long error packets, sometimes from a few dozen milliseconds to a few seconds. These errors are due to intersymbol interference, caused by the deformation of the modulation spectrum due to a process described in section 1.4.5. These errors can be reduced by using auto-adaptive methods: diversity, equalizers in the time or frequency domain. Another cause of error packets is depolarization which causes an interference between cofrequency or adjacent radio channels in cross-polarizations (see section 1.4.5).

This can be remedied by cross-polarization cancellers for the cofrequency channels and by a system of automatic transmitting power control, while controlling the power received at the extremity of the radio hop (ATPC).

Linearizer

Applications of multilevel modulations (QAM) are limited by AM–AM and AM–PM distortions in the equipments and especially in the RF amplifiers. To solve this problem, several solutions can be applied:

1. 'back off' of the transmitter output power;
2. introduction of a constant predistortion;
3. insertion of an auto-adaptive predistortion linearizer.

The first solution consists of positioning the output power below the saturation level in order that the amplifier will operate in a more linear region. This approach requires a greater output power than what is desired and the use of an amplifier with a higher power consumption. The back-offs used are 5 dB for the transistorized amplifiers and 10 dB for the TWT. To reduce these back-off values, a fixed predistortion linearizer, operating in the RF or IF bands can be used. This device, placed in front of the amplifier, has a non-linear characteristic opposed to that of the amplifier and, after adjustment of its output signal, it suppresses the non-linearity of the amplifier. The back-off reductions obtained are from 2 to 4 dB.

The non-linearities of the amplifiers vary according to temperature, characteristics of the power supplies, time, and fixed linearizers cannot compensate for these variations. The effects can be minimized by using an auto-adaptive linearizer. Figure 1.80 shows such a linearizer. The operating principle is as follows: the digital data at the input and at the output of the amplifier are compared, and a microprocessor modifies the characteristics of the linearizer to reduce the distortions between these two signals.

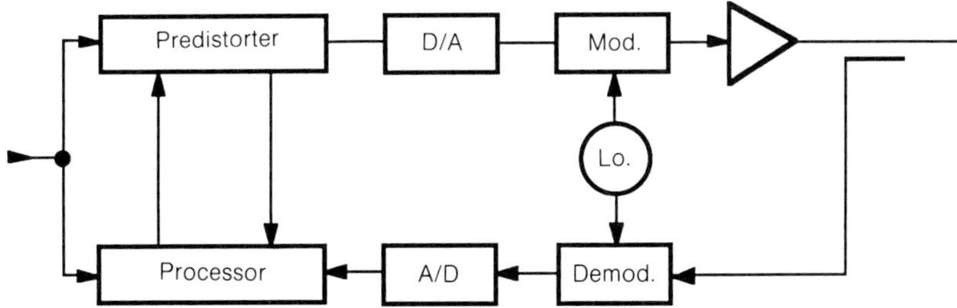

Fig. 1.80 Auto-adaptive linearizer.

Forward error correcting codes

The codes used must be simple in order to minimize the increase in the digital bit rate and the cost of the systems. Block-codes seem the most suitable to reduce isolated and random errors. The codes the most used are the Hamming, Lee, Reed–Muller and Reed–Solomon codes (Peterson). Their efficiency (number of data bits/total number of bits) is approximately 96 to 97%. The correction effects according to the code used make it possible to reduce the residual error rate from 10^{-8} to a value from 10^{-11} to 10^{-13}. It should be noted that these devices cannot correct the long error packets such as those at propagation.

Means for countering fading

The systems developed to reduce fading effects due to multiple paths can be broken down into two categories: the autoadaptive equalizers (in the frequency and time domains) and the diversity systems.

Frequency domain equalizers They operate in intermediate frequency (IF) and use two fundamental properties of the modulation spectrum: the spectrum of the signals is continuous due to the scrambling of the digital signal (scrambler–descrambler effect) and the modulation spectrum is symmetrical with relation to the carrier frequency: f_0.

The asymetry of the demodulation spectrum makes it possible to detect the effects of multiple paths and to control a correction device. Figure 1.81 shows an example of this equalizer. It uses a three coefficient transverse filter. A microprocessor controls these coefficients. The coefficients applied provide a transfer function which gives, as a function of the frequency, the following properties: the amplitude is variable, and the phase is linear.

The choice of the delay value employed in the transversal filter is used to correct the linear or quadratic distortions of the amplitude without distortion of the group delay. In practice, it is possible to associate 'linear' and 'quadratic' equalizers. On the average, the correction is identical for minimal and non-minimal phase fading.

Time-domain equalizers Time domain equalizing can be considered as the best technique for directly eliminating intersymbol interference. This time domain equalizer principle is based on the prior knowledge of certain characteristics of the signal to be corrected:

1. The existence of a time base due to a clock. A decision is made at each characteristic instant.
2. The presence of reference thresholds for assessing at the characteristic instants, the difference between the instantaneous voltage of the disturbed signal and that which should be there without disturbance. This positive or negative difference behaves like an error signal whose sign is processed by an algorithm

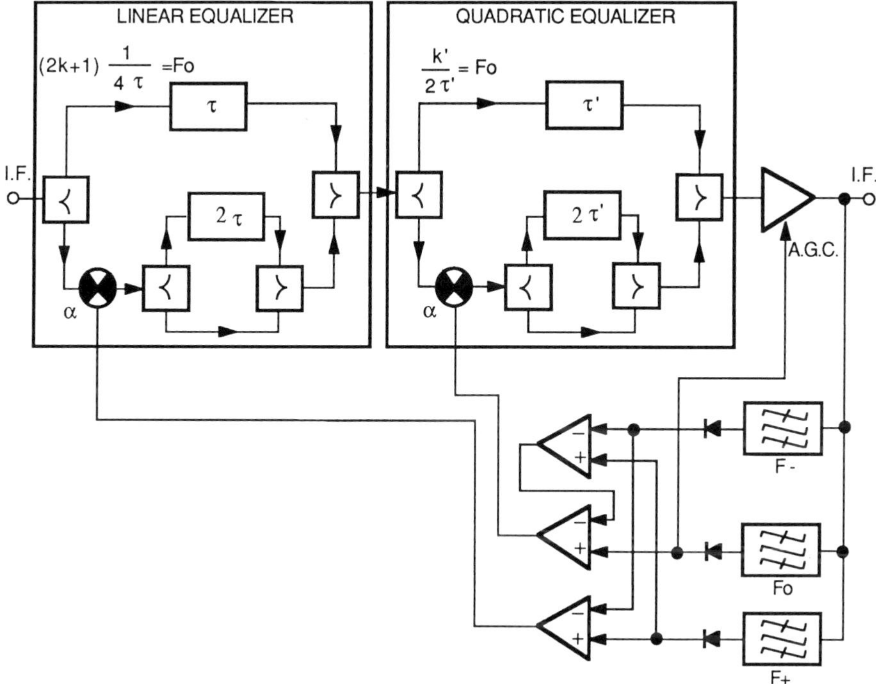

Fig. 1.81 Linear phase filter equalizer.

in order to generate control voltages of the coefficients of the direct and recursive transversal filters.

This type of equalizer is used to process the distortions produced by the amplitude variations and by the group delay at the moment of the minimal or non-minimal phase fading.

Two types of auto-adaptive equalizers have been developed. The first one operates in baseband. It uses a 3 or 5 coefficient transversal filter. Figure 1.82 shows an equalizer for a 16 QAM modulation. This equalizer is positioned after the demodulator and its performances depend upon the characteristics of the carrier recovery circuit. The influence of this device can be reduced by adding a frequency domain equalizer.

The second one operates in intermediate frequency. It comprises two direct transversal filter circuits and two recursive transverse filter circuits. It is the latter part which is effective for correcting the effect of non-minimal phase fading. Figure 1.83 shows the general diagram of this equalizer.

Diversity techniques Diversity techniques are methods which are well known for improving the quality and availability of digital microwave links in the case of propagation by multiple paths. Space, frequency and cross-band (two RF carriers

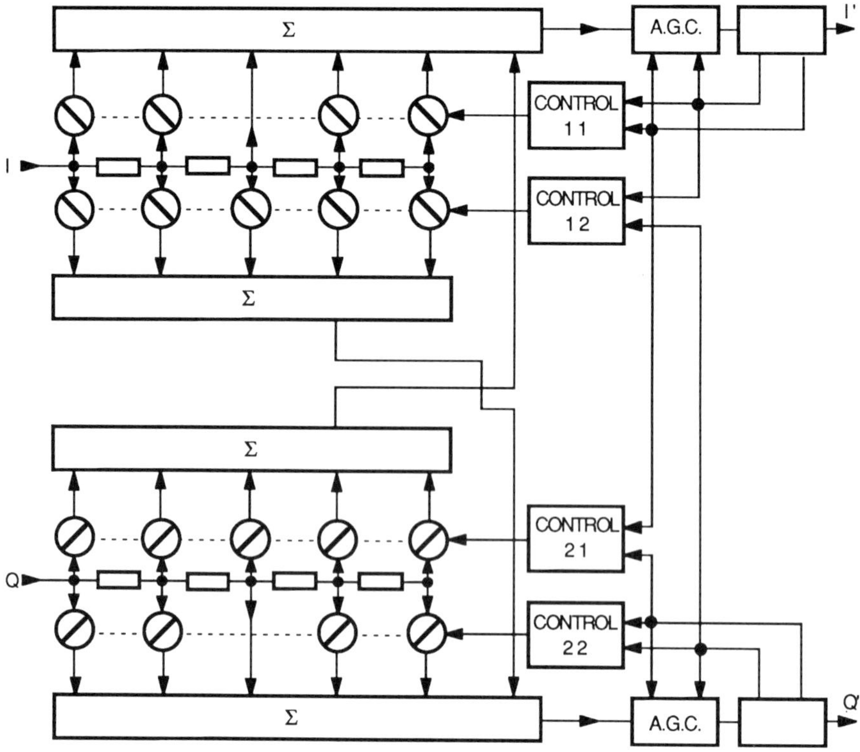

Fig. 1.82 Baseband time domain equalizer.

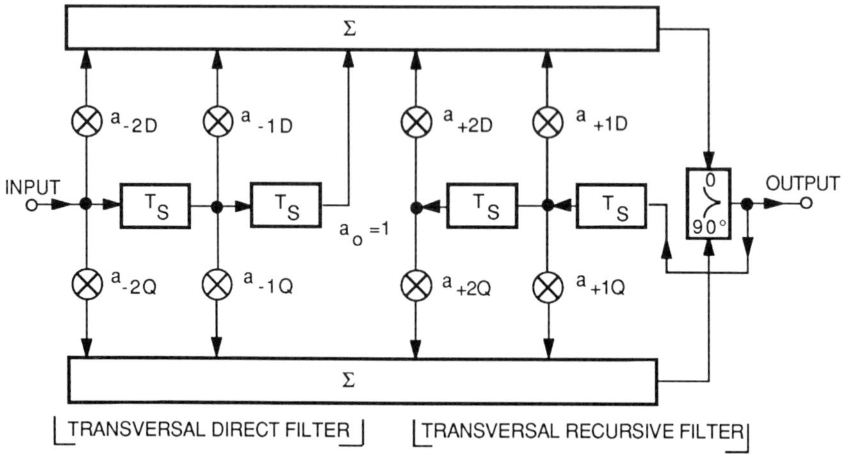

Fig. 1.83 If domain equalizer.

are situated in different frequency bands) diversity can be employed. The frequency diversities and the diversities between frequency bands are rarely employed because of the importance of the frequency bands occupied, and in particular in the high digital bit rate systems.

Space diversity uses two receiving antennas and combines the signals received. It is used to reduce the effects of flat and selective fading. The two antennas must be spaced vertically by 150 λ (wavelength). The signals are combined by one of the following devices: a hitless combiner, an IF in-phase combiner, an IF minimum dispersion combiner or a notch in-phase combiner.

Hitless switching combiner This method is used with equipment operating on 'hot standby'. Figure 1.84 shows this combiner. Generally, an error detector based on an eye diagram is used to control the switching.

In-phase combiner The process is based on an automatic rephasing of the useful signals, followed by a linear summing of them in IF. The summing of the noises is random. The combiner comprises a phase shifter and an electronic control

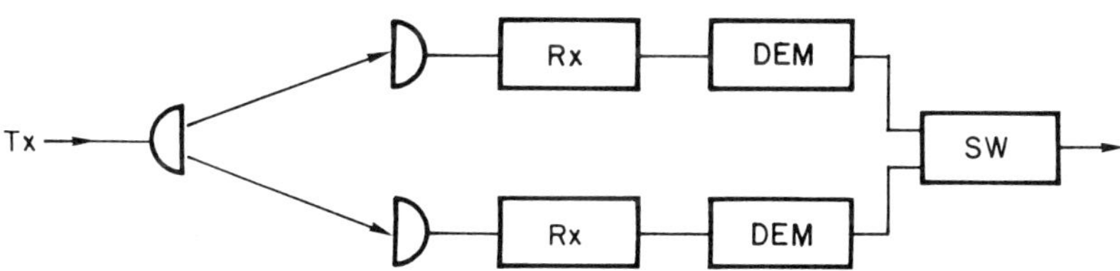

Fig. 1.84 Hitless switching combiner.

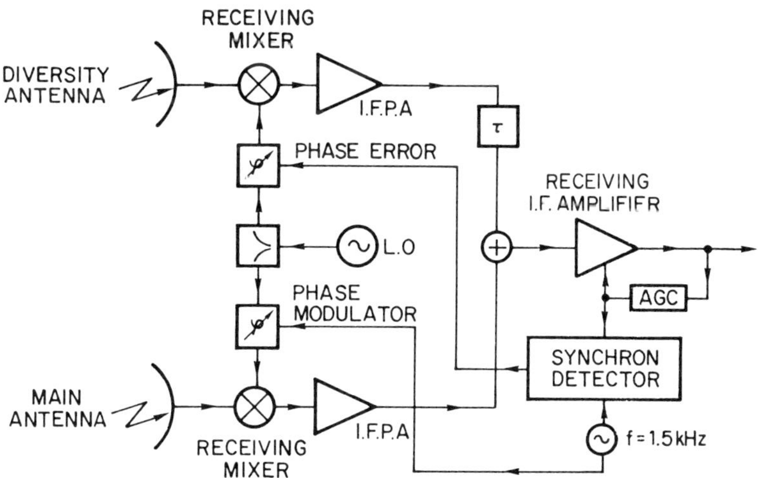

Fig. 1.85 IF in-phase combiner.

system for rephasing the signals received. An adder combines the two IF signals. Figure 1.85 gives an example of this combiner.

IF minimum dispersion combiner Generally, the disturbing signals received by the two antennas have substantially the same amplitude. One has only to combine them in phase opposition to reduce their influence. This is performed via a phase shift system, controlled by a microprocessor, based on checking the form of the sufficiently scrambled signal (filter $f-$ and $f+$). After combining, the disturbing signal is weaker than the useful signal and the fading is always at minimal phase. This combiner is shown in Fig. 1.86.

However, the useful signal can be considerably attenuated when the signals received by each antenna have the same amplitude and when the disturbing signals are negligible. To avoid this problem, the minimum dispersion combiner is automatically replaced by an in-phase combiner, when the combined signal level is less than a predetermined level.

Notch in-phase combiner This combiner is a generalization of the minimum dispersion combiner. It uses a number of detectors (>3), distributed in the IF band, to measure the amplitude variation of the signals in the passband. Based on these signals, a device, consisting of a microprocessor and a phase shift system, works to 'flatten' the signal spectrum and reduce the notch produced in the transfer function. An example of this combiner is shown in Fig. 1.87.

Cross-polarization interference canceller The devices can operate in RF, in IF or in baseband. The RF setting of these devices is complex and the baseband

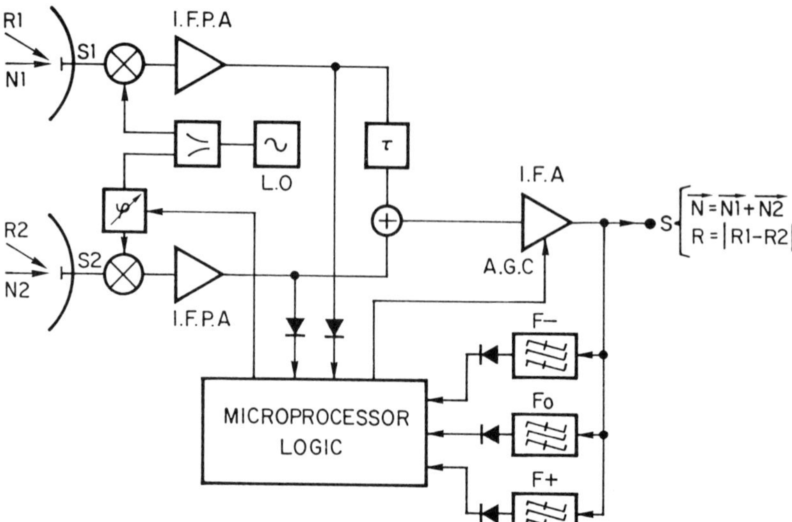

Fig. 1.86 IF minimum dispersion combiner.

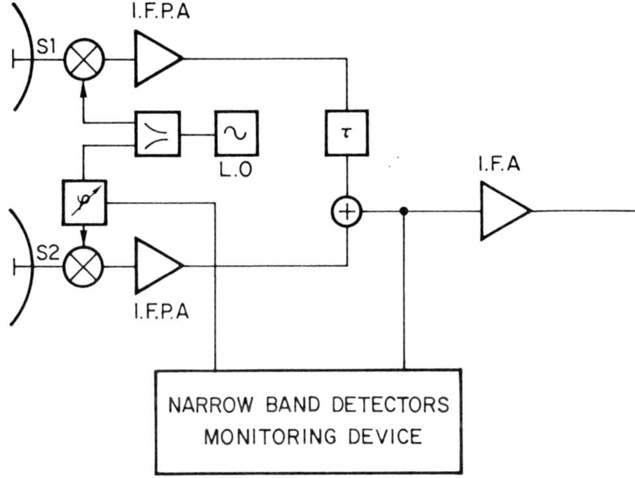

Fig. 1.87 Notch in-phase combiner.

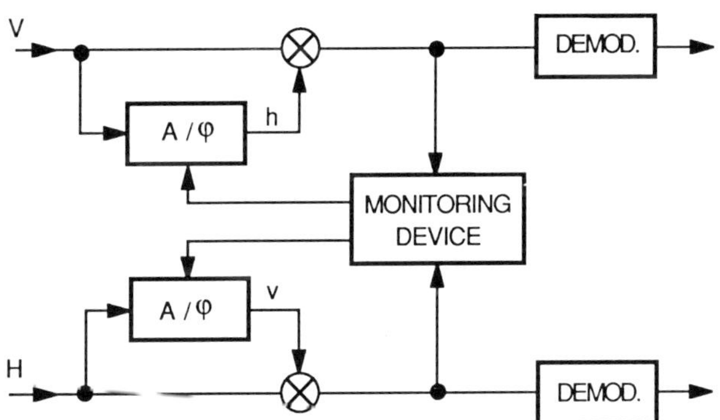

Fig. 1.88 Cross-polarization interference canceller.

suppressor requires many components, which is the reason why this equipment is generally built in IF. Figure 1.88 shows the operating principle.

Let us assume that the IF access indicated (V) has two useful signal components V, and one part of the signal (*H*), created by depolarization. The cross-polarized signal behaves like a jammer. To eliminate it, we superpose at V a fraction *h* of H, having a suitable phase and amplitude (*h* is subtracted to suppress the interference). The same operation is performed on H. The adjustments in amplitude and in phase are obtained from a control circuit. The XPD improvement is 6 to 10 dB.

Automatic transmitted power control (ATPC) On an AB link (Fig. 1.89), the ATPC is used to provide transmitting power control (station A) based on the

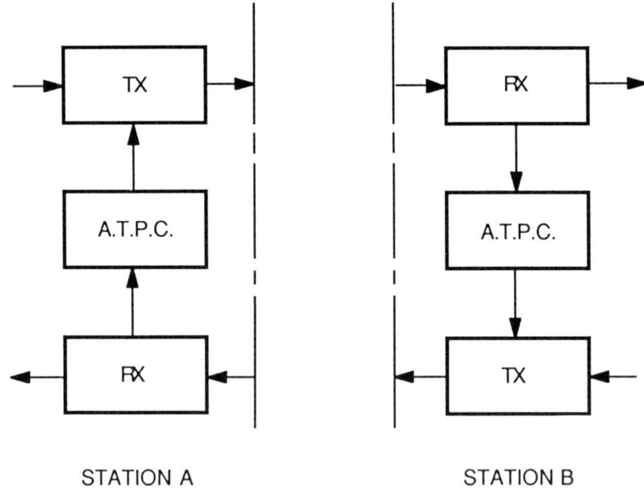

Fig. 1.89 ATPC system.

power received (station B). The transmitting power is maximal only during the deep fading. In the absence of fading, the maximal power is reduced by approximately 15 dB. The maximal power is only transmitted a low percentage of the time. This makes it possible to reduce and even to eliminate the interchannel interference on the same route or between channels radiating from the same station.

1.4.7 Predicting outages due to propagation

Among the outage calculation methods, proposed in CCIR Report 784, this example presents the prediction method based on signature curves. Digital transmission is considered as interrupted when the BER is $\geqslant 10^{-3}$. This BER must not be exceeded more than time T in any one month. This time T is given by the following relationship:

$$T = T_o \times p$$

where T_o is the observation period (one month $= 2.6 \times 10^6$ s) and p is the outage probability.

The main causes of errors due to propagation are:

1. thermal noise and its variations under the influence of flat fading;
2. signal distortion produced by selective fading;
3. noise due to interference.

Generally, the noise power due to thermal effects and to interference can be summed. In this way, the probability of interruption (BER $> 10^{-3}$) is estimated based on the probability of an outage due to flat fading (p_F) and selective fading

(p_s). Nevertheless if $p_F \cong p_s$, an interaction between noise and signal distortions develops a synergy effect. The overall outage probability can be expressed approximately by:

$$p = (p_F + p_s)k$$

where k characterizes the effect of synergy.

Flat fading outages

Outage probability is given by the following relationship:

$$p_F = \mu W_F$$

where μ is the multiple path presence factor during the operation and W_F is the probability that the received power p_r will be weaker than power p_T (threshold of the power received for a BER $= 10^{-3}$).

The probability of a deep fading is approached by a Rayleigh distribution:

$$W_F(P_r \leqslant P_T) = 1 - \exp(-P_T/P_o) \simeq \frac{P_T}{P_o}$$

where P_o is the power received for a transmission in free space. In this way, in the absence of selective fading, the ratio P_T/P_o is defined as the 'flat fading margin'

$$m_F = \frac{P_o}{P_T}.$$

The factor of the presence of fading due to multiple paths during the worst month is given by the following semi-empirical formula:

$$u = KQf^B d^c$$

where d is the radio hop length (km), f is the channel radio frequency (GHz), K is the climate dependent coefficient, Q is the coefficient dependent on the nature of the ground and B and C are coefficients dependent on the geographical location.

Under these conditions, the outage time due to a flat fading is given by the following relationship:

$$T_F = T_o \times \mu \times \frac{1}{m_F}.$$

Selective fading outages

By extending the 'margin' concept to selective fading, the 'selective fading margin' (m_s) is defined as the depth of fading which is exceeded during the same number of seconds T_s as the limit error rate (10^{-3}) in the absence of flat fading (errors due to thermal noise) (35):

$$T_s = T_o \times \mu \times \frac{1}{m_s}.$$

By using the signatures, it is possible to estimate the outage time T_s of a given equipment from the outage time T_{SR}, obtained on an experimental link. Under these conditions, we show that if m_{SR} is the margin against the selective fading of the reference equipment, margin m_s of the equipment considered is given by the following relationship:

$$m_s = m_{SR} \times \frac{S_R}{S}$$

S_R and S are surfaces under the signature of the reference equipment and of the equipment being studied, calculated according to the method defined in section 1.4.5.

In this example, the link measured by Giger and Barnett was chosen as the reference link. It has the following characteristics:

length:	42.5 km
operating frequency:	5.945 GHz
modulation:	8 PSK
digital bit rate:	78 Mbit/s
outage time due to selective fading:	725 s during the worst month
KQ =	6×10^{-7}
B =	1
C =	3.

The reference margin m_{SR} for selective fading is:

$$m_{SR} \cong 10^3.$$

The selective fading outage is:

$$T_s = T_o \times \mu \times \frac{1}{m_{SR}} \times \frac{S}{S_R}.$$

Overall outages

Overall outages are given by

$$T = k(T_F + T_s).$$

For northeastern Europe:

$$T = k \times 3.65 \times 10^{-2} f \times d^{3.5} \left[\frac{1}{m_F} + 10^{-3} \frac{S}{S_R} \right].$$

The margin m_F concerning fading must be reduced to take into account interference and, at certain frequencies, the influence of gases and of hydrometeors. The insertion of countermeasure devices in the equipment can improve the values of m_F and m_S. For the overall outage calculation, it is necessary to take the imperfections due to equipment into account.

1.5 SPECIFIC NATURE OF OVER-THE-HORIZON MICROWAVE LINKS

The tropospheric scatter or troposcatter microwave links (over-the-horizon), operate in frequency bands from 450 to 5000 MHz. They make is possible to have radio hops with lengths from 100 to 800 km. Circuits of several thousand kilometers are obtained by using several links serially.

1.5.1 Properties of the propagation medium

Figure 1.90 shows how the transmitting and receiving antennas exchange energy over the horizon via a common scatter volume. Dispersion of the incident electromagnetic radiation characterizes the over-the-horizon propagation mechanism. The signal received results from the composition of a large number of elementary vectors representing various propagation paths. Unlike the case of line-of-sight with the existence of multiple trajectories as exceptional, we find ourselves under permanent conditions of multiple paths. Consequently, the field received is modulated in amplitude and in phase. These unwanted modulations take place at the timing of the fluctuation of the heterogeneities of the scatter volume: a few hertz. Figure 1.91 gives an example of the response in amplitude A and in phase φ of a tropospheric channel.

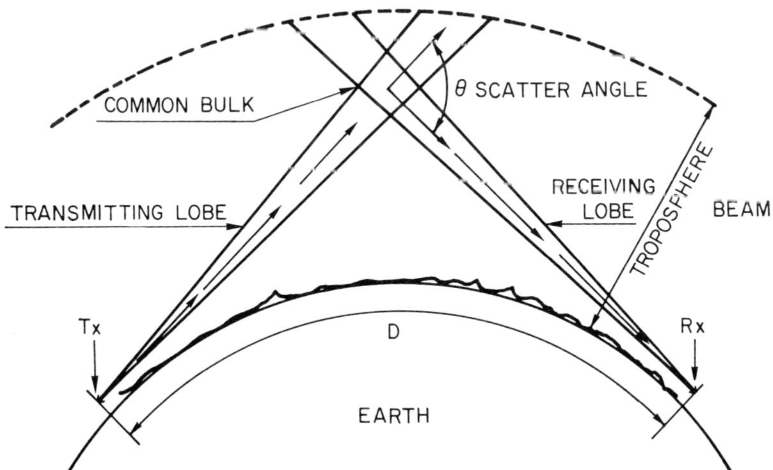

Fig. 1.90 Troposcatter link.

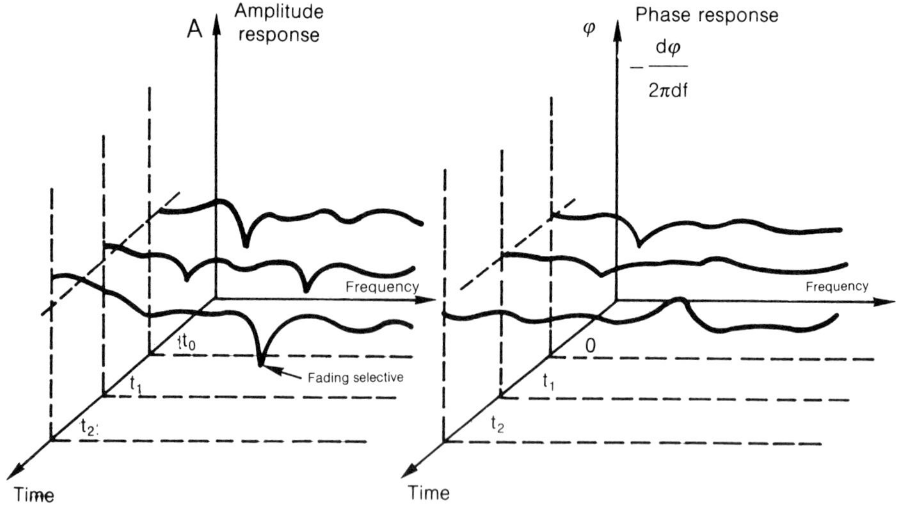

Fig. 1.91 Amplitude and phase responses.

Propagation attenuation

The following factors have been noted. At fixed path length, the propagation constant is defined by the median attenuation and by the statistical distribution of its variations as a function of time. At a given instant, the propagation attenuation varies with the frequency and limits the passband of the tropospheric channel (selective fading).

The median attenuation (50% of the time) and its long term variations depend upon the geographical and weather radio conditions encountered on the link to be set up:

1. The propagation attenuation is substantially proportional to the cube of the frequency.
2. The path profiles (distance between stations, ground obstructions, land relief, altitude of the stations) characterized by the scatter angle θ, influence the propagation attenuation value.
3. The weather conditions have a more complex effect. Their consequences are of three types.

   (a) According to the types of climates, (continental or maritime, temperate, equatorial, subtropical, desert, polar, etc.), the median attenuation and its fluctuations take different values.

   (b) Under seasonal effect, the median monthly attenuation goes through a maximum during a certain period of the year. This period is located:

      (i) in winter for temperate climates;
      (ii) in the dry season for sub-tropical climates;

Table 1.18 Troposcatter-link parameters

Frequency (GHz)	*d* (km)	L_0 (dB)	*L* (dB)
1	200–500	143	210–240
2	150–350	149	210–240
4.7	100–200	157	210–240

(iii) in the rainy season for tropical climates;
(iv) at the beginning of the summer for desert climates.

The maximal variations in the median monthly attenuation during the year are approximately 20 to 25 dB for a distance of 200 km in a temperate climate. They are lower for an equatorial climate (10 to 15 dB). The variations diminish as the distance increases. During a single day, the attenuation goes through a maximum during the afternoon. This daily effect is very important in sub-tropical climates where the amplitude of the variations reaches 25 dB. In other climates, these variations are smaller; they reach 5 dB in temperate climates and 10 dB in tropical climates. The variations diminish as the distance increases.

The short term variations of the median attenuation are characterized by a distribution of levels which obeys the Rayleigh law. In Volume 2 of this book and in CCIR, we can find complementary information and a calculation method. For example, Table 1.18 gives the order of magnitude of attenuation *L* (50% of the time) for values of θ from 15×10^{-13} to 40×10^{-3} rad. The values of *L* are greater by 50 to 100 dB than those of L_0 (free space).

Propagation selectively—coherence band

The transfer of the energy comes from the many paths having slightly different delays. The amplitude of the field received results from vector summing its many components.

Beginning with the following example where the field received comes from two scatter sources (Fig. 1.92), we determine that the amplitude of the field received as a function of the frequency goes through a maximum for $\omega = 2k\pi$ and through a minimum for $\omega(2k+1)\pi$. This approach suggests that the propagation constant presents a selectivity (Fig. 1.93).

Depending upon the frequency autocorrelation function, expressed by the degree of correlation existing between the amplitude of the fields received at the two frequencies spaced by df, this is used to determine the frequency band B_c (coherence band). From this we deduce:

$$B_c = \mathrm{d}f_c = \frac{k}{\pi\Delta}$$

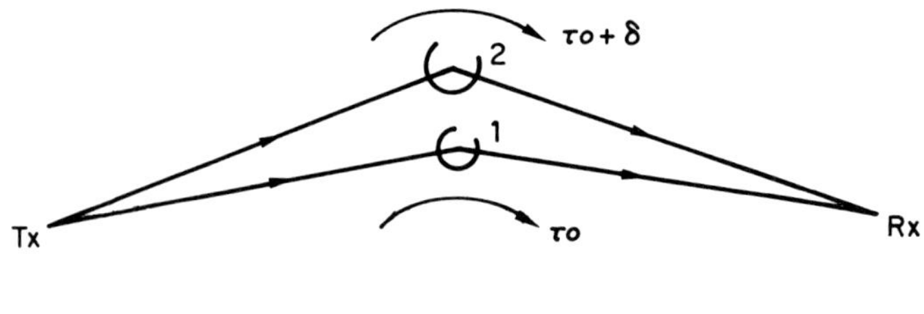

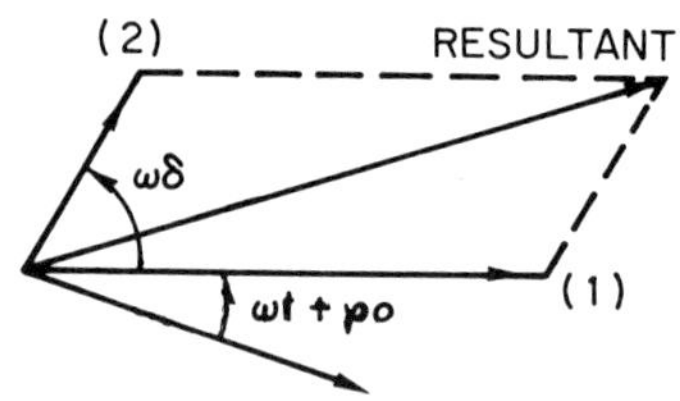

Fig. 1.92 Multipath propagation.

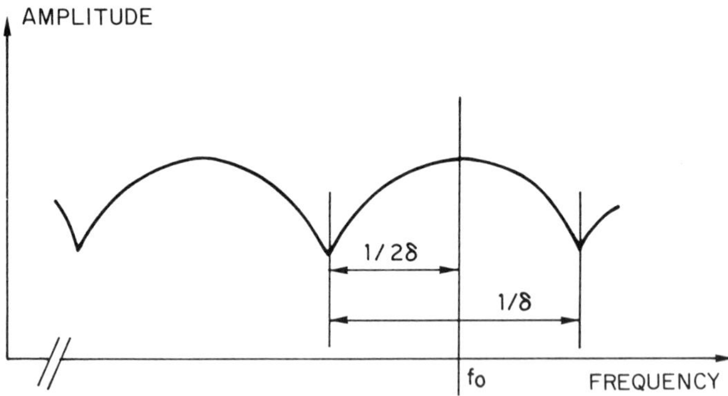

Fig. 1.93 Amplitude of the receiving field.

where Δ is the differential delay existing between the paths of the highest and the lowest altitude is $(D/8c)\alpha(\theta + \alpha)$, with $c =$ the speed of light. An experimental study (Fig. 1.94) has made it possible to determine the medium value of the coherence band B_c (50%) and its variations as a function of time.

1.5.2 Equipment characteristics

The structure of a terminal station, is similar to that of the line-of-sight microwave links. However, to compensate the effects of the propagation medium—very

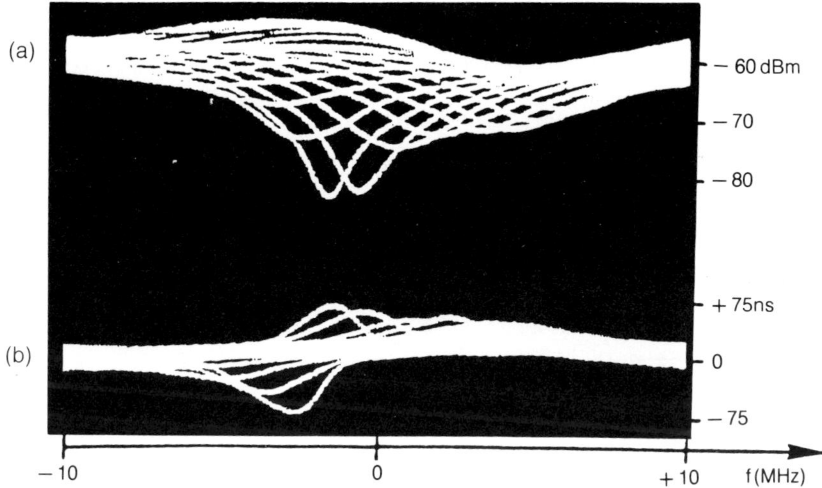

Fig. 1.94 Examples of amplitude and GD responses: (a) amplitude variation; (b) group delay variation.

substantial attenuation 170 to 240 dB, selectivity of the propagation medium—the following devices are used:

1. high gain antennas;
2. high power transmitters;
3. high sensitivity receivers;
4. systems with reception in diversity.

In the case of digital transmissions:

1. modulations with small number of states;
2. auto-adaptive equalizers;
3. self-correcting codes.

High gain antennas

These are generally in the form of large parabolic reflectors, illuminated by an external system with two polarizations. Two types of structure are used: the symmetrical parabolas of 5, 10 and 20 m in diameter, with a feed at the focus, and the asymmetrical paraboloids with a square cut of 18, 27 or 36 m per side and an offset radiator.

The characteristics of the antennas used are as follows (Table 1.19):

1. the gain (G);
2. the beam width (α).

The side-lobe attenuation must be at least 25 dB. Under these conditions, the noise temperature of the antennas is close to 200 K.

Table 1.19 Typical characteristics of troposcatter antennas

	450 MHz		900 MHz		2700 MHz		4500 MHz	
Antenna (m)	G (dB)	α (mR)	G (dB)	α (mR)	G (dB)	α (mR)	G (dB)	α (mR)
5	25	142	31	71	41	23.7	47	11.8
10	31	71	37	35.5	47	11.8	53	5.9
20	37	35.5	43	17.7	53	5.9		
30	41	23.7	47	11.8				
40	43	17.7	49	8.8				

We note that in over-the-horizon radio links, above a certain value, the gain increase is slower than the increase in the parabolic dimensions. This phenomenon, called antenna coupling losses, varies with time. In link planning, only the medium value (Fig. 1.95) of this loss is taken into consideration.

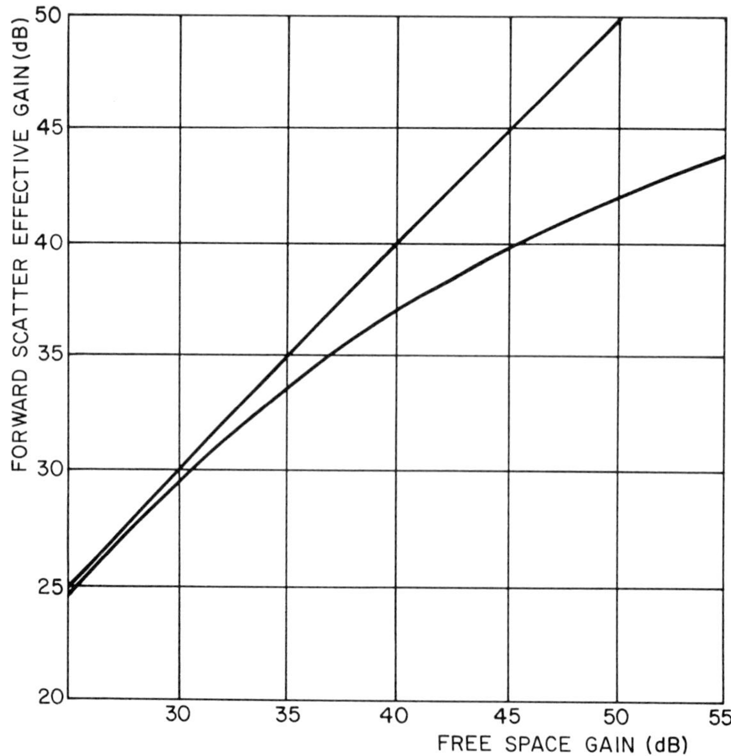

Fig. 1.95 Antenna coupling loss.

The physical characteristics of the antennas are determined by the use conditions and the enviroment. As an example, their approximate weight is as follows:

parabolic of 5 m: 250 to 500 kg
parabolic of 10 m: 1.5 to 2 tons
parabolic of 18 m: 25 to 35 tons
parabolic of 27 m: 70 to 110 tons.

Power amplifiers

The powers used in over-the-horizon microwave links are from 100 W to 100 kW. There are three types of active components:

1. transistors;
2. amplifier klystrons;
3. travelling-wave tubes (TWT).

The power transistors used operate up to 5 GHz. They have an output power of about 10 W and a gain of from 6 to 8 dB. The adaptive circuits are built using micro-electronic technology. A suitable coupling of these amplification stages provides the desired gain and power (Fig. 1.96). The output power is about 100 W from 2 to 5 GHz and the overall efficiency is 25 to 35%. The equipment is cooled by natural convection or by forced air circulation.

The amplifier klystrons are used to provide powers of:

1 kW in frequency bands 450, 900, 2500, 4500 MHz
10 kW in frequency bands 450, 900, 2500, 4500 MHz
75 to 100 kW in frequency bands 450, 900, 2500 MHz.

Their gain is approximately from 30 to 40 dB; the overall efficiency is:

20% for a transmitting power of 1 kW
30% for a transmitting power of 10 kW
35% for a transmitting power of 75 kW.

The tubes are cooled by air circulation for 1 kW and 10 kW (900 MHz only), and by water circulation at 10 kW and above.

The travelling-wave tubes are used in equipment mainly designed for military applications (possibility of fast frequency change). The powers which can be reached are approximately:

100 to 300 W in frequency bands 450, 900, 2500, 4500 MHz
1 kW in frequency bands 2500, 4500 MHz
5 to 10 kW in frequency bands 2500, 4500 MHz.

The gains obtained are from 30 to 40 dB and the overall efficiency varies from 10 to 35%. The tubes are cooled by air circulation up to 1 kW and by water

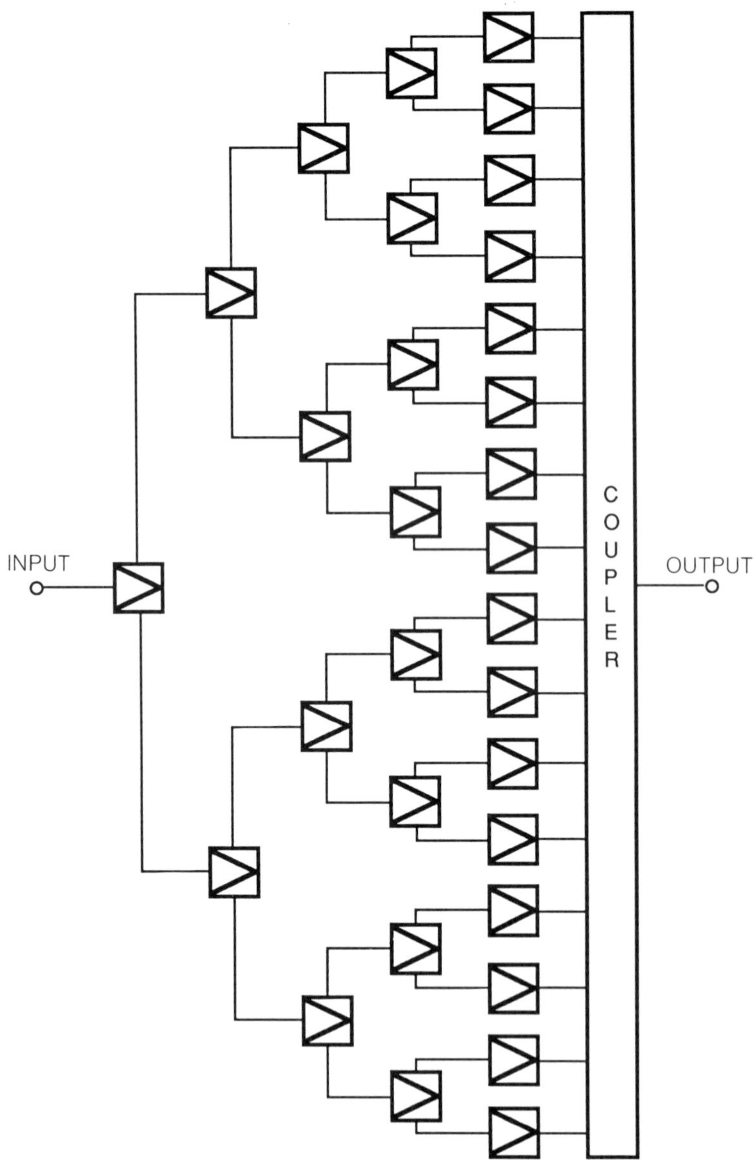

Fig. 1.96 Power transistor amplifier.

circulation above 1 kW. Tetrode and triode tubes were employed in amplifiers up to 1000 MHz. The output power did not exceed 1 kW with an overall efficiency of 35% and cooling was done by forced air.

High sensitivity receivers

The transmission quality is defined by the value of the ratio of the signal (C) to the noise (N) at the receiver input.

The noise power N is given by the following relationship:

$$N = (NF)kTB$$

where NF is the receiver noise factor, k is Boltzmann constant, T is the absolute temperature at receiver input, and B is the receiver passband. To minimize the value of N, there is an advantage in reducing NF and B.

B must remain compatible with the signals to be transmitted to avoid distortion. The minimal filtering bands are defined by the following criteria:

1. Carson for the frequency modulation of analogue signals;
2. Nyquist for the transmission of digital signals.

The noise factor of the receivers can be improved by using transistorized amplifiers positioned at the input of the receivers and sometimes, near the antenna. The gain of these devices is from 15 to 20 dB and the noise factor from 2 to 3 dB.

Diversity of reception systems

Propagation is affected by deep and fast fading. This fading follows a Rayleigh statistic. The phase shifts between the various components coming onto a single antenna affect the bandwidth of the propagation medium and produce residual intermodulations in the case of analogue signals and 'irreductible errors' for the transmission of digital signals.

These degradations can be eliminated by using two or more receivers, operating at fairly distant frequencies (frequency diversity: $\Delta f/f = 4$ to 10%) with antennas sufficiently spaced (space diversity: $e = 100\lambda$). A time domain diversity can be obtained by repeating the message at different time intervals. The structure of a quadruple space diversity link is shown in Fig. 1.97.

For links performing digital transmissions, high order diversity systems (8 to 16) are implemented. Furthermore, certain modulation methods involve using a set of n frequencies and assigning to each symbol a group of p frequencies among the n. An example of this method is shown in Fig. 1.98 (system in two frequencies among four). The reception of just one of the two frequencies is enough for the recognition of the signal received. This system provides an order 2 diversity.

The implementation of the diversity techniques requires the use of devices for combining the signals received. This combination can be performed by linear methods (analogue signals, digital signals before regeneration), or logical methods

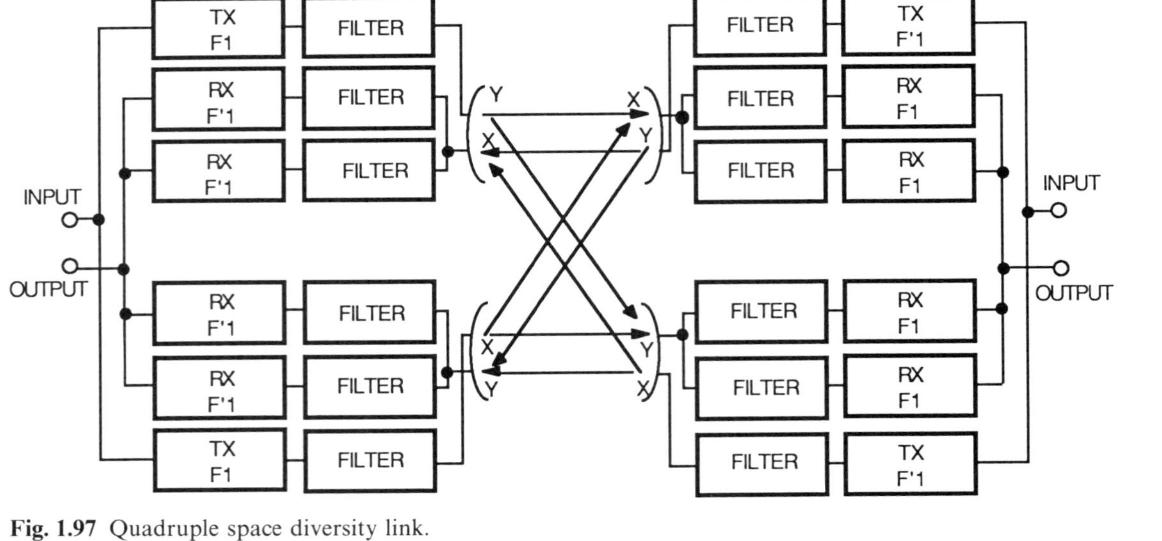

Fig. 1.97 Quadruple space diversity link.

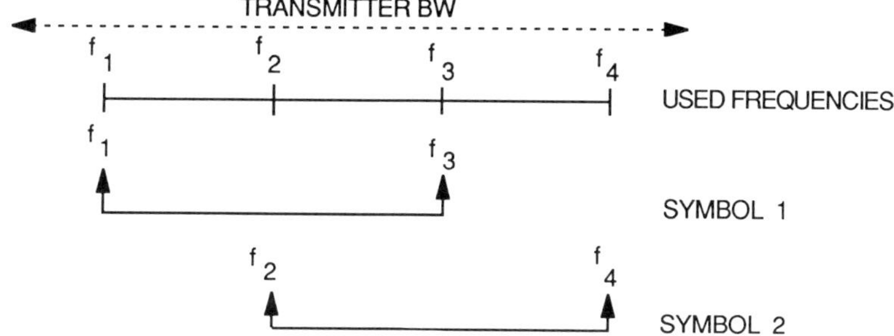

Fig. 1.98 Two frequencies among four system.

(digital signals after regeneration). There are three categories of linear combinations:

1. maximal ratio: in each receivers, the signals received are weighted by a factor which depends on the field received and of the noise level; the signals from the various receivers are then summed linearly;
2. equal gain: the signals received are summed at a constant noise level;
3. by selection: at each instant, the signal received with the highest level is selected.

The implementation of linear combination implies that no non-linear device can be installed in the reception circuit before the combiner. The combination is done in intermediate frequency or in baseband (coherent modulation processes (Figs 1.99 and 1.100).

Combination by logic processes applies to digital signals after regeneration. This consists of establishing a majority criteria on an odd number of signals. It makes it possible to obtain an efficiency which is equivalent to that of the maximal ratio combining, but a diversity order two times higher is needed.

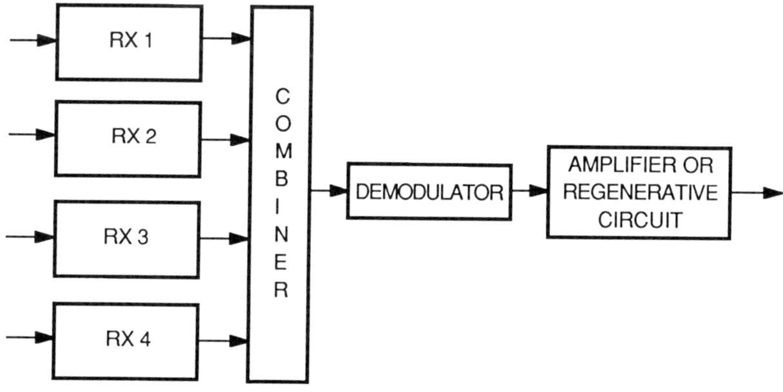

Fig. 1.99 IF combination.

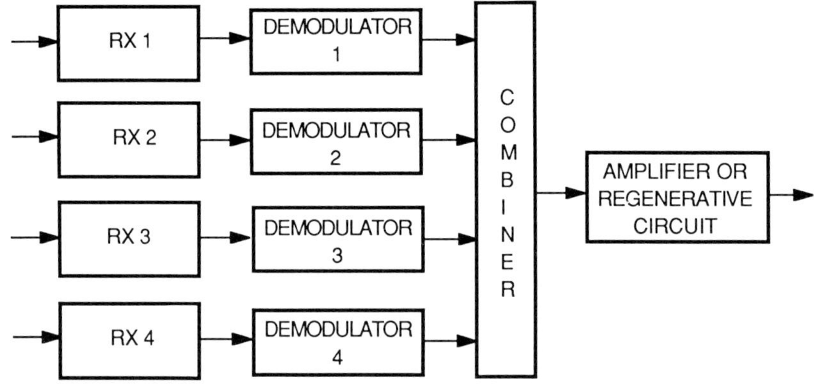

Fig. 1.100 Baseband combination.

To summarize, the following curves give performances obtained by diversity systems for:

1. the analogue signals; the diversity gain is given by the median signal-to-noise ratio increase for obtaining the same depth of fading exceeded during the same percentage ($\eta\%$) of time (Fig. 1.101);
2. the digital signals: an example is given in Fig. 1.102 for a bit rate of 2 kbit/s, a CPSK modulation and an equal gain combination.

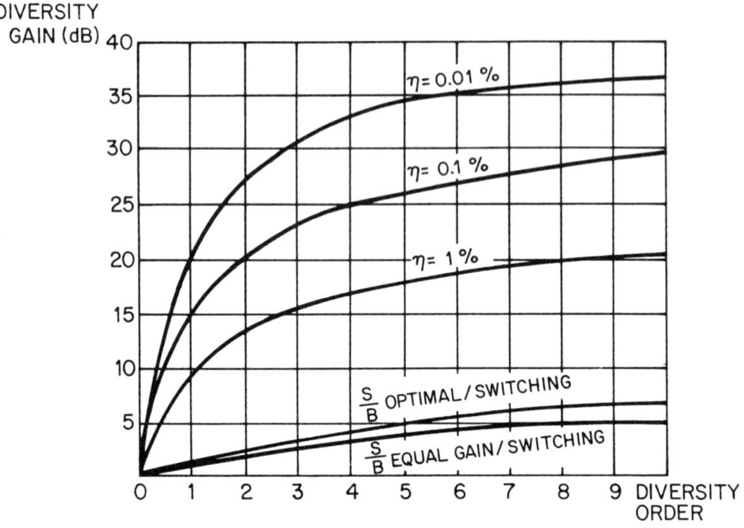

Fig. 1.101 Analogue signals diversity system performance.

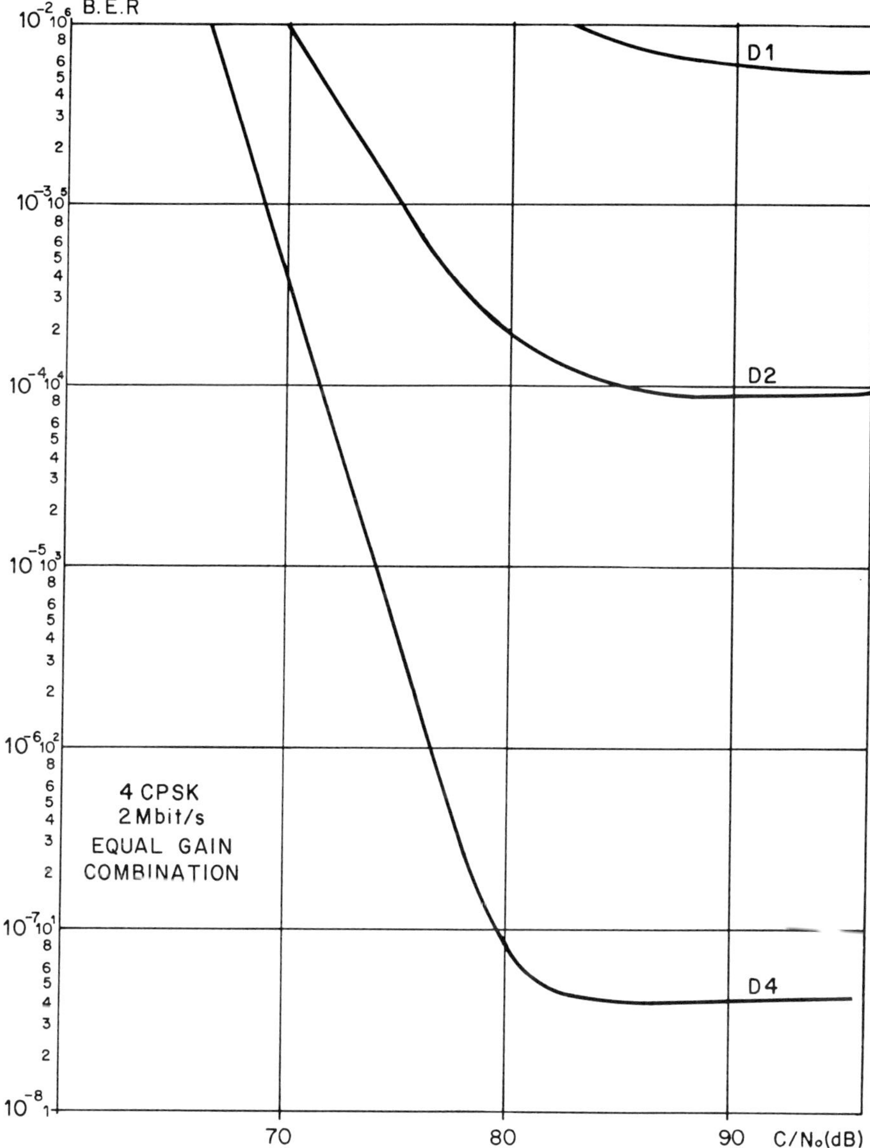

Fig. 1.102 Digital signals diversity system performance.

Forward error correcting codes

As in the case of line-of-sight microwave links, error correcting codes are used to eliminate isolated or erratic errors. The block codes are generally used. The coding system is Hamming, Lee or Reed–Solomon type.

1.5.3 Transmission quality

Analogue signals

Phase or frequency modulation is generally used for its advantageous properties, transmission equivalent stability, modulation gain, etc. The methods for calculating the signal-to-noise ratio, described in section 1.4.1, can be applied to determine the $(S/N)_{min}$.

Digital signals

Two methods are generally used for transmitting digital signals.

1. Direct modulation of a radio frequency by digital signals. This method is used for high speed transmissions (2 to 8 Mbit/s).
2. Indirect modulation where the pulses modulate a subcarrier located in the baseband, used for transmitting analogue signals. It is used for signals at low speed (50 to 9600 bit/s). The transmission of these signals is not covered here.

The transmission quality of digital signals is defined in section 1.4.4.

The origin of errors is located in the equipment, the environment and the propagation medium. The definitions of the structure of equipment and of the communication network are used to specify the limits of the first two sources of errors. Their magnitude is generally negligible. The determination of the equipment electrical characteristics, i.e. transmitting power, noise factor, modulation and demodulation process, play an essential role in the reduction of errors produced by the propagation medium. These errors essentially depend upon:

1. the signal-to-noise ratio at the receiver input during the time interval considered and in presence of non selective fading, obeying the Rayleigh law: P_{e1};
2. intersymbol interference due to selective fading: P_{e2};
3. frequency and phase variations of the carrier during the time separating two symbols: P_{e3}.

It is assumed that the total error probability P_{eo} due to the propagation medium is:

$$P_{eo} = P_{e1} + P_{e2} + P_{e3}.$$

In general P_{e3} is ignored in cases of high digital bit rates, P_{e1} is determined using the method described in section 1.4. To calculate the value of P_{e2}, we can use the procedures described by Sunde.

REFERENCES

Barnett, N. T. (1972) Multipath propagation at 4, 6 and 11 GHz, *Bell System Technical Journal*, **51**, 321–361.

Bui-Hai, N. (1978) *Antennes micro-ondes. Application aux faisceaux hertziens*, Masson, Paris.

C.C.I.R. Recommendations and Reports, green book, volumes V–IX–XII, IUT Geneva.

C.C.I.T.T., Volume III, IUT Geneva.

de Luca, O. *et al.* (1986) Auto-adaptativité dans les faisceaux hertziens numériques, *l'Onde Electrique*, **66**(2).

Emshwiller, M. (1978) Characterization of the performance of PSK digital radio transmission in the presence of multipath fading, *ICC78 conference record.*

Fagot, J. and Magne, P. (1961) *Frequency modulation theory. Application to microwave links*, Pergamon Press.

Federal communication commision Establishment of policies and procedures for the use of digital modulation techniques in microwave radio and proposed amendments to Parts 2 and 21, F.F.C. 74 – 985 docket n° 19311, Washington DC, September 1974.

Feher, K. (1981) *Digital communications—Microwave applications*, Englewood cliffs, N. J., Prentice Hall.

Gerard, P. (1981) Egaliseur autoadaptatif d'amplitude pour faisceaux hertziens numériques, *Revue technique Thomson-CSF*, **1**.

Giger, A. J. *et al.* (1981) Effects on multipath propagation on digital radio. *IEEE trans. com.*, **29**.

Giger, A. J. *et al.* (1986) Time and frequency fluctuations of microwave interference due to terrain scatter, *IEEE Global Telecommunications Conference.*

ITT (1968), Reference data for radio engineers, Howard W. Sams & Co., inc.

Lin, H. (1977) Impact of microwave depolarization during multipath fading on digital radio performance, *BSTJ*, **56**.

Magne, Ph. Faisceaux hertziens, collection Techniques de l'ingénieur, E 7520 et E 7521.

Magne, Ph. Faisceaux hertziens numériques, collection Techniques de l'ingénieur, E 7540 et E 7541.

Martin, L. (1981) Relative amplitude and delays of rays during multipath fadings, *Second international conference of antennas and propagation, York (U.K.).*

Mottle, T. O. (1977) Dual polarized channel outages during multipath fading, *BSTJ* **56**.

Murase, T. *et al.* (1981) 200 Mbit/s 16QAM digital radio system with new countermeasure technique for multipath fading, *IEEE.*

Peterson, W. W. *Error correcting codes* MIT Press.

Rummler, W. D. (1979) A new selective fading model. Application to propagation data, *Bell system technical journal*, **58 (5)**, 1037–1071.

Rummler, W. D. (1982) A simplified method for the laboratory determination of multipath outage of digital radios in the presence of thermal noise, *IEEE trans. on communications*, **30 (3)**, 487–494.

Sunde, E. D. *Communication System Engineering Theory.*

Sylvain, M. (1985) Panorama des études sur les trajets multiples, *Annales des Télécommunications*, **40 (11-12)**, 547–564.

Vigants, A. (1975) Space-diversity engineering, *Bell system technical journal*, **54 (1)**, 103–142.

Vigants, A. (1981) Distance variation of two-tones amplitude dispersion in line of sight microwave propagation, *PIEE comm.*, **18**, session 68—Propagation and digital radio.

2

Satellite links

Jean Salomon

2.1 GENERAL

2.1.1 Introduction

Most significant human progress generally results from unpredicted technological breakthroughs. Which J. Verne or H. G. Wells could have anticipated that for every man—at least, if he lives in a developed country—it would be commonplace to receive or send instantaneous oral, visual or data messages all over the world thanks to a mixture of cables (electrical and optical), radio microwaves and artificial satellite technologies?

Satellite communications are directly derived, as concerns transmission techniques, from point-to-point terrestrial links. However, they are indeed very special microwave links which use an unattended relay located very far away—usually at some 40 000 km—from the earth stations. Satellite communications have been the subject of many books and publications. In this chapter, although we shall put an emphasis on the microwave techniques proper, as applied to the satellite transponders (section 2.3) and to the earth stations (section 2.4), we shall also, more generally, outline (in sections 2.1 and 2.2) the basic principles of satellite communications links and systems.

2.1.2 Communication satellite systems

Satellite links are arranged to form communication satellite systems. A communication satellite system is composed of the space segment and the earth segment.

The space segment comprises not only the space station, i.e. the satellite, but also the tracking, telemetry and command (TTC) earth stations which provide its logistic support ground facilities. A communication satellite is composed of a space platform which carries the payload, i.e. the useful equipment (comprising the transponders and antennas).

Communication satellites can be classified according to their orbit around the earth: '12 hours' satellites have been operated for communications (the former USSR's MOLNYAs); satellites orbiting at low or medium altitude can be used for positioning, navigation, earth observation, data collection etc., and are currently receiving new attention for providing communications for mobiles.

However, by far the most commonly used orbit for communication satellites is the geostationary satellite orbit (GSO). This is a circular orbit concentric to the Earth, located on the equatorial plane with a radius of about 42 200 km, i.e. an altitude of 37 786 km. The elementary laws of dynamics show that, when launched on such an orbit, a satellite travels at an angular speed of 360° per 24 hours (geosynchronous orbit). Therefore the satellite appears theoretically motionless in relation to any reference point on the earth's surface (in fact, there always remains some minute residual drifts, often less than $\pm 0.1°$). This is a very advantageous characteristic since the satellite operates like a fixed lighthouse in the sky, thus providing permanent coverage of the specified terrestrial areas. Also, the design of the earth stations is simplified and often no antenna tracking system is required (however, in the case of a large antenna, this can be necessary to compensate for the residual satellite drift).

The earth segment is the name given to the complete set of the Earth stations which exchange information via the satellite, i.e. which transmit and receive RF carriers bearing signals to/from the satellite transponders. Therefore, the main components of an earth station are:

1. the antenna which is pointed towards the satellite and which is used both for transmitting and receiving the RF carriers, these two functions being accessed separately through a duplexer;
2. the RF transmitter, usually called the high power amplifier (HPA) system, because a high power is required to reach the distant satellite;
3. the RF receiver, the front end of which is called the low-noise amplifier (LNA), because of its ability to receive and amplify carriers from the satellite at a very low level of power without adding too much noise;
4. the so-called ground communication equipment, (GCE) which performs the functions of frequency conversion, modulation and demodulation;
5. the terrestrial interface modules, i.e. the equipment needed to transfer the information signals to the users, through terrestrial communication networks, or even directly to a terminal.

2.1.3 Utilization of the radio frequency spectrum

The radio frequencies used for satellite communications are defined by the radio regulations of the ITU (International Telecommunication Union). In order to isolate the transmitter and the receiver, both in the satellite transponders and in the earth stations, two different carrier frequencies are utilized for implementing a single, one-way, satellite communication link: the up-link (U/L) frequency for transmitting from an earth station to the satellite and the down-link (D/L) frequency for transmitting from the satellite to the earth stations. Frequency translation from U/L to D/L is performed in the satellite transponders.

The ITU allocates the frequency bands according to the types of services for which the system is intended. As concerns satellite communications, the three main types of services defined by the ITU are: the fixed satellite service (FSS), used for all types of communications between fixed users, the mobile satellite services (MSS), for communications with maritime, aeronautical or land mobile users and the broadcasting satellite service (BSS) which is reserved for the broadcasting of sound or television programs directly from very high power satellites to home terminals.

The most commonly used frequency band for the FSS is the 6/4 GHz band (the solidus mark '/' shows the separation between the U/L frequency and the D/L frequency). This band is often (mis)named as the C-band. It extends from 5.850 to 7.075 GHz (U/L) and from 3.4 to 4.2 GHz plus 4.5 to 4.8 GHz (D/L), but the parts extending from 5.925 to 6.425 GHz and from 3.7 to 4.2 GHz have been, by far, the most widely implemented up to now.

Another frequency band which is becoming widely used is the 14/11–12 GHz band, often (mis)named the Ku-band. The frequency allocations to the FSS in this band are very complex but, at present, the most commonly implemented parts of this band extend, for the U/L, from 14 to 14.5 GHz and, for the D/L, from 10.95 to 11.2 GHz, 11.7 to 12.2 GHz, 12.5 to 12.75 GHz. Higher parts of the 12 GHz band are allocated to the BSS.

For the MSS systems, the 1.6/1.5 GHz (L-band) frequencies are used (plus the 2.5 GHz band for the radio determination satellite service (RDSS)) but new frequencies should be added in the near future for coping with the expected development of these services.

Other bands are used, or will be used in the future, in particular the 30/20 GHz band (the Ka-band). The ITU not only allocates the frequency bands, but is also involved in the very complex regulatory process for allocating positions to satellites on the geostationary satellite orbit (GSO). In fact, as well as the frequency spectrum, the capacity of the GSO occupancy is limited, due not to the risk of actual collisions which are very unlikely (an angular separation of 0.01 degree corresponds to a distance of 7.4 km on the GSO), but to possible radioelectrical interference between the earth and space segments of nearby satellite systems. The result of this restricted availability of both frequency and location is that the so-called satellite communication orbit/spectrum is a nowadays a limited resource for humanity.

2.1.4 Specific characteristics

The specific characteristics of satellite communications can be listed as follows.

Coverage Figure 2.1(a) shows that communications via a particular geostationary satellite are possible within that part of the earth—called the global coverage—which is visible from the satellite; i.e. within a cone of 17.3° apex angle, (this corresponds to 15.27° apex angle as seen from the earth's centre). By 'visible',

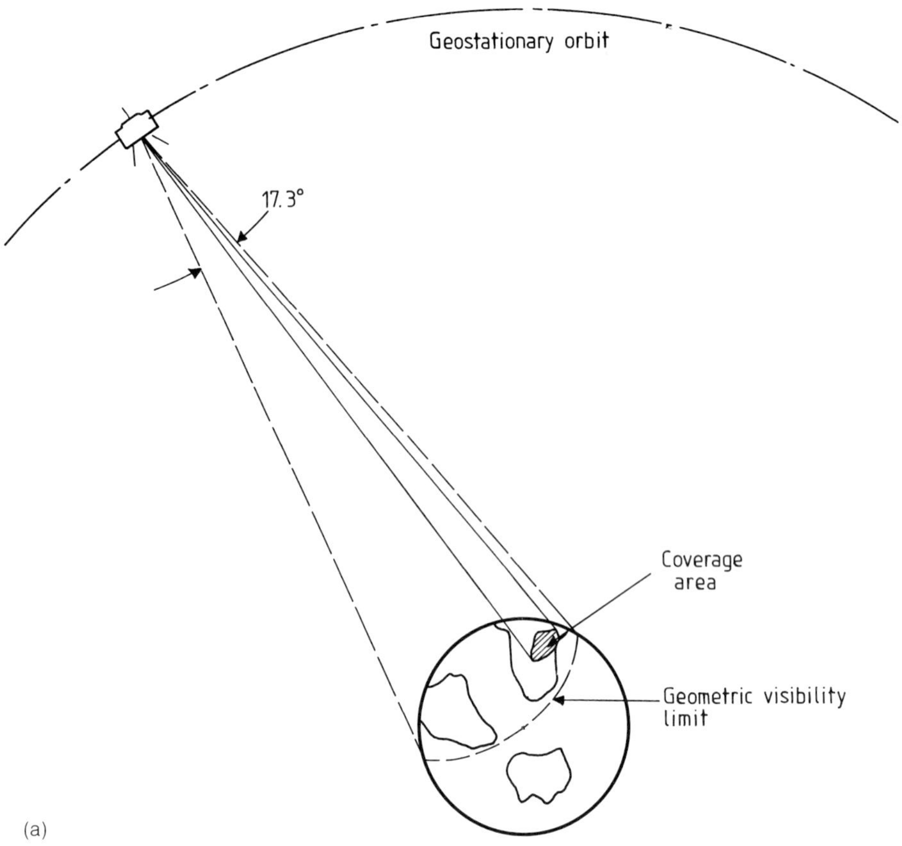

Fig. 2.1 (a) Geostationary satellite earth coverage. (b) World planisphere with GSO elevation angles. The satellite is located, for illustration, at 30°W. The pattern can be copied on a transparency and centred at any sub-satellite point.

it is meant that the cone is defined by earth–satellite rays which are at least at an angle of 5° over the local horizontal direction. Note that, in consequence, at least three satellites, spaced by about 120°, are required to cover the major part of the globe, with the notable exception of the polar regions (Fig. 2.1(b)).

In fact, in most modern satellite systems, the situation is different, as shown in Fig. 2.1(a)*, the satellite is purposely equipped with directive (spot beam) antennas in order to restrict the coverage area(s) to those regions which are actually to be served by the particular satellite communications system. This feature protects the system against interference and increases the satellite antenna gain. In fact, the most significant performance parameter of the satellite for

*The satellite is not necessarily located vertically over the coverage area. It can be shown that it is often advantageous to use oblique beams from the satellite. This increases the coverage area for a given satellite antenna directivity (i.e. gain).

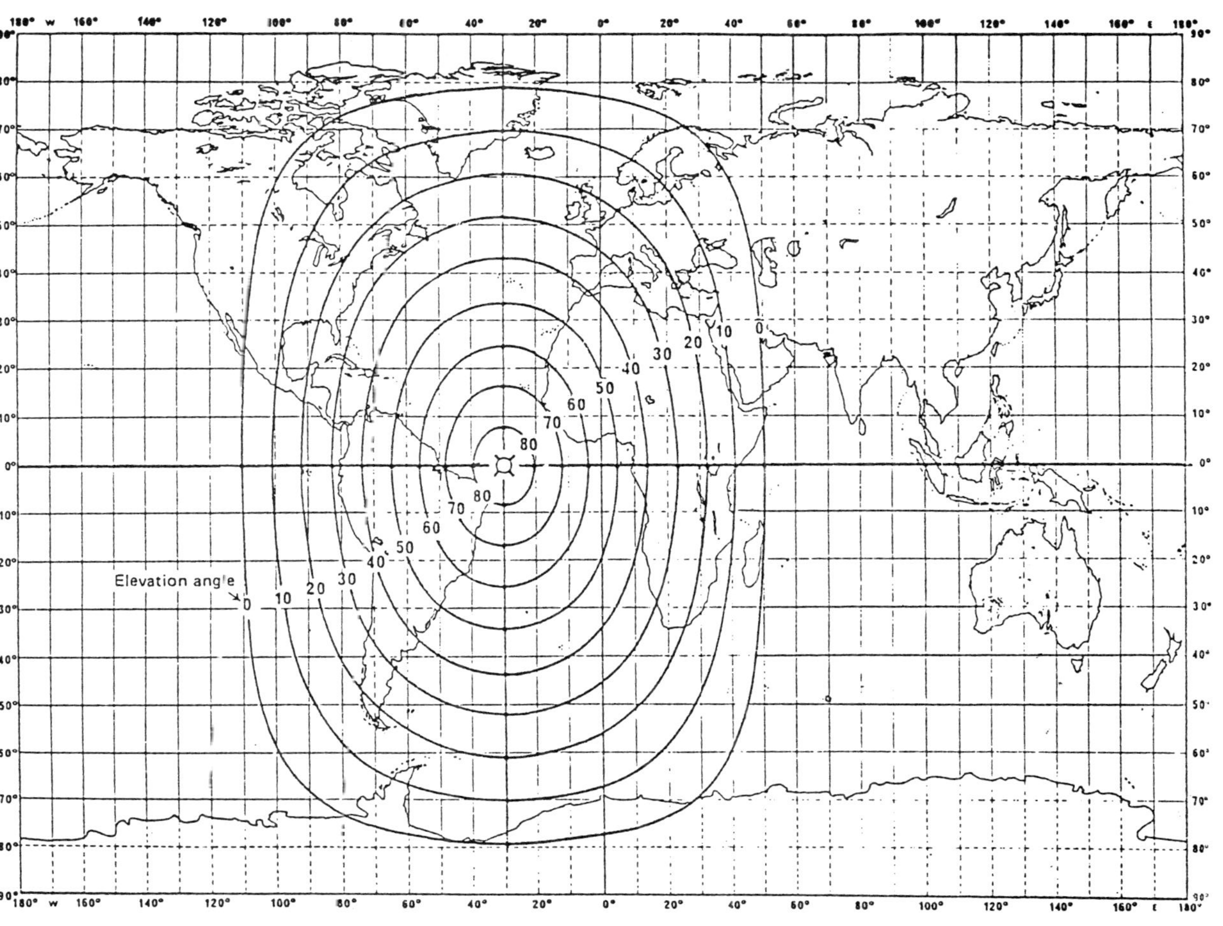

(b)

communications is its so-called equivalent isotropically radiated power (e.i.r.p.) which is the product of the transponder amplifier power by the satellite antenna gain. For example, a 20 W HPA connected to a 22 dBi gain antenna (13° beamwidth) gives a 35 dBw e.i.r.p. The higher the e.i.r.p., the smaller the earth station antenna needed.

High path loss The penalty to be paid for the very great distance of the satellite is very high free space path losses, both on the up-link and the down-link (e.g. about 200 dB at 6 GHz). Although long considered as insurmountable, these losses can now be rather easily offset by using, in particular in the earth stations, high gain antennas, very sensitive receivers (i.e. LNAs with a very low internally generated noise) and powerful transmitters (HPAs).

Multiple access Multiple access, one of the most attractive features of satellite communications, is the ability to transmit, via the same satellite transponder, several carriers, coming in from, and going to different earth stations. There are several methods for implementing multiple access, but the most important are frequency division multiple access (FDMA) and time division multiple access (TDMA):

In FDMA, a given earth station transmits, in a given RF bandwidth allocated in a part of a transponder, a carrier loaded by a multi-destination, multiplexed signal (i.e. a baseband assembly formed by a number of channels—in general telephone channels—which can be sent to various other earth stations). In consequence, in FDMA, each earth station transmits at least one RF carrier, but must be equipped to receive several carriers and, more precisely, to receive all those carriers which contain signals sent to it.

For example, if a station A has to establish telephone links with correspondents connected to stations L,M,N, station A will transmit a carrier loaded (multiplexed and modulated) by the telephone half-circuits destined for stations L,M,N and will receive the carriers from L,M,N in order to extract (demodulate and demultiplex) from them those telephone half-circuits which will form complete circuits (two-way or duplex telephone circuits) with the half-circuits sent from A.

For low-density traffic ('thin route telephony'), FDMA is often used without multiplexing, i.e. each RF carrier is modulated by only one telephone channel (half-circuit): this is called SCPC (single channel per carrier).

TDMA is the dual image of FDMA, the frequency being replaced by the time: in TDMA, all channels are carried at the same radio frequency and each station is allocated, within a periodic time 'frame', a short time interval (called a burst) during which it transmits, in digital modulation, its multi-destination, multiplexed signal.

Depending on the satellite e.i.r.p., on the earth station characteristics and on multiple access/modulation methods, the traffic capacity of a 40 MHz bandwidth satellite transponder can be typically from 300 to 1000 telephone channels (half-circuits). Advanced digital methods (DCM) can even multiply these capacities by a factor 2 to 5.

Multiple access processes may also be classified according to the channel assignment methods: the most conventional method is pre-assigned multiple access (PAMA), in which each channel is permanently established between the two corresponding users. In contrast, in demand assigned multiple access (DAMA or simply DA), all channels form a pool and a given channel is assigned between two corresponding users only for the effective duration of the call. DAMA systems, because they perform a concentration process (just like in a telephone switch), are useful in the case of sporadic traffic.

Distribution A communication satellite can be used as a 'broadcasting transmitter in the sky' for distributing one-way information signals, emitted by a single central earth station, to a multiplicity of stations (often small receiver-only (RO) stations) all over the coverage area. The more common application is the distribution or broadcasting of television programmes to 'TVRO' stations. However, the information signals can also be data signals and, in fact this is now becoming an important application of satellite communications (e.g. for information distribution from press agencies, stock exchanges, etc.).

Wideband transmission medium The bandwidth of each satellite transponder is usually at least 40 MHz and often up to 80 MHz. Modern satellites are equipped with several transponders which share the total allocated bandwidth (e.g. 500 MHz). Moreover, it is now current practice to reuse, possibly several times, this allocated bandwidth. This frequency reuse capability can be performed by two methods, which are mutually compatible: frequency reuse by beam separation, in which transponders using the same frequency bands are serving different coverage areas, and frequency reuse by polarization discrimination (also called frequency reuse by dual polarization). In the latter, two transponders using the same frequency band are connected to the same satellite antenna through two ports radiating two different, orthogonal, RF wave polarizations (for more details, see section 2.2.3 and especially Fig. 2.10). On the INTELSAT VI satellite, for example, some parts of the 6/4 GHz band are reused six times.

Serviceability and flexibility Satellite communications offer paramount operational advantages: easy and rapid installation of the earth stations, whatever their location and environment, high flexibility for changes of services, traffic plans and earth segment composition.

Propagation delay The only real disadvantage of the geostationary satellites is the propagation time between two earth stations which, due to the considerable distance of the satellite, may reach some 275 ms. In telephony, the round-trip delay of 550 ms would induce unacceptable deteriorations of the transmission quality if the echo effects could not be alleviated by the use of echo control devices, viz. the echo suppressors or, better, the more modern and efficient echo cancellers. In data communications, e.g. for signalization and for communications between computers, special protocols must be implemented for coping with the delay.

Technically, voice communications (telephony) can be transmitted by modulating the frequency of the RF carrier (frequency modulation or FM) by a single telephone channel signal bandwidth: from 300 Hz to 3.4 kHz occupying a 4 kHz baseband) or by a frequency division multiplex (FDM), i.e. a baseband assembly of multiple telephone channels (typically 12 to 972 channels). This is the conventional analogue mode of transmission.

2.1.5 Main applications and techniques

Communications (telephony)

By far, the main applications of satellite links are communications systems, especially for telephony. These are implemented in the framework of international, regional or national public (or in some cases, private) networks. Long distance international telephony traffic is currently the most important service carried by satellite links, mainly via the satellites of the INTELSAT international organization.

At present, these satellites carry approximately 2/3 of the world's international traffic. However, the utilization of satellite systems for telephony is not limited to international links and they are implemented more and more in national (domestic) networks, down to the level of the so-called rural networks. In developing countries in particular, satellite systems could prove to be the only cost effective means of bringing communications to small, isolated communities. Also to be mentioned are the very important communication services towards mobile earth stations: at present, thanks to communication satellite systems, most of the commercial ships can be connected to public telephony networks all over the world. In the near future such services should be extended to aviation and even to land mobiles.

However, the current trend is increasingly to implement digital transmission: in this mode of transmission, the (analogue) telephone signals are digitally encoded by sampling and quantization (pulse code modulation (PCM)), typically at 64 kbit/s, then usually assembled by time division multiplexing (TDM) and, finally transmitted by modulating the phase of the RF carrier (phase shift keying (PSK) modulation). More details on multiplexing and modulation will be given in section 2.2.

Communications (data and 'new services')

Apart from the conventional voice services, telegraphy, telex circuits and low bit rate data (e.g. at 50 or 75 bit/s) can be conveyed by conventional (analogue) telephone circuits. Both FDM and TDM techniques are used to transmit several such data channels (e.g. up to 46) in a single telephone baseband.

Digital transmission opens up the way to much more efficient data communications, with much higher bit rates (typically 64 kbit/s, 2 Mbit/s or even more), making it possible to implement 'telematics' and, more generally 'new communications services' (such as computer interconnection, interactive information distribution, data bank transfer, remote printing, video-conferencing, etc.). Now, satellite systems using very small and economical earth stations (often called very small aperture terminals (VSATs)), installed in the final user premises and directly connected to local computers may offer more practical and cost efficient services than terrestrial communications networks.

In the future, end-user-to-end-user, fully digital communications carrying, indifferently, voice and/or data, will be available in the framework of the so-called ISDN (integrated services digital network), the standardization of which is currently under progress at the ITU.

Television

Before the introduction of satellite communications, the only available wideband medium for distributing television programmes from the studios to the broadcasting transmitters was the terrestrial microwave link. The consequence was that long distance (e.g. intercontinental) real-time TV distribution simply did not exist.

1964 saw the beginning of satellite TV transmissions but, at that time, large and rather expensive earth stations were required, which limited the applications to the long distance part of the transmission. Since the 1970s the introduction of domestic satellite systems with higher e.i.r.p. often permits their utilization as a cost effective means of replacing terrestrial microwave networks for transmission up to the regional or local TV broadcasting transmitters. The next step took place after the early 1980s, following the availability of mass produced, low cost, small TVRO earth stations: it consists of distributing TV programs directly from powerful satellites (e.g. with e.i.r.p. greater than 35 dBW at 4 GHz and than 45 dBW at 11 or 12 GHz) down to communities such as cable TV networks (CATV), hotels, master TV antennas for large apartment buildings (SMATV) and even to isolated homes. Ultimately, direct broadcasting satellites operating in the 12 GHz band with very high power (e.i.r.p. = 55 dBW or more) and under the regulations of the BSS, should permit high quality direct home reception with very low cost stations.

Apart from these TV programme transmission, distribution and broadcasting services, satellites are currently used by TV operators for transmitting special events to the studios by the means of transportable earth stations. This is called satellite news gathering (SNG).

2.1.6 Historical overview

Table 2.1 briefly recapitulates the main milestones of the short, but highly packed history of satellite communications. Going farther back would not be very significant, although the reaction-propelled wooden pigeon of Archytas of

Table 2.1 Satellite communications historical overview

1945	(Oct.) '*Extra-terrestrial-relays*' by A. C. Clarke (Wireless World, London)
1957	(Oct. 4) Launching of the first artificial satellite (Sputnik 1, USSR). First reception of satellite signals
1960	(Aug.) Launching by the NASA (USA) of 'Echo-1', a 30 m metallized balloon in a circular orbit at 1600 km for relaying TV and telephone signals by passive reflection of radio waves
1962	Foundation of the first company devoted to satellite communications: the COMSAT corporation (USA)
	Launching by the USA of two non-geostationary, active (6/4 GHz) relaying satellites: the AT & T's TELSTAR-1 (10 July) and the NASA'S RELAY-1 (December)
	First transatlantic TV and telephony communication tests between large earth stations in USA (Andover), France (Pleumeur-Bodou) and UK (Goonhilly)
1963	(July) Launching of SYNCOM-2 (USA/NASA), the first geostationary communications satellite (7/2 GHz band, 2 W travelling wave tube) capable of relaying 1 TV programme or 300 multiplexed telephone circuits)
	First ITU Conference for establishing international regulations on satellite communications
1964	Establishment, initially by 19 nations, of the INTELSAT Organization
	TV transmission of the Tokyo Olympic games
1965	Beginning of INTELSAT operations with the launching of the first INTELSAT-I satellite ('Early Bird') and opening of commercial traffic between USA, France, Germany and UK
	Beginning of TV distribution in the USSR by the MOLNYA-1 elliptical orbit satellite (12 hours revolution)
1967–1971	INTELSAT-II (240 telephone circuits or 1 TV channel) then INTELSAT-III (1500 telephone circuits and/or up to 4 TV channels), then INTELSAT-IV (4000 telephone circuits and 2 TV channels) series of satellites; world-wide operation of TNTELSAT
1972	First domestic geostationary satellite communication system established in Canada with the ANIK-1 satellite
1974	Launching, by France and Germany of SYMPHONIE-1, the first 3-axis stabilized geostationary communications satellite
1975	First INTELSAT-IV A satellite (20 transponders with 'frequency reuse', 6000 telephone circuits +2 TV channels)
	STATSIONAR, USSR's first geostationary satellite
1976	PALAPA-1: first dedicated (nationally owned) satellite for implementing a domestic communications system in a developing country
1978	Launching of the first Ku-band satellites in Japan (BSE: experimental Broadcasting Satellite Service in the 12/14 GHz band) and in Europe (OTS satellite of the European Space Agency (ESA): experimental 14/11 GHz European regional communications system)
1979	Establishment, initially by 26 nations, of the INMARSAT Organization for maritime satellite communications

Table 2.1 *continued*

1980	First INTELSAT-V satellite (27 transponders, 12000 telephone circuits +2 TV), introducing in the INTELSAT global system: 'frequency reuse' by beam separation and by RF dual polarization, operation both 6/4 and 14/11 GHz bands and both in FDMA and TDMA multiple access
1983	First operational communication satellite in the 30/20 GHz band: CS-2 (Japan)
	First launch of the ECS (EUTELSAT-I) 14/12 GHz satellites of the EUTELSAT European Organization
1985	Entering operation by INTELSAT of its international business service (IBS) and of its global TDMA telephony network

Tarentum (4th century BC), the 13th century Chinese 'arrows of flying fire' and, more seriously, the far-sighted theoretical publications of Konstantin Tsiolkovsky – a Russian scientist – on rockets, space flights (1903) and even on the geostationary orbit (1895) could claim some antecedence. However, only the visionary paper of Arthur C. Clarke (1945) combined for the first time, the concept of the geostationary satellites – using a solar powered electric generator – with its actual application as a space relay for world-wide communications and broadcasting by the means of very high frequency radio waves. Clarke's only lack of foresight was that he did not envision unmanned space stations, and maybe that his dream could be realized in less than 20 years!

2.1.7 Existing satellite systems

As concerns communications, three types of systems are currently in service, i.e. the international, regional and domestic (national) systems:

International systems

There are only three systems which are classified as international, because they operate on a world-wide basis under the management of an international organization. the INTELSAT and the INTERSPUTNIK systems (the former being, by far, the more important) as concerns the fixed satellite service, and the INMARSAT system, as concerns the mobile satellite service.

The International Telecommunications Satellite Organization (INTELSAT) owns and operates, on behalf of its 114 participating countries ('the signatories'), a global satellite network which can be also used by any non-signatory country. In fact, INTELSAT only owns the space segment, while the earth segment remains the property of the operators (in general, the national telecommunications authorities). The space segment comprised 13 geostationary satellites operating over the Atlantic, Indian and Pacific Ocean regions in 1988. Most of these satellites were of the INTELSAT V generation, and were progressively replaced, commencing in 1989, by the INTELSAT VI series. Each INTELSAT VI satellite has a traffic capacity of 35 000 telephone circuits (and 3 TV programmes). This capacity will even be extensible up to some 120 000 circuits by implementing digital circuit multiplication (DCM) techniques. More than 1600 international earth-station-to-earth-station communication links are currently in service for providing multiplexed telephony (more than 48 000 circuits in 1988) and also data and TV transmissions (70 000 hours in 1988).

Earth stations for these international services must comply with INTELSAT specifications: they range from the main 'standard A' stations which are equipped with large, steerable (automatic tracking), 32 to 18 m diameter antennas down to small stations for the 'VISTA' rural service which are equipped with 4.5 m antennas. But this is not the whole story about INTELSAT: the organization also leases or sells transponders for domestic services and TV transmissions. INTELSAT's domestic services are currently used by more than 30 countries. Since its foundation, the INTELSAT system has completely fulfilled the needs of its users by responding to the impressive growth rate of the demand and ensuring a very high quality of service. However, INTELSAT is now facing two types of competition.

On the one hand, in the framework of the present trend of the US Administration to 'deregulate' the telecommunications services, private companies have been authorized to launch and operate satellites for international services. On the other hand, and this is much more significant, the optical fibre cable is now providing a very wideband and cost effective new transmission medium, especially for submarine links (for example, the TAT 8 transatlantic cable will offer a capacity of 8000 to 40 000 circuits).

The INTERSPUTNIK organization currently includes 14 countries which are operating 14 earth stations. The space segment is, at the moment, formed by two Soviet STATSIONAR satellites.

The International Maritime Satellite Organization (INMARSAT), with 18 signatories at present, operates a system of satellites over the three oceans to provide telephone, telex, data, facsimile and also distress and safety global communications services to ships, off-shore platforms, etc. There are already (as of January 1988) over 6000 ships equipped with ship earth stations (SES) which, via 20 coast earth stations (CES), enable connection to public communication networks all over the world. In the future, INMARSAT will provide services to smaller ships (with a new type of small, low-cost station, called 'Standard C' SES). Also INMARSAT plans to introduce aeronautical services and even land mobile services.

Regional systems

Two systems are at present in operation to provide international communications, but only on a regional basis: the EUTELSAT and the ARABSAT systems.

The European Telecommunication Organization (EUTELSAT) operates, on behalf of its 26 member countries, a system comprising, at present four

EUTELSAT-I satellites. It provides public digital telephony (TDMA) services between 15 main earth stations and also television programme transmission and business data communications (satellite multi-service system (SMS)). Moreover, several transponders are leased by various European countries, especially for TV distribution to small TVRO stations. By 1990, EUTELSAT introduced its second-generation (EUTELSAT-II) satellites which offer more capacity (16 operational transponders in lieu of 9), higher e.i.r.p. and an improved European coverage.

The Arab Satellite Communications Organization (ARABSAT) operates a system with two satellites and offers to its members public telephony, television and radio programme transmission services. It also provides domestic communications in some member countries.

Domestic systems

There are two ways for a country to establish a domestic (national) satellite communication system. The first, and simplest, is to lease or to buy some space segment (one or several transponders or even only a part of a transponder) to an operating organization. As explained above (International systems) many countries (in particular developing countries) are actually implementing their own domestic system through INTELSAT transponders. The second way is to build a complete system with its own (national) satellites. It is not possible to detail all those national systems, but a partial list of which is as follows: Australia (AUSSAT system), Brazil (BRAZILSAT), Canada (TELESAT system with ANIK satellites), China (STW satellites), France (TELECOM 1 satellites), Germany (DFS-Kopernicus satellites starting in 1989), India (INSAT satellites), Indonesia (PALAPA satellites), Italy (ITALSAT satellites starting in 1990), Japan (CS-2 satellites and other planned private systems), Luxembourg (ASTRA satellite for TV distribution, starting in 1989), Sweden (TELE-X satellite, under development), the USA (Several domestic, privately owned, systems with various satellites: AURORA, COMSTAR, CONTEL-ASC, GALAXY, G-STAR, SATCOM, SBS, SPACENET, TELESTAR, WESTAR), the former USSR (ORBITA, EKRAN and MOSKVA domestic systems, especially intended for television distribution and using the MOLNYA-3, RADUGA and GORIZON satellites). Other countries are now planning to implement national satellite systems.

2.2 BASIC PRINCIPLES

2.2.1 The basic satellite communication link

Figure 2.2 shows in its simplest form a satellite link carrying a duplex (two-way) telephone circuit: the earth station A transmits to the satellite an up-link (U/L)

*France and Germany have also started a television direct broadcasting satellite system (BSS) (TDF 1/TV-SAT). Japan also operates, at present, a BSS system (BS).

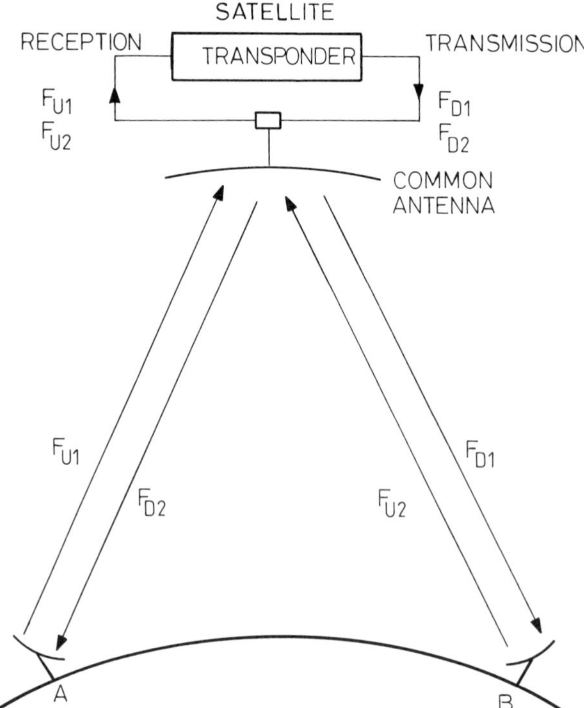

Fig. 2.2 Elementary duplex satellite link.

carrier wave (modulated by the telephone signal) at radio frequency (RF) F_{u1} (e.g. 5980 MHz). The satellite antenna and transponder system receives this carrier and, after frequency conversion from F_{u1} to F_{d1} (e.g. 5980 MHz − 2225 MHz = 3755 MHz), amplifies and re-radiates it as a down-link (D/L) wave which is received by the earth station B. For establishing the return link, B transmits an U/L carrier at another RF F_{u2} (e.g. 6020 MHz) which is received by A at the converted D/L RF F_{d2} (e.g. 6020 MHz − 2225 MHz = 3795 MHz). Note that, more generally:

1. except in the so-called SCPC (single channel per carrier) case, the carriers are modulated by an assembly of telephone signals (multiplex);
2. there are usually different antennas, on board the satellite, for the U/L (receiving antenna) and for the D/L (transmitting antenna);
3. the satellite is equipped with several transponders (the two links F_{u1}/F_{d1} and F_{u2}/F_{d2}, bearing the two half-circuits, may be transmitted through different transponders).

Anyhow, the link must be designed to provide a reliable, good quality communication, which implies that the signal transmitted by the sending earth station must reach the receiving earth station at a carrier level sufficiently higher than the signals generated by various, unavoidable sources of noise and interference. In fact, the communication quality is directly related to the carrier-to-noise ratio (C/N) of the link, i.e. the ratio of the modulated carrier received power (C) to the noise power (N) resulting from the cumulative effect of the noise sources found all over the link. The link budget calculation, which allows determination of the C/N is a very important issue and will be the subject of section 2.2.4. Before going into this calculation, some fundamental definitions and formulae for antennas and noise will be restated.

2.2.2 Definitions and formulae

Antennas

The maximum gain of an antenna (g_{max}), i.e. the maximum of the gain function or the gain in the direction of maximum radiated power (for a transmitting antenna) or maximum received power (for a receiving antenna)*, is:

$$g_{max} = \frac{4\pi A_e}{\lambda^2} \tag{2.1}$$

with

$$A_e = \eta A$$

Where λ is the wavelength ($\lambda = c/f$), c is the velocity of RF waves $= 3 \times 10^8$ ($m \cdot s^{-1}$), f is the radio frequency (Hz), A is the antenna actual (geometrical) aperture (m^2), A_e is the antenna effective aperture area (m^2), η is the antenna efficiency ($\eta \leqslant 1$; $\eta = 1$ for an ideal, lossless antenna with a 'uniform aperture illumination'. Actual antennas typically have values between 0.6 and 0.8.)

For a circular aperture with a diameter D (this is the case for earth station antennas, which are generally equipped with parabolic reflectors), g_{max} can be written:

$$g_{max} = \eta \left(\frac{\pi D}{\lambda} \right)^2. \tag{2.2}$$

These formulae, which are defined in the direction ($\theta = 0, \Phi = 0$) of the maximum of the gain function, can be generalized in any direction (θ, Φ):

$$g(\theta, \Phi) = \frac{4\pi A_e(\theta, \Phi)}{\lambda^2},$$

$A_e(\theta, \Phi)$ being the effective aperture area projected in the direction (θ, Φ).

For a (theoretical) isotropic antenna, $g(\theta, \Phi) = 1$ and:

$$A_{iso} = \frac{\lambda^2}{4\pi}. \tag{2.3}$$

*For more details, refer to the chapter on antennas in Volume 2.

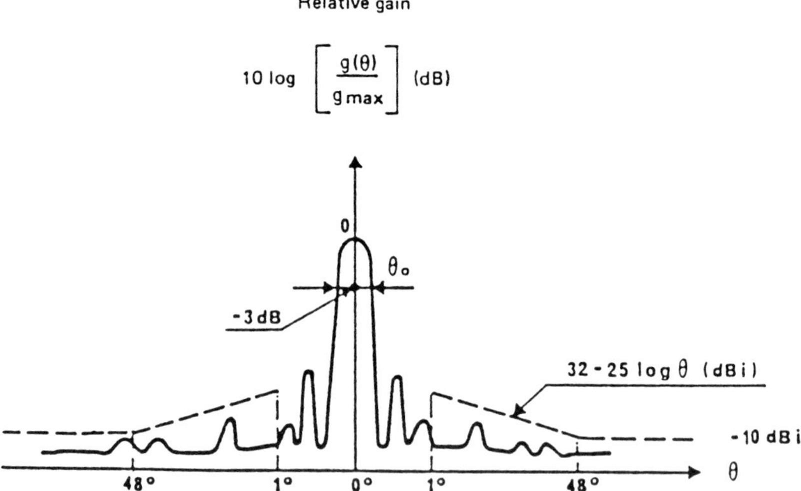

Fig. 2.3 Antenna radiation diagram.

The maximum gain (g_{max}) is often called simply 'the gain' (g) and expressed in dBi (dB over isotropic antenna) as $G(G = 10 \log g)$.

Figure 2.3 represents a cut of the relative gain function, e.g. in the plane $\Phi = 0$, this is called a radiation diagram (or a radiation pattern). The beamwidth at a given power level is proportional to λ/D (D being the aperture dimension in the cut plane; in the case of a circular aperture, D is again the antenna diameter). The beamwidth is generally defined at the half-power (-3 dB) level and this 'half-power beamwidth' can be written as:

$$\theta_0 = k\frac{\lambda}{D} \tag{2.4}$$

k depends on the aperture 'illumination law'; for high efficiency antennas, a common value is $k_\theta = 65°$.

The maximum gain can be expressed, as a function of the half-power beamwidths in the two orthogonal planes, by the following approximate formula*.

$$g_{\text{max}} = \frac{27\,000}{\theta_0 \cdot \Phi_0}. \tag{2.5}$$

Of course, in the case of a circular aperture $\theta_0 = \Phi_0$. Figure 2.4 gives typical values of the gain and the half-power beamwidth of circular aperture antennas.

*Equation (2.5) is strictly valid only for conventional beams, such as those of earth station antennas, and not for special shaped beams, such as those used for improving the coverage of satellite antennas.

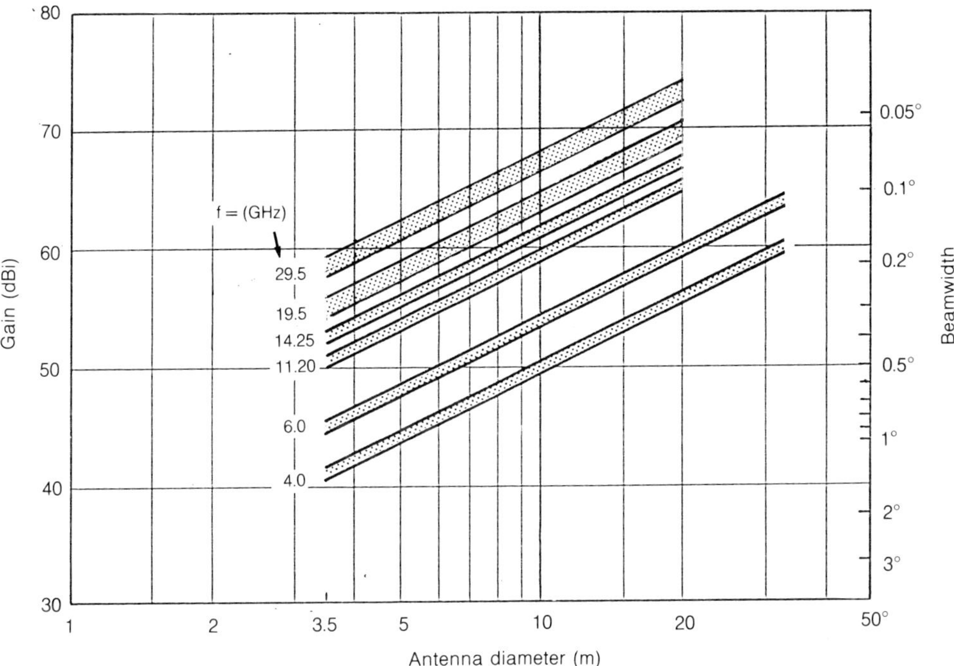

Fig. 2.4 Antenna gain and beamwidth.

Power radiated and received by an antenna

The power radiated, in a given direction, by an antenna transmitting a power p_T can be expressed by the power flux density (pfd) (in $w \cdot m^{-2}$) at the (large) distance d:

$$\text{pfd} = \frac{p_T \cdot g_T(\theta, \Phi)}{4\pi d^2}. \tag{2.6}$$

Usually, pfd is measured in the direction of the maximum gain $g_{max} = g_T$ and the product $p_T \cdot g_T$ is the equivalent isotropically radiated power (e.i.r.p.) which has already been defined in section 2.1.4 and which is the figure of merit of the station at transmission. It is expressed in watts, or more frequently, in dBW, as (e.i.r.p.) $= P_T + G_T$ (with $P_T = 10 \log p_T$ and $G_T = 10 \log g_T$).

Now, if the wave is captured at the distance d by a receiving antenna having an effective area A_{eR}, then the received power is equal to pfd$\cdot A_{eR}$, or, using (2.1) for changing A_{eR} to g_R (the receiving antenna gain):

$$p_R = p_T \cdot g_T \cdot g_R \cdot \left(\frac{\lambda}{4\pi d}\right)^2 \text{(w)}. \tag{2.7}$$

This important formula, which gives power transfer between two antennas will be widely used in the link budget calculations. Note that the so-called free space attenuation, which is the power transfer (p_T/p_R) between two isotropic antennas, can be written as:

$$l = \left(\frac{\lambda}{4\pi d}\right)^2 \qquad \text{or (in dB)} \qquad L = 20\log\left(\frac{4\pi d}{\lambda}\right). \tag{2.8}$$

For calculating d, see section 2.2.4. For example $L = 200\,\text{dB}$ at 6 GHz and at a distance $d = 40\,000\,\text{km}$.

In addition to this free-space attenuation, other losses are to be accounted for in a power transfer calculation:

1. the atmospheric propagation losses (section 2.2.3), which can vary from a few tenths of a dB at 4 GHz to tens of dBs at very high frequencies (e.g. in the 30/20 GHz band) and which depend on local rain precipitations;
2. the losses caused by offset in antennas pointing and by RF polarization mismatch;
3. the ohmic losses in the antenna feeders.

Noise power

Noise temperature definitions The thermal noise power available at the input of a matched network is given (in W) by the well-known Nyquist formula:

$$N = kTB \tag{2.9}$$

or (in $\text{W}\cdot\text{Hz}^{-1}$)

$$N_0 = kT$$

where k is Boltzmann's constant $= 1.38 \times 10^{-23}$ ($\text{J}\cdot\text{K}^{-1}$). T is the absolute temperature of the network (K), B is the 'noise bandwidth' of the network, i.e. the bandwidth of an equivalent, theoretical, rectangular filter delivering the same noise power as the actual network and N_0 is the noise spectral density ($\text{W}\cdot\text{Hz}^{-1}$).

Although strictly applicable to the thermal noise given by resistors, it is a common practice to extend the Nyquist formula (in a limited bandwidth) to other types of circuits (e.g. to amplifiers) and to other types of electrical noise, with T being the equivalent noise temperature, i.e. the temperature of a matched resistor which would deliver the same noise power.

The Nyquist formula is often expressed in dB:

$$N = -228.6 + 10\log T + 10\log B \text{ (dBW)}.$$

The noise power to be accounted for in a satellite link results from various sources, internal (receiver noise) and external (antenna noise).

Receiver noise An ideal noiseless receiving amplifier would amplify an input noise not more than the input signal (i.e. with the same gain). Due to internal

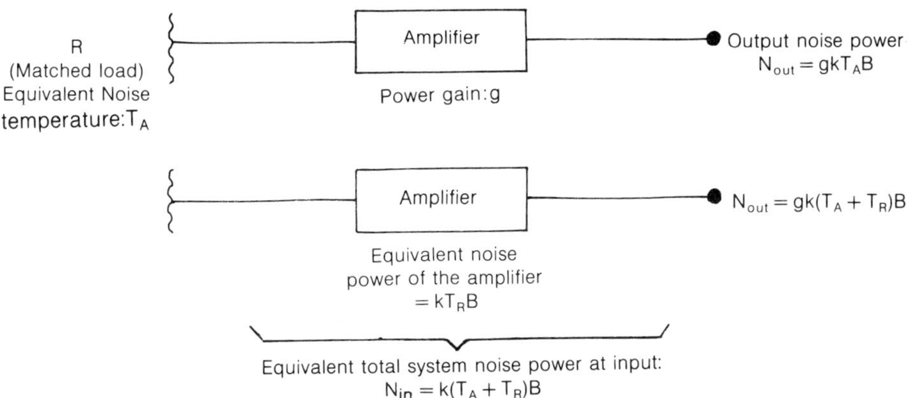

Fig. 2.5 Equivalent total system noise power.

noise, an actual receiving amplifier will bring an additional noise power which, referred to the receiver input, is expressed by an equivalent amplifier noise temperature T_R (as explained above). Common values of T_R for the low noise amplifiers (LNAs) used in modern receivers are between 30 and 150 K, depending on the frequency band and on the LNA design.

Anyhow, as represented in Fig. 2.5, the total noise temperature in (12.9), referred to the receiver input, is $T = T_A + T_R$. Note that the matched load equivalent noise temperature is here called T_A because the input of the receiver is usually connected to the (matched) antenna. T_A is thus the antenna equivalent noise temperature, the calculation of which will be explained below. Note also that the noise caused by the receiver is sometimes expressed by its noise figure F, the relation between F and T_R being: $T_R = (F - 1)T_0$, T_0 being, by convention, equal to a normal ambient temperature value of 290 K. In fact, since T_R is usually much less than 290 K, T_R is more practical to use than the noise figure in satellite communications.

Cascaded networks The receiver is actually a system composed of cascaded circuits and, more precisely of a few amplifying stages or other networks (such as the down-converter, etc.), each one having its own gain g_i and its noise temperature T_{R_i}. It can be easily demonstrated that, under such conditions the receiver noise temperature is:

$$T_R = T_{R1} + (T_{R2}/g_1) + (T_{R3}/g_1 g_2) + \cdots.$$

This formula is important because it shows that the noise contributions of the successive stages are lessened by the total gain of the preceding stages. Consequently, the RF preamplifier (LNA) must have a low T_{R1} and a high g_1.

Antenna noise temperature The noise temperature of an antenna results from the noise power collected by the receiving antenna from various external sources.

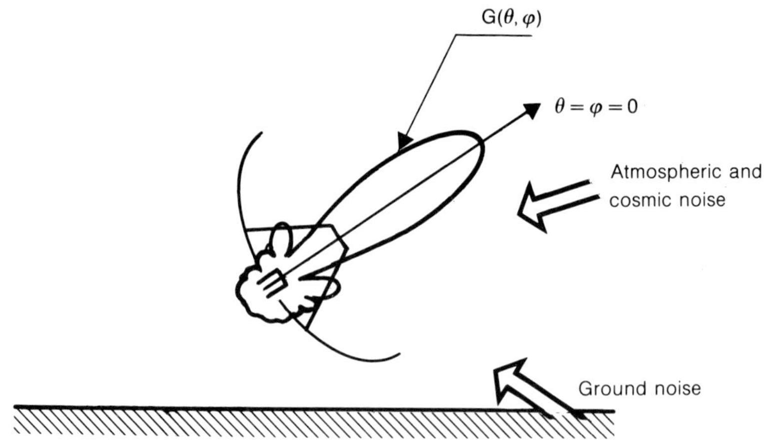

Fig. 2.6 Contributions to antenna noise temperature.

This is schematized in Fig. 2.6 (which represents an earth station antenna and its radiation diagram, in polar coordinates) and is expressed by:

$$T_{\mathrm{A}} = \frac{1}{4\pi} \iint g(\Omega)\, T(\Omega)\, \mathrm{d}\Omega \tag{2.10}$$

where $\mathrm{d}\Omega$ is the elementary solid angle in the spatial direction Ω, $g(\Omega)$ and $T(\Omega)$ are the antenna gain and the noise temperature of the noise source in the spatial direction Ω.

Terrestrial noise contributions are dominant in this formula. They are due:

1. to atmospheric attenuation*; this noise contribution decreases rapidly when the elevation angle of the antenna increases, since the higher the elevation, the shorter the length of the rays in the atmosphere;
2. to ground, because of the soil absorption noise† which is collected by the antenna; this is why earth station antennas must have low sidelobe levels in the ground directions (including in the rear directions);
3. finally, extraterrestrial noise from radio stars and from the residual cosmic noise is to be accounted for a small contribution (a few K) to T_{A}.

Figure 2.7 shows typical values of T_{A} in the case of earth station antennas (add about 5 K to these values to account for internal antenna losses). At high frequencies (over 10 GHz) a more complete calculation of T_{A} is needed to account for additional losses due to rain precipitations.

As concerns satellite antennas, they feature a high T_{A} (about 290 K) since the main lobe of their radiation diagram necessarily intercepts the earth atmosphere and ground.

* Remember that 'attenuation' or 'absorption' means a loss (equivalent to the loss in an ohmic resistor) and therefore brings a noise power contribution.

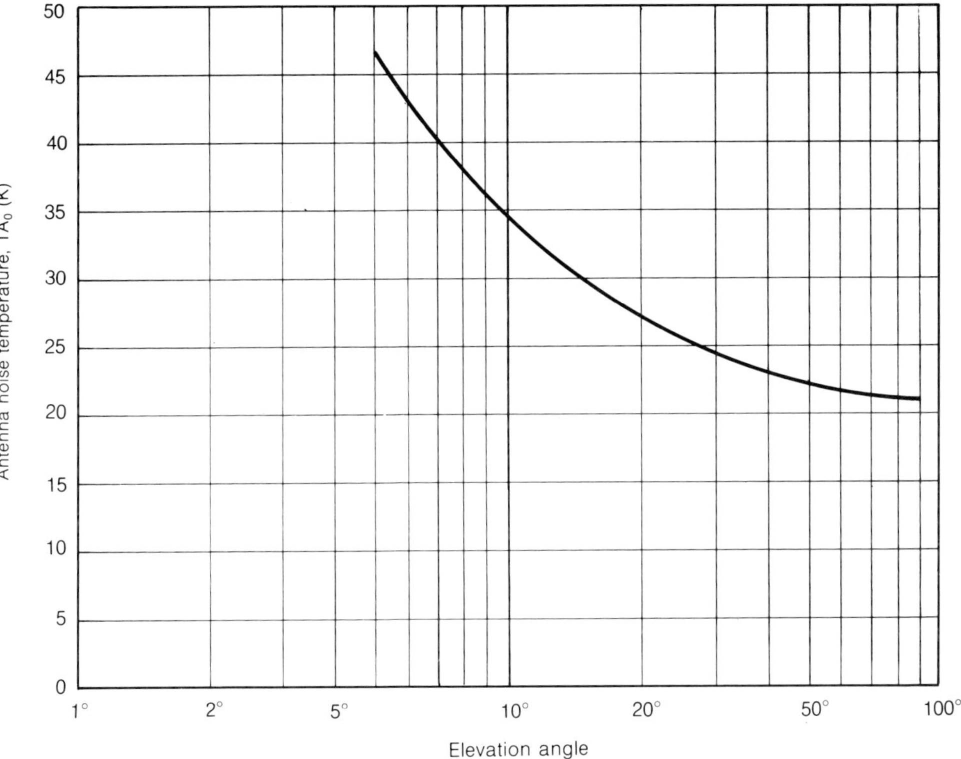

Fig. 2.7 Typical clear sky earth station antenna noise temperature.

Total system noise temperature The total noise temperature of the receiving system, including all the noise power contributions from the antenna and the receiver, and referring to the receiver input, is:

$$T = T_A/a + T_0(1 - 1/a) + T_R.$$

In this formula, a supplementary contribution is included to account for a possible loss a in the RF line (a waveguide or coaxial line) connecting the antenna to the LNA. Here, a is a power ratio (input/output, i.e. $a > 1$), often expressed in dB ($10 \log a$) and the RF line is supposed to be at the ambient temperature $T_0 = 290$ K. In fact, the earth station must be designed in order to minimize this loss (by locating the LNA directly at the antenna duplexer port or by using very low loss connections). Therefore, the loss is generally small and the following simple formula can be used (adding to T_A about 7 K per 0.1 dB loss):

$$T = T_A + T_R.$$

Figure of merit of a station at reception

Just as the equivalent isotropically radiated power (e.i.r.p.) was defined above as the figure of merit of the (space or earth) station at transmission, it will be shown in section 2.2.3 that the gain-to-noise temperature ratio (G/T) is the figure of merit of the station at reception. This is the ratio between the gain of the antenna at reception (G) and the noise temperature of the receiving system (T) (see above). The G/T is usually expressed in $\mathrm{dB \cdot K^{-1}}$ ($10 \log G - 10 \log T$).

Earth stations' G/T typical values range from $35\,\mathrm{dB \cdot K^{-1}}$ (main 4 GHz stations with a 15 to 18 m diameter antenna) to some $16.5\,\mathrm{dB \cdot K^{-1}}$ (12 GHz data transmission microstations with 1.2 m antenna). Space stations G/T at 6 GHz typically range from about $-19\,\mathrm{dB \cdot K^{-1}}$ for a global beam antenna to $-3\,\mathrm{dB \cdot K^{-1}}$ for a pencil beam (zone beam) antenna.

2.2.3 Other topics

Intermodulation in non-linear amplifiers

Most power amplifiers used in earth station HPAs and in satellite transponders exhibit a saturation phenomenon and, consequently, a non-linear behaviour when they are operated in the vicinity of their maximum available output power. If multiple carriers are simultaneously transmitted through these amplifiers, intermodulation occurs. In particular, intermodulation causes interference in the satellite transponders in the very common case where they are implemented in FDMA. More precisely, when the different carriers at RF frequencies f_1, f_2, f_3, etc. are simultaneously amplified in the non-linear amplifier, intermodulation occurs, causing unwanted carriers, called intermodulation products, to appear at frequencies:

$$f_x = m_1 \cdot f_1 + m_2 \cdot f_2 + m_3 \cdot f_3 + \cdots$$

m_1, m_2, m_3, etc. are positive or negative integer numbers and $|m_1| + |m_2| + |m_3| + \cdots$ is the order of the intermodulation product X.

The most important points concerning intermodulation noise in satellite transponders are listed below:

1. practically, only odd-order products can fall in the bandwidth of a given transponder;
2. intermodulation product power decreases with the order and only third-order products, such as $2f_1 - f_2$ or $f_1 + f_2 - f_3$, and sometimes fifth-order products are to be taken in account;
3. the number of intermodulation products increases very quickly with the number of input carriers (for example, for 3 carriers, there are 9 products and for 5 carriers there are 50).

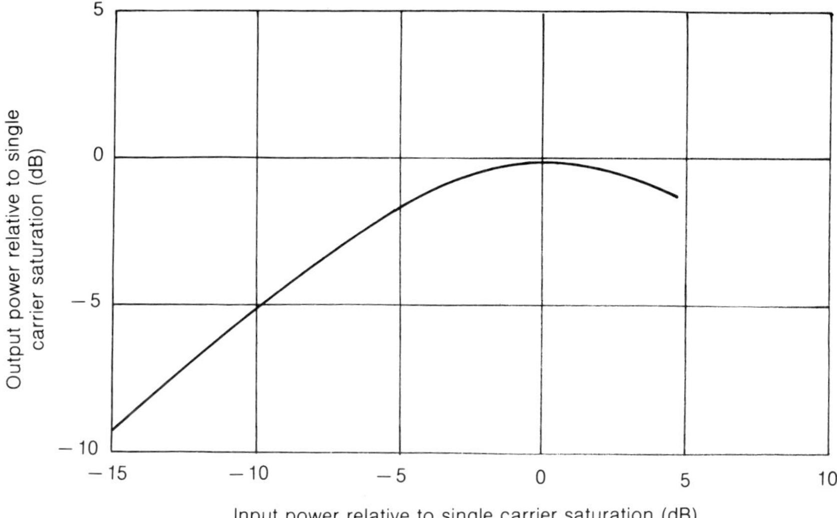

Fig. 2.8 Typical satellite TWT output/input power characteristic.

4. An example of the characteristic of a satellite amplifier is shown in Fig. 2.8. This is the output/input power transfer curve of a travelling-wave tube amplifier (TWTA). TWTAs are the most common type of satellite amplifiers, although solid state amplifiers are also frequently utilized. The curve clearly shows that the operation region near saturation is non-linear. Calculations of the intermodulation products power are performed by approximating this curve by series expansion. Calculations are feasible only for a few carriers. In the case of a great number of carriers, approximate formulae are used (this is the case for SCPC (section 2.1.4) where several hundreds of carriers are transmitted in a single transmitter). Figure 2.9 gives an example of the intermodulation power in the case of 3 carriers.
5. Most important is the selection of the operating point on the transponder characteristic. By reference to the saturation point, the operating point is measured by the corresponding input back-off $(BO)_I$ and output back-off $(BO)_O$. The input back-off is the ratio (expressed in dB) between the input power (single carrier) at saturation and the actual input power; the output back-off is the same ratio for output powers. The larger are the back-offs, the more linear is the operation and therefore the smaller is the power of the intermodulation products. Approximate formulae for intermodulation calculations are given in section 2.4.5.
6. In some circumstances, it is feasible to select the frequency plan (i.e. the frequency arrangement in the transponder) in order to minimize the interference caused to the wanted carriers by the unwanted intermodulation products.
7. Also, special pre-distortion networks, called linearizers have been developed,

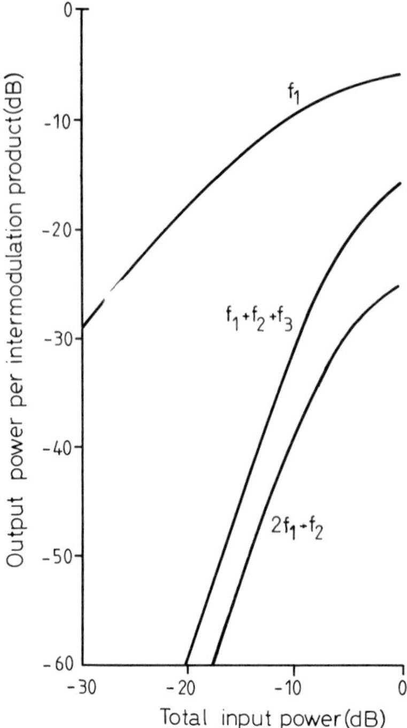

Fig. 2.9 Typical intermodulation power characteristic (three carriers).

which compensate the amplifier non-linearities and reduce the intermodulation products power.

In the case of digital carriers, the non-linear behaviour of the amplifier also causes spectral distortions (AM–PM conversion, see section 2.4.5, etc.) which induce signal deteriorations and out-of-band interference. This is why, even in the case where a single (TDMA) carrier is transmitted, it may be necessary to operate the transponder with some back-off.

Although emphasis has been put on the case of satellite transponders, similar considerations are applicable to unwanted intermodulation products radiated by earth stations using non-linear HPAs, and in particular wideband TWTs.

Frequency reuse

The principles of frequency reuse have already been explained in section 2.1.4. Figure 2.10 illustrates the two, mutually compatible, applicable methods:

Frequency reuse by beam separation, also called spatial isolation frequency reuse: in this method, the same U/L and D/L frequency bands are simultaneously used in different satellite antenna beams. The simplest example (shown in the figure) is the one implemented on the INTELSAT-IV A satellite: here a couple

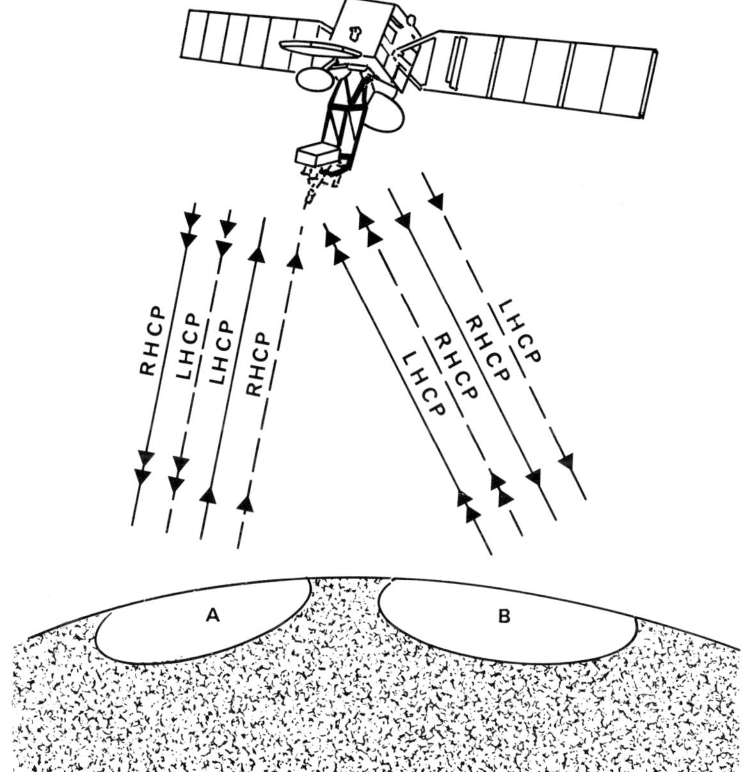

Fig. 2.10 Illustration of frequency re-use methods (quadruple frequency re-use in the INTELSAT V system). Double frequency re-use by beam separation: the same frequency bands (and polarizations) are used twice, once for transmission from zone A to zone B (lines with single arrow) and once from zone B to zone A (lines with double arrow). Double frequency re-use by dual polarization: the same frequency bands (and directions of transmission) are used twice, once (solid lines) for transmission by left-hand circular polarization (LHCP) and reception by right-hand circular polarization (RHCP), and once (broken lines) for transmission by right-hand circular polarization (RHCP) and reception by left-hand circular polarization (LHCP).

of 'hemispheric' beam transponders (eastern and western) allows signal transmissions east-to-west and west-to-east. This doubles the usable bandwidth capacity, but larger multiplication factors are possible with multiple spot-beams. This method requires no special equipment in the earth stations. On board the satellite, separate antennas or multiple feed antennas must be used with special provision for sufficient isolation: on the INTELSAT-IV A satellite, an isolation better than 27 dB is provided.

Frequency reuse by polarization discrimination also called dual-polarization frequency reuse allows doubling the usable bandwidth capacity. It is performed by using two orthogonally polarized RF waves in the same satellite antenna beam, i.e. for the same coverage area: these two polarizations may be linear (e.g. horizontal and vertical) or circular (right-hand and left-hand circular polarizations: RHCP and LHCP). This method requires that the earth stations are equipped with special feeds with 4 ports: one RHCP (or 'X' linear polarization) and one LHCP (or 'Y') port for transmission, and one LHCP (or 'Y'), and one RHCP (or 'X') port for reception. Sufficient isolation—typically more than 30 dB—must be provided between orthogonal ports. Separate antennas are often used on the satellite for the two polarizations to provide sufficient isolation.

Atmospheric propagation

It has already been explained in section 2.2.2 that atmospheric attenuation, especially due to rain, can be a significant factor at the high frequencies (especially over 10 GHz). It is not possible here to go into the details of atmospheric attenuation calculations* which depend on:

1. the elevation angle of the satellite, as seen by the earth station (Fig. 2.11), because this determines the length of the path in the rain layer;
2. the percentage of time during which a given link quality is specified;
3. the climatic zone where the earth station is located: regions with intense (e.g. tropical) rains are subject to high attenuations if percentages of time greater than, say, 99% are to be considered.

The consequences of these atmospheric effects on the satellite links may be threefold.

1. A supplementary loss factor may have to be included in the up-link ($l_{ATU} > 1$, or L_{ATU} in positive dB) and in the down-link (l_{ATD}, or L_{ATD}) budgets.
2. The earth station antenna noise temperature T_A (section 2.2.2) is increased. The complete formula including possible antenna feeder losses ($l_F > 1$, or l_F in positive dB) at atmospheric losses is as follows:

$$T_A = T_{A0}/l_F + T_0(1 - 1/l_F) + (1 - 1/l_{ATD})(T_{AT} - T_C)1/l_F \qquad (2.12)$$

   where T_{A0} is the 'clear sky' antenna noise temperature (see (2.10)), T_0 is the physical reference ambient temperature = 290 K, T_{AT} is the physical temperature of the atmosphere ≈ 270 K, and T_C is the 'clear sky' noise temperature ≈ 15 K. Note that for an approximate value of the two first terms, simply add 7 K per 0.1 dB feeder loss (L_F) to T_{A0} (given by Fig. 2.7) as already explained in section 2.2.2; the third term gives the additional antenna noise temperature (δT_A) due to the atmospheric losses for example $\delta T_A = 27$ K for $L_{ATD} = 0.5$ dB and $\delta T_A = 92$ K for $L_{ATD} = 2$ dB).

*Methods for calculating the atmospheric attenuation are given in references [1] and [2].

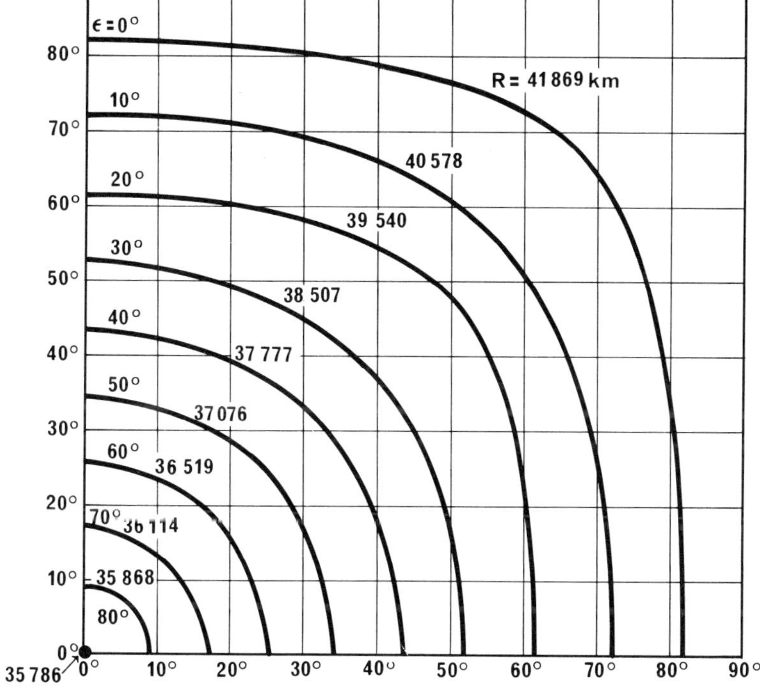

Fig. 2.11 Elevation angle ε and distance R of a geostationary satellite. Relative station longitude θ = distance between the station and the satellite longitude.

3. Rain can distort the polarization of the RF waves and this effect (rain-induced cross-polarization) can introduce interference in the satellite links implementing dual-polarization frequency reuse.

2.2.4 The link budget

Service quality objectives

It was already explained at the beginning of this section that the purpose of a link budget is to calculate the carrier-to-noise ratio (C/N) which is the first and essential parameter for evaluating the quality of the link. Given the C/N (or the C/N_0, or the C/T, which are also used in lieu of the C/N), and the modulation parameters, two types of calculations are to be carried out, depending whether the transmission uses analogue modulation or digital modulation, for evaluating the link transmission performance:

Analogue modulation In the case of analogue modulation, the link transmission performance is characterized by the baseband signal-to-noise (S/N) ratio obtained

Table 2.2 CCIR quality objectives to be met for telephony and ISDN*

Measurement conditions	*Analogue telephony* — *Noise power at reference level*	*Digital (PCM) telephony* — *BER*	*Satellite HRDP*† *forming part of a 64 kbit/s ISDN connection* — *BER*
20% of any month (1 min mean value)	10 000 pW_{op}	—	—
20% of any month (10 min mean value)	—	10^{-6}	—
10% of any month	—	—	10^{-7}
2% of any month	—	—	10^{-6}
0.3% of any month (1 min mean value)	50 000 pW_{op}	10^{-4}	—
0.01% of any year (integrated value over 5 ms)	1 000 000 pW_o unweighted	—	—
0.05% of any month 1 s mean value)	—	10^{-3}	—
0.03% of any month	—	—	10^{-3}

*ISDN: integrated services digital network.
†HRDP: hypothetical reference digital path.

Notes

1. This table refers to Recommendations 353, 522 and 614 of the ITU/CCIR (Comité Consultatif International des Radiocommunications — International Radio Consultative Committee).
2. The values quoted in the 2nd, 3rd and 4th columns must not be exceeded for more than the time percentages indicated in the 1st column.
3. In the 2nd column (analogue telephony), the specification provides for the noise level in each telephone channel, expressed in picowatts (pW = 10^{-12} W) and referred to a signal reference level of 1 mW (10^{-3} W). This is why the unit is called 'pW_o', the last 'p' ('pW_{op}') means that the measurement is made through a 'psophometric weighting network' which emulates the human's ear frequency response. It is easy to convert these units in (S/N) in dB: 10 000 pW_{op} corresponds to (S/N)p = 50 dB, 50 000 pW_{op} to (S/N)p = 43 dB and 1 000 000 pW_o (S/N) = 30 dB.
4. The 4th column refers to the so-called 'satellite hypothetical reference digital path' (HRDP) for a 64 kbit/s channel forming a part of a connection in the framework of the integrated services digital network (ISDN, see Sec. 12.1.5.2.).

Quality objectives for television

For all types of long distance television programme transmission, the CCIR specifies that the (S/N) (weighted) should be at least 53 dB for 99% of the time and 45 dB for 99.9% of the time. However, these specifications are often relaxed since satellites permit more direct transmission than terrestrial links.

at the demodulator output. The S/N ratios to be met for public communications are quoted in Table 2.2 (column 2). The relationship between the (S/N) and the (C/N) will be explained in section 2.2.5. However, it is important to note forthwith that, in the common case of frequency modulation (FM) this relation, expressed in dB, is linear whenever the (C/N) is higher than a certain value $(C/N)_{Th}$.

$(C/N)_{Th}$ is called the demodulation threshold, for $(C/N) < (C/N)_{Th}$ the (S/N) deteriorates rapidly. For conventional demodulators $(C/N)_{Th}$, expressed in dB, it is approximately equal to 10 dB. However, demodulators with an improved threshold can be designed. For these so-called threshold extension demodulators (TED), $(C/N)_{Th}$ can be lowered to 7 dB or even less.

Digital modulation In digital modulation, the link transmission performance is characterized by the bit error rate (BER) obtained at the demodulator output. The BER to be met for public communications are quoted in Table 2.2 (columns 3 and 4). The relationship between the BER and the (C/N) or the (C/T) will be explained in section 2.2.5. In the common case of phase shift keying (PSK), there is no demodulation threshold. However, for simple baseband detection, a minimum (C/T) is also often required. This minimum (C/T) can be lowered, for a specified BER, by implementing baseband processors called forward error correction (FEC) coders and decoders (codecs): these encode the normal bits (which carry the useful information) in common with additional bits and use a special decoding technique to recover the useful information with an improved BER. In conclusion, Fig. 2.12 schematizes the various steps for calculating the link budget and, thence, the link quality (including step-by-step modifications of the link parameters and of the modulation parameters). Of course, computer programs are often used to perform these calculations.

The link transmission performance is not the only factor which defines the link quality. The other important factor is the link availability (availability = 100 – unavailability in %). The unavailability is the percentage of the required time during which the link is out of service (broken or unaceptable quality, these conditions being defined by the CCIR). Practically, satellite systems prove to be of excellent availability. For example, in the INTELSAT system the average availability (1987) was 99.947% (average earth station: 99.978%, space segment: 99.994%).

Link budget: thermal (C/T)

Up-link In accordance with (2.6) and (2.7), the carrier power level (c_U) received from the earth station at the satellite receiver input is:

$$c_U = p_{ET} \cdot g_{ET} \cdot g_{SR} / l_U$$

where $p_{ET} \cdot g_{ET}$ is the earth station $(\text{e.i.r.p.})_E$ in the satellite direction, g_{SR} is the satellite antenna receiving gain in direction of the earth station, and l_U is the up-link free space attenuation (see (2.8)).

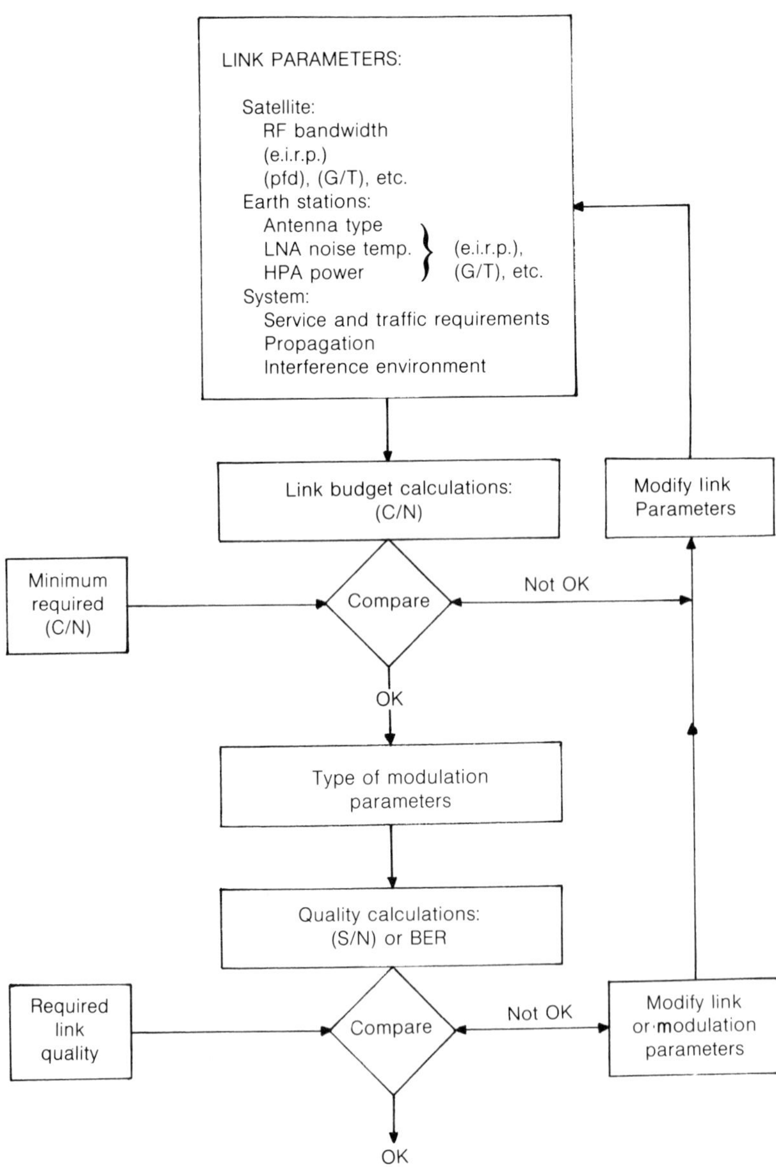

Fig. 2.12 Link budget and link quality: calculation schematic.

Therefore, the up-link (U/L) carrier-to-noise temperature ratio is (in $w \cdot K^{-1}$):

$$(c/T)_U = (e.i.r.p.)_E \cdot g_{SR}/l_U \cdot T_S$$

or:

$$(c/T)_U = (e.i.r.p.)_E \cdot (g/T)_S/l_U \qquad (2.13)$$

or:

$$(c/T)_U = (g/T)_S \cdot (\lambda^2/4\pi) \cdot (e.i.r.p.)_E / 4\pi d^2$$

or:

$$(c/T)_U = (g/T)_S \cdot (\lambda^2/4\pi) \cdot (pfd)_U$$

where $(g/T)_S$ is the satellite figure of merit at reception, $(\lambda/4\pi)$ is the isotropic antenna effective area A_{iso} (from 2.3), $(pfd)_U$ is the earth station power flux density at the satellite input (2.6)*.

All these formulae are usually expressed in $dBw \cdot K^{-1}$:

$$(C/T)_U = (E.I.R.P.)_E + (G/T)_S - L_U \quad (2.13(a))$$

$$(C/T)_U = (G/T)_S + 10\log(\lambda^2/4\pi) + (PFD)_U \quad (2.13(b))$$

with capital letters meaning dBs (like $X = 10\log x$).

We note that.

1. All calculations in this section are carried out on the basis of (C/T). Of course, it is easy to convert them into (C/N_0) (in dB·Hz), by adding $-10\log k = 228.6$ to the formulae in (C/T) (in $dBw \cdot K^{-1}$), or into (C/N) (in dB), by adding $228.6 - 10\log B$ (B is the noise bandwidth in Hz) to the same formulae.
2. For calculating L_U (and also L_D, see below), refer to the Fig. 2.11 which gives two useful parameters: the distance of the satellite and also the earth station antenna elevation angle in the direction of the satellite, both as a function of the locations of the satellite on the equatorial (geostationary) orbit and of the earth station.
3. In case of a significant atmospheric attenuation, add $-L_{ATU}$ in (2.13(a)) and (2.13(b)).

Down-link Using the same type of calculations, the down-link (D/L) carrier-to-noise temperature ratio is (in $w \cdot K^{-1}$)

$$(c/T)_D = (e.i.r.p.)_S \cdot g_{ER} / l_D \cdot T_E$$

or:

$$(c/T)_D = (e.i.r.p.)_S \cdot (g/T)_E / l_D \quad (2.14)$$

or:

$$(C/T)_D = (E.I.R.P.)_S + (G/T)_E - L_D (dBw \cdot K^{-1}) \quad (2.14(a))$$

where $(e.i.r.p.)_S = p_{ST} \cdot g_{ST}$, the satellite (e.i.r.p.) in the direction of the receiving earth station, g_{ER} is the earth station antenna receiving gain in the direction of the satellite, $(g/T)_S$ is the satellite figure of merit at reception, and l_D is the free space attenuation of the down-link.

*The maximum level (transponder saturation) of $(pfd)_U$ is often quoted as a specification of the satellite. For $(pfd)_U = (pfd)_{Smax}$, the maximum possible earth station (e.i.r.p.) can be inferred as: $(e.i.r.p.)_{Emax} = (pfd)_{Smax} \cdot 4\pi \cdot d^2$.

We note that:

1. In case of a significant atmospheric attenuation, add $-L_{ATD}$ in (2.14(a)) and use (2.12) in $(G/T)_E$;
2. $(C/T)_D$ is often the dominant factor in the total link budget.

Link budget: satellite intermodulation

The up-link and down-link contributions to the total link budget have been analysed above in the form of $(C/T)_U$ and $(C/T)_D$. These contributions are relative to thermal noise. However, the contribution of interference noise due to satellite intermodulation (section 2.2.3) must also be accounted for whenever several carriers are transmitted (in FDMA) through the satellite transponder. This supplementary contribution will be expressed by a factor $(C/T)_{SIM}$ (or $(C/N_O)_{SIM}$ or $(C/N)_{SIM}$). In the typical example of Fig. 2.9, the difference between data in abscissa and in ordinate are equivalent to the $(C/N)_{SIM}$ per intermodulation product, in dB).

Another example is the following (approximate) formula applicable to SCPC which gives the $(C/T)_{SIM}$ per carrier as a function of the number of simultaneously active carriers (N) and of the $(BO)_O$ in dB:

$$(C/T)_{SIM} = -150 - 10\log N + 2\cdot(BO)_O.$$

The farther the operating point is from saturation, i.e. the larger are the back-offs, the more linear is the operation and therefore the smaller are the intermodulation products, i.e. the larger is the $(C/T)_{SIM}$ to be considered in the overall link budget.

Practically, according to an optimization process which will be explained below, it is very often necessary to reduce the effectively available satellite output power by some back-off. In consequence, in the $(C/T)_U$ and $(C/T)_D$ calculations, the available satellite (PFD) on the one hand, P_{ST} and $(E.I.R.P.)_S$ on the other hand, are to be reduced respectively by $(BO)_I$ and $(BO)_O$.

Total link budget

Taking into account the various contributions of section 2.2.3, the total link budget can now calculated by adding the various noise powers for a given (reference level) carrier power. Consequently, the total (C/T) is obtained by adding the inverse ratios of the various contributing (C/T)s:

$$(C/T)_T^{-1} = \Sigma_i (C/T)_i^{-1}.$$

Practically:

$$(C/T)_T^{-1} = (C/T)_U^{-1} + (C/T)_D^{-1} + (C/T)_{SIM}^{-1}. \qquad (2.15)$$

(Note that this formula cannot be expressed in dBs).

In the common case of multiple carriers in the same transponder (FDMA), the maximum possible $(C/T)_T$ is obtained by optimizing the transponder operating point as follows. If the operating point is too far from the saturation, i.e. if the input and output back-offs $(BO)_I$ and $(BO)_O$ are too large, the intermodulation noise will be very low, corresponding to a large $(C/T)_{SIM}$, but the satellite e.i.r.p. will be reduced too much corresponding to a relatively poor $(C/T)_D$. If, on the other hand, the transponder is operated too near saturation, the intermodulation noise will be rather high, corresponding to a small $(C/T)_{SIM}$ and simultaneously the satellite e.i.r.p. will be high, corresponding to a large $(C/T)_D$ (and also to a large $(C/T)_U$). The optimization process is illustrated in Figs 2.12 and 2.13 and shows that, in a typical example, the optimum $(BO)_I$ is about 6 dB, corresponding to a $(BO)_O$ of about 2.2 dB.

Finally, other terms must often be added in (2.15) to account for other interference sources (e.g. from other earth stations in the network, or from the space and earth stations of other nearby satellite networks). In particular, in the case of frequency reuse, a term $(C/T)_{FR}$ should be added to account for the limited

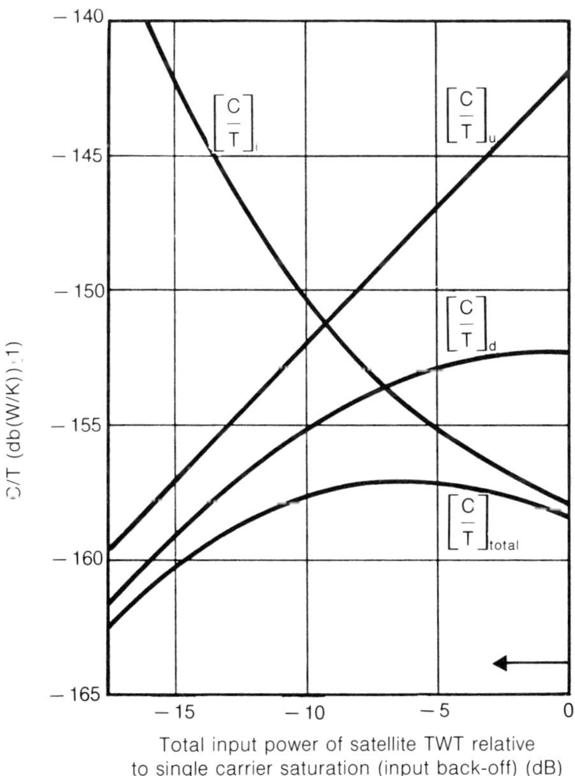

Fig. 2.13 Example of link budget optimization.

isolation between the concerned transponder and other(s) transponder(s) at the same frequency.

2.2.5 Link quality

Referring to section 2.2.4 (Quality objectives), the purpose of this section is to translate the (C/N) (or C/N_0, or C/T), as calculated in section 2.2.4, into the parameter characterizing the link quality, i.e. into the signal-to-noise ratio (S/N) in the case of analogue modulation and into the bit error rate (BER) in the case of digital modulation. Reference is also made to Table 2.2 and Fig. 2.12. Note that it is prudent practice to provide for a few dB margin in supplement to the required (C/N) (or C/N_0, or C/T) in order to take into account possible differences between theoretical and actual equipment performances and also unexpected signal degradation or interference. It should be noted that modulation, demodulation and signal processing are outside the framework of this book, the present section will give only a brief overview of these subjects.

Analogue modulation

By far, the most common type of analogue modulation is the frequency modulation (FM), because this is a constant signal envelope (constant amplitude) method which is well adapted to the non-linear power amplifiers used in satellite communications.

The conventional frequency modulation formula below gives the (S/N) at the output of the demodulator:

$$(s/n)_W = \frac{3}{2}\frac{\delta f^2}{(f_M^3 - f_0^3)}\frac{C}{kT}q \tag{12.16}$$

where δf is the amplitude of the frequency modulation deviation (Hz): this usually results from the application of a test baseband signal; f_0 and f_M are the minimum and maximum frequency of the spectrum of the baseband signal (i.e. of the signal which modulates the RF wave) (Hz) (note that $\delta f/f_M$ is the modulation index), and q is a factor accounting for the 'pre-emphasis' and the 'weighting' improvements*.

The radio frequency bandwith (B_{RF}) occupied by the modulated signal is given by the approximate Carson's rule:

$$B_{RF} = 2(\delta f_{MAX} + f_M) \tag{2.17}$$

where δf_{MAX} is the maximum amplitude of the frequency modulation deviation (Hz).

*'Pre-emphasis' is a process applied to the baseband signal before modulation in order to equalize the quality all over the frequency band at the demodulator output: the higher baseband frequencies (which should have naturally a lower frequency modulation gain), are amplified with a differential gain greater than the lower frequencies. The differential gain function is null at the cross-over frequency. Weighting is defined, for telephony in Note 3 of Table 2.2.

Multiplexed telephony In this case of multiplexed telephony (FDM/FM), it is not the overall (S/N) which is of concern, but the (S/N) in the individual telephone channels. Under this condition, (2.16) can be written as follows, in dB, after some arrangements and simplifications:

$$(\mathrm{S/N})_w = (\mathrm{C/N}) + 20\log(f_{TT}/f_M) + 10\log(B_{RF}/b) + Q \tag{2.18}$$

where $(\mathrm{S/N})_w$ is the ratio (converted in dB) of the 'test-tone' power to the weighted noise power in the upper telephone channel (i.e. the less favoured) of the multiplex. The $(\mathrm{S/N})_w$ is measured at the so-called 'zero relative level' (1 mW); f_{TT} is the rms value (amplitude/$\sqrt{2}$) of the frequency deviation given by a test-tone in one telephone channel (Hz) (the test-tone usually being at 1 kHz); f_M is often taken here as the mid-frequency of the highest telephone (Hz); b is the bandwidth of a telephone channel (≈ 3100 kHz) and Q is taken as 6.5 dB (i.e. 4 dB pre-emphasis improvement in the upper channel + 2.5 dB psophometric weighting).

Here, δf_{MAX} (which defines B_{RF} by (2.17) is the peak frequency modulation deviation resulting from the application of the multi-channel (multiplexed) baseband signal. Its relation with f_{TT} (which is taken in rms and in a single channel) is conventionally defined by the following formulae. These take into account the statistical properties of the telephone channels (effective channel occupancy, voice level distribution etc.):

$$\delta f_{MAX} = f_{TT} \cdot g \cdot l$$

where g is the peak-to-rms ratio = 4.47 (13 dB) for less than $n = 120$ channels in the multiplex and 3.16 (10 dB) for $n \geqslant 120$, and l loading factor of the multiplex given by: $20\log l = -1 + 4\log n$, for $n < 240$ channels; $20\log l = -15 + 10\log n$, for $n \geqslant 240$.

Examples: the following example is taken from INTELSAT specifications (the figures actually specified are marked*):

multiplex: 252 channels*;
allocated bandwith (in the transponder): 15 MHz*;
actual $B_{RF} = 12.4$ MHz*;
top baseband frequency $f_M = 1052$ kHz*.

This results in:

$\delta f_{MAX} = 5148$ kHz (2.16);
Multichannel rms deviation $f_{MC} = 1627$ kHz* ($= \delta f_{MAX}/g$ with $g = 3.16$);
Deviation (rms) for test tone $f_{TT} = 577$ kHz* ($= f_{MC}/l$ with $l = 2.82$).

A minimum (C/N) = 13.6 dB* is specified. Therefore, application of (2.18) gives:

$$\begin{aligned}(\mathrm{S/N})_W &= 13.6 - 5.22 + 36.02 + 6.5\\ &= 50.9\ \mathrm{dB}.\end{aligned}$$

This $(\mathrm{S/N})_w$ corresponds to the INTELSAT specification of 8200 pW^*_{0p} which refers to the CCIR specification (Table 2.2), with due account for some external interference sources.

SCPC/FM In the case of a single telephone channel ($f_M = 3400$ Hz, $f_0 = 300$ Hz), formula (2.16) can be written in dB as follows:

$$(S/N)_w = (C/T) + 127.4 + 20 \log f_{TT} + Q \tag{2.19}$$

f_{TT} results again from the application of the Carson's bandwidth rule ($B_{RF} = 2(\delta f_{MAX} + f_M)$) and from $\delta f_{MAX} = f_{TT} \cdot g \cdot l$ (here g is approximately 0.25 and l is between 8.4 and 12.6 depending on the 'clipping' specification in case of high voice level). Q is again $4 + 2.5$ dB.

Some important points are to be mentioned because of their consequences on the SCPC system design:

CHANNEL SPACING

SCPC channel frequency spacing determines the maximum possible number of channels in the satellite transponder, i.e. the maximum traffic capacity. The most usual spacings are: 45 kHz, 30 kHz and 22.5 kHz. These correspond respectively to a maximum of 800, 1200 and 1600 telephone channels in a transponder of 36 MHz bandwidth.

VOICE ACTIVATION

In the case of FDM/FM systems, due to a statistical averaging effect in the multiplex loading, the factor l 'automatically' takes into account the effective channel occupancy which is not the case in SCPC. To cope with this problem, 'voice activation' is usually provided in SCPC channels. This consists of transmitting each SCPC/RF carrier (thanks to a 'speech detector' circuit) only during the effective speech periods, i.e. only during about 40% of the time (40% is the average 'activity factor' of a telephony half-circuit). Voice activation saves the RF power in the transponder and thus allows benefit of the actual transponder traffic capacity: in fact, in the above mentioned cases of 800, 1200 and 1600 telephone channels, only an average of 320, 480 and 640 channels are simultaneously transmitted in the transponder.

THRESHOLD EXTENSION DEMODULATION

Often, in SCPC, the (C/N) required from the link budget is lower than the threshold of conventional demodulators and threshold extension demodulators (TED) are needed (see the example below and section 2.2.4).

COMPANDING

'Companding' means implementing compression–expansion. This is a process which, during transmission, reduces the dynamic range of the speech signal and which, conversely, expands this range at reception in order to restore the original voice levels. Compandors (companding circuits) are very often provided in

SCPC/FM channel units. Such circuits actually provide an additional advantage (an additional Q factor in formula (2.16) which amounts to 13 dB up to 20 dB, depending on the subjective evaluation of the voice transmission quality). This advantage is due to both an objective reason (reduction in noise power) and a subjective reason (lower noise during pauses).

Example: assume a 45 kHz channel spacing. By applying the Carson's bandwidth rule and accounting for some guard band between the channels, we take $\delta f_{MAX} = 15\,\text{kHz}$ ($B_{RF} = 36.8\,\text{kHz}$). Then, assuming $g \cdot l = 2.5$, and $f_{TT} = 6\,\text{kHz}$, (2.19) gives:

$$\begin{aligned}(\text{S/N})_W &= (\text{C/T}) + 127.4 + 75.56 + 6.5 \quad (\text{dB}) \\ &= (\text{C/T}) + 209.5 \quad (\text{dB})\end{aligned}$$

or, using companding and accounting for 15 dB companding gain:

$$(\text{S/N})_{W+COMP} = (\text{C/T}) + 224.5. \quad (\text{dB})$$

This means that, for 10 000 pW_{0p} noise power, i.e. $(\text{S/N})_{W+COMP} = 50\,\text{dB}$ equivalent link quality (Table 2.2), the total link budget should give:

$$(\text{C/T}) = -174.5\,\text{dB}\cdot\text{W}\cdot\text{K}^{-1}$$

This corresponds to (C/N) as low as 8.5 dB* which shows that a threshold extension demodulator is needed.

Television (TV/FM) From (2.16), the following formula is derived for conventionally evaluating the quality of satellite transmitted TV programmes:

$$(\text{s/n})_W = \frac{3(r_1 \cdot \delta f_{PP})^2}{f_M^3} \frac{C}{kT} q. \tag{2.20}$$

where δf_{PP} is the peak-to-peak frequency deviation resulting from a 1 V peak-to-peak tone applied at the pre-emphasis cross-over frequency; r_1 is the ratio of the nominal peak-to-peak amplitude of the luminance signal to the peak-to-peak amplitude of a monochrome composite video signal ($r_1 = 0.714$ or 0.7 depending on the TV standard: 525/60 or 625/50), and f_M is the top baseband frequency, usually taken at 5 MHz, $f_0 \ll f_M$.

Therefore, the formula can be written in dB as:

$$(\text{S/N})_W = 32.4 + (\text{C/T}) + 20\log(r_1 \cdot \delta f_{PP}) + Q$$

where $Q = (P) + (W)$. The pre-emphasis (P) and weighting (W) characteristics are defined by the CCIR and depend on the TV standards and on the measurement conditions. Referring to the pre-emphasis cross-over frequency, for a 5 MHz measurement bandwidth and for the unified weighting network (which conven-

*This (C/N) value is based on $B_{RF} = 36.8\,\text{kHz}$. In fact, the effective noise bandwidth resulting from actual fitering is somewhat smaller and (C/N) higher.

tionally accounts for the average viewer's sensitivity to the various noise spectrum frequencies), the total $Q = 14.8$ or 13.2 dB for the 525/60 or 625/50 TV standards.

Example: the calculated $(S/N)_W$ is 50.3 dB under the following conditions (From INTELSAT TV transmission standards):

Allocated satellite bandwidth: 20 MHz (INTELSAT V half-transponder),
Receiver bandwidth: 18 MHz,
TV standard: 625/50 (lines/Hz),
$\delta f_{PP} = 21$ MHz
$(C/T) = -138.7\,\text{dBw}\cdot\text{K}^{-1}$ thence $(C/N) = 17.3$ dB.

Note that there is a certain amount of 'overdeviation' due to the fact the 'Carson bandwidth' is greater than the receiver bandwidth. This could cause some 'truncation noise' under poorer conditions (lower C/T) than specified.

With regard to the TV associated sound transmission the sound (audio) programme associated with the video is usually transmitted by modulating a sub-carrier which is combined with the video baseband signal before final frequency modulation of the carrier by this (video and sound) composite signal. The sound signal results from the modulation by a high quality sound (15 kHz bandwidth) of the sub-carrier centre frequency) (in the upper part of the baseband).

Digital modulation

It has already been explained (section 2.2.4) that phase shift keying (PSK) is the most common modulation method in digital satellite transmission. This is because, again, this has a constant signal envelope method and also because it combines good bit error rate (BER) performance with simplicity of modulation and demodulation circuits.

The basic PSK is the binary PSK (BPSK or 2-PSK): it is represented by a binary code (bit 0 or 1) by the carrier phases 0 or π. In quadratic PSK (QPSK or 4-PSK), two binary codes (a two-bit character forming a symbol) are represented by four possible carrier phase states (e.g. 00 by 0, 01 by $\pi/2$, 11 by π, 10 by $3\pi/2$).

More generally, in multi-phase (M-ary) PSK (M-PSK), n binary codes forming M symbols (with $M = 2^n$) are represented by M signal elements (possible carrier phase states). Each symbol is transmitted by the RF carrier in a time duration T_S, i.e. the symbol rate (number of symbols per second expressed in Baud) is $R_S = 1/T_S$. The required signal bandwidth (Nyquist bandwidth) (in Hz) is:

$$B = \alpha \cdot R_S = \alpha \cdot R/n \tag{2.21}$$

R being the bit rate and α being a coefficient accounting for the effective spectrum width (α is often about 1.2).

There exist many other digital modulation methods: FSK (frequency shift keying), ASK (amplitude shift keying), APK (composite amplitude-phase shift keying), etc. In fact BPSK and QPSK are most commonly implemented in satellite systems, QPSK often giving the best power–bandwidth efficiency trade-off. This is because, for $M > 4$, the required power per transmitted symbol is increased for a given performance. Also BPSK and QPSK modulation and demodulation are rather simple, QPSK equipment implementing two BPSK-type channels in quadrature (the 'P' and 'Q' channels).

In the bit error rate (BER) at the demodulator output, which characterizes the transmission quality, errors can be due to inter-symbol interference, phase jitter in carrier and bit timing recovery, etc. However, the main contribution to errors is the thermal (and miscellaneous) noise resulting from the total $(C/T)_T$, as calculated by the link budget. For coherently demodulated BPSK and QPSK, the BER due to thermal noise (or equivalent) is expressed as follows by the probability of errors of a signal in the presence of additive gaussian noise:

$$P_E = \tfrac{1}{2}\mathrm{erfc}\frac{E_B}{N_0}. \tag{2.22}$$

where erfc is the complementary error function, $\mathrm{erfc}(x) = (2/\sqrt{\pi})\exp(-t^2)\,\mathrm{d}t$; E_B is the energy per bit, and $N_0 = kT$ (noise power per Hz).

This formula is represented in Fig. 2.14. For BPSK, $(E_B/N_0) = \alpha\cdot(C/N)$ since $C = E_B\cdot R$ and $B = \alpha\cdot R$. For QPSK, $(E_B/N_0) = (\alpha/2)\cdot(C/N)$ since $C = E_B\cdot R$ and $B = \alpha\cdot R/2$. This formula is valid for coherent PSK, which means that a reference signal (usually obtained from the signal itself by a carrier recovery technique) is used for demodulating the signal. However, for fully coherent detection, the phase ambiguity must also be removed. If, for simplifying the circuitry, the signal is differentially encoded (i.e. the information is represented as the difference between adjacent carrier phases), P_E is twice the value of (2.22).

Forward error correction (FEC) It has already been explained in section 2.2.4 that the required minimum (C/T) can be lowered, for a specified BER, by implementing baseband processors called forward error correction (FEC) codes which implement redundancy in the transmission as follows: to a given block of K bits (which carries useful information at a bit rate R_I), the FEC encoder associates N bits ($N > K$) which constitute a code word. The transmitted bit rate becomes R_C. Applying a knowledge of the encoding law, the FEC decoder selects the most likely information block. The ratio $r = K/N = R_I/R_C$ is called the code rate. There are two consequences of this process, which is illustrated by Fig. 2.15.

The (Nyquist) signal bandwidth required for transmission is increased and becomes:

$$B = \alpha\cdot R_S = \alpha\cdot R_C/n = \alpha\cdot R_I/(n\cdot r)\ (\mathrm{Hz}) \tag{2.23}$$

(with $n = \log_2 M$ in M-ary modulation, as explained above). For example, for transmitting $R_I = 64$ kbit/s in QPSK ($n = 2$) with a FEC coding rate $r = 2/3$, $B = (1\cdot 2)\cdot R_I\cdot 3/4 = 57.6$ kHz, in lieu of 38.4 kHz without FEC).

The BER performance is improved by a coding gain. Let us consider the BER

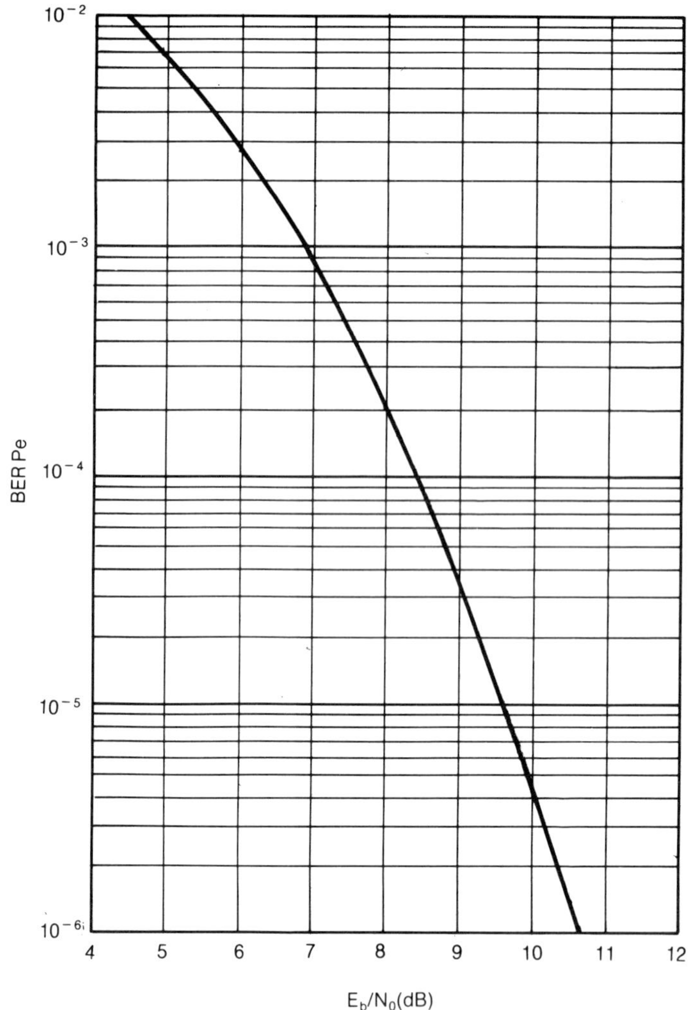

Fig. 2.14 Theoretical BER performance of coherent PSK.

curve giving the error probability on information bits (P_{EI}) as a function of (E_I/N_0), E_I being the energy per information bit and $N_0 = kT$ as usual. Typical examples of such BER curves are given in Fig. 2.16. Then, the coding gain is measured in dB by the difference, for a given BER, between the required values of the (E_I/N_0) without and with FEC (without FEC and for BPSK/QPSK, the curve is the same as in Fig. 2.12 with $E_B = E_I$ and $P_{EB} = P_{EI}$). Note that the coding gain is really a net gain because it takes into account the supplementary energy (10 log 1/r in dB) required to transmit the $N-K$ redundancy bits (in dB: $E_I/N_0 = E_C/N_0 + 10 \log 1/r$, E_C being the energy per transmitted encoded bit). Consequently, for high BERs

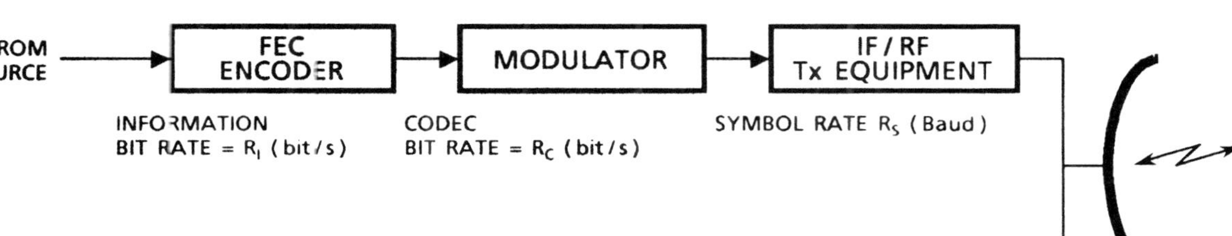

Fig. 2.15 Earth station scheme for digital communications with forward error correction.

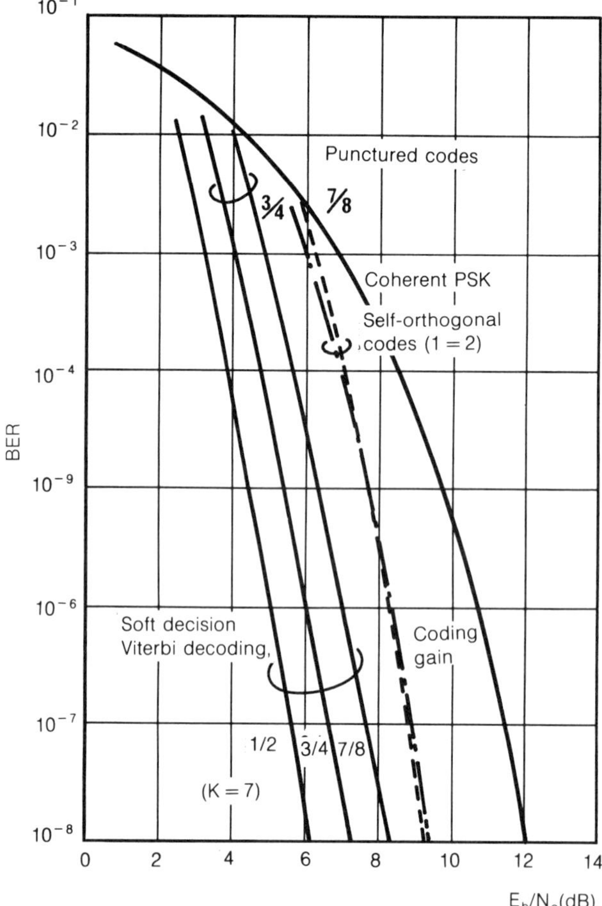

Fig. 2.16 Examples of BER performance with FEC (convolutional codes).

(over about 10^{-2}), the coding gain becomes negative because the coding advantage can no longer compensate for the increase of errors resulting from the reduction of E_I to E_C (in the transmission channel). On the contrary, the coding gain increases when BER decreases.

It is not possible here to describe the FEC processes, the various classes of codes (block codes, convolutional codes, etc.) and the decoding methods and algorithms (Viterbi maximum likelihood criterion, sequential decoding, etc. See, for example, Clarke and Cain, 1981). Note, however, that the coding gain is improved if the demodulator delivers the analogue (sampled) value of the received bits and not a simple 1 or 0. This allows decoding of the bits by a 'soft decision' process, which can commonly improve the coding gain by 2 dB compared to 'hard decision' decoders.

Introducing FEC codes in digital transmission is similar to increasing the modulation index ($\delta f/f_M$) in analogue transmission: in both cases, the required bandwidth increases and the required (C/T) decreases for a given link quality. Thanks to the availability of VLSI technology, FEC codes are now powerful tools for improving the BER of digital links or, conversely, for reducing the link parameters (earth station antenna diameter or amplifier power, etc.) for a given required BER.

2.3 COMMUNICATION SATELLITE TECHNOLOGY

2.3.1 Communication satellite construction

Classes of communication satellites

A communication satellite is comprised of a space platform (or bus) and a payload. An exploded view of a typical communication satellite is shown in Fig. 2.17. Up

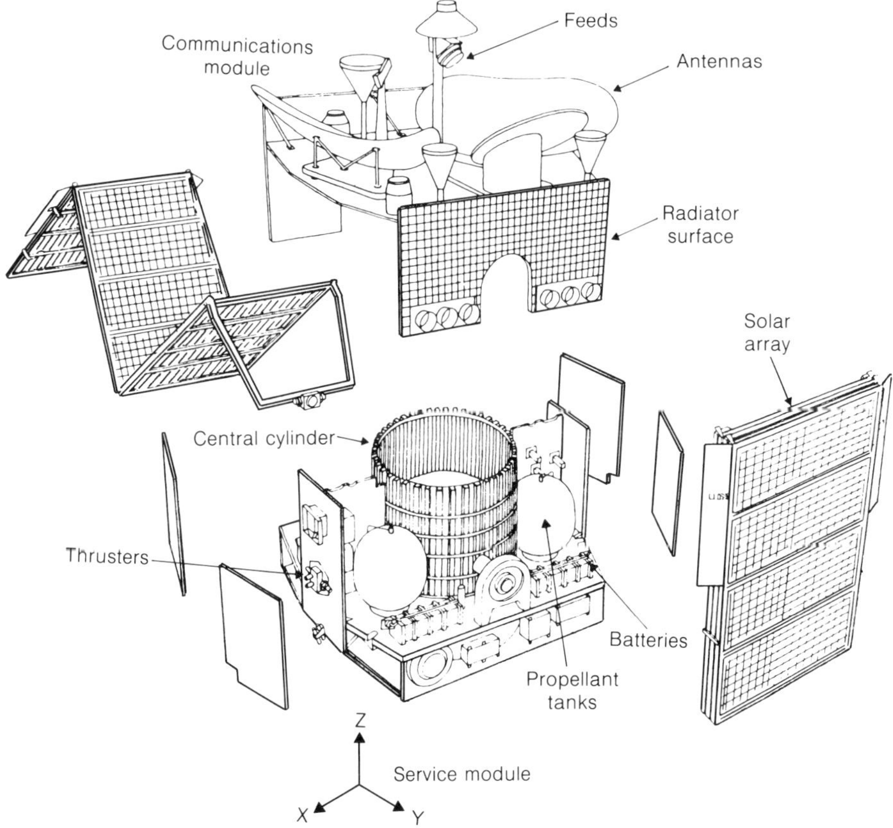

Fig. 2.17 Exploded view of a communications satellite (three-axis stabilized: TELECOM-1).

to now, two classes of communication satellites have been used: spin stabilized and body-fixed stabilized satellites.

The spin stabilized satellites The whole body of these satellites is rapidly rotated (between 30 to 120 r.p.m) around one of its principal axes of rotation. This creates a gyroscopic stiffness which maintains the body permanently in a fixed direction, relative to absolute coordinates (except variations due to perturbing torques which must be corrected as explained below). In the case of geostationary satellites, this imposed fixed direction is perpendicular to the equatorial orbit, i.e. north–south (pitch axis spinning). Spin stabilization is a simple solution, with the advantage of facilitating the thermal control because the body elements facing the sun are continuously changing. Solar cells for the power supply are located all around the cylindrical body.

The first INTELSAT satellites (INTELSAT-I and -II) were simple spin-stabilized satellites. However, such rotating platforms must carry toroidal antennas, which can radiate only a small part (about 4%) of their RF power towards the earth. This is why the technique has now been improved by using 'dual spinning', i.e. by locating the antennas (or even the whole communication payload) on a 'despun platform': such a platform is counter-rotated with one direction pointing towards the earth's centre. The INTELSAT-IV and -VI satellites are of this type (Fig. 2.18).

Note that most satellites, whatever their construction use spin stabilization during the transfer phase of their positioning on the geostationary orbit.

The body-fixed stabilized satellites (Fig. 2.17) These are often called three-axes stabilized, but this is misleading since all satellites are in fact three-axis stabilized. Here, all the parts of the spacecraft are maintained in a fixed position relative to the earth. Stabilization is also performed by gyroscopic stiffness, through utilization of an internal momentum wheel(s).

The space platform sub-systems are briefly described in this section. The communication sub-systems composing the payload are described in section 2.3.2.

The satellite structure

The primary structure, which must bear the equipment and ensure its precise positioning, whilst withstanding severe mechanical stresses—especially during launching—is generally made of aluminium alloys in various shapes: tubes, shells, stiffened frames, honeycomb panels. The secondary structure (solar generators, antenna reflectors), is usually made of composite materials (epoxy resins, carbon fibres) in order to combine lightness, high rigidity and low thermal expansion.

The attitude stabilization sub-system

Attitude stabilization is needed for keeping the satellite antenna beams in the desired direction. This is done by control loops which actuate rotations about

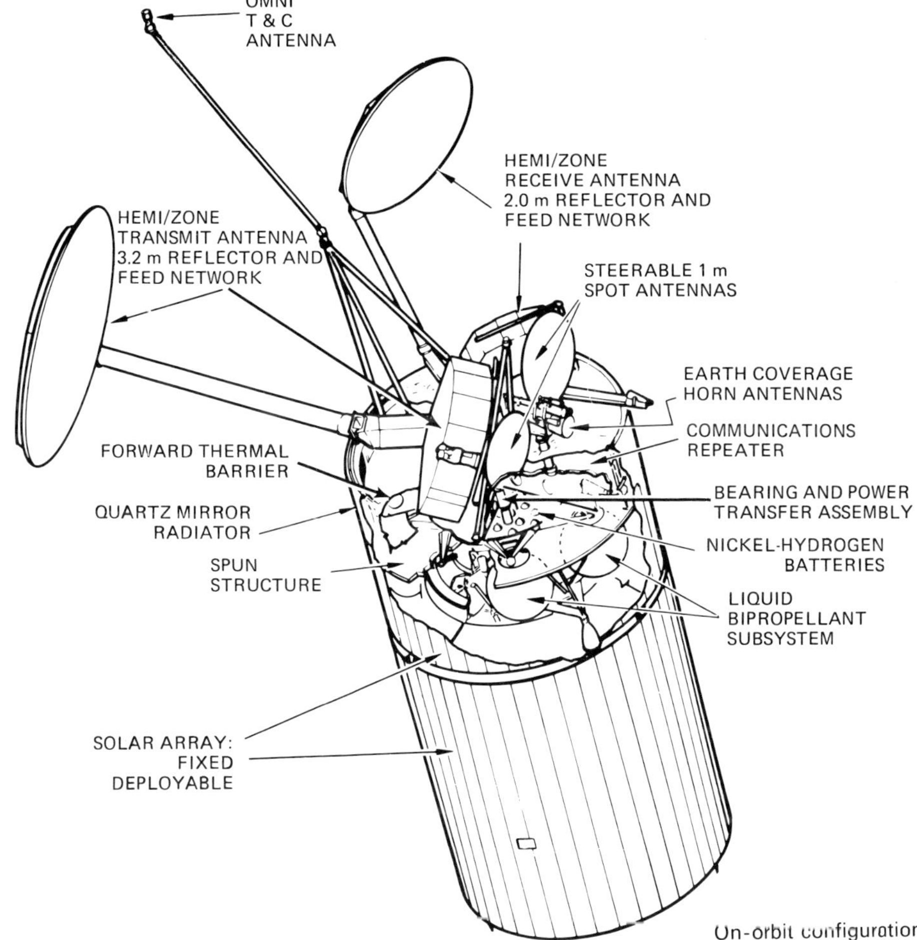

Fig. 2.18 Example of spin-stabilized satellite (INTELSAT VI).

the three reference axes (yaw: the earth's centre direction i.e. the local vertical; roll: east–west, and pitch: normal to the orbital plane, i.e. north–south) Attitude measurement is made by sensors, which can be passive or active. Most passive sensors are based on the measurement of earth's horizon (with an accuracy of about 0.05°) by infrared measurement of the difference between the earth's temperature, (about 255 K) and the sky's temperature, (about 5 K) and of the sun's direction. Active sensors which make use of a ground-based beacon are more precise and may allow easier independent orientation of multiple on-board antennas.

The satellite platform orientation undergoes north–south roll, typical drifts of about 0.02°/hour (mainly due to solar radiation pressure). The corresponding pointing errors are corrected by small thrusters, magnetic coils (which create small torques by interacting with the earth's magnetic fields) or by orientating 'solar sails'. East–west pitch errors are corrected by acting on the acceleration of the spinning elements.

The orbit control (station keeping) sub-system

Various disturbance sources tend to modify the geostationary satellite orbital position:

1. the lunisolar attraction, the effects of which (rotation of the orbital plane) are to be corrected by north or south thrust impulses;
2. the 'earth triaxiality', i.e. the non-verticality of the gravity, due to earth's equator ellipticity. This is corrected by thrust impulses in the direction of the orbit;
3. the solar radiation pressure, which decelerates the satellite in the morning and accelerates it in the afternoon (corrections also in the orbital direction).

As viewed from the earth, the satellite's residual motions shows up as a figure eight with a north–south component (orbit inclination) and an east–west (in-plane) component.

The objective of the orbit control system is to keep the satellite in a specified 'box' (or 'window') in longitude and latitude. This is done by periodically firing small thrusters at appropriate points on the orbit. Current practice limits the box to more stringent dimensions than those specified by the CCIR ($0.1° \times 0.1°$) in order to avoid the need for tracking on small earth station antennas. The propellant mass consumed by the orbit control thrusters is between 1.8% and 2.5% of the spacecraft mass per year, depending of the propellant type and technology. Therefore, the orbit control system consumption is often the main limitation to the satellite's effective life duration. Sometimes, satellite systems are used, after their specified end of life, by allowing increased orbital deviations (this is called inclined orbit utilization of the satellite).

The thermal control sub-system

The thermal control system maintains the spacecraft within the temperature range compatible with the correct operation of its various parts (e.g. -10 to $+60$ °C for electronic equipment but 0 to $+20$ °C for batteries).

In fact, the satellite is subject to severe thermal conditions. It receives solar radiation (except during eclipse periods) on one side, and radiates towards cold space on the other side (the earth thermal flux and the albedo—reflection of sun by the earth—being negligible). Also, in the case of three-axes stabilization, the side faces of the spacecraft are subject to daily solar variations (this is not encountered with spin-stabilized satellites).

The mean temperature of the satellite results from thermal balance between, on the one hand, the energy externally received (sun) and internally dissipated (e.g. travelling-wave tube amplifier) and, on the one hand, the energy radiated towards outer space.

The most used thermal control methods are:

1. the use of surface finishes, with careful selection of their absorbance and emittance coefficients;
2. the use of thermal coatings, such as multilayer 'super-insulation' blankets (made of layers of thin plastic films coated with deposited aluminium);
3. the use of heat radiators, such as rigid OSRs (optical solar reflectors) or flexible SSMs (second surface mirrors), especially for dissipating, on the north and south panels, the heat losses of the power tubes,
4. if needed, the use of active devices (especially in the case of very high power satellites): heat pipes, mechanically actuated louvres, electrical heaters.

The power supply sub-system

Up to now, communication satellites have always used arrays of solar cells as a primary source for supplying the electrical power (typically 1 to 2 kW or even more). Although body-fixed satellites needed to be folded during launching and require orientation mechanisms for permanently maximizing solar flux interception (1350 W/m^2), they provide more efficient utilization of the solar cells than spun satellites. Each solar cell (4 to 8 cm^2 p-type silicon wafers with a thin n-type layer) supplies about 50 mW power under 0.5 V (efficiency of 10 to 14%). Current solar arrays, which are composed of a series/parallel assembly of multiple cells, supply more than 60 W/m^2 (end of life).

For ensuring permanent operation, even during the eclipse periods, the satellite must be equipped with a secondary source. This provides electrical power at least to the basic control equipment and to all or part of the transponders (some satellites provide only partial communication operation during eclipses). The secondary source uses electrochemical accumulators, usually nickel–cadmium batteries (delivering 30 to 40 W h/kg). Other more efficient sources, i.e. with a higher power-to-weight ratio are under development.

Although the various equipment sets require regulating at various d.c. voltages, the solar array actually supplies an irregular power and voltage. This is why electrical conditioning (regulation, d.c./d.c. converters and protection circuits) is provided in the power supply system.

The telemetry, command and ranging sub-system

In association with the TTC (telemetry, tracking and control) earth stations, this on-board system provides the following functions:

1. reception of command signals for current satellite operations (these signals should be carefully protected against various types of interference by coding, spectrum spreading, acknowledgement procedures);

2. transmission of telemetry signals collected from the satellite sub-systems (the TTC earth stations may also use these signals as a beacon for measuring the satellite's angular position);
3. reception, conversion and retransmission of the command carrier; the satellite distance is known (ranging)—within a few metres accuracy—by measuring the phase of a set of low frequency modulation signals (tones).

For being transmitted, these signals may use carriers in the communication band. However a special transponder (in the VHF band or, in modern satellites in the 2 GHz band) usually provides operation whenever the satellite is not stabilized (positioning phase or malfunction).

The propulsion sub-system

The propulsion sub-system is comprised of the low thrust actuators which are used for the attitude and orbit corrections (see above) and the apogee motor, with a very high thrust, which provides the velocity increment needed, at the apogee of the transfer orbit, for injecting the satellite on the geostationary orbit.

Such a process is, in fact, the most cost efficient for launching geostationary satellites: it consists of launching first the spacecraft at some 200 km altitude. From this point (perigee), the spacecraft is placed on an elliptical orbit (the transfer orbit) with an apogee at 36 000 km and an inclination close to the latitude of the launching site. The apogee motor is fired to produce, from the apogee, the required circular orbit in the equatorial plane. Note that, when launching is made by the space transportation system (STS), (Space Shuttle), a supplementary motor, the perigee motor, is needed for injecting the satellite into its transfer orbit at the time that the shuttle crosses the equatorial plane from its parking orbit, a circular orbit at some 290 km altitude.

2.3.2 Communication satellite payload

General

A typical overall block diagram of a communication satellite payload is shown in Fig. 2.19. It is composed of the following parts.

1. A wideband sub-system for receiving the bulk of the up-link (U/L) carriers (e.g. at F_U in the 6 GHz band). This includes one U/L antenna with an overall coverage (as shown in the figure) or a few U/L antennas, each one with a partial coverage (spot beam antennas); and one wideband receiver per U/L antenna. Each wideband receiver comprises an input passband filter, a low-noise amplifier, a down-converter with a local oscillator (e.g. at 2.225 GHz) for converting the U/L carrier frequencies (F_U) into down-link (D/L) carrier frequencies (F_D) and an amplifier.

2. One input demultiplexer (IMUX) per U/L antenna. This is a divider equipped with filters. It has one wideband input port and N narrower band

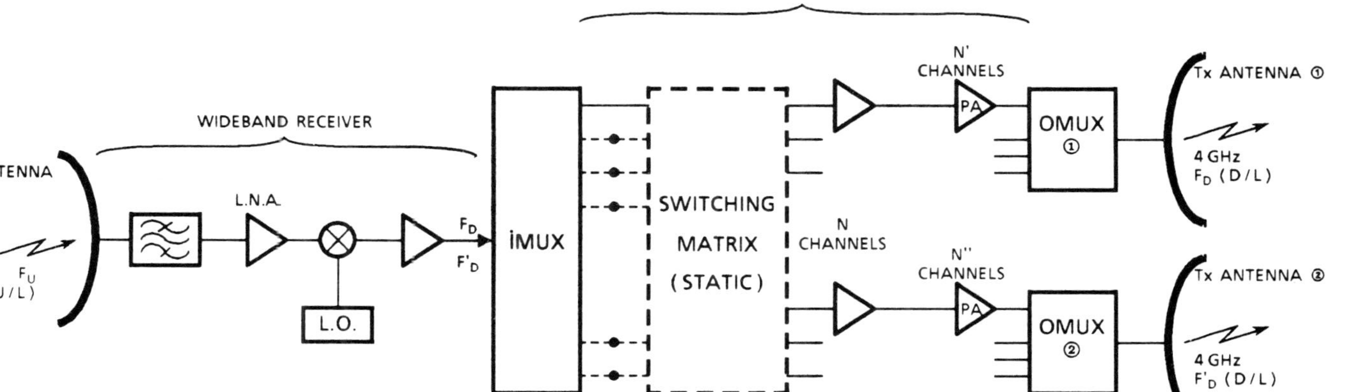

Fig. 2.19 Overall block diagram of a typical communication payload. Rx, reception; Tx, transmission: LNA, low noise amplifier; LO, local oscillator; IMUX, input demultiplexer; OMUX, output demultiplexer, PA, power amplifier.

output ports. Its function is to separate the various carriers by their respective frequencies and to assign them to N different amplifier chains, usually called channels.

3. A channelized sub-system which amplifies separately these N channels, up to the required D/L power level: each channel equipment forms one transponder proper (or repeater) with a bandwidth of 40, 70 or 120 MHz (typical values).

4. One or several output multiplexer(s) (OMUXs): each OMUX (e.g. N' input ports and one output port) combines one group of N' channels and is associated with a separate D/L beam (i.e. a separate earth coverage) for transmitting the D/L carriers to their destined earth stations. Each D/L beam is radiated, either by a separate D/L antenna, or by a single multi-beam (i.e. multi-feed) D/L antenna.

5. A switching matrix (not provided in the simplest satellites) for distributing the channels to the various D/L beams. The static switching matrix shown in the figure allows modifications of the channel assignments by telecommand orders from a TTC earth station. It is not to be confused with the dynamic switching matrices which are able to operate in millisecond times on some modern satellites (e.g. INTELSAT VI).

Of course, actual communication payloads are, in general, much more complex. In particular redundancy is always provided for ensuring a specified communications availability: wideband receivers are very often doubled (1 + 1 redundancy) as well as some power amplifiers and other critical equipment. In the channelized sub-system, $n + m$ type redundancy is often sufficient: this means that m spare equipment units are provisioned (usually with automatic switching) in case of a malfunction of any of the n active units ($m < n$).

The figure shows a transparent satellite, which is the case for most satellites under current operation. This is a 'black box' which simply retransmits the received carriers (after frequency translation and amplification) without modifying in any way the signals conveyed by these carriers. In the future, satellites will implement active payloads, with beam switching (as mentioned above, this is already the case in INTELSAT VI satellites), and/or signal baseband processing and regeneration (see below).

Two OMUXs (and the corresponding channels), if connected to two well-separated (isolated) beams, can be operated in the same frequency band: this mode of operation is called frequency reuse by beam separation (section 2.1.4). Also, two OMUXs in the same frequency band can be connected to a single beam, i.e. to a single antenna, through the two access ports of a dual-polarized feed: this is the dual-polarization frequency reuse mode of operation.

Some satellites are equipped with multiple payloads: for example, the French TELECOM 1 satellite carries one 6/4 GHz payload (2×40 MHz + 2×120 MHz transponders), one 8/7 GHz payload and one 14/12 GHz payload (6×36 MHz), the Indian INSAT and the Arab league ARABSAT satellites includes one 6/4 GHz payload for communications and one 6/2.5 GHz payload for direct TV broadcasting (plus, in INSAT, a 400 MHz data relay repeater). Other satellites include interconnected repeaters in different frequency bands: for example, the INTELSAT V and VI satellites are equipped with 6/4 and 14/11 GHz interconnectable transponders, which means that the following up/down links are possible: 6/4 GHz, 14/11 GHz, 6/11 GHz, 14/4 GHz.

Protection against mutual interference and, more generally electromagnetic compatibility problems impose severe constraints in the design of multiple payload satellites. For the sake of illustration, Fig. 2.20 shows the very complex arrangement of the INTELSAT VI satellites. In fact, these are the most sophisticated satellites ever designed. To complete this figure, here is a list of the INTELSAT VI communication payload hardware:

16 receivers (6 GHz band)
4 receivers (14 GHz band)
39 driver amplifiers (4 GHz band)
50 input filters
20 up-converters (4/11 GHz)
42 travelling-wave tube amplifiers (TWTAs) (4 GHz band)
15 solid-state power amplifiers (SSPAs) (4 GHz band)
20 travelling-wave tube amplifiers (TWTAs) (11 GHz band)
50 output filters.

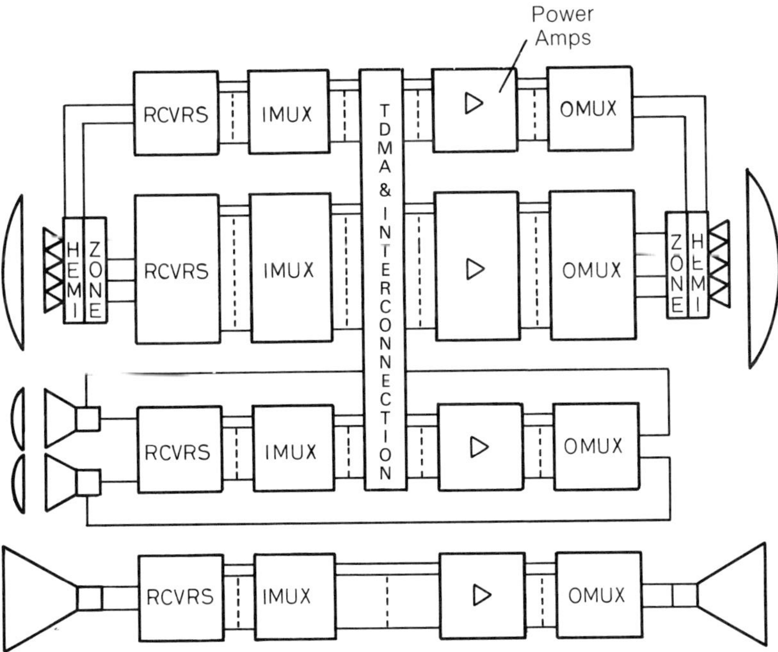

Fig. 2.20 INTELSAT VI satellite payload.

The antenna sub-system

Main characteristics It would be of little interest to describe here the simple types of antennas (e.g. toroidal) used in early satellites. In fact, modern communication satellite systems performance largely relies on the RF characteristics of their complex 'antenna farm' (i.e. all the antennas grouped on the fixed satellite platform, should it be dual-spinned or body fixed). The most important of these characteristics are as follows.

The radiation beam shape which, ideally, should give a 'footprint' on the earth tailored to the desired coverage, thus maximizing the transmit and receive gains and therefore maximizing the satellite e.i.r.p. and G/T in the wanted zone while minimizing unwanted interference sent or received to/from outside.

The sidelobe radiation level which should be as low as possible for the same reason of minimizing unwanted interference. However, recommendations on satellite antenna sidelobes have not yet been issued by the CCIR, except in the case of the broadcasting satellite service (BSS).

The RF polarization (either linear or circular) which should be as pure as possible especially in the case of dual polarization frequency reuse implementation. Polarization purity may imply, firstly, that the level of the unwanted polarization in the radiation beam should be low and, secondly that the two orthogonal polarizations, if radiated by the same antenna (through two access ports) should be well isolated. The required figures for minimum discrimination and isolation are generally about 30 dB.

In the case of circular polarization (e.g., right-hand—RHCP), the amount of the RF field in the unwanted polarization (e.g., left-hand—LHCP) is related to the ellipticity of the (imperfect) circular polarization by the following formula (which is similar to the relation between the reflection coefficient and the voltage standing wave ratio in a transmission line):

$$i = \frac{ar + 1}{ar - 1} \qquad \text{or, in dB} \qquad I = 20 \log i$$

where i is the isolation (or discrimination), i.e. the ratio of the wanted polarization to unwanted polarization RF fields and ar is the axial ratio, i.e. the ratio of major to minor axes of the polarization ellipse.

This is the difficult task of the satellite antenna designer to find the best compromise, for each specific application, between compliance to these RF characteristics and, at the same time, to the mechanical requirements, and in particular, to volume and weight specifications.

Normal beam antennas It will be explained below how shaped beam antennas can be designed for fitting complex geographical coverages. However, very often, a part of the transponders is assigned to the coverage of simple circular or elliptical zones of the earth. This can be performed by antennas herein called normal beam antennas. The first and simplest case is the global beam antenna, which radiates a conical beam with a 17.5° beamwidth, thus giving the maximum possible coverage of the earth, as seen from the geostationary satellite. This is usually a conical horn, the performance of which (symmetry of revolution of the radiation beam, sidelobes level and polarization purity) can be improved by various technologies: corrugated horns (special annular grooves inside the inner wall), and horns with longitudinal fins, multimode excitation, etc.

The coverage of narrower zones of the earth is performed by the so-called spot beam antennas (usually provided with orientation capability). These are generally reflector antennas, constructed with a paraboloidal reflector illuminated from its focus by a feed (or primary source, usually a horn)*. Circular beams are generated by reflectors with a circular contour and elliptical beams by reflectors with an elliptical contour (see gain and beamwidth formulae in section 2.2.2).

The utilization of dual-reflector systems may facilitate the mechanical design by improving the location of the feed. Dual-reflector systems usually consist of a main paraboloidal reflector and of a sub-reflector (hyperboloidal in Cassegrain antennas and ellipsoidal in Gregory antennas). Another dual-reflector technique uses two cylindro-parabolic reflectors with their generators in perpendicular planes.

Also, an asymmetrical configuration of the reflector(s) (offset antennas) can be used both for improving the mechanical design and for avoiding the aperture blockage caused by the primary source (and, possibly, by the sub-reflector). For example, Fig. 2.21 shows an offset Cassegrain antenna. Here the main reflector is an off-axis part of the paraboloidal surface (and the same for the hyperboloidal sub-reflector). Due to their inherent assymetry, offset reflectors introduce some radiation defects, in particular as concerns polarization purity. However this can be improved by special techniques, such as modified reflectors (section 2.4.2).

Note that the earth coverage characteristics of a satellite antenna (in association with its transponder), i.e. its e.i.r.p. and G/T, are usually given at the edge of coverage. This corresponds to a certain level of the antenna beamwidth: although half-power (-3 dB) is used, it is a better practice to specify the edge of coverage at about -4.3 dB, which can be shown to be an optimum level. This can be explained as follows. Let us assume a normal beam with a gain (dBi) G_{MAX} at the beam centre and $G(\alpha)$, N dB below G_{MAX}, in a given direction α (from this centre) at the edge of coverage. $G(\alpha) = G_{MAX} - N$: for N too small (e.g. $N = 2$ dB), G_{MAX} will be relatively low (because the antenna beamwidth will be too large); on the other hand, for N too large (e.g. $N = 6$ dB), G_{MAX} will be relatively high (narrow antenna beamwidth). To calculate the optimum N, $g(\alpha)$ (power ratio) can be (very well) approximated by the following formulae, in the case of a

*The term 'illuminated' by the feed actually refers to transmission. However the antenna operates similarly at reception where the feed 'collects' the incoming radiation after its reflection on the paraboloid.

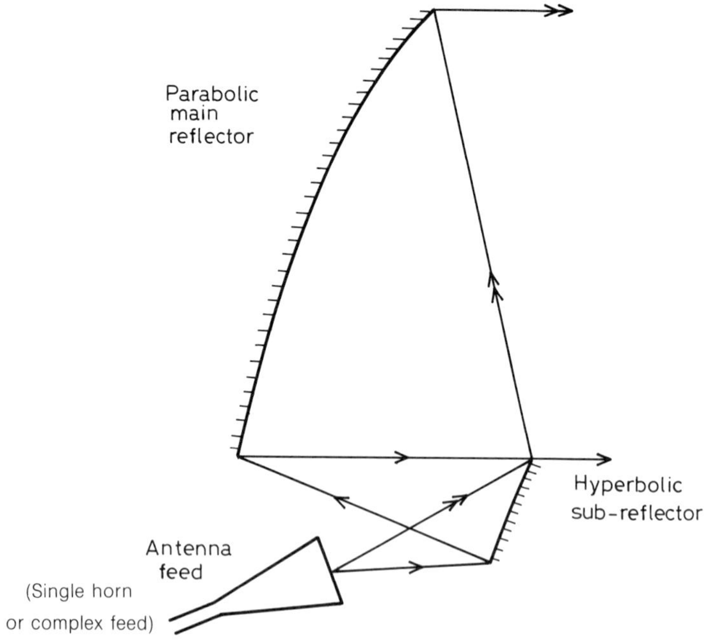

Fig. 2.21 Offset Cassegrain antenna.

circular spot beam:

$$g(\alpha) = g_{\text{MAX}} \cdot \exp\{-k(\alpha/\alpha_0)^2\}$$

where α_0 is the 3 dB half-beamwidth, $k = \log 2/\log \mathrm{e} = 0.693$, and $g_{\text{MAX}} = C/\alpha_0^2$ (see Equation (2.5)). Taking the derivative of $g(\alpha)$ versus α_0 and making $\mathrm{d}g(\alpha)/\mathrm{d}\alpha_0 = 0$ gives the optimum $\alpha = \alpha_0/\sqrt{k} = 1.2\alpha_0$ and $N = 10 \log \mathrm{e} = 4.34$ dB.

Multiple spot beam antennas If a reflector (or a dual-reflector) antenna is illuminated by several feeds properly located in the focal plane—near the focus—each feed port will be connected to a separate, independent, spot beam. However, it should be noted that the out-of-focus beams are deteriorated by a 'coma aberration' which depends on the ratio f/D (f = focal length, D = antenna diameter) and, of course of the offset angle (the angle between the focus and the feed directions, as seen from the reflector's apex). The larger is f/D, the larger the acceptable offset angle. Also the beams cannot be too contiguous, which means also that there is a minimum distance to be kept between the feed apertures (in the usual case of horn feeds, this minimum distance is imposed by the horn aperture physical dimension): in fact, the aperture centres (phase centres) of two feeds should be distant by an angle (as seen from the reflector's apex) at least equal—in radians—to λ/D (λ is the wavelength), for a proper isolation between the two beams (of course, isolation is automatically obtained if the beams are operated at different frequencies or in orthogonal polarizations).

Shaped beam antennas It is now assumed that the multiple beams, as described above, are no more independent but are fed in common, thus forming a single complex beam which results from the in-phase addition of the component RF fields (note that, in this case, the component beams can be contiguous): the single complex beam formed by this method is called a shaped beam which can be tailored to a specified coverage of the earth (see examples in Fig. 2.22). The

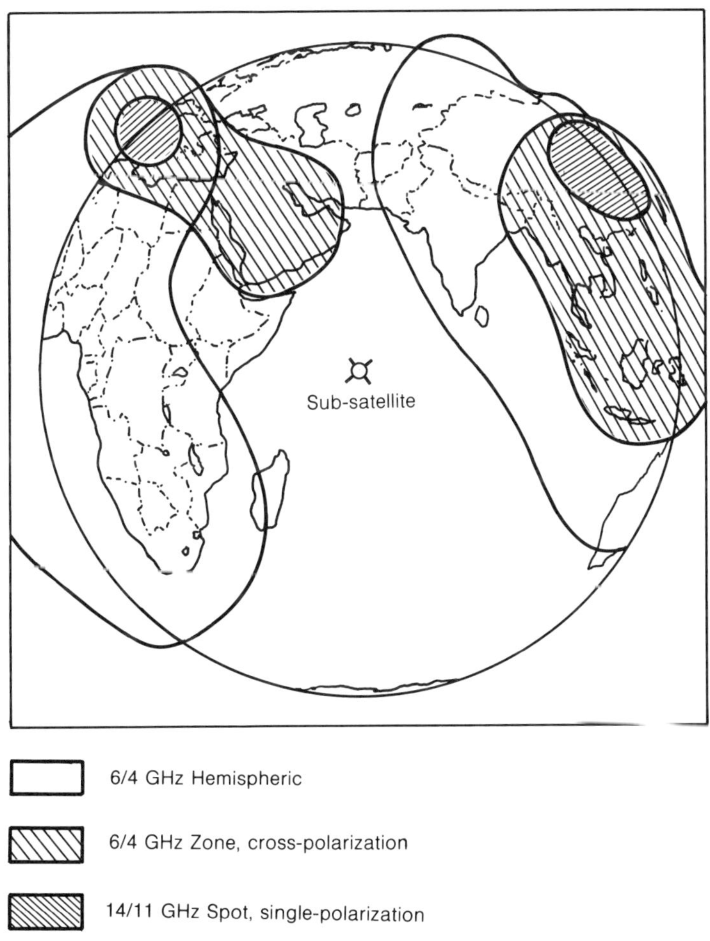

Fig. 2.22 INTELSAT V satellite: typical coverages. Satellite location: 61.4 °E. The 14/11 GHz spot beams are steerable and may be moved to meet traffic requirements as they develop. A global beam coverage is also provided.

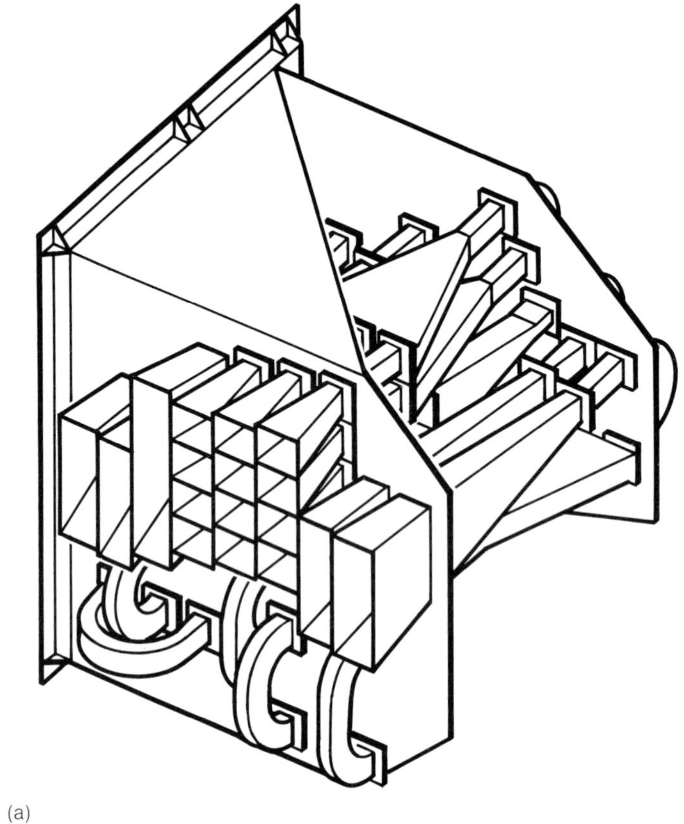

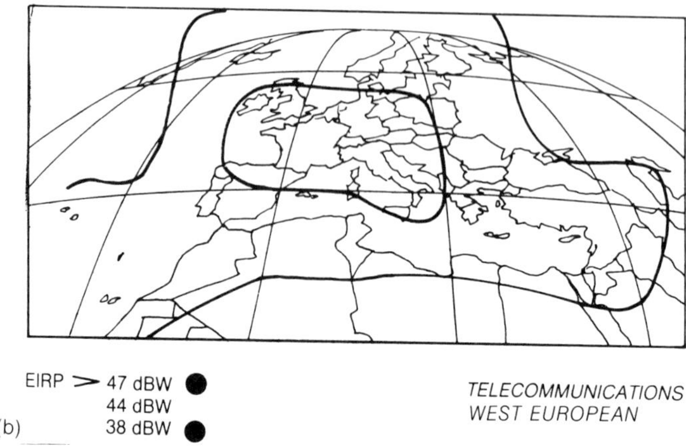

Fig. 2.23 (a) An example of a complex satellite antenna feed with (b) the corresponding coverage (EUTELSAT II satellite antenna).

in-phase addition of the RF fields is performed by connecting the component primary sources to the transponder input port through a beam forming network, which is a passive power divider (or combiner if considered at reception) with phase shifters for in-phase tuning. It is even possible to control the amplitudes and phases of the beam forming network from the TTC earth stations for changing the earth coverage on request. Altogether, the beam forming network and the component primary sources can be seen as constituting a single complex antenna feed. From this concept come the following remarks.

The field on the radiating aperture of the complex feed (feed illumination) is to be modelled as a 'filtered image' of the coverage of the earth. This should be self-evident in terms of pure geometrical optics. However, physical optics (i.e. diffraction laws) show that, due to its limited dimensions (compared to the wavelength), the antenna reflector acts on the antenna diagram as a low-pass filter with a cut-off frequency λ/D: the larger the reflector, the more precise can be the beam shaping (the slope of the diagram at the beam edge—or roll-off—is determined by the width of the beam components at the edge, i.e. λ/D).

The 'imaging transfer' between the complex feed illumination and the antenna radiation diagram (or the earth's coverage) can also be demonstrated by Fourier transformation, resulting from the fact this transfer results from two successive Fourier transform operations. The first converts the feed illumination into the feed radiation diagram (or reflector illumination) and the second converts this reflector illumination into the antenna diagram.

Of course, a precise beam shaping also requires many component primary sources: the number of sources in a given plane should be approximately $\theta \cdot D/\lambda$, θ (in radian) being the angular width of the coverage in this plane. For example, the INTELSAT V satellite hemizonal 6/4 GHz antenna comprises 88 component sources (horns) feeding a 2.44 m reflector (INTELSAT VI: 147 sources with a 3.2 m reflector). Using the specified earth coverage as an input, computer programs are available to perform the design of the complex antenna feed. Such programs can even be coupled with other programs performing mechanical analysis, thermoelastic analysis etc. in order to allow the complete radioelectrical and mechanical antenna system design. Figure 2.23 shows a shaped beam antenna with a complex, multi-horn feed and summarizes the calculation results.

Antenna feeds, polarizers As explained above, most satellites use microwave horns as a feed or as component primary sources. In the case of circularly polarized beams, a polarizer (often a circular waveguide with a quarter-wavelength plate) is connected to each horn. However, other types of sources can be used, such as dielectric radiators or helices (which directly radiate circularly polarized waves). Also, an antenna can be operated in linear polarization, the linear field being converted to circular polarization by an external polarizing plate as explained below.

Polarization (and frequency) sensitive surfaces Surfaces which selectively perform transmission, reflection and/or polarization conversion of microwave radiation, depending on the polarization, and, possibly on the frequency of the incident plane wave, find many utilizations in satellite antennas. These surfaces can be made of grids, viz. regular arrays of metallic wires or stripes (width longitudinal to the propagation direction)—often embedded in a dielectric foam material—or of a lattice of waveguide cells.

If the metallic elements (separated by an interval e), are set parallel to an incident linear polarization, they act as reactive elements and, for $e < \lambda/2$, they can totally reflect this polarization. In consequence, such a surface can be used to transmit and 'filter' a linearly polarized wave or to reflect it. This selective function can be used, either for separating two orthogonally polarized waves in a common frequency band, or for separating waves at various frequencies. Frequency sensitive surfaces are called dichroic surfaces: for example, the surface acts as a reflector at lower frequencies, e.g. 4 GHz, and is transparent at higher frequencies, e.g. 11 GHz).

An example of orthogonal polarization separation in the same frequency band is given in Fig. 2.24: here two different reflectors composed by two orthogonally polarized grids are superposed on a single mount and are illuminated by two nearby located, orthogonally polarized (linear polarizations) feeds. The front reflector reflects one polarization (e.g., horizontal) and is transparent for the other polarization (e.g., vertical) while the rear one reflects this latter polarization. Two different beams, with two different directions are formed in this way.

Polarization sensitive surfaces, with the metallic elements oriented with a 45° angle versus the incident linear polarization, are also used for converting pol-

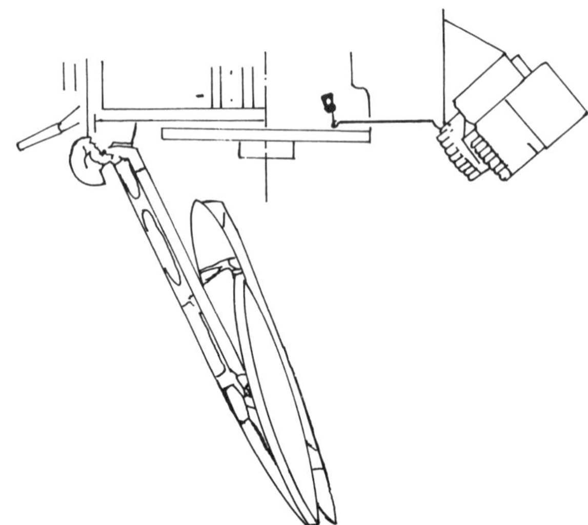

Fig. 2.24 Double-polarization, double-reflector satellite antenna.

arization. For example, a quarter-wavelength plate, converting linear polarization to circular (or conversely) can be designed using the fact that the wavelength into the plate is different for the two polarization components of the wave, parallel and perpendicular to the wires or stripes, therefore introducing a $\pi/2$ phase shift between these components. A reflector with an eighth wavelength thick polarizing surface operates similarly. Also, a reflector with a quarter-wavelength thick polarizing surface is a polarization rotator, which, for example, converts an horizontal polarization into a vertical polarization. This is used in Cassegrain antennas to avoid the aperture blockage caused by the sub-reflector: in such an application the sub-reflector reflects the primary source waves but is transparent to the secondary waves after their polarization rotation on the main reflector.

The wideband receiver

The low-noise amplifier currently uses field-effect transistors (FETs) in the 6 GHz band (typical noise figure $NF = 2$ dB). At higher frequencies, the new high electron mobility transistors (HEMT) are now available and offer good performance ($NF \approx 3$ dB) up to the 14 GHz band ($NF \approx 4$ dB) and even up to the 30 GHz band ($NF \approx 6$ to 7 dB).

The down-converter is designed for conversion losses around 5 dB and high rejection of unwanted frequencies (harmonics etc.). The local oscillator (power ≈ 0.1 W) uses a crystal controlled oscillator followed by multipliers, or a microwave dielectric cavity oscillator with crystal reference control. It must feature good long term temperature frequency stability ($\approx 10^{-6}$) and low phase noise (e.g. less than 60 dB below the carrier at 100 Hz from the carrier frequency).

Note that the above description refers to single conversion operation. Double conversions may be needed in the case of very large bandwidth systems or whenever the local oscillator frequency or its harmonics fall into the useful bandwidth. In the case of double conversion, the wideband receiver down-converter provides a first intermediary frequency and second converters are provided in the channelized sub-system. The wideband receiver—with a typical overall gain of 50 to 60 dB—must be operated in a linear region in order to minimize intermodulation.

Note that the various components used in the receiver must not only feature high performance, but also, just like all satellite components, be specially selected for compliance with 'space qualification' specifications.

The input demultiplexers and the output multiplexers

Each receiving antenna is connected, through a wideband receiver, to one input demultiplexer (IMUX) for channelization. Figure 2.25 shows a quite common IMUX arrangement: here, since losses do not really pose problems, the utilization of an input 3 dB coupler and of cascaded circulators provides a simple solution

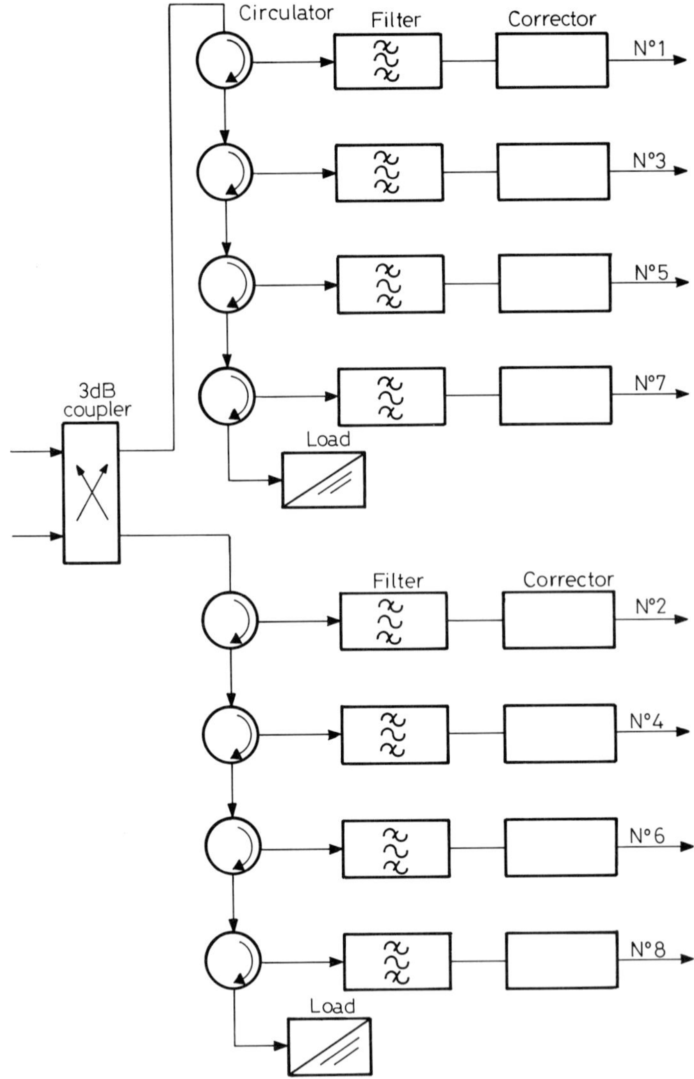

Fig. 2.25 Satellite input demultiplexer (IMUX).

with good VSWRs and channel isolations. Furthermore, the 3 dB coupler permits, on the one hand, to connect two wideband receivers to its input ports for 1 + 1 redundancy and, on the other hand, to 'interleave' the channel splitting, i.e. to separate channels No. 1, 3, 5, etc. (numbering by increasing rank of centre frequencies) through the filters connected to one of the two coupler output ports and channels No. 2, 4, 6, etc. through the filters connected to the other port, thus lessening the filter slope performance requirements.

Each transmitting antenna is connected to a group of channels through an output multiplexer (OMUX). The OMUXs are more bulky and more difficult to design than the IMUXs: this is because, here, losses must be kept low to avoid thermal problems and significant radiated power reductions. This is why, in OMUXs, filters are generally mounted as lateral arms on a 'waveguide manifold' so as to successively pass through or short-circuit the input carriers incoming from the power amplifiers of the connected channels. Fig. 2.26(a) is a schematic diagram of an half-OMUX: here again, two manifolds are used, one connecting the odd-numbered channels, the other connecting the even channels (not represented). The two manifolds are connected to the same antenna through a matched (lossless) T junction. Figure 2.26(c) shows the actual realization (INTELSAT VI 4 GHz band OMUX) and Fig. 2.26(b) the selectivity measurements through the various channels.

IMUX and OMUX microwave filters are to comply with severe specifications as concerns their amplitude and group delay frequency responses (filter masks) in order to avoid multiple paths in the MUX (out-of-band steep slopes) and signal distortions (in-band flat response): these specifications are met by implementing multiple pole filters (often in association with amplitude- and phase-equalizers). The frequency responses are of the minimum phase variations types: Chebyshev (specified level of out-of-band 'sidelobes' or of in-band amplitude oscillations) and elliptic (well-defined transmission zeros). As concerns technological realization, dual-mode cavities are widely used. In modern designs, size and weight are reduced by inserting dielectric resonators inside the cavities. The need for simultaneous low weight, mechanical stability and low thermal expansion often implies the utilization of exotic materials, such as thin invar or carbon fibres (with silver or gold metallization). Special care must be applied to the surface finishes in order to avoid 'multipactor' effects (high voltage breakdown in space vacuum).

The power amplifiers

Each channel is equipped with its own power amplification chain with a gain of about 55 dB or even more. The main components of this chain are:

1. an attenuator, controlled from the TTC earth stations, for remotely tuning the overall gain;
2. a driver amplifier (solid-state);
3. the power amplifier, which is the transponder output stage. Although this is often a travelling-wave tube amplifier (TWTA), solid-state power amplifiers (SSPAs) are now available with output levels up to about 30 to 50 W at 4 GHz and 10 W at 11/12 GHz. SSPAs tend more and more to be preferred to TWTAs because of to their lower total weight and their better linearity. Of course, tubes remain needed for the higher frequencies (e.g., 20 GHz) and for very high power (e.g. up to 200 W for 12 GHz direct broadcasting satellites).

e1
e3
e5
e7
e9
Rg=1
K'1
K'3
K'5
K'7
K'9
1
3
5
7
9
K1
K3
K5
K7
K9
Rc=1
output
short-circuit
L9
L7
L5
L3
L1

(a)

0dB
-50
-60
-25
-30
25
30

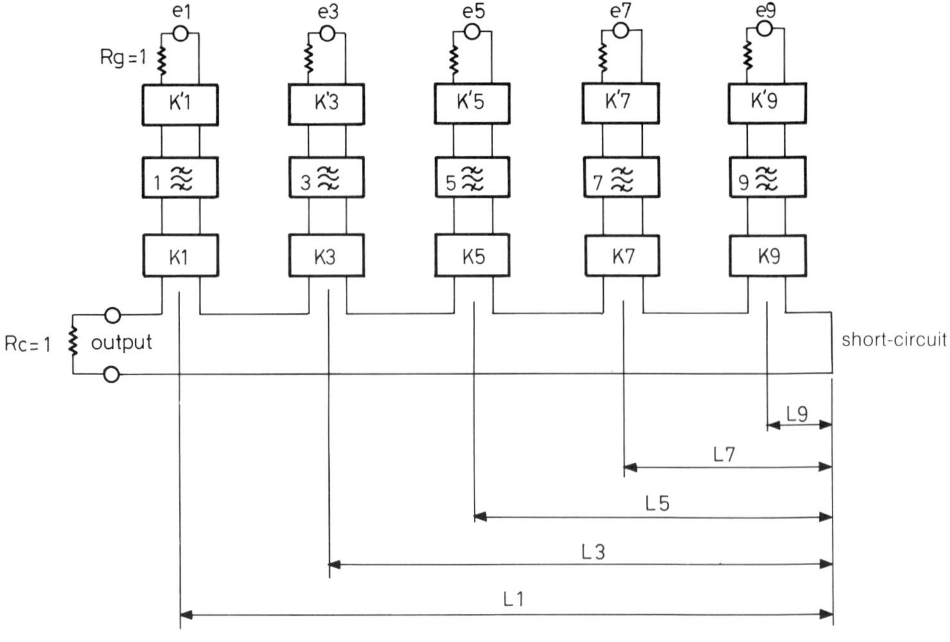

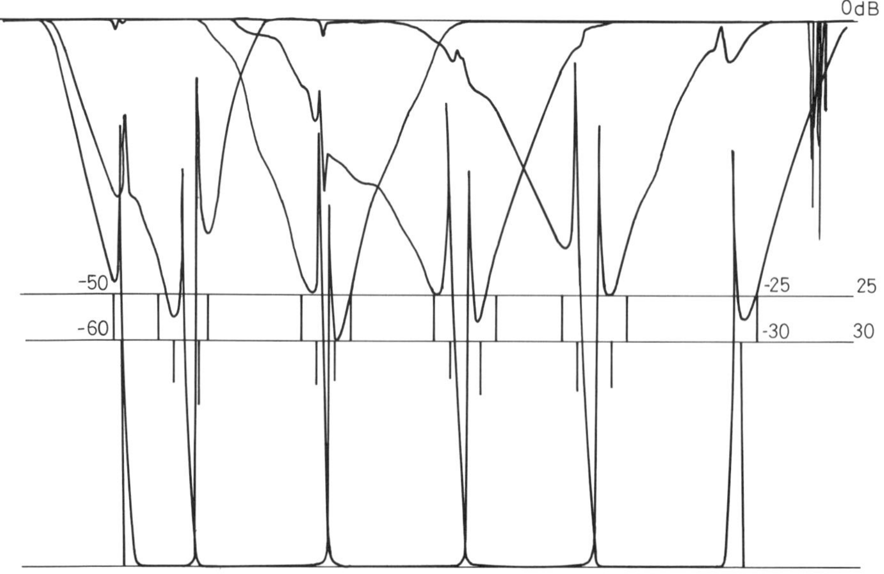

(b)

(c)

Fig. 2.26 (a) Block diagram of the half-OMUX: *ei*, transponder power amplifiers; K*i*, *i*, K′*i*, couplers/matching devices and microwave cavity filters; Li, equivalent to $(2k + 1)$ quarter wavelengths at frequency F*i*; Rg, Rc, matched impedances at input and output. (b) measured selectivity of the half-OMUX (attenuation vs. frequency). (c) A typical satellite output multiplexer: half-OMUX–odd-numbered channels (INTELSAT VI, 4 GHz).

As already explained, the most important characteristic of the power amplifier is its linearity. Problems caused by intermodulation (see sections 2.2.3 and 2.2.4) can be improved by inserting a linearizer at the power amplifier input: this is a pre-distortion device which (partially) compensates for the power amplifier non-linearities, thus minimizing intermodulation (in multiple carrier—i.e. FDMA—operation) and AM–PM conversion (specially in digital transmission*). A typical example is shown on Fig. 2.27.

Active payloads

Up to now, only transparent satellite payloads have been considered and, in fact nearly all existing communications satellite operate in this way. However, in the

*AM–PM conversion are phase variations (phase modulation) which appear in amplifiers near saturation in the presence of amplitude variations (amplitude modulation). In digital transmission, this causes spectral distortions of the signal. In particular, even for well filtered input signals, out-of-band signals re-appear at the power amplifier output ('sidelobe regrowth', see also section 2.4.5)

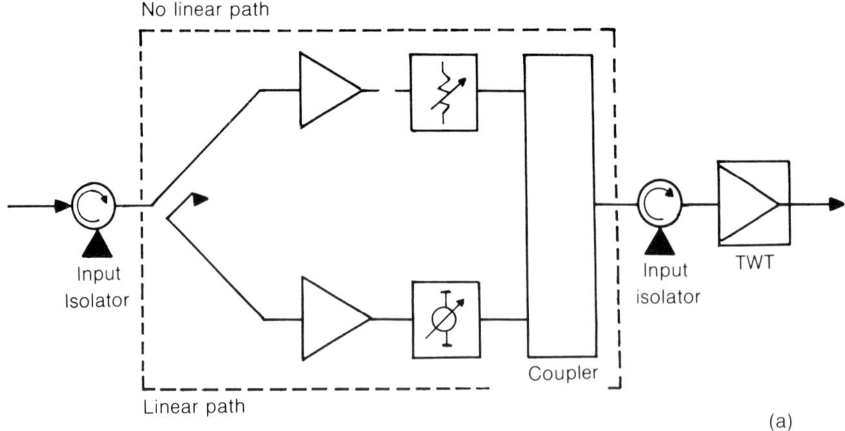

future, much more sophisticated systems, using active payloads, should be implemented, with main applications in digital satellite communications networks.

Dynamic switching matrices The microwave dynamic switching matrix already implemented in the INTELSAT VI satellites will be briefly described as an illustrative example: this switch operates in connection with time division multiple access (TDMA), thus providing an on-board time-domain switched TDMA system (SS/TDMA). This system allows dynamic interconnection, in millisecond times, of any two beams among six different satellite beams. The operation is controlled by an on-board control unit with a programmable memory clocked by a stable time reference. A given earth station, located in a given beam footprint, synchronizes its transmission by reference to the on-board switching sequence, in order that its information bursts reach the satellite at the precise times—inside the TDMA frames—when its own beam is actually connected to the beam serving its correspondent's earth station.

Regenerative transponders On-board regeneration consists of demodulating the up-link carriers in the satellite receiver and remodulating the down-link carriers by the recovered baseband signals before their transmission through the power amplifiers and D/L antennas. In itself such an operation presents several advantages due to the separation of the up- and down-links and to the signal regeneration before remodulation: the overall link bit error rate (BER) is no longer fixed by the total (E_B/N_0) (section 2.2.5) but becomes the sum of U/L and D/L BERs, non-linear distortions are reduced, the transponders can be operated at saturation etc., the net result being an improvement of the link budget and therefore, a possible reduction of satellite and/or earth station parameters (e.i.r.p., antenna size).

However, much more significant are the various possibilities offered by on-board switching and/or processing the digital baseband signals.

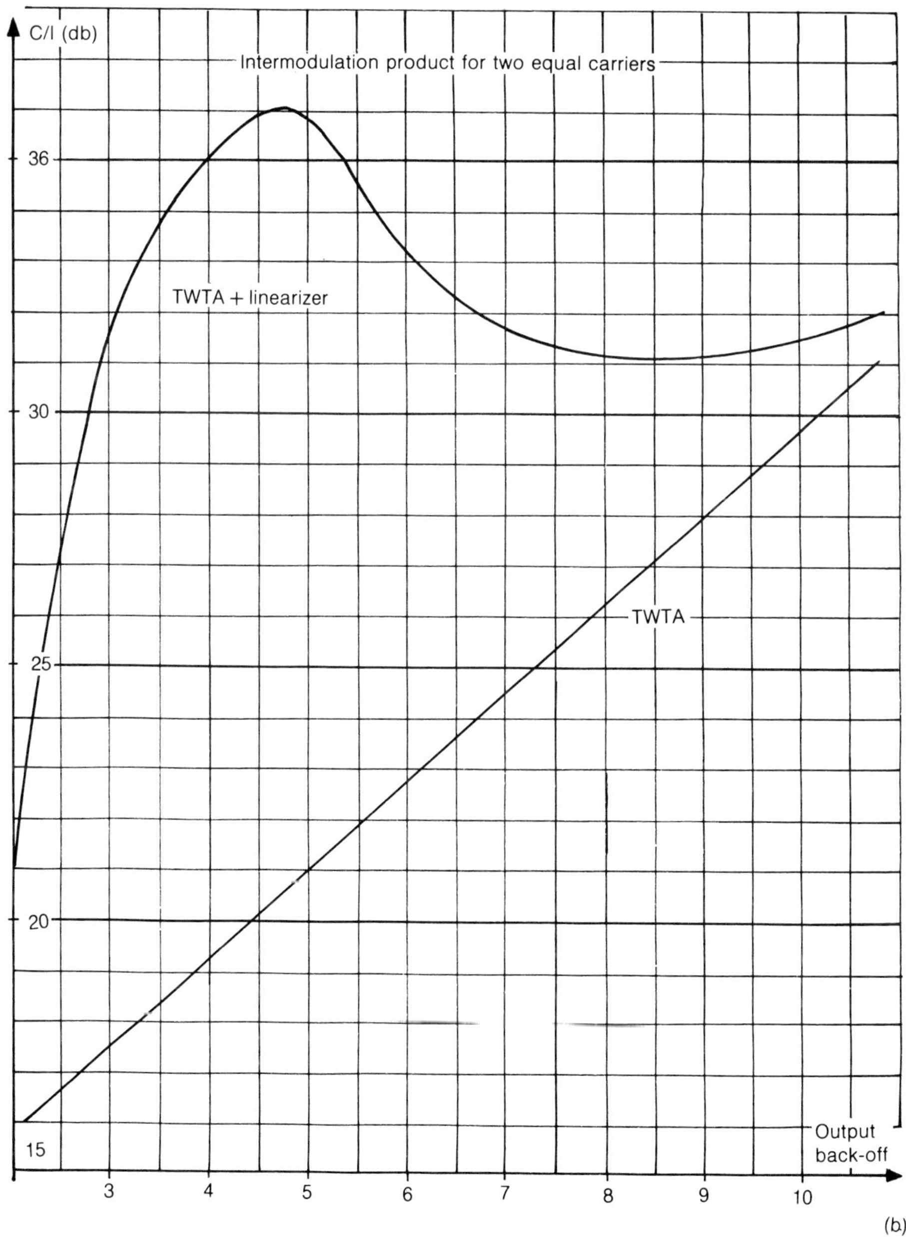

Fig. 2.27 Example of a linearizer: (a) block diagram; (b) typical results.

High speed baseband switching matrices Such matrices can be constructed which are much simpler and more efficient than microwave matrices (as concerns size, power consumption etc.).

Bit stream processing Should allow adaptation of the multiple access schemes as a function of the type(s) of traffic involved in the communications system and, consequently, to minimize the earth station costs. This can be done by storing, in on-board memories, the signals incoming from the various U/Ls (possibly at different bit rates) and by multiplexing and modulating them for transmission to the various D/Ls with the appropriate digital format conversion (store and forward methods). A promising example is to use various FDMA/TDM/PSK (continuous wave) for the U/L carriers and TDMA/TDM/PSK (burst mode) for a common D/L carrier. Such a scheme should minimize the earth stations e.i.r.p. requirements and optimize the satellite bandwidth and power utilization.

Forward-error-correction (FEC) This can be implemented on the remodulated bit streams in order to improve the D/L budgets or to compensate for D/L losses, in the higher frequency bands, during rain periods.

In conclusion, although the best-adapted, most cost-effective fields of application of these new satellite systems are still under discussion, as is the problem of their overall reliability, much effort is currently being made with these advanced techniques, both in R and D (wideband multi-carrier demodulators, highly integrated on-board processing circuitry, synchronization, demand assignment and network control methods, system studies, etc.) and in the practical development and realization of pre-operational satellites (Italian ITALSAT, US NASAs ACTS, etc.).

Inter-satellite links (ISL)

Inter-satellite links between geostationary satellites should be introduced in future global satellite communications systems, due to their attractive features. ISLs will allow direct interconnection (i.e. by avoiding a double, or even a triple hop) of earth stations covered by different satellites. For example, remote earth stations using satellite No. 1 in a domestic system could access directly to international service via an ISL connecting this satellite to satellite No. 2. More generally, ISLs can be seen as a means for extending the maximum coverage of a single satellite (40% of the earth's surface).

ISLs can increase the flexibility in positioning the satellites on the geostationary orbit with, *inter alia*, the following benefits: possible interference reduction, increase of the earth stations elevation angle (especially for very high frequency—e.g. 30/20 GHz—operation where atmospheric losses must be minimized).

Two types of ISLs are currently envisaged: microwave and optical links.

Microwave ISLs These should use the ITU's allocated frequency bands (section 2.1.3), which, for this type of service, are: 22.55 to 23.55 GHz, 32 to 33 GHz, 54.2 to 58.2 GHz and 59 to 64 GHz. Tracking between two satellites for operating the microwave ISL proves to be rather simple as well as the acquisition process which could be provided by command from an earth station. However microwave ISLs appear to suffer from a necessarily bulky and power consuming on-board implementation (e.g. an antenna up to 2 m diameter and several hundreds watts RF power).

Optical ISLs However, these appear attractive due to the very high gain provided even by small antennas. A consequence is also the low laser diode power required (e.g. less than 1 W). The drawback is the extreme narrowness of the beam (e.g. less than 1 md° for a 10 cm mirror at 1 μm wavelength) and therefore, the difficult problems posed by acquisition, then by tracking the beam direction between the two satellites. Another advantage is the large available bandwidth for signal transmission, but the simplest modulation method (on-off keying of the laser diode) implies on-board regeneration. Another problem to be mentioned is the operation in the presence of sunshine.

2.4 EARTH STATIONS

2.4.1 General

'Earth station' is a general term, covering all types of a satellite communications terminal located on the earth's surface. This extends from the high traffic international centres, which comprise one or several big antenna(s) (e.g. 32 m to 15 m diameter), multiple very high power amplifiers (e.g. 3 kW RF power) and a large complex of communication and control equipment, down to the domestic television receive only stations (the TVROs with 3 m to 0.5 m antennas) or to the small data communication terminals (the VSATs (very small aperture terminals)) which comprise of only an antenna (e.g. 1.5 m to 0.6 m) and a PC-size equipment rack directly connected to a user data workstation or computer. An earth station is generally installed on land (fixed location, but possibly mobile, e.g. on a truck), but it can be also installed on a ship (ship earth station (SES)) or even on an aircraft for the—future—aeronautical service. Table 2.3 summarizes some important types of earth stations.

A brief description of an earth station has been already given in section 2.1.2. A general, very simplified, block-diagram is shown in Fig. 2.28. In this section, the main earth station microwave subsystems will be described, viz.:

1. the antenna system, including the associated equipment: duplexer, polarizer, tracking equipment, etc.;
2. the RF receiver, i.e. the low noise amplifier (LNA);
3. the RF transmitter, i.e. the high power amplifier (HPA);
4. the RF/IF converters, i.e. the up- and down-converters (U/C, D/C).

Table 2.3 Typical earth stations under current operation

Type	*Band*	*#Applications* *@ Analogue/Digital* *◆ Multiple access/ multiplexing/modulation*	*Main performance specifications*	*Construction and equipment*
INTELSAT Standard A[a]	C	#International high traffic: *Telephony and data:* @Analogue: ◆FDMA/FDM/FM @Digital: ◆SCPC/QPSK ◆FDMA/TDM/QPSK +FEC (IDR) ◆TDMA/TDM/QPSK (120 Mbit/s) *Television:* @Analogue: ◆FDMA/FM	#(G/T)O $\geqslant 35\,\mathrm{dB \cdot K^{-1}}$* #$G_R \approx 54.8$ dBi #$T_R < 50$ K #$G_T \approx 57.3$ dBi #Dual polarization frequency reuse (RHCP + LHCP) ar $\leqslant 1.06$* #Tracking accuracy $\approx 0.03°$ #E.I.R.P. (dBw): ◆SCPC ≈ 61 ◆FDM: From ≈ 66 (24 channels to $\approx$ 85 (972 ch.) ◆IDR ≈ 55 to 85 ◆TDMA ≈ 85 ◆TV ≈ 85	#Antenna diam > 15 m #Dual-reflector: shaped Cassegrain #High perf. 4-port duplexer-polarizer #Automatic tracking: monopulse or step #Ant. pedestal: AZEL #LNA: uncooled parametric or cooled FET #HPA: TWT or klystron (e.g. 3 kW) #Multiple U/Cs, D/Cs #Communication equipment: FM or QPSK (IDR) modems, TDMA terminal
INTELSAT Standard B[b]	C	#International medium capacity traffic: *Telephony and data:* @Analogue: ◆FDMA/CFDM/FM @Digital: ◆SCPC/QPSK ◆FDMA/TDM/QPSK +FEC (IDR)	#(G/T) O $\geqslant 31.7\,\mathrm{dB \cdot K^{-1}}$* #$G_R \approx 51.6$ dBi #$T_R < 55$ K #$G_T \approx 54.6$ dBi #Single polarizations: Tx:LHCP or RHCP.Rx:RHCP or LHCP ar $\leqslant 1.06$* #Tracking accuracy $\approx 0.045°$ #E.I.R.P. (dBw): similar to Standard A[c]	#Antenna diam ≈ 11 m #Dual-reflector: shaped Cassegrain #High perf. 2-port duplexer-polarizer #Auto. Tracking: step tracking #Ant. Pedestal: AZEL #LNA: uncooled parametric or colled FET

		Television: @Analogue: ◆ FDMA/FM		#HPA: TWT or klystron (e.g. 3 kW) #Multiple U/Cs, D/Cs #Communication equipment: FM or QPSK (IDR) modems
INTELSAT Standard C and EUTELSAT main telephony/ TV stations	Ku	#International high traffic: *Telephony and data*: INTELSAT: @Analogue: FDMA/FDM/FM @Digital: IDR EUTELSAT: @Digital: ◆ TDMA/TDM/QPSK (120 Mbit/s) *Television*: @Analogue: ◆ FDMA/FM	#(G/T) O $\geqslant 37 + X$ dB·K^{-1}* #$G_R \approx 60.3$ dBi #$T_R < 150$ K #$G_T \approx 62.3$ dBi #Single polarizations: linear orientable (orthogonal T_x/R_x) #Tracking accuracy $\approx 0.03°$ #E.I.R.P. (dBw): similar to Std A[c]	#Antenna diam > 11 m #Dual-reflector: shaped Cassegrain (often BWG feed) #Auto. tracking: monopulse or step #Ant. pedestal: AZEL #LNA: cooled FET (HEMT) #HPA: mostly TWT (e.g. 2 kW), klystron #Multiple U/Cs, D/Cs #Communication equipment: FM or IDR modems or TDMA
INTELSAT Standard D ('VISTA' service)	C	#Low density telephony traffic (rural communications) *Telephony*: @Analogue: ◆ SCPC/FM with companding	Standard D1 (small remote stations) #(G/T) O $\geqslant 22.7$ dB·K^{-1}* #Single polarizations: Tx:LHCP or RHCP, RX:RHCP or LHCP ar $\leqslant 1.3$* #E.I.R.P. (dBw): 51.7 (RX:D2) 55.6 (RX:D1)[c]	#Antenna diam ≈ 4.5 m #Dual-reflector: Cassegrain #2-port duplexer-polarizer #Fixed-pointed #LNA: Uncooled FET #HPA: SSPA (e.g. 10w) #Single U/C, D/C #Communication terminal: SCPC modems (Channel units)

Table 2.3 (*Continued*)

Type	*Band*	*#Applications* *@ Analogue/Digital* *◆ Multiple access/ multiplexing/modulation*	*Main performance specifications*	*Construction and equipment*
			Standard D2 (central and main stations) #Antenna System, LNA, HPA: performances and equipment similar to Standard B (except ar $\leqslant$ 1.09) #Communication equipment: SCPC channel units	
INTELSAT Standard E ('IBS' service)	Ku	#Business communications @ Digital: ◆ FDMA/TDM/QPSK + FEC (IBS)	Standard E1 (small, on premises stations) #(G/T) O $\geqslant 25\,\mathrm{dB\cdot K^{-1}}$* #Single polarizations: linear orientable (orthogonal T_x/R_x) #E.I.R.P. (dBw) $\approx$ 51 (RX:E2) per 64 kbit/s[c]	#Antenna diam $\approx 3.5\,\mathrm{m}$ #Dual-reflector Cassegrain or front fed parabolic #Fixed-pointed (weekly correction) #LNA: uncooled FET #HPA: SSPA (e.g. 5 W) #Communication terminal: QPSK (IBS) modems
			Standard E2 (medium size, 'teleport' stations) Performance and equipment similar to Standard E1 stations except: #(G/T) O $\geqslant 29\,\mathrm{dB\cdot K^{-1}}$ (antenna $\approx 5.5\,\mathrm{m}$) #HPA: SSPA or TWT	

			Standard E2 (main stations) Performance and equipment similar to Standard E2 stations except: #(G/T) O $\geqslant 34\,dB \cdot K^{-1}$ (antenna $\geqslant 8$ m) #Auto.tracking #HPA: TWT
INTELSAT Standard F	C	#Business communications	Standard F stations are similar to Standard E but operate in C-band: #F1: (G/T) O $\geqslant 22.7\,dB \cdot K^{-1}$ (antenna ≈ 4.5 m) #F2: (G/T) O $\geqslant 27\,dB \cdot K^{-1}$ (antenna ≈ 7.5 m) #F3: (G/T) O $\geqslant 29\,dB \cdot K^{-1}$ (antenna ≈ 9 m)
Typical stations for VSAT systems	Ku or C	#Business communications between a central station ('hub') and remote microstations (VSATs) @ Digital: Hub to VSAT *links*: ◆ FDMA + CDMA/TDM/BPSK(+FEC) or: ◆ FDMA/TDM/BPSK (+FEC) (typically $\leqslant 256$ kbit/s) *VSAT to hub links*: ◆ FDMA + CDMA/TDM/BPSK(+FEC) or: ◆ FDMA/TDMA/TDM/BPSK (+EFC) (typically $\leqslant 64$ kbit/s)	Central station ('hub'): similar to standard E2/F2 stations Microstations ('VSATs'): #'Outdoor unit': • Antenna (typically 1 m to 2 m): front fed parabolic • LNA: uncooled FET } integrated in antenna feed • SSPA: 3 to 10 W } integrated in antenna feed • U/C, D/C #'Indoor unit': • Demodulator (+FEC decoder) • Modulator (+FEC encoder) • Signal processor • Data interface unit } Personal computer-size unit

Table 2.3 (*Continued*)

Type	*Band*	*#Applications* *@ Analogue/Digital* *◆ Multiple access/ multiplexing/modulation*	*Main performance specifications*	*Construction and equipment*
INMARSAT Standard A ship earth station	L	#Public telecommunications with ships: *Call request & telex*: @ Digital: ◆ Similar to VSAT systems *Telephony*: @ Analogue: ◆ SCPC/FM with companding	#(G/T) O $\geq -4\,dB\cdot K^{-1}$* #Circular polarizations #E.I.R.P.: 37 dBw	#'Above-deck equipment' • Antenna: parabolic (≈ 1.2 m) under a radome • 3-axes stabilized pedestal #'Below-deck equipment' • LNA & RX equipment • HPA & TX equipment • Telephone and telex sets

[a] Previous Standard A (and C) specifications asked for higher performance, viz. $G/T \geq 40.7\,dB\cdot K^{-1}$ ($39\,dB\cdot K^{-1}$ for Standard C). This resulted in the utilization of very large antennas (≈ 32 m for Standard A, 18 m for Standard C) equipped with 'beam waveguide' (BWG) feeds.
[b] Standard B stations are less expensive than Standard As but the 'space segment' cost is higher since they require more RF power from the satellite.
[c] The E.I.R.P. to be transmitted by an earth station is (approximately) inversely proportional to the (G/T) of the receiving earth station (this is indicated in the table e.g. by: (RX:E2), meaning that the receive station is a Standard E2). In digital transmission, the E.I.R.P. is directly proportional to the transmitted bit rate. Note that, in the INTELSAT system, C-band stations (e.g. Standard A or B) can be connected to Ku-band stations (e.g. Standard C), thanks to 'cross-strapping' (frequency translation) in the satellite.
* Indicates specified mandatory performance.
Frequency bands: See Sec. 12.1.3. C-band = 6/4 GHz, Ku-band = 14/12–11 GHz, L-band = 1.6/1.5 GHz.

Abbreviations:

FDMA : frequency division multiple access
TDMA : time division multiple access
CDMA : code division multiple access — See section 2.1.4
SCPC : single channel per carrier
FDM : frequency division multiplexing
CFDM : FDM with voice channel companding
TDM : time division multiplexing — See sections 2.1.5 and 2.2.5
FM : frequency modulation
PSK : digital modulation phase shift keying (BPSK: 2-phase, QPSK: 4-phase)
IDR : IDR (intermediate data rate) is INTELSAT designation INTELSAT for digital transmissions in FDMA/TDM/QPSK (+ FEC: forward error correction) with bit rates between 64 kbit/s and 44 Mbit/s.
IBS : INTELSAT business service for private data transmissions (computer communications, video-conferencing, etc.): FDM/TDM/QPSK (+ FEC) with bit rates between 64 kbit/s and 8 Mbit/s.
G/T : figure of merit at reception. The specification is given as: $G/T = (G/T)_o + 20 \log F/F_0$ ($dB \cdot K^{-1}$) (F frequency, F_0 mid-bandwidth frequency) (X additional down-link attenuation due to local rain statistics).
G_R : antenna gain at reception (down-link band) (dBi)
T_R : receiver noise temperature (K)
G_T : antenna gain at transmission (up-link band) (dBi)
E.I.R.P. : equivalent isotropically radiated power = $10 \log (G_T \times \text{HPA power})$ (dBw)[c]
T_x/R_x : transmission/reception
RHCP/LHCP: right hand/left hand circular polarization
ar : axial ratio (see section 2.3.3, footnote)
AZEL : azimuth-over-elevation antenna pedestal
LNA : low-noise amplifier
FET : field-effect transistor
HPA : high power amplifier (TWT (travelling-wave tube), klystron or SSPA (solid-state power amplifier), i.e. transistorized)
VSAT : (very small aperture terminal) conventional designation of microstations for direct business data communications.

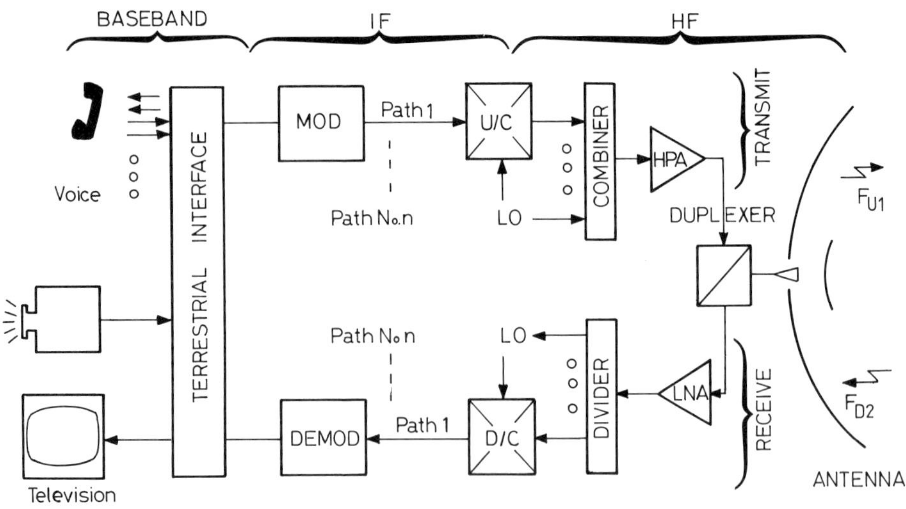

Fig. 2.28 Simplified block diagram of an earth station.

Note that the LNA, the HPA and the U/Cs and D/Cs are generally redundant, i.e. equipped with stand-by units (through automatic switching systems) in order to provide the required reliability.

2.4.2 The antenna system

Radioelectrical design

Definitions and formulae of interest here are given in section 2.2.2. Also, the reader is referred to Volume 2 of this work for the basic antenna principles.

Specific requirements for radioelectrical performances of earth station antenna systems are:

1. wideband operation: performances listed below must be complied with both in the down-link (e.g. 3.625 to 4.2 GHz) and up-link (e.g. 5.850 to 6.425 GHz) bands;
2. large antenna gain, in order to contribute to a high figure of merit (G/T) at reception and to a high equivalent isotropically radiated power (e.i.r.p.) at transmission;
3. minimum antenna noise temperature, also to contribute to a high G/T;
4. low side lobe radiation level, in order to minimize both the sensitivity to received interference signals and the emission of interfering signals (caused by the earth station off-beam e.i.r.p.),

5. high polarization purity, for avoiding reception or transmission of cross-polarized interference signals, especially in the case of dual polarization frequency reuse satellite systems (section 2.1.4).

Table 2.4 below recapitulates these requirements and their consequences on antenna design.

Table 2.4 Earth station antenna radioelectrical performances and design

Performance	*Antenna design*
Large antenna gain	Large reflector antennas High antenna efficiency (60 to 80%), whence: Quasi-uniform aperture illumination Frequent utilization of 'shaped' dual-reflector antennas (e.g. Cassegrain) High precision manufacture (reflector rms accuracy $\sigma \leqslant \lambda/500$ viz. 1 mm for 6 GHz)
Minimum antenna noise temperature*	Low radiation level in ground direction: Dual-reflector antennas Low primary source and sub-reflector 'spill-over' radiation Low losses between primary source and LNA: LNA directly connected to duplexer (whence use of dual-reflector or offset reflector) or use of very low loss waveguides or use of 'beam waveguide' feeds (for very large antennas)
Low side lobes radiation level: $G \leqslant 29-25 \log \phi$ (dBi)†	Low primary source and sub-reflector 'spill-over' radiation High performance primary source, e.g. corrugated horns High precision manufacture Minimum obstacles to radiation (reduction of sub-reflector and support arms mask effect)
High polarization purity‡	Scalar-type primary source, such as corrugated horns In the case of circular polarization: high performance polarizers In the case of dual polarization frequency reuse: high performance four-port duplexer–polarizer

*Refer to sections 2.2.2 and 2.2.3.

†The CCIR stipulates that the gain of 90% of the side lobe peaks should not exceed this limit. ϕ is the off-axis direction. This formula applies to large antennas (D/λ) 150. For more details, see CCIR Recommendation 580.

‡As shown in Table 2.3, two types of RF wave polarizations are commonly used in satellite communication systems: circular polarization (CP) and linear polarization (LP). Often, the up-link (U/L) and down-link (D/L) polarizations are orthogonal, i.e. right hand and left hand circular polarizations (RHCP/LHCP) or two right-angle oriented LPs, e.g. horizontal and vertical. However, the two orthogonal polarizations are simultaneously used, both on U/L and D/L when dual polarization frequency reuse is implemented (section 2.2.3). Polarization purity is defined in section 2.3.2.

Main types of earth station antennas

Nearly all earth stations are reflector antennas with a circular aperture (circular contour). Although a simple paraboloidal reflector can be sufficient for small earth stations such as TVROs, the required performance usually calls for dual-reflector (Cassegrain or Gregory) antennas*. Another major advantage of dual reflector systems is that the feed (primary source) is located near the reflector vertex, therefore allowing easy connection of the equipment, with short waveguide runs. For mechanical construction reasons, large antennas use a rotationally symmetrical geometry. In the case of smaller antennas, an asymetrical design (offset antennas) often gives better performance. Two techniques specific to earth station antennas will be described below: modified reflectors and beam waveguide (BWG) feeds.

Modified reflectors This technique (also called reflector shaping), which is applicable to dual-reflector antennas, is illustrated by Fig. 2.29 for the case of a Cassegrain antenna. In fact, the utilization of two reflectors allows optical stigmatism to be maintained while bringing out a supplementary degree of freedom in the design. Here, the purpose is to make more uniform the antenna aperture illumination (and, consequently to improve aperture efficiency, e.g. from 0.6 to 0.8) by modifying the shape of the subreflector in such a way as to increase the concentration of the rays towards the edge of the main reflector. Such a modified subreflector no longer has the theoretical shape of an hyperboloid. Therefore, and in order to maintain a constant path length (FA + AB + BD) between the phase centre of the primary source (horn) and the aperture, it is also necessary to modify the shape of the main reflector (which no longer has the theoretical shape of a paraboloid). In terms of diffraction, this means an uniform phase on the antenna aperture.

Beam waveguide (BWG) feeds A BWG (Fig. 2.30) is an assembly of focusing elements, usually reflectors, between which near-field radiation is propagated. Such a technique allows remote location of the primary source (horn) of a reflector antenna and to convey (at transmission) the primary radiated field up to the reflector system focus with very low losses, typically 0.1 dB (the same process taking place at reception in the reverse direction). As shown in the figure, thanks to intermediary—plane—mirrors, it is also possible to keep the primary source in a fixed position while rotating the reflector system. This is a very practical arrangement in the case of large, bulky antennas (e.g. $D/\lambda > 250$) because the whole reception and transmission equipment (LNAs and HPAs) can be conveniently housed in a room under the antenna pedestal and directly connected to the primary source (via the duplexer), therefore avoiding losses from rotary joints and waveguides.

*All these terms have been defined for satellite antennas in section 2.3.2.

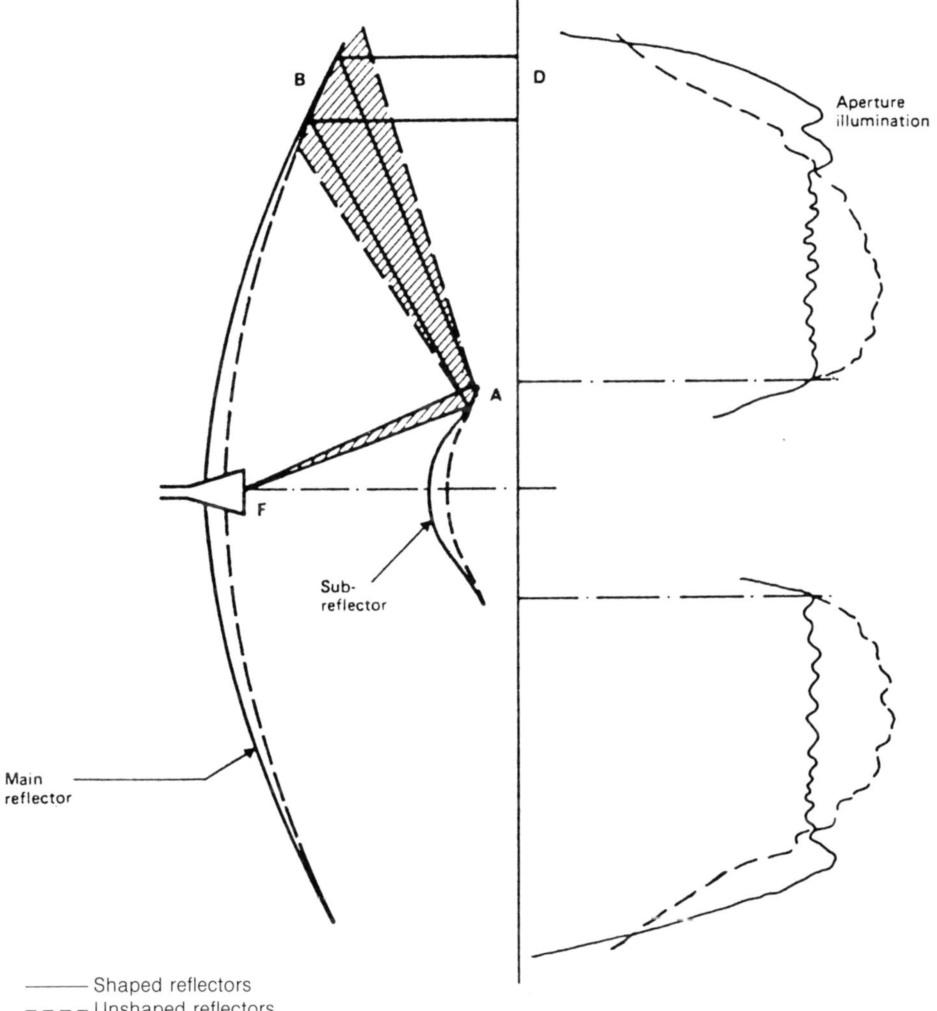

Fig. 2.29 Field distribution of Cassegrain antenna.

Feed systems

The feed system of an earth station antenna (Fig. 2.31) is composed, not only of the primary source (with, possibly, a BWG), but also, possibly, of a tracking mode coupler (TMC, in the case of monopulse tracking (section 2.4.2)) of a duplexer with two ports (transmission T_x and reception R_x) or with four ports (two T_x and two R_x in the case of dual polarization frequency reuse) and of a polarizer system (in the case of circular polarizations).

The most commonly used type of primary source is the corrugated horn (already

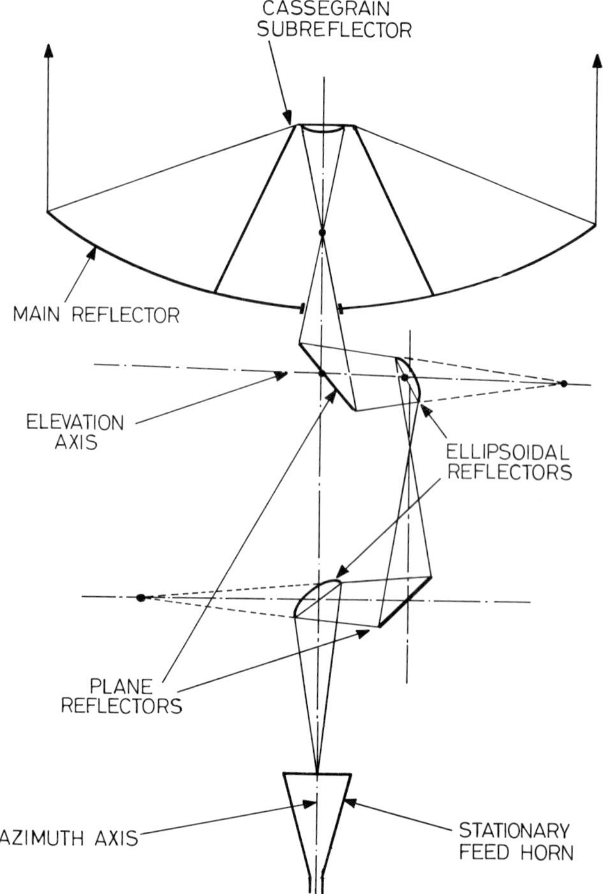

Fig. 2.30 Beam waveguide antenna feed.

cited for satellite antennas): this is a conical horn in which annular grooves are cut in the inner wall. The depth of the cut being $\approx \lambda/4$, the inner wall is equivalent to an impedance $Z = \infty$ for all propagation modes. This results in an optimum, rotationally symmetrical illumination on the horn aperture with a uniformly polarized electric field: actually, the corrugated horn (a special case of the so-called scalar-type sources) provides, in a wide frequency band, an excellent primary radiation pattern for all polarizations.

The tracking mode coupler (TMC) will be described with tracking systems (see below).

The duplexer is composed of orthomode junctions (OMJ) These are usually composed of one circular waveguide and of two, orthogonally oriented, rectangular waveguides. In the simplest systems where transmission (T_x, up-link) and

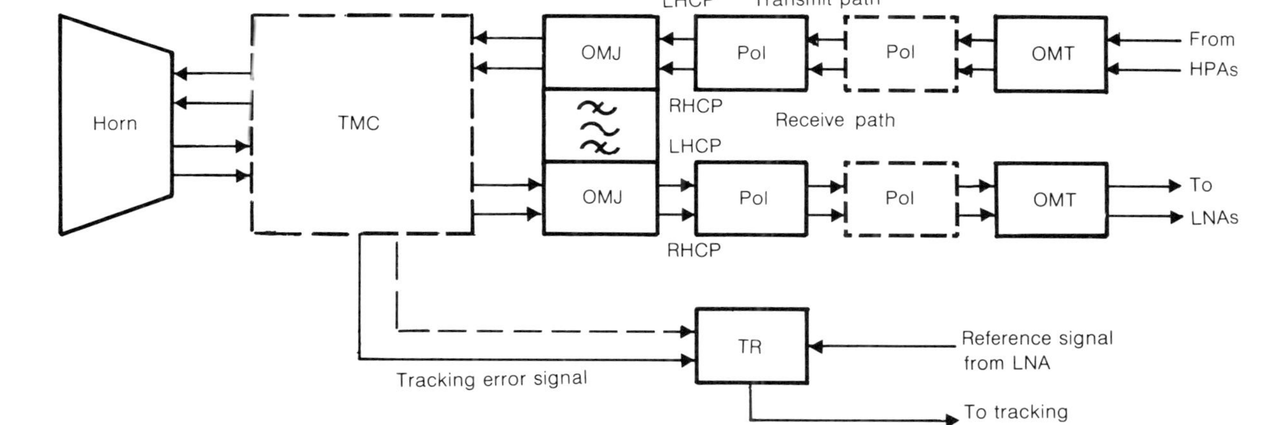

Fig. 2.31 Typical antenna feed block diagram (frequency re-use). TMC, tracking mode coupler; OMJ, orthomode junction; Pol, polarization; OMT, orthomode transducer; TR, tracking receiver, LHCP, left-hand circular polarization; RHCP, right-hand circular polarization.

reception (R_x, down-link) are operated not only at different frequency bands (e.g. 6 and 4 GHz), but also in two orthogonal polarizations, such a junction is sufficient to separate T_x from R_x signals. However, in dual polarization frequency reuse systems, frequency filters combine with the OMJ(s) to separate the four different, orthogonally polarized, T_x and R_x channels.

The polarizer system is usually composed of quarter-wavelength plates (dielectric or metal fin array) located in circular waveguides. The operation of quarter-wavelength plates has already been explained in section 2.3.2 and is illustrated by Fig. 2.32. Fig. 2.32(a) shows how an input linearly polarized (LP) electrical mode (E_{in}) is converted by the plate in two rectangular components (E_{out}), the one parallel to the plate being delayed by $\pi/2$. The sum of these two components is a circularly polarized wave, RHCP or LHCP, depending on the $\pm 45°$ orientation of the plate. Of course the operation is reciprocal, CP being converted to LP at reception. Fig. 2.32(b) similarly shows how a half-wavelength plate with an orientation α changes, thanks to a π delay, the orientation of a LP field by 2α.

Although elliptical polarization (EP, the most general description of an RF electric field) is not normally implemented in satellite communications, various defects can slightly distort CP or LP into EP. The most common defect is caused by propagation through rain, due to the oblate shape of rain drops which induces differential phase and attenuation.

In a dual polarization frequency reuse feed, an EP field (e.g. nearly RHCP) is converted at reception, not only into a (wanted) copolarized component (e.g. RHCP), but also into a small (unwanted) cross-polarized component (e.g. LHCP) which causes interference in the orthogonal channel (e.g. LHCP). Polarization compensation devices have been proposed (although there are very few examples in actual operation). In the 6/4 GHz band, rain causes no significant differential attenuation, so that phase compensation devices, based on quarter-wavelength

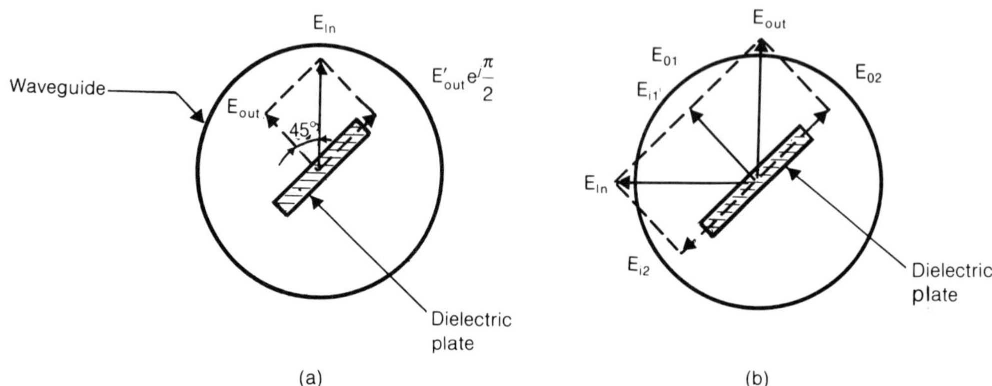

Fig. 2.32 Operation of a polarizer, (a) for circular polarization and (b) for linear polarization.

and half-wavelength plates could be used. At high frequencies, differential attenuation would complicate the compensation.

Mechanical design

The above-described antenna (reflector(s) and feed system) is supported by a pedestal (or antenna mount). In order to provide for orienting the antenna beam in the satellite direction, this pedestal is usually comprised of two orthogonal movable axes, each one equipped with a drive subsystem. Except in the case of small antennas, these subsystems are motorized and servo-controlled by a tracking system (see below).

Figure 2.33 shows a modern INTELSAT Standard A earth station (see first case in Table 2.3). In this 'compact' version, the antenna is built on the top of

Fig. 2.33 An INTELSAT Standard A earth station antenna (Azel mount).

the equipment room. The Cassegrain antenna, with a 16 m main reflector, is mounted on a 'king post', 'AZEL', pedestal which is comprised of a large ball bearing allowing the servo-driven azimuthal rotation and a jackscrew actuator for allowing the required, servo-driven, elevation movement. Other types of mounts are:

1. wheel-and-track (also AZEL, where the whole antenna system rotates on a large circular rail);
2. *X–Y* (with two orthogonal axes for limited steerability around any central *Z* direction, see Fig. 2.34);
3. polar mount (with a 'right ascension axis'—or 'hour axis'—parallel to the earth's axis and a 'declination axis' for allowing a limited movement orthogonally to the geosynchronous orbit);
4. three-axis mount (for stabilized ship earth stations), etc.

Figure 2.35 shows the limiting case of a very small antenna for a 'VSAT' station, where only manual orientation in a fixed direction is required: here, the whole RF equipment (T_x/R_x) is located on the antenna.

The antenna reflector(s) must be constructed with high accuracy and stiffness in order to achieve the specified performance (especially as concerns the antenna gain), even under extreme environmental conditions. The gain loss δG due to

Fig. 2.34 An INTELSAT Standard E earth station antenna (X–Y mount).

residual, uniformly distributed, surface errors, is given by the following formula:

$$\delta G = 10 \log [\exp - (4\pi\sigma/\lambda^2)]$$
$$\approx - 686(\sigma/\lambda)^2 \quad \text{(dB)}$$

where σ is the rms manufacturing tolerance and λ the wavelength.

Tracking systems

Large antennas (in terms of D/λ) must be equipped with a tracking system in order to keep the antenna beam axis pointed towards the satellite, in spite of residual satellite movements, of mechanical loads on the antenna (effects of wind and weight) and, possibly, of atmospheric propagation effects (at higher frequencies).

Fig. 2.35 A VSAT earth station antenna (complete outdoor unit, including RF transmitter and receiver).

The following (approximate) formula gives the maximum diameter D of antennas where no tracking system is required:

$$D/\lambda \leqslant 18^\circ\sqrt{n}/(\mathrm{SSK} + \mathrm{APE})$$

where SSK is the satellite station-keeping (degrees), i.e. the maximum north–south (N/S) and east–west (E/W) satellite movements; APE is the antenna pointing error (degrees), which can be due to factors such as misalignment of mechanical axis and primary source, deformations of the antenna and pedestal by wind, gravity, thermal effects, etc.; n is the acceptable antenna gain loss (dB). λ is the wavelength (m).

For example, SSK $= \pm 0.07^\circ$ (due to 0.05° N/S and 0.05° E/W), APE $= 0.05^\circ$ and $n = 1$ dB results in $D/\lambda \leqslant 150$, e.g. $D \leqslant 7.5$ m at 6 GHz ($\lambda = 5$ cm.) or $D \leqslant 3.1$ m at 14.5 GHz ($\lambda = 2.07$ cm).

Note that this formula can easily be derived from the fact that the shape of a radiation pattern in the vicinity of its axis is approximately parabolic (in dB) (see formula in section 2.3.2). Of course, even in such a case of (small) non-tracking antennas, at least the capability for a manual orientation should be provided.

In the case of medium-size antennas and/or of more significant satellite movements (SSK), a simple solution is program tracking. This consists in actuating the antenna by motors which are controlled by a program permanently giving the actual satellite position. This program can either be derived from local orbit calculations or from data provided by a control station.

Now, in the case of large antennas a complete automatic tracking system is to be provided. This consists of a drive and servo subsystem equipped with gear or jackscrew drive mechanisms and motors. These are fed by electrical currents derived from servo-loops, the error signals of which are generated by a tracking receiver.

By processing a special carrier received from the satellite (the so-called satellite beacon), the tracking receiver generates error signals, i.e. an azimuth signal and an elevation signal which are proportional (with $\pm$ sign) to the angular difference between the satellite direction and the antenna axis (maximum gain direction). Two types of tracking systems are used: monopulse and step-track.

Monopulse is the more sophisticated, but also the more reliable and precise tracking method. It is derived from radar technology and consists of extracting the error signals directly from the antenna feed, through special output ports called the azimuth and elevation 'difference ports'. This is illustrated by Fig. 2.36 which shows a four primary horns feed system (only two horns are shown in this plan cut). The antenna now generates three antenna diagrams, through three ports: the conventional antenna diagram, called the reference diagram (also used for normal telecommunications signals) is available at the 'sum' (Σ) output port of the hybrid coupler (e.g. a magic tee).

The 'difference diagrams' are available at the two 'difference ports' (Δ, one only shown in the figure) of the coupler. In fact, the four horn feed represented in

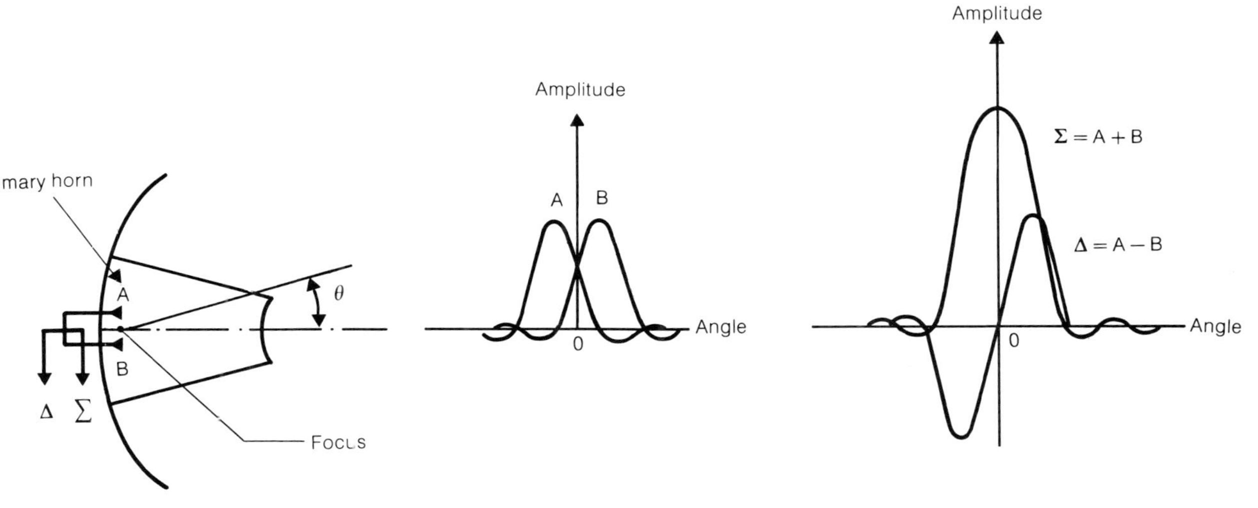

Fig. 2.36 Monopulse tracking system operation: (a) the principle of the multi-horn tracking system; (b) patterns of each horn; (c) superposed pattern.

Fig. 2.36 is never used in modern antenna design because it cannot provide good radiation diagrams and gain. It is replaced by a single multimode horn which propagates not only a fundamental mode (e.g. the TE_{11} circular waveguide mode), but also higher odd modes (e.g. TM_{01}, TE_{01}, etc.). These modes are extracted by a tracking mode coupler (TMC, see Fig. 2.31). The fundamental mode generates the conventional 'reference diagram' and the higher odd modes generate the azimuth and elevation difference diagrams. The corresponding signals are coherently detected by the Σ signal in the tracking receiver (a typical block diagram is shown in Fig. 2.37).

Monopulse systems feature a very high tracking accuracy (e.g. 0.01° with a 32 m Standard A antenna, see formula below), even in the presence of intense wind gusts, to which they are able to react instantaneously. However monopulse feed, TMC and tracking receiver are rather complex, especially when associated with dual polarization frequency reuse antenna system. This is why, except in the case of very large antennas (e.g. $D/\lambda \geqslant 400$), the simpler and less costly step-track system is usually preferred. This uses a 'climbing the hill' servo-system. In this method which does not require difference signals, the antenna is steered by small progressive angular increments (steps). If the received (beacon) signal decreases, as compared to the previous step, the step-track processor will command the antenna to be steered in the opposite direction. Of course, such a non-instantaneous method is not able to cope with rough wind gusts, nor with possible satellite beacon short term instability.

The tracking accuracy is defined as the residual error angle between the antenna beam axis and the direction of arrival of the satellite received signal under automatic tracking operation. It is limited by many factors, in particular by servo-defects (backlash etc.), by wind effects and also, fundamentally, by the tracking receiver thermal noise. The effects of wind, which are much more relevant in non-monopulse systems, are usually specified under two conditions: condition 1 where full antenna performance must be kept (e.g. 13 m/s wind with 20 m/s gusts), and condition 2 where antenna performance, though degraded, remains sufficient for operation (e.g. 20 m/s wind with 27 m/s gusts). The tracking error (rms) due to the receiver thermal noise is given by the following formula (strictly valid for monopulse tracking):

$$(\delta\theta)_{\text{Thermal noise}} = (1/s)\cdot\sqrt{kTB/P_{\text{R}}}$$

where s is the slope of the difference diagram ($s \approx 0.4\cdot D/\lambda$, in rad^{-1}, where 0.4 is a typical value for the odd mode coupling efficiency (for example $s = 170$ rad^{-1}, or 3 degrees^{-1} for a 32 m antenna at 4 GHz), kTB is the noise power ($k = 1.38 \times 10^{-23}$ J·K^{-1}, T is the total noise temperature in the error channel of the tracking receiver, B is the error integration bandwidth, i.e. servo-bandwidth, e.g. 1 Hz). P_{R} is the power received from the satellite beacon in the reference channel of the tracking receiver.

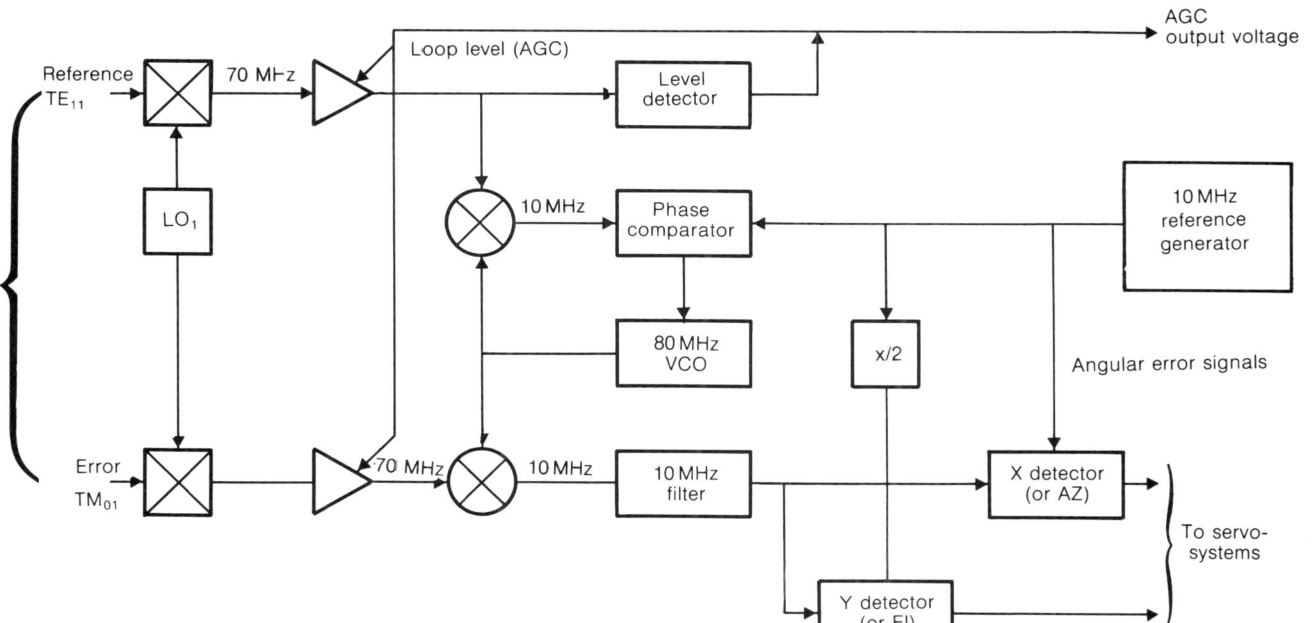

Fig. 2.37 Block diagram of a monopulse tracking receiver.

2.4.3 The low-noise amplifier (LNA)

The low-noise amplifier (LNA) is the input stage of the receiver subsystem of the earth station. As already explained (section 2.2.2), it must feature a very low intrinsic noise temperature and a gain high enough to 'mask' the noise contributions of the subsequent stages. In the 1960s, the first satellite communications earth stations (in particular the big 'horn antennas' of Andover in the USA and Pleumeur–Bodou in France) were using cryogenic maser LNAs. Then came the era of cryogenic parametric amplifier LNAs. By cryogenic, it is meant that the complete amplifier is cooled down to some 20 K (−253 °C) by a gaseous helium system. Such a low physical temperature allows us to obtain impressively low noise temperatures (e.g. 15 K at 4 GHz). Thanks to the technology progress in diode and transistor design, much simpler, less bulky, more reliable and cheaper solutions are now used (even if their performance is a little lower). In fact all current LNAs use either parametric amplifiers or FET (field-effect transistors), both uncooled or slightly cooled by thermoelectric (Peltier diodes) systems. The latter ones (FET) are progressively replacing the former ones (parametric).

Parametric low-noise amplifiers Parametric amplification is obtained by driving an inversely biased semiconductor diode by a local oscillator (the 'pump') at a very high frequency F_P. The diode is called a varactor, because its capacitance varies with the applied voltage. It is easy to demonstrate that, under these conditions, the applied RF input signal (at a frequency F_S, with $F_P \gg F_S$) 'sees' the circuit as a negative resistance and is therefore amplified with a very low intrinsic noise (since, theoretically, there are only reactive components). Note that an image ('idler') signal is generated, at a frequency $F_I = F_P - F_S$, and must be rejected. Note also that the amplified output signal is reflected at the input signal and must therefore be separated by an RF ferrite circulator. A typical parametric amplifier is shown in Fig. 2.38.

The gain–bandwidth product of a parametric amplifier stage is a direct function of the bandwidth of the signal and idler RF circuits. An approximate formula is:

$$\sqrt{G}\cdot B = 2/(1/B_S + 1/B_I)$$

G being the gain (power ratio) and B_S, B_I being the bandwidths of the signal and idler circuits.

The following formula gives the noise temperature T_E of the parametric amplifier (which is due to the varactor residual resistivity):

$$T_E = (1 - 1/g)\cdot\frac{T_P\cdot Q_S^2\cdot(F_S/F_I)^2 + 1}{Q_S^2\cdot(F_S/F_I)^2 - 1}$$

where g is the gain of the amplifier, T_P is the physical temperature of the varactor (K) and Q_S is the dynamic quality factor of the varactor at the signal frequency F_S ($Q_S = F_{CO}/F_S$, F_{CO} being the varactor cut-off frequency).

The formula shows that low T_Es are obtained with a high cut-off frequency

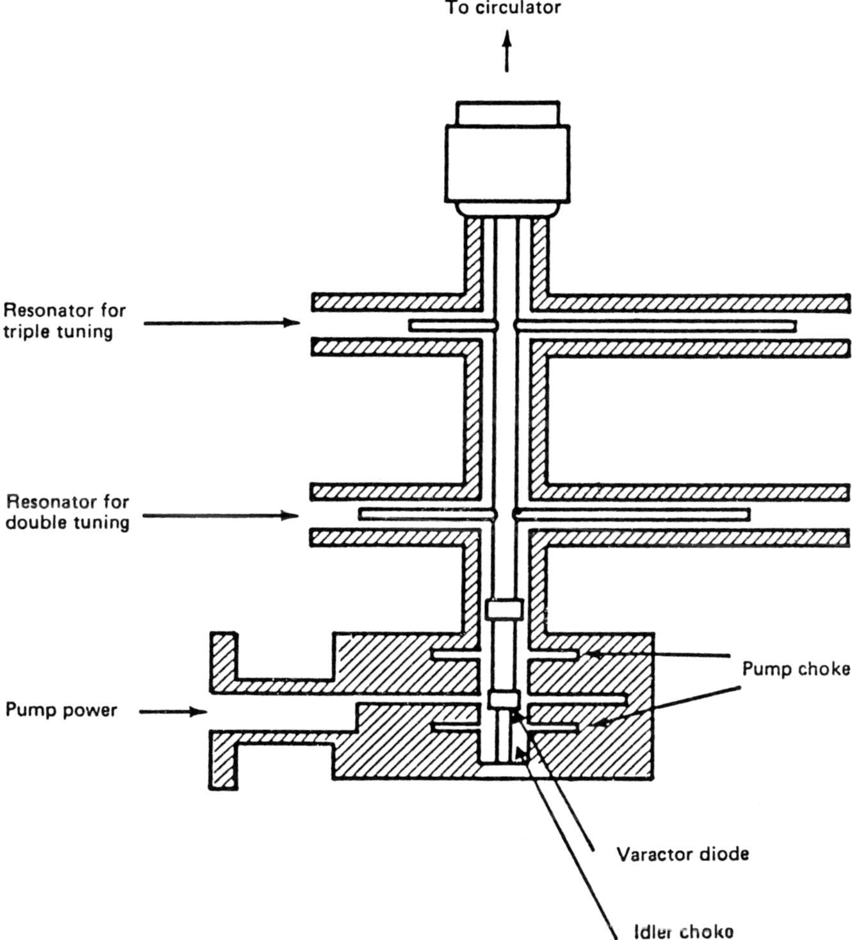

Fig. 2.38 Parametric amplifier.

varactor and with a high pump frequency (resulting in a high F_I). Varactors with F_{CO} as high as 1000 GHz are available and pump oscillators, using Gunn diodes up to some 60 GHz are commonly implemented. If required, further improvement can be obtained by operating at a lower T_P, e.g. by thermoelectric cooling.

Field-effect transistors (FET), low-noise amplifiers Due to their shot noise (and other noise mechanisms), bipolar transistors cannot be used in LNAs over 1 GHz. On the contrary, the—mostly thermal—noise of FETs can be reduced by using high electron mobility semiconductor material (gallium arsenide (GaAs) is the most common) and sub-micron fabrication technologies. A typical LNA FET is made of a 0.3 μm thick n-type GaAs layer epitaxially grown on a semi-insulating GaAs substrate. Electrodes (source, drain and Schottky-barrier gate) with very small geometries (e.g. 0.25 μm gate length) are used. FET LNAs can be manufactured either in hybrid form with discrete components or, in the case of mass production, in the form of MMICs (microwave monolithic integrated circuits). Although GaAs FETs are now the 'workhorse' of LNAs, having replaced parametric amplifiers at least up to the Ku-band, they are themselves progressively replaced by the new high electron mobility transistors (HEMT) (Smith and Swanson, 1989).

In fact, HEMTs are directly derived from FETs technology and can be easily exchanged with FETs in LNAs. HEMTs differ from conventional FETs by the semiconductor zone where electrons flow from source to drain (when a voltage is applied to the gate electrode). In a FET, the electrons flow in a channel layer which is doped by donor ions and are therefore scattered by these ions. In a HEMT, a potential well is created, in a GaAs buffer zone, by a AlGaAs/GaAs interface. The electrons form a 'two-dimensional electron gas' travelling in undoped GaAs. This reduces scattering and increases mobility and velocity of the electrons. HEMTs exhibit lower noise and higher frequency performance (HEMTs can be used up to, at least 100 GHz).

Table 2.5 LNAs: typical noise temperatures

Frequency band	*LNA type*	*Cooling*	*Typical noise temperature* (K)
C-band (3.7 to 4.2 GHz)	Parametric	Cryogenic	15
		Thermoelectric	35
		No cooling*	50
	FET or HEMT	Thermoelectric	50
		No cooling*	75
Ku-band (11.7 to 12.2 GHz)	Parametric	Cryogenic	15
		Thermoelectric	80
		No cooling*	100
	FET or HEMT	Thermoelectric	120
		No cooling*	200
Ka-band (17.7 to 19.5 GHz)	Parametric	Cryogenic	50
		Thermoelectric	200
	FET or HEMT	Cryogenic	100
		Thermoelectric	200

* In uncooled systems, a temperature stabilization process is usually provided.

Table 2.5 recapitulates the noise temperature performance of current LNAs. Note that in the case of very small stations (TVROs, VSATs), the LNA is often associated with the down-converter (D/C) in a single unit—called a low-noise converter—which is directly connected to the antenna source.

2.4.4 Measurements of noise temperatures and antenna G/T

In this section, some types of measurements, which are specific to low-noise antennas and receivers and which are common practice for earth stations, are explained. Fig. 2.39 represents a common block-diagram for these measurements, which are based on the 'Y factor' method (this term being explained below).

Noise temperature of LNAs (T_R) The LNA noise temperature is measured by comparing: (1) the noise power received when the LNA input is connected to a matched load at a well defined cold physical temperature ('cold load', usually at the liquid nitrogen temperature $T_{CL} = 77.3$ K) with (2) the noise power received when the LNA input is connected to a matched load at a well defined ambient physical temperature ('warm load', e.g. at the reference temperatue $T_0 = 290$ K).

In case (1), the received power is:

$$P_{R1} = kTB \approx k(T_{CL} + T_R)B$$

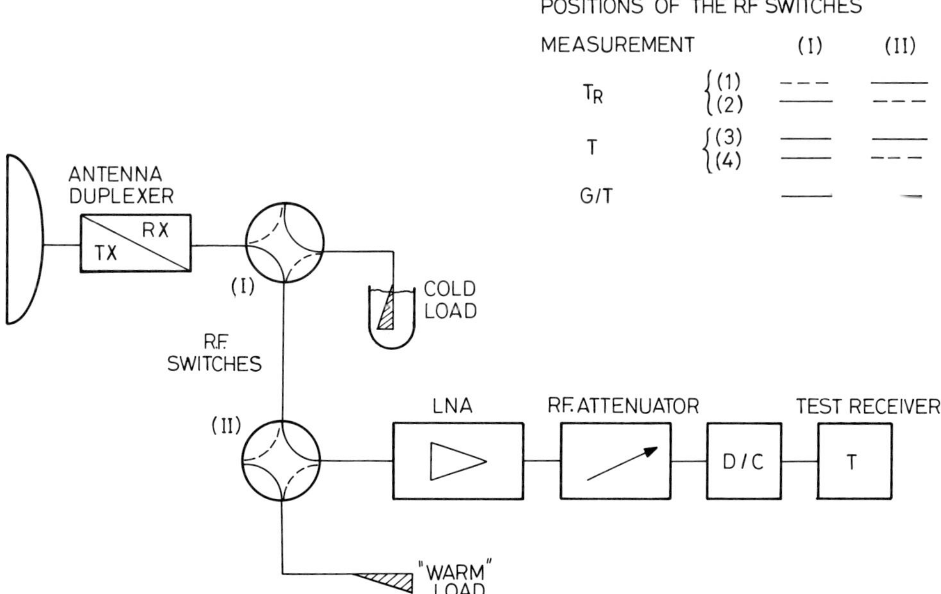

Fig. 2.39 Measurement of T_R, T and G/T.

(see section 2.2 with $T_A = T_{CL}$ and B being the actual measurement bandwidth of the test receiver). In case (2), the received power is:

$$P_{R2} = k \cdot (T_0 + T_R) \cdot B.$$

In fact, an attenuator is set at a (dB) value $10 \log Y = 10 \log (P_{R2}/P_{R1})$ in order to adjust these two powers at the same deviation on the RF test receiver voltmeter. Whence T_R as a function of the Y factor:

$$T_R = (T_0 - Y \cdot T_{CL})/(Y - 1).$$

Total system noise temperature (T) The total system noise temperature, which represents all the noise contributions in the earth station at reception (antenna: T_A and LNA: T_R) has been defined in section 2.2.2. It is measured by comparing: (3) the total noise received by the earth station through the antenna with: (4) the noise received when the antenna is replaced by a warm load, e.g. at $T_0 = 290$ K.

In case (3), the antenna being pointed towards a 'cold' region of the sky (i.e. a 'normal' region where there is no radiosource), the received power is

$$P_{R3} = kTB \approx k \cdot (T_A + T_R) \cdot B.$$

In case (4), the received power is:

$$P_{R4} = k \cdot (T_0 + T_R) \cdot B.$$

With the attenuator set at: $10 \log Y = 10 \log (P_{R4}/P_{R3})$, T is measured as: $T = (T_0 + T_R)/Y$.

G/T measurement The G/T of the antenna with its LNA is measured by using as a distant power source a radiosource, also called a 'radiostar', i.e. an astronomical source of RF power. The most significant radiosources (Cassiopeia A and Taurus) are well known: their radiation can be considered as white noise in the earth station receiving bandwidth and their received power flux (S in $W \cdot m^{-2} \cdot Hz^{-1}$) has been very accurately calibrated. More precisely, the measurement consists of comparing the total (noise) power received in case (5), where the antenna is pointed towards a 'cold' region of the sky, with the noise power in case (6) where the antenna is pointed towards the radiosource.

In case (5): $P_{R5} = kTB$ and in case (6):

$$P_{R6} = [k \cdot T + (1/2) S \cdot (G\lambda^2/4\pi)] \cdot B$$

where $(G\lambda^2/4\pi)$ is the antenna effective aperture are (section 2.2) and where the factor (1/2) is due to the fact that the radiosource emission being unpolarized, only half of its power is received in the normal antenna polarization. Again, with $Y = P_{R6}/P_{R5}$, the measured (G/T) is given by:

$$(G/T) = (Y - 1) \cdot 8k\pi/(S \cdot \lambda^2).$$

More precisely, the (G/T), which is usually expressed in dB, is given by:

$$(G/T)(dB) = 10 \log [(Y - 1) \cdot 8k\pi/(S \cdot \lambda^2)] + K_1 + K_2 + K_3 + K_4 + K_5$$

with S (in $W \cdot m^{-2} \cdot Hz^{-1}$) $= 1061 \cdot 10^{-26}$ for the CASSIOPEIA A radiosource and $S = 685 \cdot 10^{-26}$ for the TAURUS radiosource.

The term K represents various correction factors which account for: K_1 the atmospheric attenuation as a function of the elevation angle, K_2 the angular width of the radio source (as seen by the antenna beamwidth), K_3 the decreasing of the radiosource flux with the elapsed time ($K_3 = 0$ for Taurus), K_4 the variation of the flux with the (RF) frequency and K_6 for the polarization (in fact, $K_6 = 0$ for Cassiopeia A and Taurus). The values and formulae for these factors will be found in Ref. [8].

2.4.5 The high power amplifier (HPA)

General

The high power amplifier (HPA) is the output stage of the earth station equipment, just before emission by the antenna. More precisely, an individual HPA is a part of the HPA system, which may comprise:

1. an input combiner (represented on Fig. 2.28);
2. the individual HPAs proper: only one is shown in the figure, but even in the simplest stations, at least a (1 + 1) redundant configuration, i.e. one active and one stand-by HPAs are provided. In high traffic capacity stations, more than one active HPA may be needed, operating either in parallel or in different frequency bands;
3. an output combiner (not represented on the figure), if there is more than one HPA, for coupling the HPAs to the antenna duplexer.

Table 2.3 above gives typical data on the equivalent isotropically radiated power (e.i.r.p.) which are required for the transmission of some usual types of carriers by some usual types of earth stations.

A general relation giving the HPA dimensioning, i.e. the required output power of a HPA (P_{HPA}, measured at saturation) is given below (in dBw), in the general case of multiple carrier operation:

$$P_{HPA} = 10 \log \left[\sum_{i=1}^{n} (\text{e.i.r.p.})_i \right] - G_T + (BO)_O + L_T \quad (\text{dBw})$$

where (e.i.r.p.)$_i$ is the (e.i.r.p.) required for the ith carrier, note (e.i.r.p.) is a power expressed in w, (E.I.R.P.) in dBw) G_T is the antenna gain (in the transmission band) (dBi), $(BO)_O$ is the HPA output back-off required for sufficiently linear operation (dB below saturation, see section 2.2.3) and L_T is the total losses between the HPA output port and the duplexer input port (this includes waveguide losses and, possibly, output combiner coupling loss) (dB).
In the particular case of n equal carriers, this becomes:

$$P_{HPA} = (\text{E.I.R.P.})_i + 10 \log n - G_T + (BO)_O + L_T \quad (\text{dBw}).$$

For example an INTELSAT Standard B earth station ($G_T = 54.6$ dBi) must transmit, through a single TWT HPA, three 'IDR' type carriers, one at 2.048 Mbit/s (i.e. a primary order multiplex of 30 + 2 PCM channels at 64 kbit/s) and two at 384 kbit/s, requiring respectively 66 and 59 dBw E.I.R.P.s. For keeping the emission of intermodulation products below the specified power level (see below), computation and tests show that a minimum output back-off $(BO)_O = 5$ dB is needed. Accounting for some waveguide losses and for a 3 dB coupler (used as an output combiner) gives $L_T = 3.5$ dB. Therefore, the TWT should have a saturated power of:

$$P_{HPA} = 10 \log [4 \cdot 10^6 + 2 \times 7.9 \cdot 10^5] - 54.6 + 5 + 3.5$$
$$= 21.37 \text{ dBw}$$

i.e. ≈ 137 w (i.e. a 150 W tube).

Selection of the amplifier type

Two types of microwave tubes are used in earth station HPAs: travelling-wave tubes (TWT) and klystrons. In the case of small stations, e.g. for rural telephony (Standard D1) for data transmissions (VSATs), solid-state (FET) amplifiers are now common practice. Also, in the case of microwave tubes HPAs, FETs are used as a first stage (preamplifier).

Travelling-wave tubes (refer to Volume 1) Thanks to their specific wideband performance (e.g. 500 MHz at C-band), TWTs provide the means for amplifying in a single HPA, with the specified gain and group delay uniformity, all the carriers to be transmitted by an earth station, whatever be their frequency. Just as for satellite transponders power amplifiers (see section 2.3.2), the insertion of a linearizer at the TWTs input allows operation of the tube nearer saturation, i.e. with a smaller back-off, even in the case of multi-carrier transmission. A very wide range of earth station TWTs is currently available.

Some typical features of these TWTs are summarized below:

1. output power: C-band (5.850 to 6.425 MHz): from 40 W to 13 kW,
   Ku-band (14 to 14.5 GHz): from 15 W to 3 kW,
   Ka-band (27.5 to 30.5 GHz): from 25 W to 1 kW;
2. gain: typically between 35 dB and 50 dB;
3. slow wave structure: lower power TWTs use a helicoidal slow-wave structure (helix); for high powers (e.g. more than 3 kW in C-band, but much less at higher frequencies), coupled cavity structures are needed;
4. focusing: all low or medium power TWTs (up to 1 kW or even more) are now constructed with PPM (periodic permanent magnet) focusing; only very high power ones are equipped with electromagnetic coils;
5. cooling: nearly all modern TWTs can be cooled either by simple conduction or by forced air; Only the highest powers may need liquid cooling;

6. efficiency: TWTs efficiency is rather low (often ≈ 10%). Modern techniques such as depressed collector operation and multiple collectors provide the best efficencies by reducing collector heat dissipation;
7. power supply: power supply is rather complex in the case of a TWT HPA (even with PPM focusing): three different high voltages are needed (anode, helix and collector); also, means must be provided for protecting the helix during switching-on the power supply (until voltages are nominal). Typical figures for a modern 500 W TWT are: 6 kV/1 mA for the anode, 10 kV/15 mA max (well regulated) for the helix, and 5.5 kV/400 mA for the collector.

Klystron tubes (refer to Volume 1) Klystrons are intrinsically relatively narrow band tubes. Klystrons for earth station HPAs are usually constructed so that their instantaneous bandwidth covers one satellite transponder frequency band, viz. about 40 MHz (or even 80 MHz) in the C-band and 80 MHz in the Ku-band. However, this instantaneous bandwidth can be set in the frequency band of any transponder (by mechanically tuning all the cavities of the interaction space). Often, this tuning device can be remotely controlled.

Notwithstanding these bandwidth limitations, klystrons are very frequently used for HPAs. This is because of the following advantages which often make klystrons more economical and easier to operate than TWTs:

1. high efficiency (35% or more);
2. very simple implementation: permanent magnet focusing and forced air cooling are possible up to at least 3 kW power. The power supply needs only one high voltage (between cathode and collector or body, e.g. 8.5 kV/1 A for a 3 kW klystron);
3. long service life (30 000 or 40 000 hours);
4. capability to operate at reduced power (with lower consumption).

The preferred fields of applications of klystron HPAs are:

1. medium or high power: typically, from 700 W to 3.5 kW (or even more; in fact, klystrons are not common at a power less than about 500 W);
2. transmission of one carrier (e.g. TV) or several carriers (e.g. SCPC) towards a single, well-defined transponder;
3. transmission of a few (n) FDMA carriers towards different transponders. Often, in such a case, the power of a single medium power TWT could be insufficient and the utilization of n (active) klystron HPAs (+m in stand-by, with $m(n)$ may prove to provide a more reliable and economical solution.

Solid-state power amplifiers FET (GaAs) with more than 5 W at C-band and 2 W at Ku-band are currently available. Even more power is possible by paralleling two or more FETs. Such solid-state amplifiers exhibit significant advantages when only a relatively low power HPA is required: they are very economical and feature a low power consumption, a high reliability and also better linearity than tube HPAs.

Impairments in HPAs, non-linearity effects

Apart from their output power, gain and bandwidth characteristics, the transmitting system of an earth station and, in particular of the HPA must comply with other performance specifications. These include:

1. gain and time delay variations versus frequency;
2. residual amplitude modulation;
3. harmonic generation;
4. unwanted and noise emissions;
5. intermodulation;
6. amplitude modulation to phase modulation (AM–PM) conversion.

Gain and time delay variations must be kept within specified limits inside the signal bandwidth in order to avoid signal distortions, although delay variations can be compensated by inserting time delay equalizers (TDE) in the transmission chain at intermediate frequency (IF). Also, in a multi-carrier operation, an effect of gain versus frequency variations with AM–PM conversion is to cause intelligible cross-talk between carriers.

Residual amplitude modulation is produced mainly by the HPA power supply ripple. Its effect is to induce distortion noise through the satellite transponder (due to its own AM–PM conversion).

Microwave tubes generate harmonics at significant levels (typical second harmonic levels are $-30\,\mathrm{dB}$ for klystrons and up to $-10\,\mathrm{dB}$ for helix TWTs). Harmonic filters are often inserted at the HPA output in order to lower these levels down to -50 or $-60\,\mathrm{dB}$.

Unwanted and noise emissions transmitted by an earth station throughout its operating bandwidth may cause interference to other systems. In particular, electron fluctuations in the tube generate noise at RF, even in the absence of input signals. This effect can be very significant when multiple stations, in TDMA systems, are transmitting, on a time shared basis, the same carrier.

The two last effects, intermodulation and (AM–PM) conversion, are due to the non-linearity of the HPA characteristic (output versus input signal amplitude and phase). They have already been explained in section 2.2.3.

Intermodulation The intermodulation characteristics of a microwave tube are usually specified by its manufacturer as the level of the third-order intermodulation products generated by two reference carriers at the same output power level.

In an actual application, with two carriers at frequencies F_I and F_J and at powers p_I and p_J, the following approximate formula gives the output power (in W) of each third-order intermodulation product (at frequencies $2F_\mathrm{I} - F_\mathrm{J}$, with: $I = 1$, $J = 2$, or $I = 2$, $J = 1$):

$$(\mathrm{IM})_\mathrm{IJ} = \frac{(\mathrm{IM})_\mathrm{O}}{p_\mathrm{O}} \frac{p_\mathrm{I}^2 \cdot p_\mathrm{J}}{p_\mathrm{O}^2}$$

where $(IM)_O$ is the output power of the third-order intermodulation products generated by the two reference carriers at an output power p_O. Or, in dB:

$$(IM)_{IJ}(dBW) = -D_3 - 2(p_O) + 2(p_I) + (p_J)$$

where the powers are expressed in dBW and where $D_3 = -10\log(IM)_O/p_O$.

For example: for a TWT with 1.3 kW saturated output power, $D_3 = 26$ dB for two 60 W (p_O) reference carriers (manufacturer's data). If now $p_1 = 200$ W and $p_2 = 150$ W (i.e. with an output backoff $(BO)_O = 10\log(1300)/(200 + 150)$ or 5.7 dB), then:

$$(IM)_{12} = -26 - 2(17.8) + 2(23) + (21.8) = 6.2\,\text{dBW}$$

and:

$$(IM)_{21} = -26 - 2(17.8) + 2(21.8) + (23) = 5\,\text{dBW}.$$

In the case of three carriers (or more), such a calculation must be carried out for each carrier pair. However, as already explained in section 2.2.3, there is a second type of third order intermodulation products at frequencies such as $F_1 + F_2 - F_3$, $F_1 + F_3 - F_2$ and $F_2 + F_3 - F_1$. The (approximate) formula for these products is:

$$(IM)_{123} = \frac{4(IM)_O}{p_O} \frac{p_1 \cdot p_2 \cdot p_3}{p_O^2} \qquad \text{(in W)}$$

Or, in dB:

$$(IM)_{123}(dBW) = 6 - D_3 - 2(p_O) + (p_1) + (p_2) + (p_3).$$

Note that these products are dominant, due to the 6 dB factor.

These formulae are valid only for carrier powers not too far from the reference p_O and as a first order approximation. More precise calculations are rather complex and need computer programming.

Finally, all these calculations are made for pure (unmodulated) carriers. In the actual case of modulated carriers (either analogue or digital modulation), a power spectrum spreading factor must be accounted for, thus reducing the actual level of the intermodulation products.

(AM–PM) conversion effects in digital transmission AM–PM conversion occurs when an amplitude variation (i.e. modulation) induces, through non-linear effects a phase variation (i.e. modulation). In TWTs, AM–PM at saturation is about 7°/dB. It is less in klystrons (about 4°/dB). Theoretically, AM–PM should not affect digitally encoded, phase modulated (BPSK or QPSK) RF carriers because of their constant amplitude envelope and it should be possible to operate the HPA at saturation in the case of a single digital carrier (e.g. a high bit rate – 120 Mbit/s TDMA carrier). However, this is not quite true due to filtering effects: filtering limits the signal bandwidth B ($B \approx 1.2$ R is usual for BPSK and $B \approx 0.6$ R for QPSK, R being the bit rate).

Due to this limitation, the signal amplitude is no longer constant and AM–PM conversion occurs, with two types of degradations: the bit error rate (BER) is increased and the output signal spectrum is enlarged (with unwanted radiation outside the assigned bandwidth, see note in section 2.3.2). Finally, even in this case, some back-off must be provided in HPA operation: $(BO)_O \approx 3$ dB is usual).

2.4.6 The up- and down-converters (U/C, D/C)

The telecommunications (baseband) signals modulate a carrier at an intermediate frequency (IF). The IF is typically at 70 MHz, 140 MHz or 1 GHz, depending on the signal bandwidth and on the general equipment design. Following the modulator, an up-converter (U/C) translates the IF signal into the RF signals (e.g. at 6 GHz or 14 GHz) to be transmitted by the earth station. Conversely, the received RF signals (e.g. at 4 GHz or 11 GHz) are applied to a down-converter (D/C) which translates them at the IF before their demodulation in the demodulator. Two types of converters can be used: single and dual frequency conversion U/Cs (or D/Cs).

Frequency conversion is performed by heterodyning the signal, in a non-linear semi-conductor mixer, with the wave generated by a local oscillator (LO). In addition to one (single frequency converters) or two (dual frequency converters) mixer(s) and LO(s), U/Cs and D/Cs comprise RF and IF filters, IF amplifiers (often with automatic gain control) and also group delay equalizers (GDEs), for compensating the time delay variations of the receive—or transmit chain—in the signal bandwidth).

Figure 2.40 shows three classical types of D/Cs (the block-diagrams would be similar for U/Cs). Figure 2.40(a) is a single frequency conversion D/C. It is very simple but it does not provide frequency agility: for changing the operating RF frequency, it is necessary to tune the LO and to mechanically adjust the microwave bandpass filter centre frequency.

Figure 2.40(b) is the most common type of D/C: it is a dual frequency conversion D/C with full frequency agility. Changing the operating RF only requires tuning the first LO, which can be electrically (digitally) and remotely controlled if this LO is frequency synthesized. The second part of the D/C (at 1st and 2nd IF) is frequency fixed. The 1st IF is chosen at higher centre frequency than the RF total frequency range (e.g. 500 MHz) in order to allow the rejection (by a wideband microwave filter shown on the figure) the 'image' signals, i.e. the spurious signals which should be symmetrical on a RF frequency scale versus the useful signals.

Figure 2.40(c) is another type of dual frequency conversion D/C: it is similar to Fig. 2.40(a) as concerns frequency changing. However, the power divider (or combiner for U/Cs) and the adjustable filters operate at a lower frequency (e.g. around 1 GHz). This is a very attractive solution for small stations, e.g. for television reception (TVROs) since the RF section can be integrated, in the antenna feed, with the LNA (forming a LNC: low noise converter). The whole bulk of the receiver channels are then easily connected to the indoor unit (comprising the divider and the multiple second D/Cs) by a coaxial cable (e.g. at 1 GHz).

The local oscillators (LOs) of the converters must feature high long term

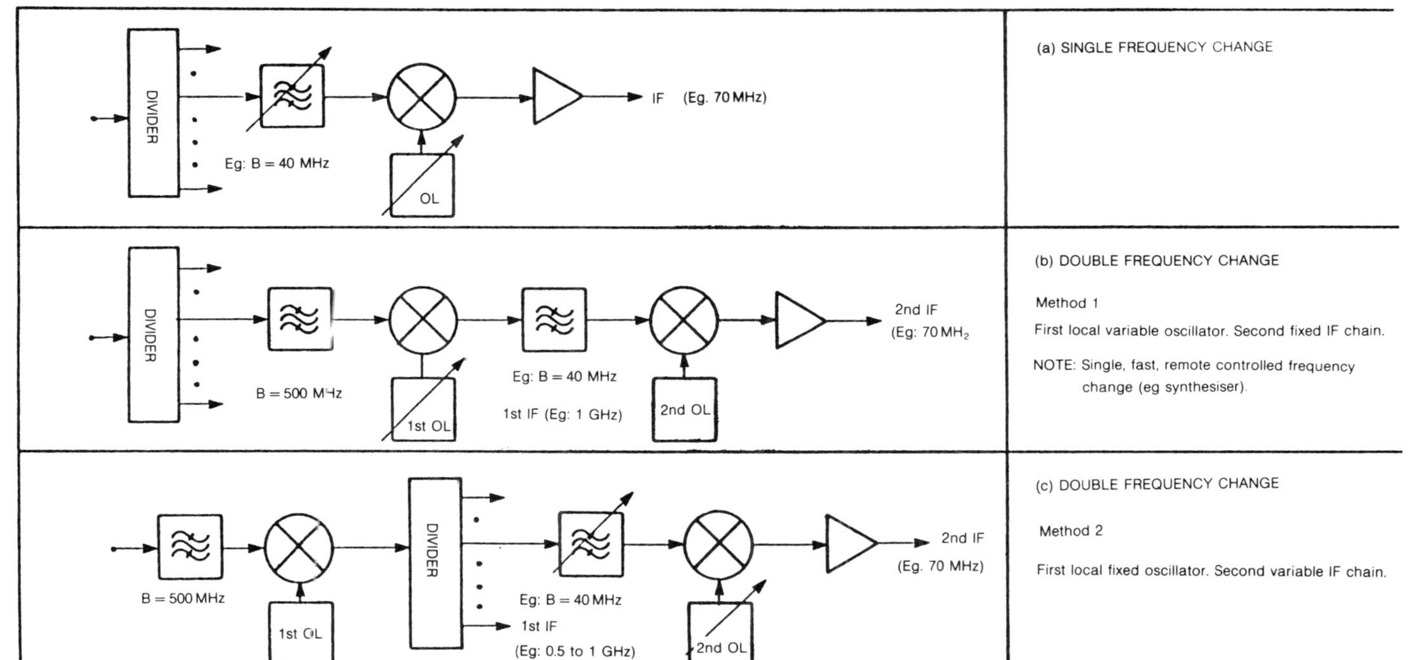

NOTE: The diagrams above show down converters.
The diagram for up converters would be similar.

Fig. 2.40 Classical types of D/Cs block diagram.

frequency stability (ranging from $\pm 10^{-5}$ to $\pm 10^{-8}$) and low phase noise characteristics. These are especially stringent in the case of multiple narrow-band voice or data signals sharing a common satellite transponder, viz. in the case of single channel per carrier (SCPC) transmission or reception.

2.5 CONCLUSIONS AND PROSPECTS

It is a general opinion that satellite communications will face increasing competition from other transmission media and, in particular, from optical fibre cables in their main current applications, i.e. long distance high capacity traffic transmission, and that, in consequence they should orientate their future development efforts towards new, innovative fields where their specific characteristics will ensure them continuing opportunities and will guarantee their share in the ever growing telecommunications market.

Fixed telephony and data transmissions

Even accounting for the optical fibre cables, satellite communications will continue to carry a big part of the international traffic and also of the domestic (especially rural) traffic in developing countries. Direct data distribution and transmissions, based on microstations (VSATs) networks, should take a growing part in the private, business communications market.

Television

The importance of satellites for the distribution of television programmes, either towards community (CATV, SMATV) or individual antennas, should be confirmed in the future. However, the best opportunities should, at last, occur through an actual developing of the broadcasting satellite service (BSS), allowing direct-to-home (DTH) distribution of high definition television (HDTV) programmes. Also to be mentioned here are the new satellite digital audio broadcasting (DAB) applications, which should allow broadcasting of fully digital, multiplexed, high quality, audio programmes to pocket-size receivers over very large regions.

Mobile services

This chapter has been mainly devoted to satellite communications and systems for fixed applications (i.e. implemented in the framework of the 'fixed satellite services'). However, it may be the mobile satellite services which have the most exciting prospects. In fact, mobile services are now expanding rapidly, both for maritime (maritime mobile satellite services (MMSS)) and land mobile applications (land mobile satellite services (LMSS)).

Maritime mobile satellite services These are already very well developed, thanks to the satellites of the INMARSAT International Organization. At present, more than 10 000 vessels are equipped with ship earth stations (SES) of the Inmarsat-A or Inmarsat-C types, which, through coast earth stations, allow direct interconnection of their users to the public telephone networks, whatever be their location on the globe. Inmarsat-A stations remain rather complex and costly (1 m antenna on a three-axes stabilized platform and under a radome) and are intended mainly for ships over 30 000 tons. Note that transportable versions of Inmarsat-A stations are also available, and widely used, for news reporting and other instant emergency communications from anywhere on the earth. New opportunities are now offered by Inmarsat-C SESs which feature compact, cheap equipment. Using a fixed helix antenna with a low gain, they are limited to the transmission of text (telex) and data messages at 600 bit/s, but this is quite sufficient in most cases. However, new types of small SES are currently under development. Using low bit rate (LRE) codes (e.g. at 4.8 kbit/s) voice, they will enable telephony transmission with an acceptable transmission quality.

Land mobile satellite services These should play, in the medium term, a significant role in the mobile communications market. Already, Inmarsat-C stations are available for providing global communication means to terrestrial vehicles, especially trucks. But other systems are now appearing: for example, the Omnitracs system (USA), and Euteltracs in Europe, are implementing small stations with a tracking antenna under a radome. The main marketing target is to provide truck transportation companies with traffic control capabilities. The stations are very easy to install and they operate via non-specialized satellites in the Ku-band (e.g. via Eutelsat satellites). Of course, they also only provide two-way text messages. In the vehicle these messages, sent by an operation centre, appear on a small console. In order to save the satellite power, the messages are actually routed sequentially by a 'queuing' process. Moreover, this system provides the operation centres with permanent localization of their vehicles, using differential distance measurement through two satellites.

More generally as concerns localization proper, the GPS (global positioning by satellite) US military system is worth mentioning: This system implements—all over the earth—21 dedicated (non-geostationary) satellites orbiting at some 17 600 km altitude. Operating by measuring distances simultaneously with three satellites (triangulation) through control earth stations, they provide, anywhere, precise localization (longitude, latitude, altitude) to very simple (nearly pocket-sized) and low-cost receivers. The nominally quoted accuracy is 18 m. However, due to a special coding process, this accuracy is accessible only to military usage and for civilian applications, the accuracy is currently restricted to some 100 m.

As concerns communications with aeroplanes, the aeronautical mobile satellite services is, at present, still in its infancy but dedicated systems should certainly enter operation in the future and there are great expectations of Inmarsat in this applications field.

On the long term, it is probable that the future of satellite communications for mobiles will largely rely on non-geostationary satellite systems, with low altitude or elliptical orbits (LEOs, low earth orbit satellites). This is because geostationary satellites are not very well adapted for mobile services, especially for telephony applications: firstly, the geostationary orbit distance implies relatively high power transmission from the earth, which is difficult for small earth terminals; then, multiple hops will generally be needed which means that long propagation delays are to be expected. Last, but most important, the elevation angle towards the satellite must be kept high to allow continuity of the communication link whatever obstacles may be on the line-of-sight, i.e. especially in the urban areas.

Very spectacular research projects are currently being carried out in this field of mobile communications using LEO satellites. To be cited in particular is the IRIDIUM system which should implement a global cellular network using 77 small satellites orbiting on 11 low altitude (760 km) polar orbits. By this means, one should always find, everywhere, a nearly vertical satellite, enabling him to establish, with his personal terminal, a communication with any other terminal in the world by finally relaying the link up to the satellite which will be, at the moment, straight over his correspondent. Other similar projects are under way (Inmarsat 'Project 21' with a mixing of geostationary and LEO satellites, Globalstar, etc.). However, in addition to the technical (and also regulatory) difficulties to be solved, there remains the big problem of frequency allocations with which the ITU World Administrative Radio Conference (WARC 92) has only temporarily dealt.

There are many other applications of satellite communications with the mobiles, such as paging services. More generally, satellites will certainly participate in the future, long term, global networks (UMTS: universal mobile telephone system and UPT: universal personal telephone) which are the subject of current intensive studies in the ITU: such networks will bring off completely new concepts in public telecommunications whereby telephone numbering will be attached to the subscriber proper (and no more to a location), therefore allowing communications to be established from anywhere without even knowing the current position of the called correspondent in the world.

Miscellaneous applications and projects

Other applications can be mentioned as well, although they are not necessary carrying telecommunications services: military satellites, earth photography by satellites for meteorology (METEOSAT) and for terrestrial resources exploration (SPOT, LANDSAT), etc.

For the future, very original satellite systems, based on entirely new designs have been cited (Petton, 1991): satellite clusters, space platforms, optical communication satellites, dipole satellites for low frequency direct broadcasting, very low altitude satellites with station-keeping through microwave beam energized propellers, geostationary satellites with a feeder cable from the earth surface, etc.

REFERENCES

[1] "Handbook: Satellite Communications (Fixed Satellite Service)" CCIR, International Telecommunication Union, Geneva 1988.
[2] CCIR Report 564? (See 12.1.1 Page 1).
[3] Clarke, A. C. (1945) "Extra Terrestrial Relays", *Wireless World.*
[4] Clark, G. C. and Cain, J. B. (1981) *Error coding for digital Communications*, Plenum Publishing Corp., New York.
[5] Marcuvitz, N. (1951) *Waveguide Handbook*, MIT Series.
[6] Matthai, G., Young, L., Jones, E. M. T. (1980) *Microwave Filters, Impedance-matching Networks and Coupling structures*, Artech House.
[7] Smith, P. M. and Swanson, A. W. (1989) HEMTs-Low Noise and Power Transistors for 1 to 100 GHz, *Applied Microwave*, 63–72.
[8] INTELSAT Document BG/T 30–32 (18/10/78).
[9] Pelton, J. N. (1991) Communication via satellite: the next 100 years, *Via Satellite*, **VI** (9).

3

Low and medium power translators and transmitters

Claude Cluniat

In order to illustrate problems occurring when developing TV translators and transmitters, some recent achievements made in one company (LGT) are presented in this chapter. First is described how to optimize the input design of a television translator in order to get a constant noise factor, independent of input level. Second, the development of television transmitter modulation stages. Lastly, the optimization of the output design of a television translator or transmitter and the enhancing of the level of transistorization by correcting third-order non-linearities and minimizing coupling losses of the power amplifiers.

3.1 OPTIMIZATION OF THE INPUT DESIGN OF A TELEVISION TRANSLATOR

3.1.1 Brief review: what are the design possibilities?

1. single or double frequency change;
2. common or separate amplification of the vision and sound channels.

There are two types of retransmission employing single frequency change: retransmission with demodulation and single-frequency change retransmission. Retransmission with demodulation, referred to by the German term 'ballangfang', is obtained by juxtaposition of a sensitive receiver and a television transmitter.

Interconnection is implemented in the baseband. This solution is ruled out due to the introduction of additional distortion from demodulation (fourth-order distortion, etc.) and non-linear distortion due to modulation (Fig. 3.1).

Single-frequency change retransmission eliminates these defects. In this case, the input signal F_1 is changed by means of a local frequency F. The output frequency F_2 is equal to the sum or the difference between F_1 and F (Fig. 3.2),

$$F_2 = F_1 \pm F.$$

Amplification and filtering functions are performed totally in RF, which intro-

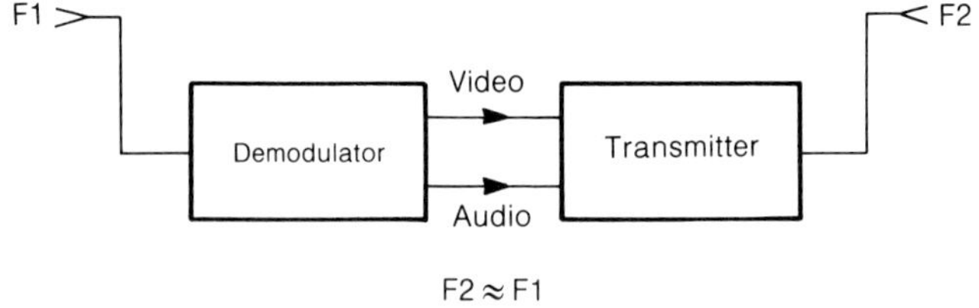

Fig. 3.1 Translator with demodulation.

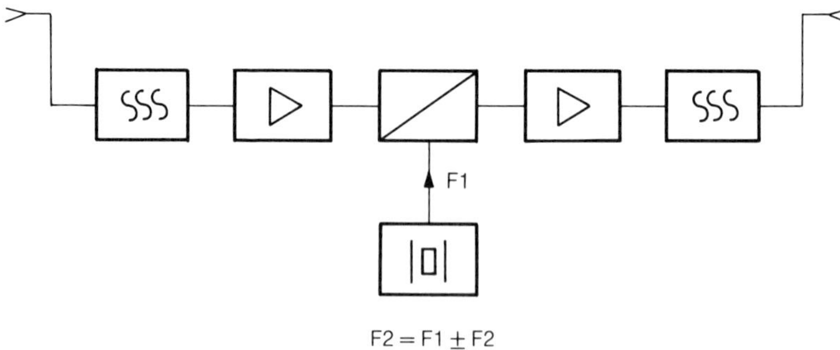

Fig. 3.2 Single frequency change translator.

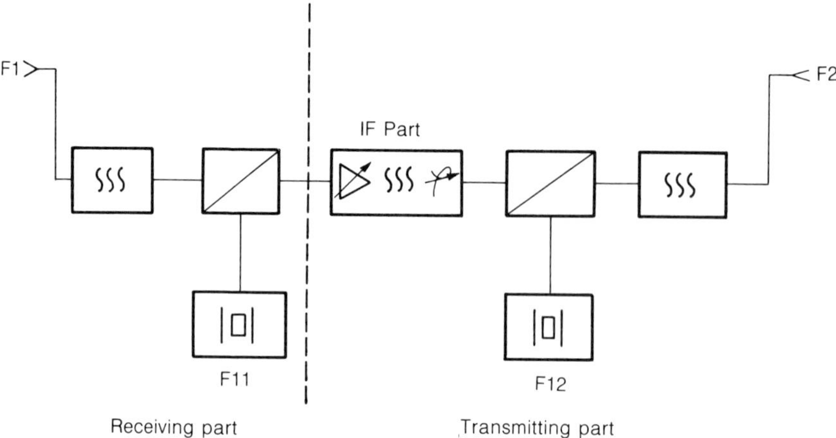

Fig. 3.3 Double frequency change translator.

duces limitations concerning stability and selectivity; therefore, it is often necessary to resort to retransmission by double frequency change (Fig. 3.3).

The important functions are, in this case, all performed in intermediate frequency:

1. generating the amplitude–frequency charateristic (filtering and selectivity);
2. linearization of the phase–frequency characteristic (correction of the group delay time);
3. amplification;
4. gain control.

The amplified and converted signals are of different types: the vision carrier is amplitude modulated whereas the sound carrier is frequency modulated. Simultaneous transmission of these two carrier signals introduces the phenomenon of transmodulation between the two channels due to amplification non-linearities, especially in the power stages. This problem can be resolved by separating the two channels; power amplification then has to be duplicated, thus constituting considerable extra expense.

Other problems appear: phase distortion, signal separation and combination instabilities are introduced. Simultaneous channel amplification, while neutralizing transmodulation phenomena by the use of non-linearity correction circuitry, turns out to be the preferred solution.

3.1.2 Optimization of antinomic couple noise factor/input stage linearity

The importance of the constant noise factor is that when the input level increases by n decibels, the signal-to-noise ratio is improved by n decibels at the translator output. This means that any antenna gain improvement or transmitter power increase leads to improved vision quality.

The theoretical expression for the noise factor at the input of the filter connected before the mixer is:

$$F_{\mathrm{Tm}} = F_1 \cdot L_{\mathrm{c}}(F_{\mathrm{FI}} + t - 1)$$

where F_1 is the filter loss, L_c is the mixer diode conversion loss, t is the diode noise temperature and F_{FI} is the intermediate frequency noise factor (Fig. 3.4).

The literal expression for the noise factor becomes:

$$F_{\mathrm{TA}} = F_1 F_2 + \frac{F_{\mathrm{Tm}} - 1}{G_2}$$

where F_{TA} is the total noise factor, F_1 is the input filter loss, F_2 is the noise factor and G_2 is the preamplifier gain.

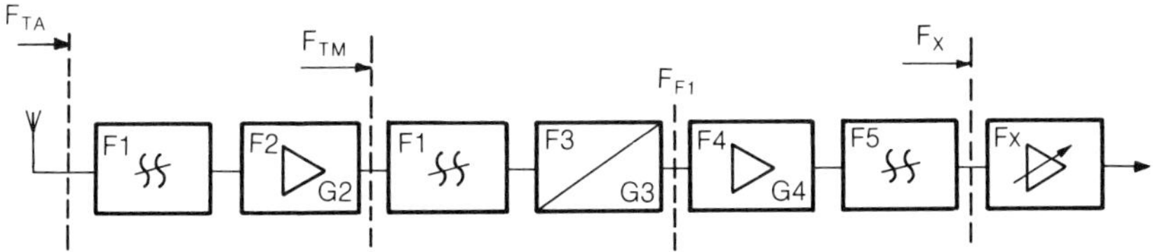

Fig. 3.4 Input schematic.

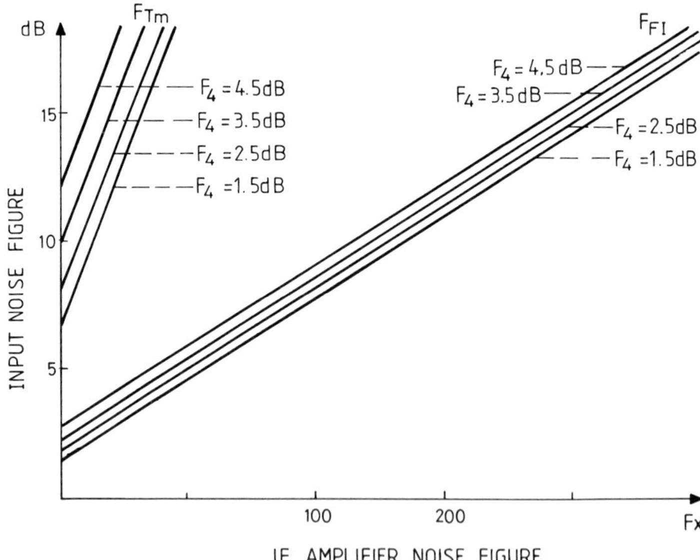

Fig. 3.5 Input noise figure versus IF amplifier noise figure.

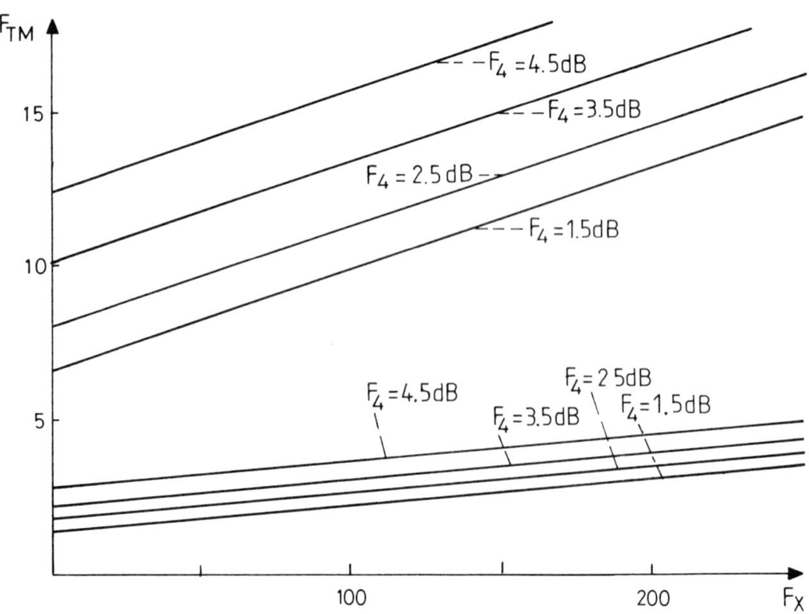

Fig. 3.6 Low noise preamplifier after mixer.

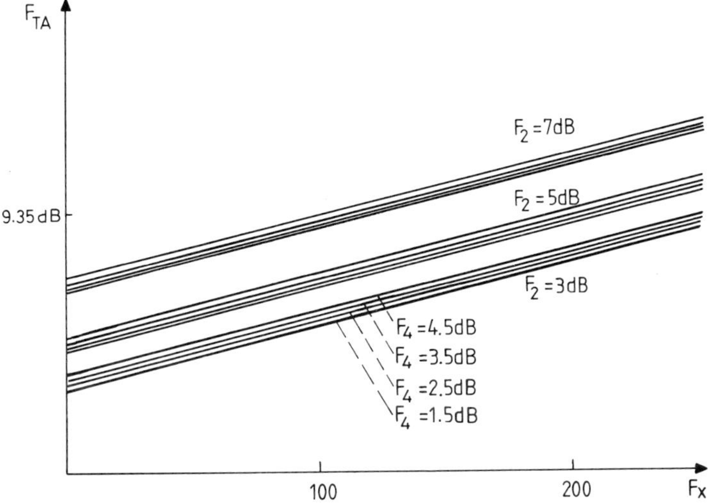

Fig. 3.7 Low noise preamplifier before mixer.

The graphs of Figs 3.5, 3.6 and 3.7 show that the use of a low-noise amplifier stabilizes the overall noise figure especially when placed in front of the mixer. The slope of the graphs $F_{Tm} = f(F_X)$ and $F_{TA} = f(F_2)$ decreases from 25×10^{-2} (Fig. 3.5) to 3.2×10^{-2} (Fig. 3.7) with a low-noise IF preamplifier placed before mixing and 2.5×10^{-2} (Fig. 3.6) with a UHF front-end low-noise amplifier. This represents a considerable improvement in the input noise figure stability (ratio = 10).

INTERMEDIATE FREQUENCY VARIABLE AMPLIFICATION

Gain control can be implemented in several ways:

1. by varying the current of one or more transistors;
2. by gain control at one or two points, inserting a variable attenuator between stages;
3. by alternately controlling the gain of several stages with a variable feedback.

The results obtained in these three cases will be very different as far as input noise figure stability and linearity of the amplitude–frequency input stages are concerned. In fact, the control must be implemented at the highest possible RF level in order to isolate the noise figure variations, inherent in gain variation, from the translator input noise figure.

The first solution is undesirable because of variations of the IF stage characteristics due to the change in transistor biasing and to linearity degradation following current variations.

The diagram corresponding to the second solution is shown in Fig. 3.8. The

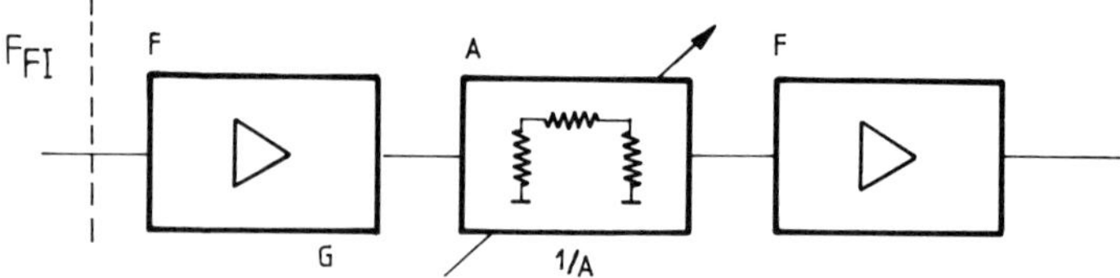

Fig. 3.8 Single point gain control.

total noise figure is:

$$F_{\mathrm{FI}} = F + \frac{A-1}{G} + \frac{F-1}{G(1/A)} = F + \frac{F \cdot A - 1}{G}.$$

For example for $F = 3\,\mathrm{dB}$, $G = 10\,\mathrm{dB}$, and $A = 40\,\mathrm{dB}$, which corresponds to a control dynamic range of ratio 100; we obtain a total noise factor of: $F_{\mathrm{FI}} = 0.2A + 1.9 \approx 33\,\mathrm{dB}$ (graph of Fig. 3.11).

This noise figure could be improved by increasing G. However, since attenuator A is a diode attenuator, its maximum operating level is around 250 mV, which limits the value of this parameter to the 10 dB level.

To reduce noise figure variations, it is necessary to envisage employing gain control at several points (Fig. 3.9). In this case, the total noise figure becomes:

$$F_{\mathrm{FI}} = F + \frac{\sqrt{A}-1}{G} + \frac{F-1}{GA^{-1/2}} + \frac{\sqrt{A}-1}{G^2A^{-1/2}} + \frac{F-1}{G/A}.$$

That is:

$$F_{\mathrm{FI}} = F + \frac{1}{G}\left[\frac{1+\sqrt{A}}{G}\right](F\sqrt{A}-1).$$

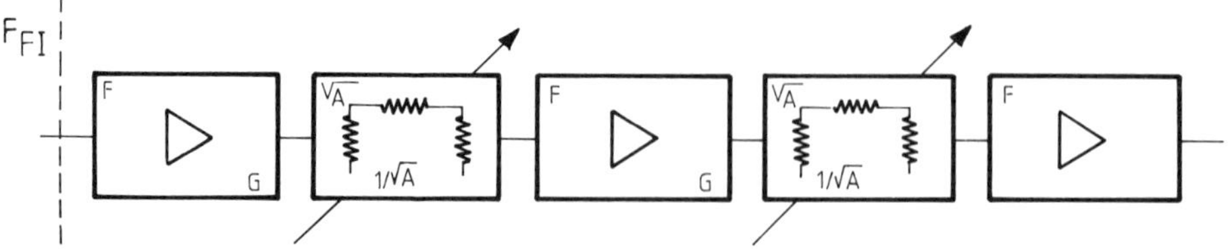

Fig. 3.9 Several point gain control.

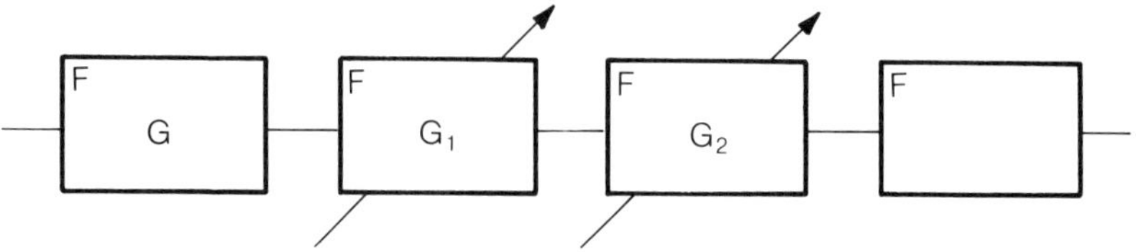

Fig. 3.10 G_1 and G_2 synchronously controlled.

The total noise figure variation F_{FI} will be much lower than in the first case.

$$F_{FI} = 2 \times 10^{-2} A + 1.9 \times 10^{-1} \sqrt{A} + 1.9$$

(See Fig. 3.11). Still for $F = 3$ dB, $G = 10$ dB and $A = 40$ dB, $F_{FI} = 23.09$ dB, which represents an improvement of 10 dB compared to single-point gain control, but is still much too high.

The last solution consists of alternately controlling the gain on two stages. The operating point of the transistors is constant (fixed biasing). The feedback increases with increase of input level, which means that the stage linearity is improved.

Taking the simplified case where gains G_1 and G_2 can be synchronously controlled without affecting the noise figure (Fig. 3.10), and with $0 \leqslant G_1$ and $G_2 \leqslant 20$ dB, we will have:

$$F_{FI} = F + \frac{F-1}{G} + \frac{F-1}{G \cdot G_1} + \frac{F-1}{G \cdot G_1 \cdot G_2}.$$

As

$$G_1 = G_2 = \frac{G_{1\max}}{\sqrt{A}}$$

$$F_{FI} = F + \left(\frac{F-1}{G}\right) \times \left(1 + \frac{1}{G_1} + \frac{1}{(G_2)^2}\right)$$

whence

$$F_{FI} = F + \frac{F-1}{G}\left(\frac{A}{(G_{1\max})^2} + \frac{\sqrt{A}}{G_{1\max}} + 1\right)$$

that is:

$$F_{FI} = 10^{-4} A + 10^{-2} \sqrt{A} + 3$$

$$F_{FI} = 7\,\text{dB}$$

for $F = 3$ dB, $G = 10$ dB, $G_{1\max} = 20$ dB and $A = 40$ dB.

OPERATIONAL ANALYSIS OF THE GAIN CONTROLLED STAGE (FIG. 3.12)

The transistor is connected in common-emitter configuration. The temperature stabilizing resistor (R_{12}) is connected in series with an HF feedback resistor R_{13}. The impedance of inductor CH_2 is of the same order as the resistance value of R_{12}. Similarly, R_{48} determines the load in the collector circuit (resistance offered to the following stage). Resistor R_{14} provides voltage feedback which stabilizes input and output admittances. The emitter feedback circuit is formed by the differential resistance of the diode connected in parallel with resistor R_{13}, which limits the stage gain variation dynamic range. When the gain control voltage (U_{cag}) increases, diode D_1 conducts and the gain of the stage increases. Conversely, if the absolute value of the gain control voltage decreases, the gain of the stage

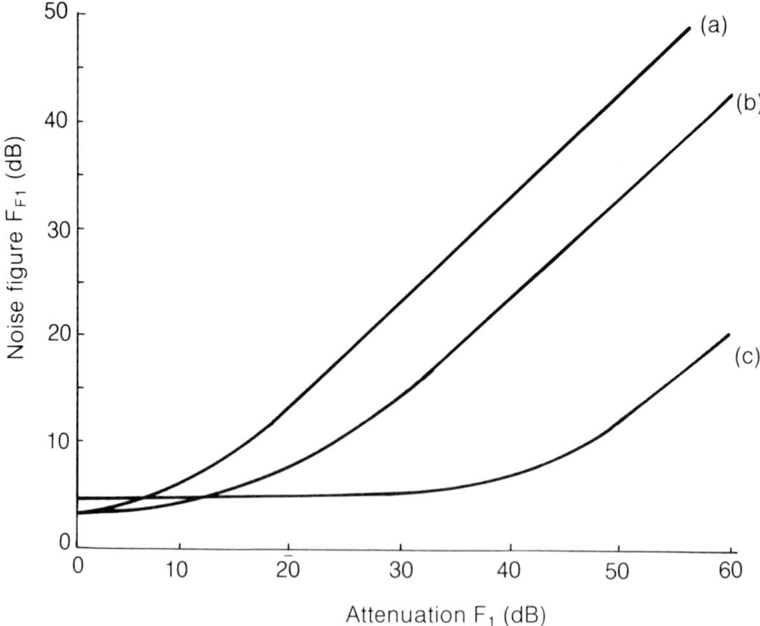

Fig. 3.11 Noise figure versus attenuation for different amplifier configurations; (a) single point gain control, (b) several point gain control, (c) synchronous control of G_1 and G_2. These correspond to Figs 3.8; 3.9; 3.10.

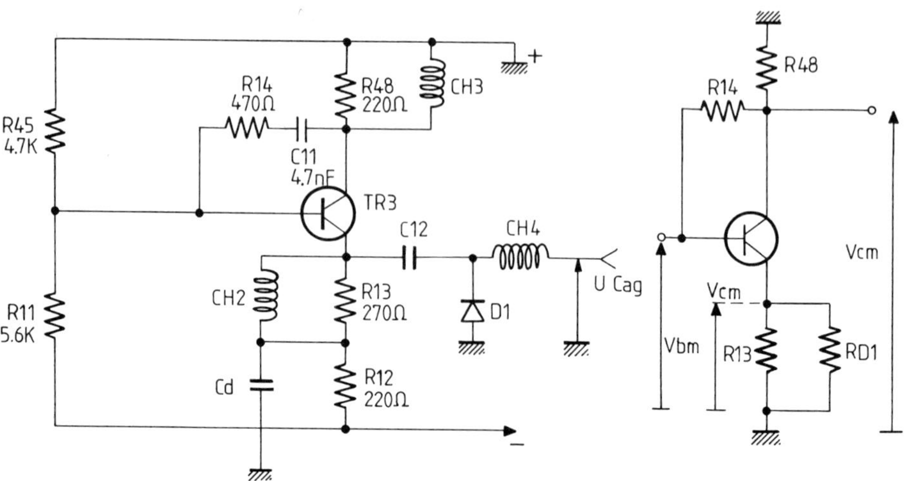

Fig. 3.12 Operational analysis of the gain controlled stage.

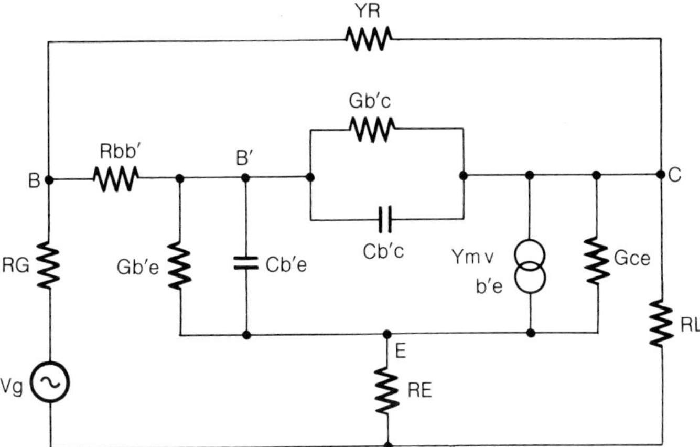

Fig. 3.13 Equivalent circuit. RE is a function of the diode differential and therefore of the AGC voltage.

decreases. The differential resistance of diode (b) varies with the AGC voltage and determines of RF gain of the stage.

The equivalent circuit is shown in Fig. 3.13.

V_g:	input generator	
RG:	input generator source resistance	
$R_{bb'}$:	base spreading resistance	
$G_{b'e}$:	conductance	
$C_{b'e}$:	capacitance	elements of the π equivalent circuit
$G_{b'c}$:	conductance	
$C_{b'c}$:	capacitance	
Y_m:	admittance of the transistor internal slope	
G_{ce}:	conductance	
RE:	emitter feedback resistance	
YR:	base-collector feedback admittance	
RL:	load resistance.	

RF FRONT END

The optimization of the input noise figure depends on four main factors:

1. low-noise VHF or UHF amplification before mixing;
2. double radiofrequency input filtering (Fig. 3.14);
3. image rejection mixing (Fig. 3.15);
4. active gain controlled IF amplifiers.

The preamplifier may be implemented by means of a thin-film microelectronic

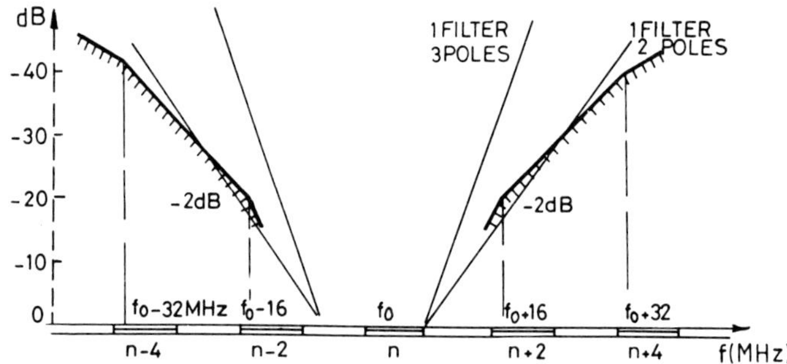

Fig. 3.14 Double RF input filtering.

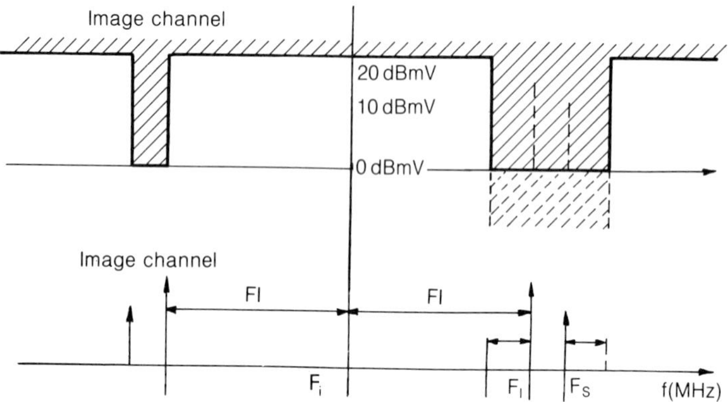

Fig. 3.15 Image rejection.

wide-band low-noise amplifier which enables all frequencies between BI and band V to be transmitted, channel input selection being achieved by double filtering.

The mixer is the image-rejection type. The main advantage of this type of mixer lies in the facto that the intermediate filter connected between the input preamplifier and the mixer proper can be simplified or even eliminated without degrading the overall noise factor.

GENERAL CHARACTERISTICS—EXAMPLE

RF preamplifier:

Operating frequency:	40 to 960 MHz
average gain:	14 dB
return loss—	
input:	⩽ 20 dB
output:	20 dB
typical noise factor:	> 3 dB
Maximum input level:	< 30 mV
intermodulation:	−70 dB for 30 mV input.

RF/IF mixer:

Operating frequency:	VHF and UHF
conversion loss:	6 to 7 dB
return loss	
RF input:	$\leqslant 20$ dB
IF output:	< 20 dB
Maximum input level:	$\geqslant 100$ mV

IF preamplifier:

Operating frequency:	intermediate frequency
maximum gain:	70 dB
dynamic control range:	$\geqslant 56$ dB
noise factor:	Maximum gain = 3.3dB
	gain reduced by 50 dB is <9dB
maximum output level:	0 dB for intermodulation of $\leqslant 70$ dB.

The preamplifier is preceded by a surface-wave IF filter as shown in Fig. 3.4. It is comprised of four transistor–amplifier stages: the input stage and the output stage are constant gain, the two intermediate stages are gain controlled by the use of PIN diodes connected as HF feedback in the emitter circuit. This set-up makes it possible to have a noise factor which is practically constant throughout the dynamic gain range of the preamplifier and to obtain an improvement in the linearity when the input level increases (increasing of the feedback).

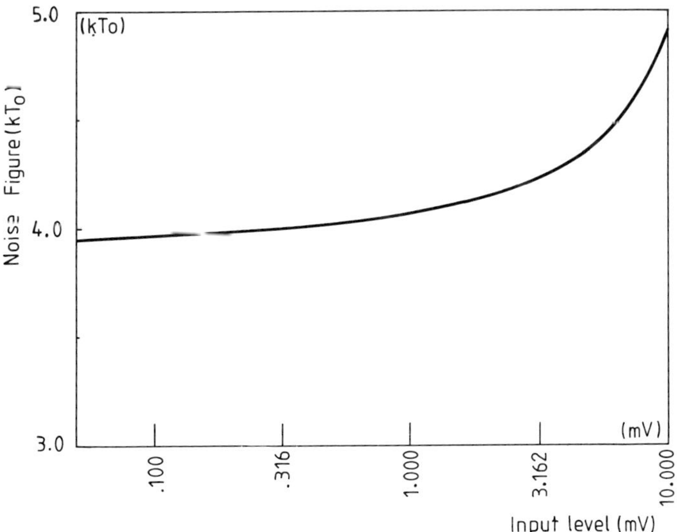

Fig. 3.16 Noise figure versus input level.

The expression for the preamplifier's noise factor is given by the formula:

$$NF_{\mathrm{dB}} = 2.3 \times \exp\left(\frac{-0.74G + 51.8}{G}\right)$$

where G is the gain of the preamplifier. The expression is valid for $20\,\mathrm{dB} \leqslant G \leqslant 70\,\mathrm{dB}$. The permitted level of input is between 50 μV and 30 mV without degradation of characteristics, both from the standpoint of intermodulation and noise. In fact, the noise factor is practically constant from 50 μV to 10 mV and does not exceed 10 dB for an input level of 30 mV (Fig. 3.16).

3.2 DEVELOPMENT OF TELEVISION TRANSMITTER MODULATION STAGES

Three main developments will be discussed: surface-acoustic-wave filters, non-linearity correction and wide-band output converters. The modulation stages can be classed under two categories. The modulation assembly includes the vision part (negative amplitude modulation) and the sound part (frequency modulation) (Fig. 3.17).

3.2.1 Vision and sound IF modulation

Due to the use of the most recent hybrid microelectronic and diffused circuits, the modulator, in spite of its complex functions can be accommodated in a very small 3/25th 'camac' drawer. In addition to excellent linearity and stability, the quality of the vision modulator is notably characterized by its high noise immunity in the extraction of the sync pulse.

The basic sync pulses provide the origin of the clamp, sync strip, sync regeneration and test line gate pulses. They are generated in a comparator which can sample the pulses coming from a very noisy low-level signal with extreme accuracy. The spurious noise present on the input video frequency signal has practically no further influence on the quality of the picture transmitted by the transmitter.

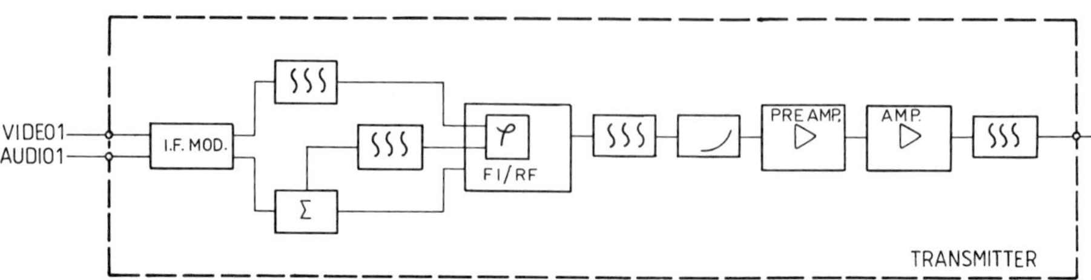

Fig. 3.17 Transmitter modulation stage.

BASIC SPECIFICATIONS

<u>Video section</u>

Input impedance	75 Ω
return loss	40 dB (to 6 MHz)
output impedance	50 Ω
video input level	0.7 V ± 6 dB
sync input level	0.3 V ± 6 dB
hum rejection	26 to 36 dB
noise conversion	−23 dB
back porch level	±0.5%
video frequency response	±0.2 dB until 6 MHz
differential gain	0.5%
Differential phase	0.5%
50 Hz square wave response	0.5% K
bart tilt	0.3% K
2T pulse shape	0.5 K
2T preshoot or overshoot	0.5%
20 T response	0.5% K
output level	300 mV
weighted noise	⩾67 dB

<u>Sound section</u>

input impedance	600 Ω balanced
output impedance	50 Ω
input level	−6 to +18 dB
frequency response	±0.5 dB (30 Hz to 100 kHz)
distortion	$<0.4\%$ 50 kHz deviation
pre-emphasis	50 or 75 μs
output level	200 mV
IF oscillator standard frequency stability	1×10^{-6} year

<u>Mechanical specifications</u>

Standard camac housing 3/25 wide.

In addition, this circuit is capable of separating sound-in-sync from the video signal, when the sound is transmitted in code in the sync pulses. The synchronous pulse of test lines 17 and 330 samples the white bar level and feeds an AGC circuit which regulates the luminance level for a variation of ±6 dB of input program amplitude. In the case of a loss of test line signal, an analogue memory holds the gain of the circuit constant at the level of the last test sampled, for some 24 hours. The original synchronization is stripped and replaced by a signal, correctly calibrated in time and amplitude. The video frequency clamping reduces the LF noise to 36 dB while still keeping a very low noise conversion (26 dB). The energy dispersed signal is thereby removed from the video wavefront.

The twin path white limiter, operating from 0 to 1 MHz, uses an original system which does not affect differential gain, even below limit level; in particular, the phase of the colour subcarrier is not affected by limiting. Modulation is performed in a symmetrical mixer which is not affected by variations of local frequency level.

The sound modulator is basically made up of a voltage controlled oscillator which generates the intercarrier frequency modulated by the audio frequency program.

In order to minimize the effects of interference, due to the colour subcarrier (intermodulation), the sound carrier is synchronized with the line frequency. This process allows the disturbance to be synchronized, and improves the resulting picture considerably.

3.2.2 IF vision corrector (Fig. 3.18)

Traditionally, the amplitude–amplitude characteristic is formed in a vestigial-sideband filter, with group delay time corrected by five to nine correction cells. These cells, in fact, have two functions: first, the linearization of the frequency phase characteristic, and second, the precorrection of the group delay time characteristic (1/2 correction or full correction depending on the standard).

International schedules of conditions show increased severity in selectivity and out-of-band protection. This has led us, initially, to produce for North America and South Africa, for instance, a twin filtering system (eight poles per filter) to which it is necessary to add an IF rejector, three group delay time correction cells required to linearize the phase and five cells to implement the precorrection.

The surface-acoustic-wave filter proves to be a simple method of overcoming this problem. This solution has been adopted for the standards M–N (North and South America) and B–G (Europe). Design study contracts with the United Kingdom and internal development work within the Thomson-CSF group, leads us to expect a generalization of this solution as concerns both the transmitter and the translator.The group delay time precorrection cells are produced in conjunction with a video-frequency low-pass filter. This improves the stability of the amplitude–frequency characteristics. See the graph of the group delay time precorrected low-pass filter in Fig. 3.19 for the B–G standard (Europe) and Fig. 3.20 for the M–N standard (North and South America).

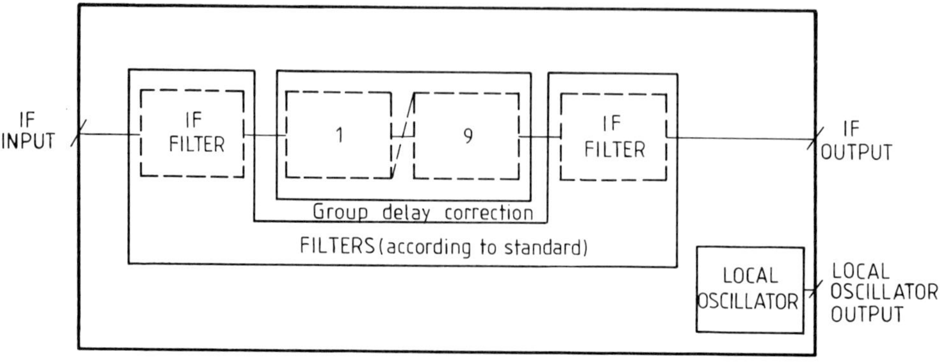

Fig. 3.18 IF vision corrector.

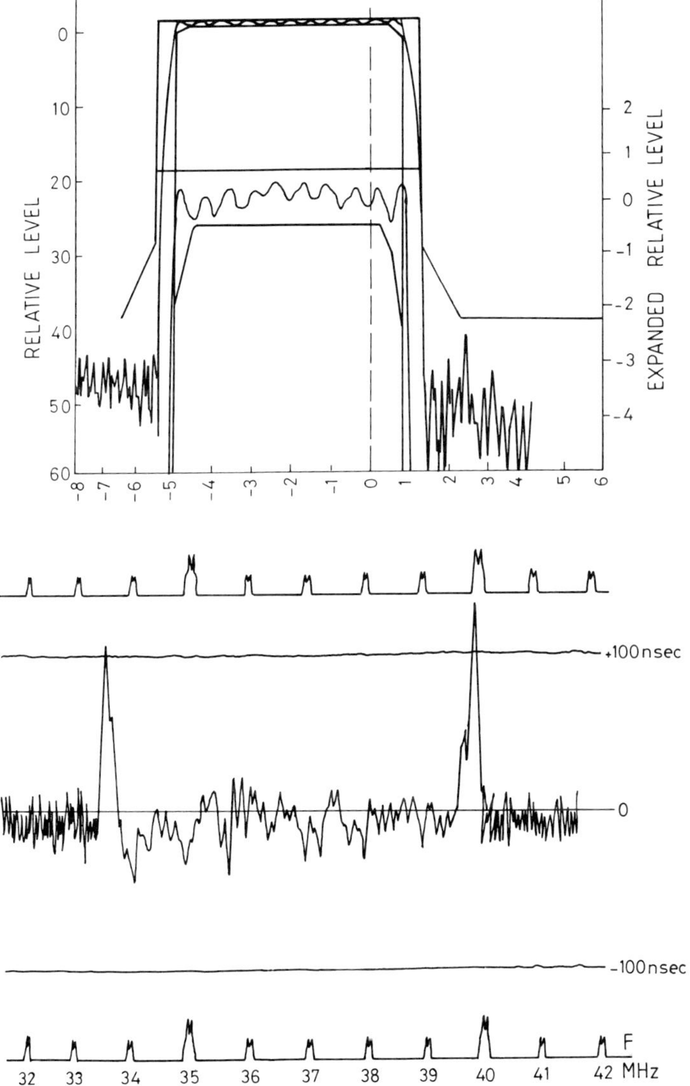

Fig. 3.19 Standard B–G graph of amplitude versus frequency with group delay time pre-correction.

3.2.3 Non-linearity correction of the vision channel

The power stages of a television transmitter introduce a large number of defects which can be suppressed either directly in VHF or UHF in the case of common amplification of the vision and sound channels (see following paragraph), or in

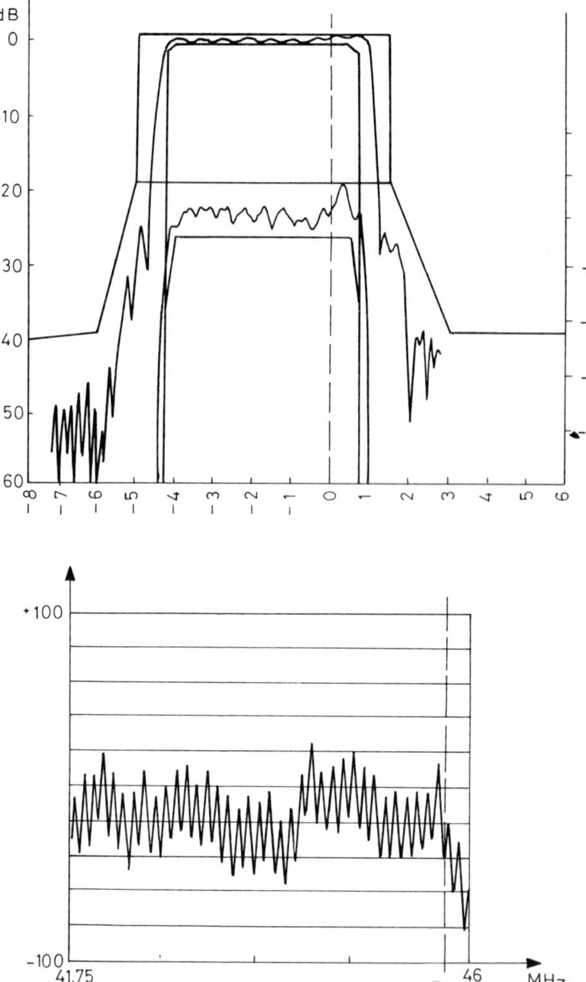

Fig. 3.20 Standard M–N graph of amplitude versus frequency with group delay time pre-correction.

the modulation stages. The non-linearity corrections of the vision channel are of two types as follows.

In video-frequency

Low-frequency corrections These corrections must work on the double band part of the modulation spectrum without affecting the single-sideband part. In other words, this circuit has to linearize the 'staircase type' video-frequency signal without, however, affecting the differential gain.

Differential phase corrections On the contrary, the phase–amplitude characteristic can be corrected without affecting the gain and the low frequency linearity. Its efficiency is better than 25%.

Non-linearity correction of the vision channel in intermediate frequency

Incidental phase correction This type of correction is indispensable when the power stages generate significant non-linearities as in the case of equipment featuring AB class biased transistor power stages or very powerful equipment coupled to a klystron transmitter.

The range of the amplitude–amplitude characteristic up to its saturation power produces phase modulation of the carrier at the rate of the so-called 'differential phase' incident video-frequency modulation signal. As a consequence, this produces a considerable degradation of the signal-to-noise ratio of the frequency modulated sound channel when it is demodulated as per the general 'intercarrier' process, the reference being taken at the level of the vision carrier.

Cross-modulation correction 'Cross-modulation' is a term meaning the percentage of sound carrier amplitude modulation due to transfer of vision modulation to the sound in the power stages. This amplitude modulation is for the most part removed by RF intermodulation corrections (see later). A ripple value less than 10% is generally accepted. Certain administrations impose main transmitter cross-modulation values which must be less than 1%. A special circuit is then required.

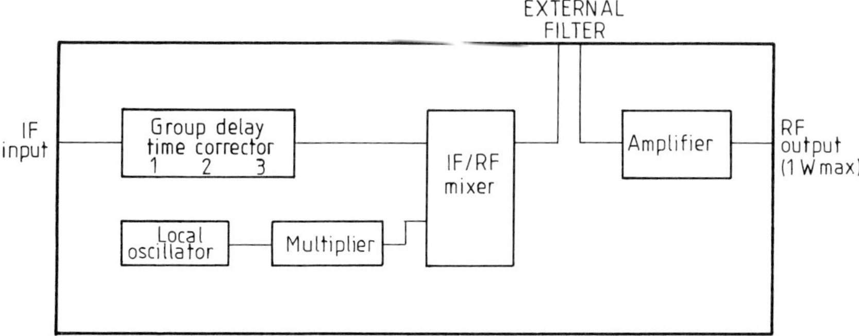

Fig. 3.21 Output wideband converter.

3.2.4 Output wideband converter (Fig. 3.21)

This subassembly requires no adjustment and provides all the functions of the standard converter, furthermore it incorporates a local frequency channel. It is made up of the following units:

1. a high level wideband IF–RF mixer (VHF and UHF);
2. a plug-in high stability thermostatically controlled oscillator;
3. a multiplier continuously adjustable by potentiometer;
4. a high-gain output amplifier permitting an output level of 1 W.

BASIC CHARACTERISTICS

Input frequency range IF:	30 to 45 MHz (IF)
output frequency range:	40 to 860 MHz
input level IF:	150 mV
output level:	1 W maximum
possible group delay correction:	0 to 300 ns
frequency stability:	2×10^{-7} month.

3.3 OPTIMIZATION OF THE OUTPUT DESIGN OF A TELEVISION TRANSMITTER OR TRANSLATOR AND ENHANCING THE LEVEL OF TRANSISTORIZATION

This optimization is achieved by correcting third-order non-linearities and by decreasing coupling losses of the power amplifiers (regrouping by Chebyshev impedance transformation).

3.3.1 Analysis of the distortions generated in power amplifiers

Intermodulation and cross-modulation When two signals are applied simultaneously to an amplifier, the non-linearity of the latter generates intermodulation and cross-modulation products. The level of these products determines the maximum operating level of the amplifier.

The measurement of intermodulation in television transposer equipment, with frequency modulated sound, is generally made by using three signals having the following levels with respect to peak sync power:

$$P_y = P_o - 8\,\text{dB}$$

$$P_s = P_o - 7\,\text{dB}$$

$$P_{sc} = P_o - 17\,\text{dB}$$

where P_o is the peak sync power level, P_y is the vision carrier level, P_s is the sound carrier level, and P_{sc} is the colour subcarrier level; at any frequency between 0 and 5.5 MHz. The level of the intermodulation products relative to the peak sync power, P_o, expressed in dB, gives the intermodulation of the measured amplifier. A limit of better than -51 dB is generally accepted under these measurement conditions.

The amplitude of the synchronous vision signal which appears on the sound carrier expressed as a percentage of the sound carrier is the cross-modulation.

A mathematical analysis of non-linearity, using a simplified model representing the law of the amplifier to be precorrected, shows that there is a relationship between intermodulation and cross-modulation. This means that if the intermodulation can be corrected, the cross-modulation will also be corrected, provided that the intermodulation generator law is representative of the law of the output amplifier.

It can be shown that if a signal of the form:

$$e_1 = A \cos a + B \cos b + C \cos c \tag{3.1}$$

is applied to an amplifier, the performance characteristic of which is represented by the following series:

$$e_s = K_1(e_1) + K_2(e_1)^2 + K_3(e_1)^3 \tag{3.2}$$

where A, B, C are the peak values of signals having frequencies given by:

$$a = 2f_a t$$

$$b = f_b t$$

$$c = f_c t$$

and where e_1 is the instantaneous voltage and K_1, K_2, and K_3 are complex numbers which represent the gain, amplitude distortion and the phase distortion of the amplifier respectively.

In the case of cross-compression, the gain at a given frequency decreases, when the level of a carrier at another frequency increases. It is clear that for well

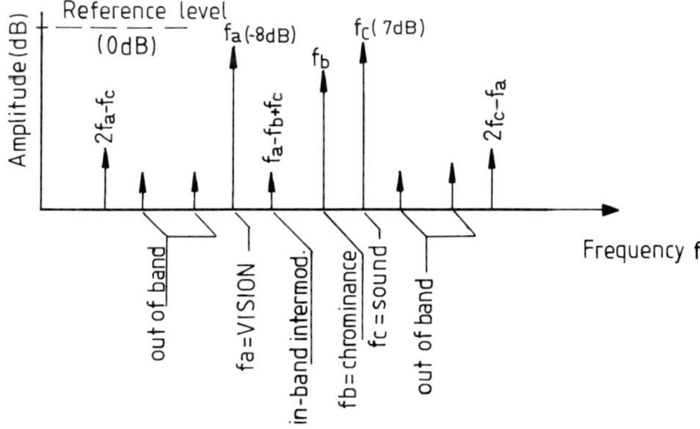

Fig. 3.22 Effect of third-order non-linearity.

defined A, B, C levels, the amplitude and phase of all components are determined by the third-order coefficient K_3.

The two components producing in-band intermodulation and vision to sound cross-modulation are:

$$\tfrac{3}{2}K_3 A\cdot B\cdot C\cdot\cos(a+b+c) \tag{3.3}$$

and

$$\tfrac{3}{2}K_3\cdot C\cdot A^2\cdot\cos c. \tag{3.4}$$

Effect of third-order non-linearity is shown in Fig. 3.22. Coefficient at frequency f_a is:

$$K_1 A + (\tfrac{3}{4})K_3 A^3 + (\tfrac{3}{2})K_3 AB^2 + (\tfrac{3}{2})K_3 AC^2$$

The levels of the signals at frequencies f_b and f_c can be calculated with identical expressions obtained by circular permutation.

Differential gain Differential gain may be defined as the change in level of a subcarrier between black level and white level. In the case of Standard L* for example, the vision signal may be synthesized by the combination of the three signals having the following relative levels:

vision carrier	$A = -4\,\text{dB}$
video frequency sidebands	$B = -9\,\text{dB}$
chrominance signal	$C = -26\,\text{dB}$

The sum of these three signals is approximately equal to the video frequency peak level. An analysis of the preceding paragraph reveals nine components at the same frequency as the input component.

The components which refer to the chrominance signal are those which represent differential gain since it is the variation of the chrominance level between black and white which defines G_{diff}.

$$G_{\text{diff}} = 1 + \left(\frac{3}{2}\cdot\frac{K_3}{K_1}\right) \times (1 - (0.3)^2)M^2 \tag{3.5}$$

where M is the peak white level.

Therefore our hypothesis also suggests that differential gain is affected by third order non-linearity components.

Differential phase Differential phase ϕ_{diff} is the peak-to-peak variation in the phase of the colour subcarrier from black level to white level. It can be shown that this change of phase is due to the third order factors.

$$\phi_{\text{diff max}} = \arctan\left[\left(\frac{3}{2}\cdot\frac{K_3}{K_1}\right)(1 - 0.3^2)M^2\right] \tag{3.6}$$

*Standard *L* is the French standard. The amplitude modulated sound is generally amplified separately from the vision.

Note that expressions (3.5) and (3.6) show that there is a relationship between third order intermodulation products, differential phase and differential gain. A single limit for intermodulation (61.5 dB for standard L) can therefore also determine the limit for the two parameters G_{diff} and ϕ_{diff}.

One obtains:

1. first-order components at a frequency identical to the input frequency;
2. second-order components representing a change in the average level of the harmonics; linear combinations of frequencies taken two by two (of the form $a + b$);
3. third-order components in a range of frequencies three times the input frequencies (of the form $3a$ and $2a + b$);
4. four components generated by combinations of three frequencies (of the form $a + b + c$).

This last group of third-order components contain components at frequencies identical to the input frequencies. When coefficient K_3 is positive, these components add to the first order output coefficients, which increase the gain and increase the level (expansion); when K_3 is negative, these components subtract from the first order output coefficients, reduce the gain, (compression). If K_3 is positive, the three components give 'self-expansion' if the level of the components is increased. If K_3 is negative the components give 'self-compression'. In a similar manner, six components produce 'cross-modulation' or 'cross-expansion'. Third order-components may be of the following forms:

(a) $\frac{1}{4}\cdot K_3\cdot A^3\cdot \cos 3a$
(b) $\frac{1}{4}\cdot K_3\cdot A^2\cdot B\cdot \cos(2a+b)$
(c) $\frac{3}{2}\cdot K_3\cdot A\cdot B\cdot C\cdot \cos(a \pm b \pm c)$
(d) $\frac{3}{4}\cdot K_3\cdot A^3\cdot \cos(a)$
(e) $\frac{3}{4}\cdot K_3\cdot AB^2\cdot \cos(a)$

3.3.2 Non-linearity correctors

The principle of operation of this sub-assembly is to insert in series with the stage requiring correction a circuit producing third-order distortion of the same amplitude but in opposite phase.

Assuming the gain K_1 of the network to be unity:

$$U_1 = U_e(1 + K'_3 U_e^2) \tag{3.7}$$

$$U_s = K_1 U_1(1 - K_3 U_1^2) \tag{3.8}$$

Replacing U_1 in (3.8)

$$U_s = K_1 U_e(1 + K'_3 U_e^2)[1 - K_3 U_e^2(1 + K'_3 U_e^2)^2]. \tag{3.9}$$

A graph of equation (3.9) $U_s = f(U_e)$ for an amplifier having six values of precorrection in the range $K'_3 = 0.05$ to 0.1 (linear axes) is shown in Fig. 3.23. In this way

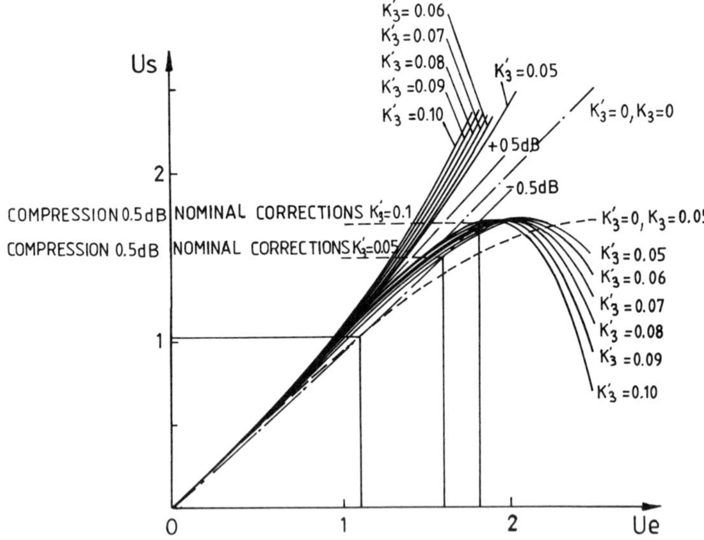

Fig. 3.23 Output *U* – versus input *U* – for six values of precorrection k'_3.

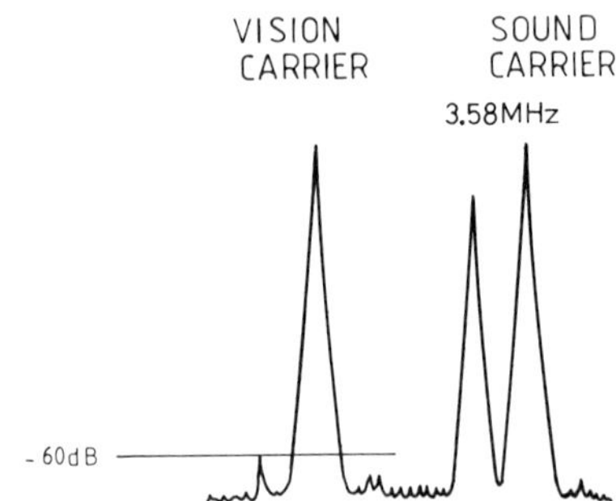

Fig. 3.24 Signal-to-noise with non-linearity correction.

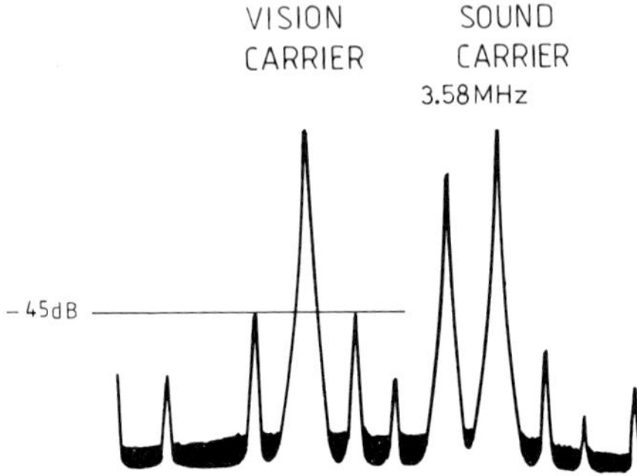

Fig. 3.25 Signal-to-noise without non-linearity correction.

for a given overall performance the output amplifier can be operated at twice normal power level. (compare Fig. 3.24 with correction to Fig. 3.25 without).

3.3.3 Amplifier assemblies

They have the following common factors.

1. Wideband, stripline techniques and the use of two transistors with splitting and combining at the input and output by means of 3 dB couplers.
2. Input and output matching to allow unlimited interchangeability and grouping (VSWR <1.2).
3. A transistorized current regulator, ensures that the amplifier is non-sensitive to temperature variations.
4. The transistors in the different paths are fed from two separate power rails to ensure program continuity.

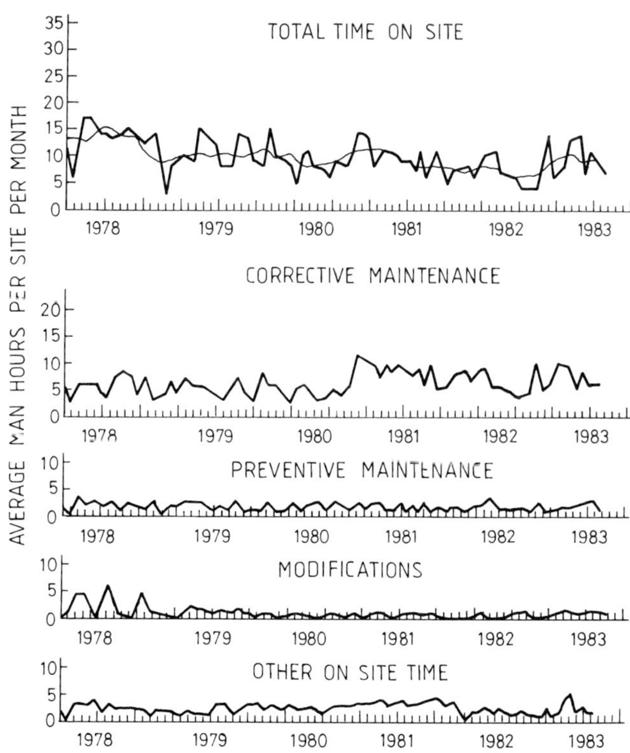

Fig. 3.26 Reliability and maintenance of UHF Mk1 tube equipment.

3.3.4 Improvement of reliability

Two types of equipment, of equal output power, are compared: the first has a transistorized driver, with a single power amplifier tube (LGT 50 V Mkl, UHF), and the second is completely transistorized and incorporates a UHF linearity corrector (LGT 50S UHF).

The graphs of Fig. 3.26 and 3.27 show the cumulative on-site time necessary to maintain the two types of equipment. They are expressed as man hours per station per month and are a good indication of the relative reliability of the two types of equipment. Average on-site maintenance time for 50 V MI LGT UHF transposer (y_1) and RUHF 50S LGT transporator is (y_2):

$$y = \frac{1}{n} \sum_{i=1}^{n} y_i$$

$y_1 = 11.09$ man-hours/site/month (Fig. 3.26) and $y_2 = 1.07$ man-hours/site/month (Fig. 3.7). The improvement ratio y_1/y_2 is greater than 10.

(We are indebted to the UK's Independent Broadcasting Authority for supplying the curves shown in Figs 3.26 and 3.27.)

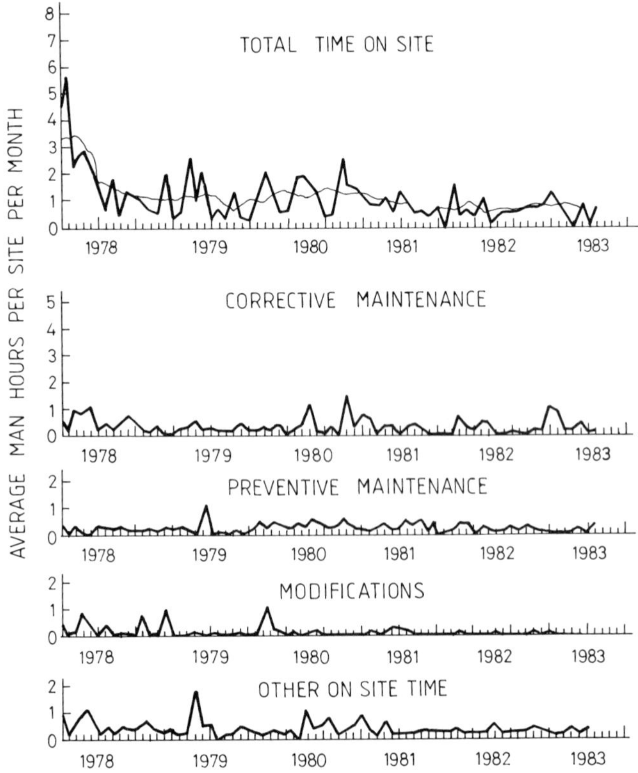

Fig. 3.27 Reliability and maintenance of solid state equipment.

3.3.5 Overall characteristics of transmitters

Frequency range:	VHF or UHF
power supply voltages:	117 V, 208 V, 220 V, 240 V } ±20%
power consumption:	1.2 KVA Max. (6 × 10 W programs)
cos φ:	0.9.

VISION TRANSMITTER

Carrier frequency stability:	10^{-6} per year
RF output impedance:	50 Ω
video input level:	1 V peak to peak onto 75 Ω
video AGC efficiency:	±6 dB with VITS
signal-to-noise ratio (vision transmitter section):	S/N weighted: 60 dB S/D weighted: 60 dB
clamp 30 dB or more for energy dispersal signal synchronized with vertical synchronization	Im ⩽ −54 dB
intermodulation products:	vision −8 dB colour subcarrier −16 dB sound } below peak sync power.

SOUND TRANSMITTER

AF input level:	0 dBm ± 6 dB balanced or unbalanced 600 Ω
signal-to-noise ratio (sound transmitter section):	S/N weighted = 70 dB (40 Hz to 15 kHz).

3.4 CONCLUSION

By carrying transistorization to the limit, the reliability of the equipment has been improved, in various practical realizations:

1. 500 W at band IV/V (Fig. 3.28);
2. 1 kW at band III (Fig. 3.29 VHF, Fig. 3.30 UHF);
3. considerable improvement in efficiency (Fig. 3.31): the insertion of a corrector doubles the efficiency of an equipment;
4. improved performance characteristics: at twice the normal power, the intermodulation products can be improved by 6 dB;

Fig. 3.28 500 W UHF transposer (passive reserve driver stages with automatic change over).

Fig. 3.29 2 × 1 kW VHF transposer (active reserve configuration).

Fig. 3.30 1 kW UHF transmitter.

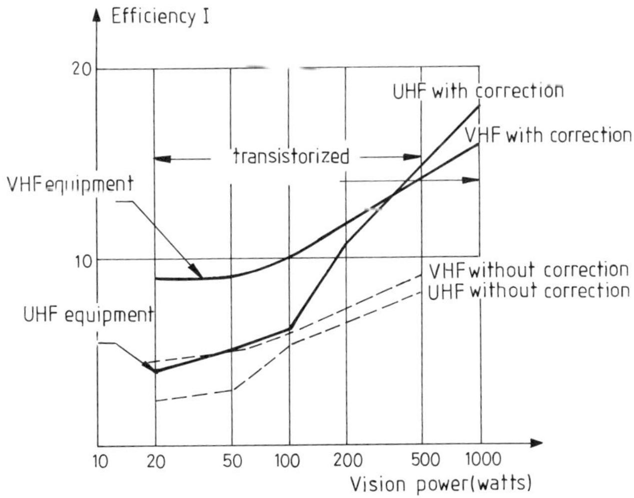

Fig. 3.31 Efficiency versus output power for different equipment types.

Fig. 3.32 100 W UHF transmitter.

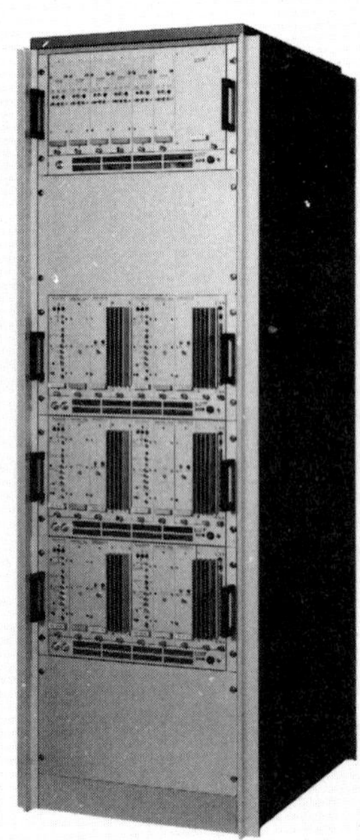

Fig. 3.33 Satellite television multiprogram reception and transmission station.

5. reduced maintenance costs: for equivalent output power, equipment employing a corrector used less than half the number of transistors in the final amplifiers;
6. single-rack, 100 W UHF or VHF transmitter (Fig. 3.32);
7. integrated equipment for satellite TV reception and terrestrial broadcast (Fig. 3.33).

APPENDIX 3.A The frequency spectrum and broadcasting channels

Table 3.A.1 Broadcasting bands

Low frequency (long wave)	(LF)	160–225 kHz	(1875–1176 m)	AM radio
Medium frequency (medium wave)	(MF)	525–1605 kHz	(571–187 m)	
Band I	(VHF)	41–68 MHz	(Channels 1 to 5)	
Band II	(VHF)	88–108 MHz		FM radio
Band III	(VHF)	174–216 MHz	(Channels 6 to 13)	405 line black and white television
Band IV	(UHF)	470–582 MHz	(Channels 21 to 34)	
Band V	(UHF)	614–854 MHz	(Channels 39 to 68)	625 line colour/black and white television

Table 3.A.2 Television channels, nominal carrier frequencies and wavelengths

Channel No.	*Frequency* (MHz)	*Vision wavelength* (m)	*Wavelength* (ft)	*Sound frequency* (MHz)	*Channel No.*	*Frequency* (MHz)	*Vision wavelength* (m)	*Wavelength* (ft)	*Sound frequency* (MHz)
Band I					*Band V*				
1	45.00	6.6621	21.8571	41.50	39	615.25	0.4873	1.5987	621.25
2	51.75	5.7931	19.0062	48.25	40	623.25	0.4810	1.5781	629.25
3	56.75	5.2827	17.3317	53.25	41	631.25	0.4749	1.5581	637.25
4	61.75	4.8549	15.9283	58.25	42	639.25	0.4690	1.5386	645.25
5	66.75	4.4913	14.7351	63.25	43	647.25	0.4632	1.5196	653.25
					44	655.25	0.4575	1.5011	661.25
Band III					45	663.25	0.4520	1.4830	669.25
6	179.75	1.6678	5.4719	176.25	46	671.25	0.4466	1.4653	677.25
7	184.75	1.6227	5.3238	181.25	47	679.25	0.4414	1.4480	685.25
8	189.75	1.5799	5.1835	186.25	48	687.25	0.4362	1.4312	693.25
9	194.75	1.5394	5.0504	191.25	49	695.25	0.4312	1.4147	701.25
10	199.75	1.5008	4.9240	196.25	50	703.25	0.4263	1.3986	709.25
11	204.75	1.4642	4.8038	201.25	51	711.25	0.4215	1.3829	717.25
12	209.75	1.4293	4.6893	206.25	52	719.25	0.4168	1.3675	725.25
13	214.75	1.3960	4.5801	211.25	53	727.25	0.4122	1.3525	733.25
					54	739.25	0.4077	1.3377	741.25

Band IV									
21	471.25	0.6362	2.0872	477.25	55	743.25	0.4034	1.3233	749.25
22	479.25	0.6255	2.0523	485.25	56	751.25	0.3991	1.3092	757.25
23	487.25	0.6153	2.0186	493.25	57	759.25	0.3949	1.2955	765.25
24	495.25	0.6053	1.986	501.25	58	767.25	0.3907	1.2819	773.25
25	503.25	0.5957	1.9544	509.25	59	775.25	0.3867	1.2687	781.25
26	511.25	0.5864	1.9239	517.25	60	783.25	0.3828	1.2558	789.25
27	519.25	0.5774	1.8942	525.25	61	791.25	0.3789	1.2431	797.25
28	527.25	0.5686	1.8655	533.25	62	799.25	0.3751	1.2306	805.25
29	535.25	0.5601	1.8376	541.25	63	807.25	0.3714	1.2184	813.25
30	543.25	0.5518	1.8105	549.25	64	815.25	0.3677	1.2065	621.25
31	551.25	0.5438	1.7843	557.25	65	823.25	0.3642	1.1947	829.25
32	559.25	0.5361	1.7587	565.25	66	831.25	0.3607	1.1832	837.25
33	567.25	0.5285	1.7339	573.25	67	839.25	0.3572	1.172	845.25
34	575.25	0.5212	1.7098	581.25	68	847.25	0.3538	1.1609	853.25

1. Frequencies for each channel are nominal and polarization is either horizontal or vertical.
2. Offset operation is used on UHF and VHF; on UHF it is either 0, +5/3, or −5/3 of line frequency; on VHF non-standard multiples of 1/12 of line frequency are used.
3. Carrier frequency tolerances on UHF are + or − 500 Hz.

Table 3.A.3 Correspondence between channel numbers and assigned frequencies for the 12 GHz satellite broadcasting band

Channel No.	*Assigned frequency* (MHz)	*Channel No.*	*Assigned frequency* (MHz)
1	11 727.48	21	12 111.08
2	11 746.66	22	12 130.26
3	11 765.84	23	12 149.44
4	11 785.02	24	12 168.62
5	11 804.20	25	12 187.80
6	11 823.38	26	12 206.98
7	11 842.56	27	12 226.16
8	11 861.74	28	12 245.34
9	11 880.92	29	12 264.52
10	11 900.10	30	12 283.70
11	11 919.28	31	12 302.88
12	11 938.46	32	12 322.06
13	11 957.64	33	12 341.24
14	11 976.82	34	12 360.42
15	11 996.00	35	12 379.60
16	12 015.18	36	12 398.78
17	12 034.36	37	12 417.96
18	12 053.54	38	12 437.14
19	12 072.72	39	12 456.32
20	12 091.90	40	12 475.50

Note: UK channels are 4, 8, 12, 16 & 20 orbit position 31 °W polarization left hand circular.

Table 3.A.4 Proposed broadcast satellite parameters for the frequency band 11.7 to 12.5 GHz

Type of modulation	FM	
Number of lines	625	
Sound sub-carrier frequency	6 MHz	
Peak–peak deviation	13.3 MHz	
Peak deviation of sound sub-carrier	50 kHz	
Receiver equivalent rectangular noise bandwidth	27 MHz	
Angle of elevation	15°	40°
Luminance signal-to-unweighted noise ratio for 99% worst month	34 dB	33 dB
Sound signal-to-weighted noise ratio for 99% of worst month	51 dB	50 dB

Table 3.A.6 Television systems used in various countries

Country	Systems used in bands			
	I/III		IV/V	
Algeria (Algerian Democratic and Popular Republic)	B, E/PAL	(13)(16)	G*, H*/PAL	(13)(16)
Germany, Federal Republic of (as was)	B/PAL		G/PAL	
Netherlands Antilles	M			
Saudi Arabia, Kingdom of	B			
Argentine Republic	N			
Australia	B/PAL	.(22)	*/PAL	
Austria	B/PAL		G/PAL	.(1)
Belgium	C, B/PAL	.(17)	H/PAL	
Brazil, Federative Republic of	M/PAL		M/PAL	
Bulgaria, People's Republic of	D/SECAM		K/SECAM	
Burundi, Republic of	K1*	.(16)	K1*	.(16)
Canada	M/NTSC		M/NTSC	
Cameroun	K1*	(15)(16)	K1*	(15)(16)
Central African Republic	K1*	.(16)	K1*	.(16)
Cyprus, Republic of	B		H*	(2)(18)
Colombia, Republic of	M		M*	
Congo, People's Republic of the	K1*	.(16)	K1*	.(16)
Korea, Republic of	M			
Ivory Coast, Republic of the	K1*	.(16)	K1*	.(16)
Cuba	M		M	
Dahomey, Republic of	K1*	.(16)	K1*	.(16)
Denmark	B/PAL		G*	
Egypt	B	.(16)	G*, H*	(14)(16)
Group of Territories represented by the French Overseas Post and Telecommunications Agency	K1			

Table 3.A.5 Characteristics of the principal television systems

Parameter	System code									
	A	M	(N)	B	C	G (H)	I	D K (K1)	L	E
Lines per picture	405	525	(625)	625	625	625	625	625	625	819
Field frequency (Hz)	50	60	(50)	50	50	50	50	50	50	50
Line frequency (Hz)	10125	15734	(15625)	15625	15625	15625	15625	15625	15625	20475
Video bandwidth (MHz)	3	4.2		5	5	5	5.5	6	6	10
Channel bandwidth (MHz)	5	6		7	7	8	8	8	8	14
Nearest edge of channel relative to vision carrier (MHz)	1.25	−1.25		−1.25	−1.25	−1.25	−1.25	−1.25	−1.25	+/−2.83
Sound carrier frequency relative to vision carrier (MHz)	−3.5	4.5		5.5	5.5	5.5	6	6.5	6.5	+/−11.15
Width of vestigial sideband (MHz)	0.75	0.75		0.75	0.75	0.75 (1.25)	1.25	0.75 (1.25)	1.25	2
Vision modulation polarity	Positive	Negative		Negative	Positive	Negative	Negative	Negative	Positive	Positive
Sound modulation FM	AM	FM +/ −25 kHz		FM +/ −50 kHz	AM	FM +/ −50 kHz	FM +/ −50 kHz	FM +/ −50 kHz	AM	AM
pre-emphasis (μs)		75		50	50	50	50	50		

1. Figures quoted are nominal.
2. Data in brackets refers to system code shown in brackets at top of column.
3. For further information reference should be made to Report 624 (Rev. 76) of the interim meeting of CCIR Group 11, Geneva 1976.

Spain	B		G	.(2)
United States of America	M/NTSC		M/NTSC	
Ethiopia	B*	.(16)	G*	.(16)
Finland	B/PAL		G/PAL	
France	E		L/SECAM	
Gabon Republic	K1*	.(16)	K1*	.(16)
Ghana	B*, G*	.(16)	G*	.(16)
Greece	B*		G*	.(3)
Guinea, Republic of	K1*	(15)(16)	K1*	(15)(16)
Upper Volta, Republic of	K1*	.(16)	K1*	.(16)
Hungarian People's Republic	D/SECAM		K/SECAM	
India, Republic of	B			
Indonesia, Republic of	B*			
Iran	B		G	
Ireland	A, I/PAL	.(4)	I*	
Iceland			G*	(2)(5)
Israel, State of	B		G*	.(6)
Italy	B/PAL		G/PAL	
Jamaica	N			
Japan	M/NTSC		M/NTSC	
Jordan, Hashemite Kingdom of	B		G*	
Kenya, Republic of	B*	.(16)	G*, I*	.(16)
Kuwait, State of	B		G*	.(19)
Liberia, Republic of	B*	(8)(16)	H*	(8)(16)
Libyan Arab Republic	B*	.(16)	G*	.(16)
Luxembourg	C		L*	.(2)
Malaysia	B		G*	
Malawi	B*	(10)(16)	G*	(10)(16)
Malagasy Republic	K1*	.(16)	K1*	.(16)
Mali, Republic of	K1*	.(16)	K1*	.(16)
Morocco, Kingdom of	B		H*	

Continued

Table 3.A.6 *continued*

Country	Systems used in bands			
	I/III		IV/V	
Mauritius	B			
Mauritania, Islamic Republic of	K1*	.(16)	K1*	.(16)
Mexico	M			
Monaco	E		L*	
Niger, Republic of	K1*	.(16)	K1*	.(16)
Nigeria, Federal Republic of	B	.(16)	I*	.(16)
Norway	B/PAL		G*	.(3)
New Zealand	B/PAL			
Uganda, Republic of	B	(9)(16)	G*	(9)(16)
Pakistan	B			
Panama, Republic of	M			
Netherlands, Kingdom of the	B/PAL		G/PAL	.(21)
Peru	M		M	
Poland, People's Republic of (as was)	D/SECAM		K/SECAM	
Portugal	B		G	
Portugese Overseas Provinces	I*	.(16)	I*	.(16)
German Democratic Republic (as was)	B/SECAM		G/SECAM	
Rhodesia	B	.(10)	G*	.(10)
Romania, Socialist Republic of (as was)	D		K*	.(2)
United Kingdom of Great Britain and Northern Ireland	A		I/PAL	
Rwanda, Republic of	K1*	.(16)	K1*	.(16)
Senegal, Republic of the	K1*	.(16)	K1*	.(16)
Sierra Leone	B	(11)(16)	G*	.(16)
Singapore, Republic of	B		G*	.(20)
Somali Democratic Republic	B*	.(16)	G*	.(16)
Sri Lanka (Ceylon, as was), Republic of	B			

South Africa, Republic of	I*	.(16)	I*	.(16)
Sweden	B/PAL		G/PAL	
Switzerland, Confederation of	B/PAL		G/PAL	.(7)
Surinam	M			
Tanzania, United Republic of	B*I*	(12)(16)	I*	(12)(16)
Chad, Republic of the	KI*	.(16)	KI*	.(16)
Czechoslovak Socialist Republic (as was)	D/SECAM		D/SECAM	
Overseas Territories for the international relations of which the Government of the United Kingdom of Great Britain and Northern Ireland are responsible	B*I*	.(16)	I*	.(16)
Overseas Territories of the United Kingdom in the European Broadcasting Area			H*	.(2)
Togolese Republic	KI*	.(16)	KI*	.(16)
Turkey	B		G*	
Union of Soviet Socialist Republics (as was)	D/SECAM		K/SECAM	
Uruguay, Oriental Republic of	N			
Venezuela, Republic of	M			
Yugoslavia, Socialist Federal Republic of (as was)	B/PAL		G/PAL	
Zaire, Republic of	KI*	(15)(16)	KI*	(15)(16)
Zambia, Republic of	B*	(10)(16)	G*	(10)(16)

(This table is reproduced from Annex 1 of CCIR report 624-Rev. 76-Study Group 11. It has been edited to account for political changes)
* planned (whether the standard is indicated or not); not yet planned, or no information received.
./ the abbreviation following the stroke indicates the colour transmission system in use (NTSC, PAL or SECAM)
(Figures in brackets refer to the following notes)

Notes:

1. Austria reserves the right to the possible use of additional frequency modulated sound carriers, in the band between 575 and 675 MHz in relation to the picture carrier.
2. The indications and notes are based on indications and notes given in Chapter 2 of the 'Technical data used by the European UHF/VHF broadcasting conference'.
3. No definite decision has been taken about the width of the residual sideband but this country is willing to accept the assumption that for planning purposes the residual side band will be 0.75 MHz wide.

Table 3.A.6 Notes *continued*

4. System 1 will be used at all stations. In addition, during a transition period, transmissions on system A will be made from Dublin and Sligo stations.
5. This country does not at present intend to use band IV and V but accepts the parameters given in the table under 'Standard G' as television standard in bands IV and V.
6. No final decision has been taken about the width of the residual side band but for planning purposes this country is willing to accept the assumption of a residual side band 1.25 MHz wide.
7. The Swiss administration is planning to use additional frequency-modulated sound carriers, in the frequency interval between the spacings of 3 and 6 MHz in relation to the picture carrier, at levels lower than or equal to the normal level of the sound carrier, for additional sound tracks or for sound broadcasting.
8. Liberia accepted for planning purposes standard B or II but reserves the right to adopt Standard M.
9. Uganda is already committed to standard B in band III. Standard G is planned for IV and V although further consideration will be given to other standards when band IV and V are to be commissioned.
10. Indications for Malawi, Rhodesia and Zambia are based on indications for Rhodesia and Nyasaland Federation given in the Final Acts of the African UHF/VHF Broadcasting conference Geneva 1963. Standard B is now in use in band I: no final decision is taken regarding systems to be used in bands III, IV and V.
11. Sierra Leone now uses Standard B but reserves the right to use any other standard compatible with the plan.
12. Tanzania, the indications are based on indications for Tanganika and Zanzibar given in the Final Acts of the African VHF/UHF Broadcasting Conference, Geneva 1963. It is intented to use standard B in bands I and III. Although Standard I is planned for bands IV and V further consideration will be given to the use of standards G and H.
13. Algeria reserves the right to change later.
14. The Arab Republic of Egypt is now studying the adoption of either standard G or H for bands IV and V.
15. In Cameroon, Zaire and Guinea, planning has been based on standard KI, but they reserve the right to use any other standard compatible with the plan when they introduce television.
16. The indications and notes 10 to 17 are based on indications and notes given in the Final Acts of the VHF/UHF African Broadcasting Conference, Geneva 1963.
17. Belgium will use standard C in bands I and III until April 1977, after which standard B will be used.
18. Cyprus is already committed to the use of standard B in band III. Standard II is envisaged for use in bands IV and V although further consideration will be given to the possible use of other standards when stations operating in bands IV and V are to be commissioned.
19. In Kuwait, if the services are called upon to broadcast in a second language, the frequency between 55 MHz and 65 MHz could be used to provide an additional frequency modulation sub carrier.
20. Singapore reserves the right to use additional frequency modulation sound channels in the band between 5.5 and 6.5 MHz in relation to the picture carrier, for additional sound channels for sound broadcasting.
21. Some existing transmitters operate with a residual sideband up to 1.25 MHz. For the future, only transmitters with a residual sideband of 0.75 MHz are foreseen.
22. Australia uses nominal modulation levels as specified for system I.

4

Radar systems

Michel-Henri Carpentier

4.1 THE HISTORY OF RADAR

A radar system (radar for *ra*dio *d*etection *a*nd *r*anging) is a system which transmits electromagnetic waves in a given part of space, receives the waves reflected by the various 'targets' existing there, and processes them in order to detect them and (except in rare examples) to determine some of their characteristics. Those characteristics could vary: it could be the horizontal position of the targets, their height, their speed and possibly their shape.

4.1.1 Before 1935

The word *radar* is in fact the name officially adopted by the US Navy in November 1940, which is relatively late, since radar systems already existed some time before, but were called differently i.e. in the USA: radio echo equipment, and in France: DEM for *détection éléctromagnetique.*

Contrary to what has very often been written, radar activity was not suddenly implemented by one person (namely Sir Watson-Watt), even if Watson-Watt was essential in the evolution of the radar activity. Radar activity existed progressively, resulting from the actions of many people, who made radar what it is today. And of course many people today contribute to its continuing evolution.

For example, in 1886, Hertz made the basic observation that there was no difference between light and electromagnetic waves (except for the wavelength), and as early as 1904 Hulfsmeyer described his 'telemobiloscope' transmitting 'electric waves reflected by metallic objects on the sea (that is ships)... to inform the captain on the bridge... of the position of an approaching ship... so as to avoid accidents...' (Electrical Magazine 1904, Phillips, 1978). Marconi, during the course of a lecture given at the Institute of Radio Engineers in New York in 1922, referred to the fact that 'it should be possible to design apparatus by means of which a ship could radiate... a divergent beam of rays... which rays, if coming across a metallic object such as another... ship, would be reflected back to a receiver... and reveal the presence of the other ship...' (Phillips, 1978).

Between 1922 and 1927, some scientists performed experiments on electromagnetic detection, such as, Taylor and Young (Naval Research Laboratory) with a 5 m wavelength, M. Mesny and P. David with a 1.8 m wavelength. In June 1930 L. A. Hyland obtained, by chance, the detection of a plane crossing a beam of 'radio waves' (9 m wavelength) and that led the NRL (Taylor, Young and Hyland) to experiment systematically until 1934 into the possibility of radio-detection in metric wavelengths, detecting planes up to 80 km. At the same time (1934), on the other side of the Atlantic, David was experimenting on a similar system for detecting planes up to 10 km, after which 30 bistatic CW 'DEM' systems were installed in France before the Second World War.

In 1934, CSF people Ponte and Gutton installed a bistatic CW system in microwaves (16 cm wavelength) on the ship Oregon to detect icebergs, while GEMA systems (Germany) obtained detection up to 10 km with 50 cm wavelength.

All those systems used CW transmission, and the distance was measured only indirectly (parallax). Using metric wavelengths, detection up to 100 km was rapidly possible, but using microwaves (wavelength less than 1 m) was more difficult as microwave technology did not yet really exist.

4.1.2 Since 1935; the pulse radar

In 1935, following the famous memorandum by Watson-Watt, the British simultaneously used two essential tools: metric wavelengths (as many others), and transmission of regularly spaced short pulses allowing for the direct measurement of the (radial) distances of the targets, by measuring the time difference between transmission and reception.

Installation of the equipment of the Chain Home in 1937, working at around 10 m wavelength (with a range around 150 km), was proven to be really efficient in detecting possible enemy raids: similar equipment was then made at shorter wavelengths to be installed on board of ships and planes. When the USA decided to join their forces with those of the UK, the pulse radar benefitted from a tremendous effort: at the end of the Second World War, the 'Allied' powers had deployed a complete catalogue of radars of all types, working at wavelengths between 20 m and a few centimetres. At the same time, the German catalogue, not supported at the same level by their authorities, was not so complete and not so efficient.

It should be noted that the action by French scientists Gutton and Berline in the field of decimetric resonant segment magnetrons 'greatly improved by the introduction of a large oxide cathode'. As indicated by Dr Megaw (1946), 'in spite of his recommendations of the use of thoriated tungsten... the oxide cathode has been preferred "by them".' Sixteen cm magnetrons, which had already given pulse powers of the order of 1 kW, were brought to Wembley by Dr Ponte of the CSF. This was the starting point of the use of the oxide cathode in practically all subsequent (allied) pulsed transmitting valves, and as such, a significant contribution to British radar. The date was the 8 May, 1940.

The typical pulse radar equipment as it was built during the Second World War and as it is sometimes still built today, was as follows:

1. a transmitter emitted regularly spaced short and powerful pulses, radiated in space by an antenna concentrating maximum transmission in a relatively angularly reduced zone of space;
2. after transmission, the antenna (generally the same) was used for reception, then connected to the receiving system by means of a duplexing device;
3. the receiver mixed the received signals with a local frequency, in order to shift down the received frequency to a more convenient lower frequency (intermediate frequency (IF)) at which amplification and filtering was achieved. (Use of amplification in IF instead of amplifying and filtering at zero frequency was necessary to reduce the importance of internal noise. Today the use of low-noise RF amplifiers before shifting down the received frequency could change the architecture of similar future radar equipments.) After IF amplification and filtering, the envelope of the received pulses was obtained by detection, allowing for range measurement by comparison of their time of occurrence compared to the time of transmission.

4.1.3 The angular measurement

Angular measurement was obtained by using a directive antenna and measuring its angular position when the received signal was maximum. Basically, three types of radars systems were used: the panoramic surveillance ('early warning') radar systems, the height-finders, and the tracking radars.

The panoramic radars used an antenna of large horizontal span and reduced vertical height, rotating around a vertical axis, with a beam narrow in azimuth, but like a fan in elevation, allowing for the measurement of the azimuths of the targets, in a so-called 2D arrangement.

When a target was detected (and its horizontal coordinates measured), if the height of the target was requested (for interceptor aircraft guidance), another radar was used for that purpose. This radar used an antenna reduced in horizontal span but large in height, moving up and down around an horizontal axis in a 'beam nodding' mode. This antenna, directed approximately horizontally towards the target, obtains the measurement of the elevation of the target, completing the (3D) measurement of three coordinates of the target. Many beam-nodding height finders were needed if many targets had to be controlled at the same time. In some cases, instead of moving up and down all the antenna system, the reflector was kept fixed (in elevation) but the beam was moved by moving the actual position of the primary feed in front of the reflector by using a convenient arrangement (Robinson or Foster arrangement).

In case a permanent accurate angular measurement of the target direction was needed (to ensure guidance of weapons), a circular antenna was used with its axis approximately directed towards the target, while an offset primary feed rotated in front of it, delivering a conical motion of a pencil beam. Resulting amplitude modulation of the signal received from a target not in the axis allowed for the measurement of the angular position of the target, and generally a tracking feedback organization tried to reduce to a minimum the amplitude modulation to keep the target as close as possible to the antenna axis.

Obviously, the angular measurement was only valid if the signal received from the target was constant, in such a way that the amplitude modulation was only coming from the motion of the beam compared to the target direction. Any natural or artificial fluctuation of the power reflected by the target introduces an error in the angular measurement, which could be drastic, especially in case of a voluntary fluctuation by an uncooperative target.

This is the reason why, in the 1950s, monopulse angular measurements were introduced. Basically, in a monopulse arrangement, the antenna provides (at least on reception) several different beams (equivalent to consecutive beam positions in former systems) at the same time, and comparison (made on the same received pulse) of respective received signals (in phase and/or in amplitude) allows the possibility of angular measurements with fewer problems coming from the target fluctuation. In case of a tracking radar, three beams are enough to completely determine the angular position of the target (even if for convenience four are often used). In the case of a panoramic rotating surveillance radar, beams stacked in elevation could be used: the pair of consecutive beams where the signals are the highest is first determined and then a finer evaluation is made by comparing the signals received in those two beams. Air defence radars organized according to that principle began to be used by the end of the 1950s (FPS 7 and FPS 27 in USA, THD 1955 in Europe, for instance) under the name of 'stacked-beam 3D radars'.

It should be noted that such stacked-beam arrangements have only been possible to design and build with fairly reasonable qualities (regarding the spurious radiation in particular) because powerful enough computer-aided design of the antenna became feasible.

4.1.4 Pulse compression and coded radars (Carpentier, 1988)

In a conventional pulse radar, quality of the range measurement is improved when the duration of the transmitted pulses is reduced, but at the same time, the bandwidth of the (IF) receiver must be widened in the relevant way, thus increasing the power of the parasitic noise accompanying the useful signals, and consequently obliging increases in the transmitted peak power.

Improving the quality of the range measurement was thus leading to a proportional increase of the transmitted peak power, with all the consequences on the detectability of the radar, and on the volume and the weight of the transmitter resulting from the increase of the peak power and the relevant increase of the high voltage. That was the immediate result from the use of only amplitude (pulse) modulation, as opposed to the possible use of phase–frequency modulation within the transmitted pulse. The tremendously successful use of magnetrons in radar systems was probably the reason for not implementing any phase modulation in transmitted signals, since magnetron-type tubes were practically unable to be conveniently phase-controlled during the transmission. Other types of valves could have been used but the high performance of magnetron radars probably prevented the scientists from implementing other architectures.

During the Second World War or immediately after, some scientists (Cauer and Kronert, Dickie) had presented the possibility of using frequency modulation on transmission, associated with a relevant filtering on reception as it was on some radar altimeters. But it is probably the book published in 1950 by Woodward which really drew attention to the interest of radars transmitting frequency–phase modulated pulses.

The first radar of that type was probably developed by the end of 1950s in the Naval Research Laboratory (USA), in which the transmitted pulses were linearly frequency modulated while a dispersive delay line was used on reception, exactly matched to that frequency modulation. This means that a copy of the transmitted signal after crossing the dispersive delay line contained no frequency modulation, and gave a signal duration approximately equal to the inverse of the spectral width of the transmitted signal Δf, achieving a time compression by a ratio equal to $T_p \Delta f$ (T_p being the pulse duration), and also a multiplication of the signal-to-noise ratio by $T_p \Delta f$ (assuming that the bandwidth was Δf before crossing the dispersive delay line).

In the beginning of the 1960s similar devices had been made (such as the Macbeth radar in France), in which the transmitted signal was binary phase-coded according to a pseudo-random sequence and correlators were used on reception achieving correlation with a replica of the transmitted signal, providing similar final results. Use of pulse compression or equivalent correlation-type receivers is now generalized in most of the military radar systems.

4.1.5 Doppler filtering

At the early beginnings of the radar activity (before the Second World War), especially when the systems were working in continuous wave, some scientists (David) had designed radars systems in which measurements concerning target speeds were achieved by means of Doppler frequency evaluation, but nothing really operational was implemented from that.

During the Second World War, Doppler processing was patented (by Busignies working with ITT in the USA), designed and implemented with the only purpose of discriminating radially moving targets from fixed echoes, in order to cancel the returns from the last ones, in the so-called moving target indication (MTI) fixes.

In those systems the basic job is to compare, from return to return, the phase-difference between transmission and reception. If this difference remains constant the target is assumed to be stationary and then the relevant returns are cancelled, if not they are kept. So the main problem was to store the relevant information from pulse to pulse, that is over a period of time around a millisecond. In the beginning the relevant storage was obtained with a quicksilver delay line, followed by quartz delay-lines, all analogue devices delivering spurious signals (out of time) and not fully compatible with a modification of the interval between pulses, while such a modification (a 'wobulation' of the pulse repetition frequency) is necessary to avoid stroboscopic ambiguities coming from the fact that the pulse repetition frequency (prf) is generally much smaller than the Doppler frequency-shift. In this case any targets whose Doppler frequency-shift is equal to zero modulo the prf are cancelled together with the fixed echoes (the Doppler frequency-shift is twice the ratio between the radial velocity v_R and the wavelength λ).

On the other side, the importance of the spurious signals delivered by those analogue (quicksilver or quartz) delay lines limited to 20 dB the quality of the MTI devices (defined as the improvement of the ratio 'power of mobile targets/power of fixed targets' in the linear zone of—the dynamic range—of the fix). Using storage-tubes (memory tubes) as offered in the 1960s by CSF, was a way to have aperiodic storage devices but unfortunately a too heavy and expensive one.

The use of digital devices (CMOS shift-registers) to store the information was, in the middle of the 1960s, the best way to cope with those problems in the French CSF for three main reasons: because of the accuracy inherent to many digital devices (provided the analogue to-digital encoding be accurate and fast enough), the intrinsic aperiodicity of CMOS shift registers, and because of the possibility of more easily controlling the shape of the filtering curves with digital devices. Fixed echo rejection became more and more efficient with digital MTIs and compatible with more sophisticated filtering functions.

Present evolution is towards the use of specific digital signal processors, using relatively classical parallel computing architectures and achieving better performance with less accuracy requested for the coefficients used in computing. Of course, in any case, performance in Doppler measurement depends on the intrinsic phase stability obtained during transmission for the transmitter and for the various requested sources of frequency.

At present, some combat aircraft radars where a high repetition frequency mode is used, extreme quality of phase stability is obtained, achieving excellent performance in clutter (parasitic echoes mainly coming from the earth) rejection.

4.1.6 Electronic scanning

Most of the antennas used in radar systems up until the end of 1950s were derived from optical systems, using a primary feed in front of a reflector. They had many disadvantages.

1. Direct radiation of the primary feed not reflected by the reflector 'spills over' it, introducing parasitic radiation (spill-over sidelobes) generally in a direction not far from perpendicular to the axis of the antenna, which is particularly bad in the case of a radar installed on a mobile platform (a ship for instance—see section 4.2.13). Spill-over sidelobes obviously disappear when using an array antenna.

2. Moving the main radiation from one direction to another one distant from the first requires mechanical movement and takes time. As a result, it is practically impossible to share the radiation of transmitted joules in order to match with the situation by sending more joules in difficult (or interesting) directions than in easy (or non-interesting) ones.
3. Deeply changing the shape of the antenna pattern is generally impossible. As a result several radar equipments are necessary if several functions are required.
4. Controlling the parasitic radiation is impossible. As a result, if parasitic radiation is important in the direction of a jamming system (or in a direction of heavy clutter), practically nothing could be done to cope with that.

Electronically steered phased-array antennas do not have those troubles: the direction of the main radiation may be deeply changed quasi-instantaneously, the antenna pattern may be completely changed extremely rapidly, and some means exist to control the configuration of sidelobes to cope with jamming problems.

This is the reason why designing and building electronically steered phased-array antennas for radar systems was a dream since the 1950s. Problems were various and mainly economical, the price of the necessary phase-shifts and the relevant arrangement was high, and in the case when the antenna is very large and or if the instantaneous bandwidth (the bandwidth of the transmitted signal) is very large (to obtain good range resolution—see section 4.2), phase shifters are not enough and controlled delay lines are needed.

The first operational example of electronically phased-array antennas is probably the radars installed on the USS aircraft carriers built at the end of the 1950s (USS Enterprise). Presently, at least in military radars, the advantages coming from 'electronic scanning' are enough to compensate for the relevant increase in complexity: electronic scanning in a vertical plane to replace the stacked-beam arrangements in air defence radars, electronic scanning in elevation as well as in bearing in most of the modern weapon systems. Regarding the combat aircraft, the necessity of a very low level of parasitic radiation (section 4.2.13) requires a very good accuracy in the control of the phase of the phase shifters which increases the difficulty, and that is the reason why complete electronic scanning is not yet operationally used today (for the moment) in combat aircraft radar systems.

4.2 GENERAL DESCRIPTION OF RADAR SYSTEMS

4.2.1 Basic principles derived from the theory of radar systems (Carpentier, 1988, Woodward, 1950)

The basic theory of radar systems is based upon the following assumptions.

First of all, the noise accompanying the 'useful' return from a target is gaussian (it has a gaussian amplitude distribution): gaussian means that the noise has the maximum disorder. If the actual noise (coming from the internal noise of the radar receiving system, or from an outside source such as natural or artificial jammers) is not gaussian, using a so-called ideal receiver as defined later could be non-optimum, which means that the actual performance in that case could be better than expected from the conclusion of a theory only valid, strictly speaking, when the noise is gaussian.

Second, the target is supposed to be a point target, that means a target whose radial dimension is zero. The ideal receiver is proven, in that case, to be optimum regarding the detection and the range localization of that point target. If the target is definitely not small compared to the range resolution of the system, the 'ideal receiver' could not be ideal and, for instance, better results would be obtained by using a filter matched not to the transmitted signal, but to the signal as modified (in phase, in amplitude, in duration) by the target. A consequence of that is also that improvement of range resolution could be obtained by sacrificing detection performance.

Third, the theory of the 'ideal receiver' assumes that there is, a priori, no particular place or zone where the presence of a target is significantly more likely than elsewhere. If this is not the case, radar receivers should have to take that into account (which is practically achieved in tracking radars receiving a 'designation' from a surveillance radar, by limiting the zone under consideration to the zone where the target could be).

Under the above assumptions, results of the radar theory are as follows. Let us call $S(t)$ the transmitted signal (duration T) and $\phi(f)$ the associated Fourier transform. Let us call $y(t)$ the received signal (sum of a possible useful signal returning from a target and of (gaussian) noise $n(t)$). Ideal reception is obtained by computing

$$C(t_o) = \int y(t)_x S(t - t_o)\,dt$$

if the noise power is constant in the interesting frequency band (where there are the spectral components of a possible return): white noise.

The probability that there is a target (a return) at a distance $ct_o/2$ (c is the speed of the light), which means that there is a return received t_o after transmission, is proportional to $C(t_o)$. Using a threshold for $C(t_o)$, above which we assume that there is a return, will then fix a false alarm probability P_f and hence a detection probability P_d.

Calling R the energetic signal-to-noise ratio given by

$$R = \frac{2E}{N_o}$$

where E is the energy of the received (useful) signal and N_o is the spectral power density of the parasitic (white) noise (the ratio between the power of the noise and the bandwidth of positive frequencies that it occupies), the decision threshold being K times the standard deviation of $C(t_o)$ in the absence of an useful signal,

P_f and P_d are given by

$$P_f = \frac{1}{2\pi}\int_K^{+\infty} \exp\left(-\frac{v^2}{2}\right) dv$$

$$P_d = \frac{1}{2\pi}\int_{K-\sqrt{R}}^{+\infty} \exp\left(-\frac{v^2}{2}\right) dv.$$

As a consequence if P_f and P_d are fixed (ultimately by the customer, directly or not), that fixes the value of R. We usually say the quality of detection depends only on R, instead of saying—more correctly—that the quality of detection determines R. That means that improving the quality of detection (decreasing P_f or increasing P_d) could be obtained only by increasing R.

Figure 4.1 shows the graph of P_d versus R for $P_f = 10^{-3}$, 10^{-5}, 10^{-10} (R is given in decibels, by taking $10\log_{10} R$). In case of a transmitted signal $S(t)$ of constant power during its duration T frequency modulated in such a way that its spectrum be practically limited to a bandwidth Δf, the power signal-to-noise S/N will be, if the noise has been limited to the band Δf, given by

$$\mathrm{S/N} = \frac{E/T}{N_o \Delta f}$$

$$\mathrm{S/N} = \frac{R}{2T\Delta f}.$$

In that case R is twice the power signal-to-noise ratio, multiplied by $T\Delta f$ (later called the pulse compression ratio). $R/2$ is also the power signal-to-noise ratio at the output of the ideal receiver.

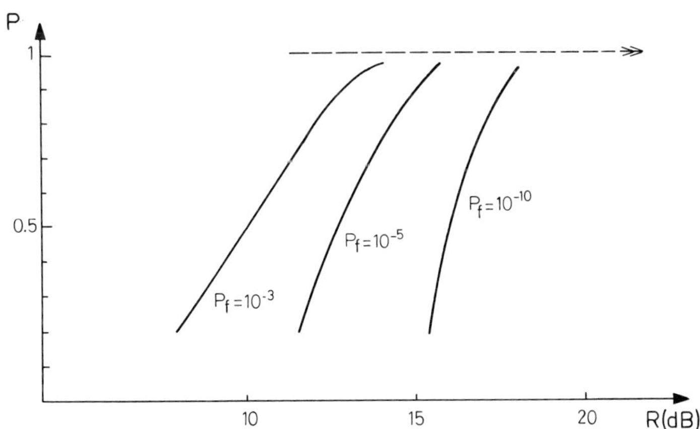

Fig. 4.1 Detection probability versus energy signal-to-noise ratio R (non-fluctuating target).

One major difference between a a classical pulse radar and a coded (or pulse compression) radar is the fact that, in the first case, because the transmitted signal was not frequency modulated $T\Delta f$ was close to 1, while in coded radars $T\Delta f$ could be very large.

To use an ideal reception is to decide that there is a return for a value of t_o for which $C(t_o)$ is above the threshold and maximum. Existence of noise $n(t)$ generally results in a position of the maximum which is not the real position of the return but close to it (if R is large enough). It is proven that the error made in deciding that the position of the return is that of the maximum of $C(t_o)$ is random (of course), gaussian, with zero mean value and with a standard deviation (Woodward's formula)

$$\frac{1}{2\pi B \times R^{1/2}}$$

in which B is the second order moment or the gyration radius of $|\phi(f)|^2$ (the power spectrum of $S(t)$), that is to say that

$$B^2 = \frac{\int (f - f_o)^2 \times |\phi(f)|^2 \, \mathrm{d}f}{\int |\phi(f)|^2 \, \mathrm{d}f}$$

where f_o is the central frequency of the transmitted signal.

That means in particular that for a given quality of detection, the error in range measurement is proportional to the increase of B. Since B is generally close to $\Delta f/3$ (between $\Delta f/2$ and $\Delta f/4$), for a given quality of detection the error in range measurement is proportional to the inverse of Δf. So, to obtain a good detection, R needs to be high enough, which means—since the noise spectral density generally does not really depend on the radar designer but from the state of the art in technology or from the jamming people—that E needs to be high enough. E directly depends on the energy transmitted during the time T of detection. And a conclusion is that in most of the cases the quality of detection essentially depends on the number of joules transmitted during the time T of detection. For a given number of transmitted joules, the larger T is, the smaller the relevant power is, the more difficult is the detection of the radar transmission by other people, and the smaller are the price and volume of the radar transmitter.

Accuracy of range measurement basically depends on transmitted bandwidth Δf. Clearly a good solution is when T is large as well as Δf, which implies that $T\Delta f$ be large. On the other side, on first approximation, range resolution is found to be equal to $(c/2) \times (1/\Delta f)$ or close to it.

Practically there are two basic means which are used to make an 'ideal reception'. The first is to work according to the above formula, which means to achieve the multiplication of the received signal by various replicas of the transmitted signal separated in time by the range resolution (by $1/\Delta f$) and to integrate each product during the time T. In fact to be complete, it would be required for each distance under consideration not only to multiply the received signal by one replica of the transmission, but by various replicas shifted in frequency by all the possible Doppler shifts expected for possible targets (separated by $1/T$). Practically, in general, there is only one multiplication by distance, but the signal out from each multiplication is filtered by a low-pass band filter (cut-off frequency of $1/T$) to obtain the integration for non- (or slowly) moving targets, and by equivalent contiguous bandpass filters centred around $1/T, 2/T, \ldots$ up to the useful maximum (Fig. 4.2). In general a useful return gives significant results in two adjacent 'Doppler filters' for two adjacent 'range gates', the real (in range and Doppler) position being obtained, if requested, by interpolation (looking for a barycentre). It has to be noted that the approximation of ideal reception which consists of using identical benches of Doppler filters for each distance is not valid if the transmitted signal is (pseudo) randomly phase-coded in which case multiplication for each distance by the various replicas of the transmitted signal shifted by all possible Doppler-shift frequencies (separated by $1/T$) is necessary.

The second means for making an ideal reception is to apply the following mathematical identity:

$$C(t_o) = \int y(t) \times S(t - t_o)\,dt \equiv \int Y(f) \times \phi^*(f) \times \exp(2\pi i f t_o)\,df$$

where $Y(t)$ is the Fourier transform of the received signal and $\phi^*(f)$ is the

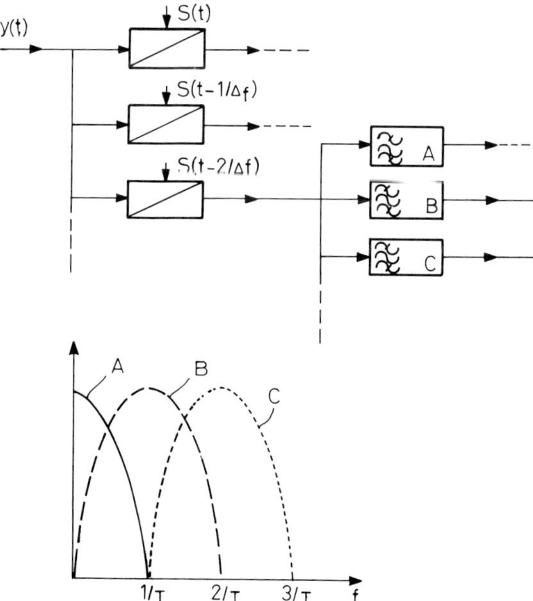

Fig. 4.2 Block diagram of a correlation radar.

conjugate of the Fourier transform $\phi(f)$ of $S(t)$. The identity means that if the signal $y(t)$ crosses a filter whose transmittance is $\phi^*(f)$, it will get out a signal whose Fourier transform is $Y(f) \times \phi^*(f)$, and which eventually is $C(t)$. A filter whose transmittance is $\phi^*(f)$ is called a matched filter and delivers at the output the wanted $C(t)$ when crossed by $y(t)$.

One main characteristic of a matched filter is that if it is crossed by a Dirac impulse (a very short pulse), at the output it delivers $S(-t)$, which is a replica of the transmitted signal $S(t)$ but time-reversed. It is the same thing to say that a matched filter could be defined by its impulse response which has to be identical to the transmitted signal but time-reversed.

Another characteristic is that if $S(t)$ is phase (or frequency) modulated, which implies that $\phi(f)$ is a complex function of f, if a replica of $S(t)$ (such as a useful signal) crosses the matched filter, it will deliver a signal with a Fourier transform equal to $\phi(f) \times \phi^*(f) = |\phi(f)|^2$ which is a real function of f. The conclusion is that a matched filter completely removes the frequency (phase) modulation of any useful incident signal which crosses it. This is why a useful signal of duration T and spectrum width Δf, after crossing the matched filter gives a (non-frequency modulated) signal whose duration is around $1/\Delta f$, ensuring not only a multiplication by $T\Delta f$ of the power signal-to-noise ratio but also a shortening of the useful signal by a ratio of $T/(1/\Delta f) = T\Delta f$ (pulse compression ratio).

But it must not be forgotten that in fact the role of the pulse compression filter is the same as that of a correlator: the signal obtained at the output of the pulse compression filter when crossed by an useful signal also has the shape of the autocorrelation function of $S(t)$: that means that if the duration of the zone where the compressed signal is significant is around $1/\Delta f$, the total duration of the compressed signal is twice ($2T$) the duration of $S(t)$, the main (central) compressed signal being surrounded by spurious ones called sidelobes. Here again, even if it is officially necessary to use several filters matched not only to $S(t)$ but also to $S(t)$ frequency-shifted by all the possible Doppler shifts (separated by $1/T$), in practice most often (and particularly when using chirp or quasi-chirp signal, see section 4.2.10) only one matched filter is used, followed by a bank of Doppler filters.

Above, the noise accompanying the useful return has been assumed to be white. In cases when the noise is not white, a 'whitening' filter has to be used (whose effect is to deliver a white noise at its output from the actual noise), followed by a filter which has to be matched not to $S(t)$ but to $S(t)$ as modified by crossing the whitening filter. The result is that when the input noise has crossed the whitening noise and the new matched filter, most often it is not at all white at the output. Practically, in most cases the whitening filter is only followed by a normal matched filter, for reasons of convenience and simplicity (moreover it should be indicated that in most cases, completely ideal reception, in case of a coloured noise, cannot be theoretically and *a fortiori* practically implemented).

Another important remark regarding ideal reception, which is particularly important in the case of surveillance radar systems, is that normally the receiver has to be matched, in fact, not to the transmitted signal, but to the signal expected from a point target, taking into account the amplitude modulation coming from the motion of the antenna beam.

Problem: as an example, let us consider a panoramic radar with an antenna of 9.54 m in span, rotating at 6 rpm, the transmitted signal being a succession of pulses with 10 μs duration, repeating at 1 kHz. Each pulse is linearly frequency modulated between 9990 MHz and 10010 MHz. The antenna is uniformly illuminated.

1. What is the shape of the received signal to which the receiver has to be matched?
2. What is the shape of the corresponding Fourier transform?

Solutions.

1. (See chapter on antennas, Volume 2.) The amplitude of the signal received from a non-fluctuating target varies like $\sin^2(9.54\pi\theta/\lambda)/(9.54\pi\theta/\lambda)^2$ versus the azimuth θ (in radians). Since $\theta = 2\pi t/10$ the amplitude of the signal received from a non-fluctuating target varies like ($\lambda = 0.03$ m) (Fig. 4.3):
$$\frac{\sin^2(627.7t)}{(627.7t)^2}.$$
2. If the transmitted signal was a non-frequency modulated CW signal (at 10 000 MHz), the expected received signal would only be amplitude modulated by the former law and the shape of the Fourier transform would be the Fourier transform of that law, known as being an isosceles triangle, the base of which is 200 Hz. Replacing a CW signal by pulses repeating at 1 kHz, leads to repeating the triangle in frequency every 1 kHz (Fig. 4.3).

In fact since each pulse is linearly frequency modulated between 9990 and 10010 MHz and not amplitude modulated, it could be considered that each frequency has the same duration (linear frequency modulation) and the same power (no amplitude modulation), hence the same energy. That implies that the envelope of $|\phi(f)|^2$ is constant over Δf (and zero outside since there is no frequency below 9990 MHz of above 10010 MHz). Then the shape of the envelope of $|\phi(f)^2|$ and then of $\phi(f)$ is constant between 9990 and 10010 MHz and close to zero outside. In fact it must be noted that the notion of stationary phase (or instantaneous frequency) implicity used is only valid if the modulation index is high enough which is in practice, the case here $\Delta f/(1/T) = T\Delta f = 200$.)

So the total spectrum of a received signal is composed of roughly 20 000 identical triangles regularly spread over 20 MHz centred on 10 000 MHz. But phase varies from triangle to triangle. If we assume that an elementary pulse lasts between $t = -5\,\mu$s and $t = +5\,\mu$s, instantaneous frequency varies like $f = f_0 + 2 \times 10^{12}t$, which means that the phase varies like
$$\varphi = 2\pi \times 10^{12}t^2$$
(in order that $(1/2\pi)(\mathrm{d}\varphi/\mathrm{d}t) = 2 \times 10^{12}t$)

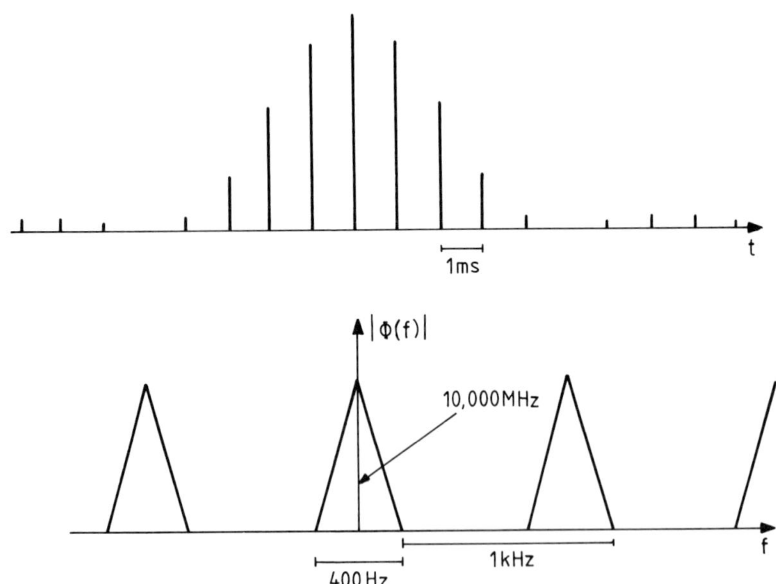

Fig. 4.3 Example of signal received by a panoramic radar.

or

$$\varphi = 2\pi \times 10^{12} \frac{(f - f_o)^2}{4 \times 10^{24}}$$

$$\varphi = \frac{\pi}{2} \times 10^{-12} \times (f - f_o)^2.$$

(The phase of the first triangle from the central one is 1.6 microradians, the phase of the second one is 6.3 microradians,..., the phase of the 10 000th one is 157 radians!)

4.2.2 About the parasitic noise

The parasitic noise $n(t)$ could be natural or artificial. Natural noise is the addition of the noise collected from outside by the antenna plus the noise generated by the receiver itself. (Natural) antenna noise depends on what antenna is looking at.

In the case of an antenna designed for space telecommunications (via satellite) which is specially designed to have little parasitic radiation in direction of the earth, antenna noise is small which means that the noise collected by the antenna has a power N_A given by

$$N_A = kT_A \Delta f$$

where k is Boltzmann's constant ($k = 1.38 \times 10^{-23}$ J/K), Δf is the bandwidth (of the receiving chain) T_A is the so-called antenna temperature, and T_A is small (a few tens of degrees Kelvin).

In the case of a panoramic ground surveillance radar antenna designed to detect low flying targets, the power radiated by the antenna in the direction of the earth (at an average temperature of 300 K) is greater, perhaps 35%, thus on reception the antenna temperature will be around 100 K.

Natural noise coming from the receiver depends on the nature of the receiver: it is given by

$$N_R = kT_R\Delta f$$

in which T_R is the receiver noise temperature. In case of a receiver using an RF low-noise amplifier before shifting-down the frequency in the mixer, T_R could be 170 K for instance. After the Second World War, when no low-noise RF amplifier was used (nor existing) and when the receiver noise was produced mainly in the mixer, relevant receiver noise temperature was frequently around 3000 K.

Total noise power in that case is

$$N_B = kT_A\Delta f + kT_R\Delta f = k(T_A + T_R)\Delta f = kT_B\Delta f.$$

If $T_A = 100$ K and $T_R = 170$ K, T_B (temperature of the receiving system) is 270 K and we usually say that the receiving system noise figure is

$$N_B = \frac{T_B}{290} = 0.93 \quad \text{or} \quad -0.3\,\text{dB}\,(10 \times \log_{10}(0.93)).$$

$$\text{If } T_A = 100\,\text{K and } T_R = 3000\,\text{K then } T_B = 3100\,\text{K}$$

$$N_B = \frac{T_B}{290} = 10.7 \quad \text{or} \quad 10.3\,\text{dB}.$$

Now, unfortunately, very often the receiver noise figure is defined by $N_R = 1 + (T_R/290)$ (the formula which would give the noise figure of the receiving system with an antenna temperature of 290 K), which gives the following results. If $T_A = 100$ K and $T_R = 3000$ K, then $N_R = 11.7$ (10.7 dB while $N_B = 10.3$ dB). If $T_A = 100$ K and $T_R - 170$ K, then $N_R = 2$ dB while $N_B = -0.3$ dB because in fact T_A is much smaller than 290 K.

But parasitic noise could be artificial and for instance it could be produced by a jammer installed at some distance from the radar system. In that case the noise power depends on the power of the jammer, on the gain of the antenna jammer in the direction of the radar system (both things are mixed when speaking of an effective emitted radiated power (e.e.r.p.) for the jammer, which would be the power radiated by an omnidirectional jammer achieving the same result (for the radar), at the distance of the jammer, on the antenna gain of the radar in the direction of the jammer.

Problem: let us consider a 1 kW CW jammer covering uniformly the frequency band 2800 to 3200 MHz, installed onboard a plane at 150 km from a radar system (high enough to be in line-of-sight from it), by means of an antenna with 6 dB gain in the direction of the radar system. The radar system has an antenna gain of 42 dB when directed towards the jammer, and an average parasitic radiation of – 50 dB compared to the main gain (sidelobes).

1. what is the e.e.r.p. of the jammer?
2. what is the noise temperature coming from the jammer when the radar antenna is directed towards it?
3. same as question (2) but when the antenna is looking at another (very different) direction.

Solutions

1. 6 dB is 4 times, so the e.e.r.p. is 1 kW × 4 = 4 kW.
2. Density of power (in W/Hz or joules) is given by

$$\frac{4000\text{ w}}{400\,000\,000\text{ Hz}} \times \frac{1}{4\pi(150\,000)^2} \times \frac{G\lambda^2}{4\pi}$$

with $G = 42$ dB, $\lambda \simeq 0.1$, which is usually obtained by the following computation (in 'decibels')

4000	+36 dB	
400 000 000		–86 dB
4π		–11 dB
$(150\,000)^2$		–103.5 dB
G	+42 dB	
λ^2		–20 dB
4π		–11 dB
	+78	–231.5 = – 153.5 dB

The result is 4.8×10^{-16} J. Dividing by Boltzman's constant (1.38×10^{-23}) it comes to about 32 000 000 K (then the natural noise could be ignored).
3. Dividing by 10^5 (–50 dB), it comes to 320 K (to be added to the natural noise).

4.2.3 Radar block diagram

Figure 4.4 represents a relatively complete and general block diagram of radar equipment (referring to Fig. 1 of Chapter 1, Volume 1). Generation of an IF (intermediate frequency) (phase modulated in case of a coded radar) is performed in 1.1. and 1.2: basically 1.1 is a stable oscillator (at IF) followed by a phase/frequency modulator 1.2 (e.g. a dispersive delay line). The signal out of 1.2 is mixed in 2 with the signal provided by a stable local oscillator (at transmission frequency shifted by the IF) and then amplified in 4 before feeding the antenna 14 via duplexing system 5.

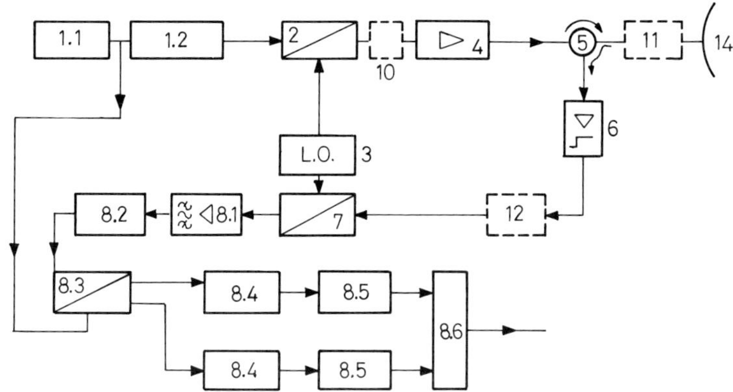

Fig. 4.4 General block diagram of a radar equipment.

The received signal RF is amplified (and limited in case of too powerful a signal) in 6 (the RF amplifier is generally a low noise one) and then mixed in 7 with the local oscillator signal (mixer 7 contains an IF preamplifier). The IF signal obtained at the output is amplified and filtered with accuracy in the case of a conventional radar, with less accuracy in the case of a coded radar where the essential matched filtering as done in 8.2 (e.g. pulse compression device), before being shifted down to zero frequency (in fact at the Doppler frequency) by the mixer 8.3. Mixer 8.3 is in fact an I and Q mixer in most cases (digital Doppler filtering) or a single side-band mixer in the case of analogue Doppler filtering. If 8.3 is not such, a problem will occur with blind phases as was the case in most of the radar systems using analogue MTI (mobile target indication) devices (section 4.2.13). The 2 outputs of 8.3 (in case of I and Q, the output in-phase with the IF reference and the output in Quadrature with the reference) are Doppler filtered in 8.4 and 8.5. 8.4 is a rejecting filter which only provides rejection of the strong clutter returns, while 8.5 is a bank of Doppler filters, such as represented in Fig. 4.2. Detection (in I and Q, recovering a vector from the two known coordinates I and Q) is performed in 8.6.

In the case of a passive electronic scanning phased-array, many phase shifters are introduced in 11 between the duplexer and the radiating elements of the antenna, which are generally reciprocal and used on transmission and reception. In the case of active electronic scanning phased-arrays, phase-shifting in transmission is placed in 10 (and not in 11), associated with another one in 12, and there are possibly as many transmission–reception modules (with 4, 10, 5, 6 and 12 boxes) as radiating elements.

Let us recall that in the case when the antenna size is not small enough compared to the range resolution of the radar, the control of the antenna pattern has to be achieved via controlling delays and not phases, and in that case delay-shifters must be added to the phase-shifters (or may even replace the phase shifters). For instance if the antenna size is 3 m and Δf is 3 MHz, range resolution is $c/2\Delta f = 50$ m and phase shifters could be used, while if the antenna size is 30 m and $\Delta f = 100$ MHz, range resolution is 1.5 m and (a combination of phase shifters plus) delay shifters have to be used.

When digital Doppler filtering is achieved, that means that between 8.3 and 8.4 an analogue to digital encoding is made, 8.4 and 8.5 being digital as well as 8.6. But on the other side, evolution of digital circuitry is such that in more and more cases matched filtering will be able to be achieved in digital circuitry also, in which case box 8.2 is placed after 8.3 (after A/D encoding).

In pulse radars, the duplexing function shuts the receiver during transmission, causing the radar to be blind during transmission periods: in the case of continuous wave (CW) radars, that cannot be accepted, of course, and two antenna systems are used: one for transmission, one for reception. Sufficient decoupling is then necessary between both antennas, which is achieved by using antennas with very small parasitic radiation in the direction of each other. In the case of simpler radars where the transmitted signals are not phase modulated, self-oscillators could be used on transmission (using magnetrons, or triodes with cavities, or transistors) and the radar system uses an automatic frequency control (AFC) such as indicated in Fig. 4.5. A small part of the transmitted pulse is then mixed with the local oscillator signal, the reference IF thus found is compared to the theoretical IF and the difference is used in a loop to change the local oscillator frequency to cancel the difference. In that case the local oscillator has to be a system in which frequency output is electronically controlled (by a voltage): 'reflex klystrons' were used in the past for that purpose, but they are replaced today by solid-state devices (incorporating varactors or equivalent). In that case, simplifications could be made: box 8.2, of course, does not exist, but banks of Doppler filters could also be eliminated (reducing the radar performance, particularly in sensitivity).

The use of coded signals is still possible with self-oscillators, but in that case phase modulation needs to be placed between box 4 and the duplexer (box 1.2/10 of Fig. 4.5): in that case an ultimate simplification could be added by using an IF equal to zero (see Fig. 4.6 where the transmitted signal is assumed to be phase-modulated continuous wave (PMCW) or frequency modulated continuous

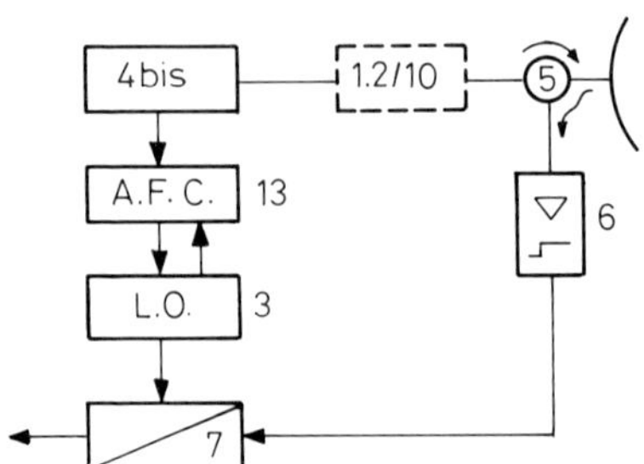

Fig. 4.5 Use of an AFC (automatic frequency control).

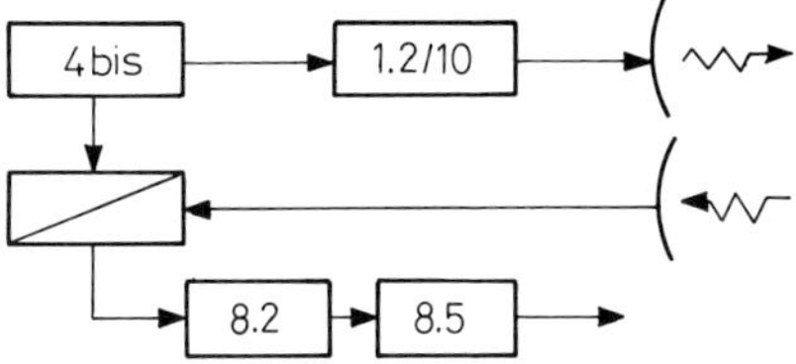

Fig. 4.6 Homodyne reception.

wave (FMCW)). In Fig. 4.6 the noise figure coming from the mixer working around zero frequency will be bad, but such an organization is valid when there is no problem of sensitivity (altimeters, proximity fuses, etc.).

If working in IF is necessary (for instance in the case of radars without RF low noise amplifiers when a relatively good noise figure is requested), in fact, noise produced by a mixer working around zero frequency is large compared to the noise with a similar mixer working around an IF of several megahertz. In fact working in IF is still convenient today to introduce some processing (such as some CFAR described below or monopulse measurements), however the AFC is generally not enough since it produces only an approximate stability of the IF frequency received.

In fact Doppler processing, even if only used for MTI action, requires, in addition, that consecutive pulses received from a fixed target give IF signals with a constant phase, which is not obtained with a self-oscillator transmitter whose phase changes randomly from pulse to pulse (except if it is phase-triggered by a very stable frequency, which is used in some radars). In that case the system represented in Fig. 4.7 is used. At each transmission a leakage of the transmitted pulse is sent to a second mixer, where it is mixed with the local oscillation: the IF obtained has the same phase reference as the transmission, and is used to trigger the phase of an IF 'coherent oscillator' COHO (the triggering requiring several IF periods). Then the output of the COHO is used as the output of 1.1. in Fig. 4.4.

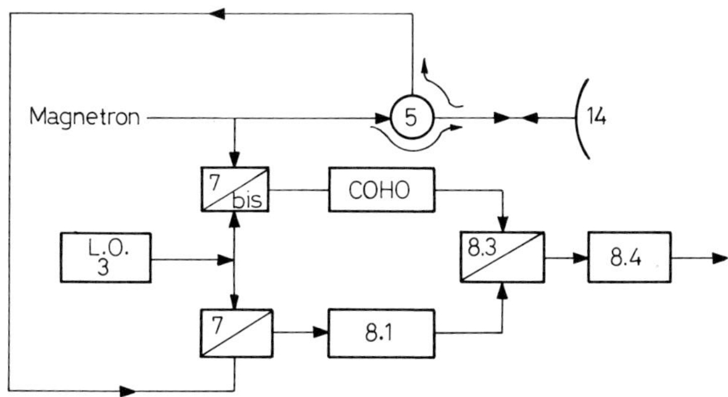

Fig. 4.7 Use of a COHO in an MTI system.

4.2.4 About antennas

Most of the basic antenna structures have been described in the relevant chapter of volume 2. One particular structure specific to surveillance radars must be indicated in addition: the antenna structures used to deliver a cosecant-squared antenna pattern.

The basic coverage requested for most of the surveillance radar systems is as indicated in Fig. 4.8, where between B and C, detection at constant height is required. In that case range OM which varies like the root of the (power) antenna gain is equal to $AB/\sin\theta = AB \times \csc\theta$, which means that the ideal solution is to have between θ_o and θ_{max}, the antenna gain varying proportionally to $1/\sin^2\theta = \csc^2\theta$.

The means to obtain that with a reflector structure are indicated basically in Fig. 4.9, for instance, by deviation from the simple parabolic reflector: in fact real

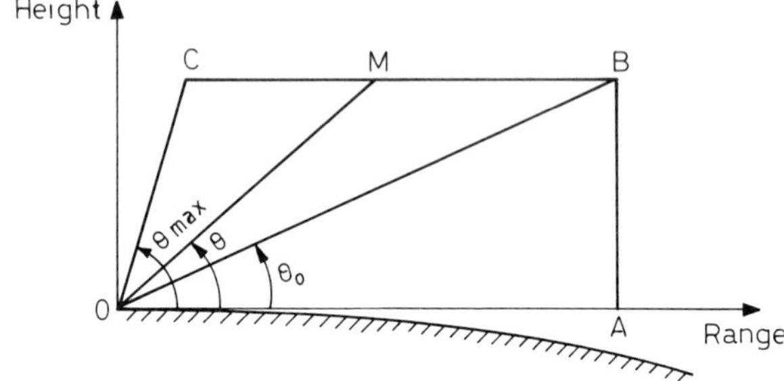

Fig. 4.8 Coverage requested from a surveillance radar.

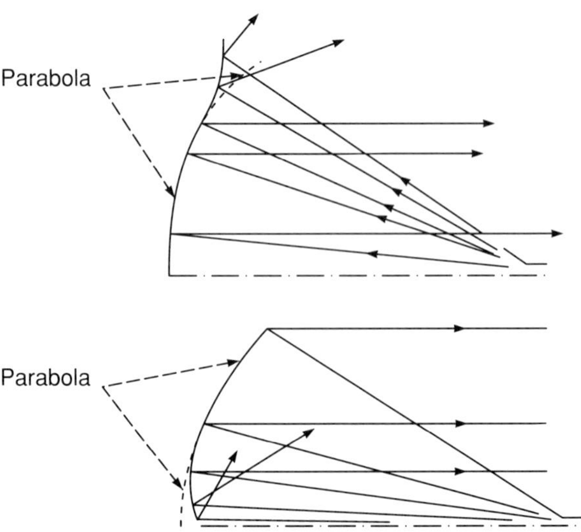

Fig. 4.9 'Double curvature' antennas.

situations are more complicated, since the geometrical optic considerations represented are not really valid with the reduced relative height of the antenna (compared to the wavelength) and real shapes are computed from those basic ones.

4.2.5 About transmitters

Many details have been given in the relevant chapters (of Volume 1): all types of transmitters are used in radar systems. Only basic notions have to be recalled here.

First, in radar systems the peak power introduces only operational problems (easy detection by the enemy, high voltage and relevant large volumes and weights) while mean power gives the performance. Evolution tends more towards the use of duty cycles relatively close to one (phase or frequency coded pulses long compared to the pulse repetition period).

Long pulses have another advantage: the relevant reduction of peak power is clearly associated with a reduction of the high voltage of the transmitting tube, but also long phase-modulated pulses are relevant to the increase of the pulse-transformer ratio, thus decreasing the voltage in most of the transmitter. An associated problem is the fact that the radar is blind for the short ranges (receiver shut during transmission): that could be corrected by transmitting, at the end of the long pulse, a classical small pulse for detection at short ranges.

Second, in radar systems evolution is towards the use of wider and wider bandwidths, particularly to allow for frequency agility in a wide band to oblige the jammers to spread their power in the widest possible band (see 4.2.9 Problems). The use of travelling-wave tubes is relevant to that evolution.

Third, better and better Doppler filtering is required for clutter elimination and target identification, which requires increasingly better phase stability during transmission. O-type tubes (TWT or klystrons) are better adapted to that requirement than the M-type tubes (magnetron, or cross-field amplifiers (CFAs)).

Fourth, multimode radars are requested more with various pulse lengths (and various pulse repetition frequencies (prf)). The use of TWT with the incorporation of a control grid are convenient for that purpose.

Fifth, the use of active arrays with a large number of transmitting–receiving (T/R) modules provides a lot of advantages which are explained later, a reason for their progressive introduction in radar systems.

4.2.6 About receivers

Low noise RF amplifiers

The use of low-noise RF amplifiers associated with a protection device (to protect the mixer) is becoming quite common. TWTs, triodes, tunnel diode amplifiers, parametric amplifiers have been used in the past for that purpose. At the present time, semiconductor amplifiers (mainly gallium–arsenide) are now preferred in most cases, providing a very reduced noise, and a simple structure requiring only a low voltage.

Sensitivity time control

At short ranges from a radar, large returns could be found (e.g. from close buildings, ship superstructures, etc.) which require a reduced sensitivity of the radar receiver for those distances in the relevant directions. RF controlled solid-state attenuation devices are used for that purpose between antenna and mixer, possibly associated with the low-noise RF amplifier.

Lin-log IF amplifier structure

IF amplifiers are usually composed of a cascade of amplifiers such as represented in Fig. 4.10 together with the individual characteristic output voltage versus input of one elementary amplifier. The graph given in the bottom of Fig. 4.10 explains why such an amplifier is called lin-log.

The overall graph 0 versus I for the four stage amplifier is given by the continuous line of the graphs represented in Fig. 4.11 where an obvious saturation

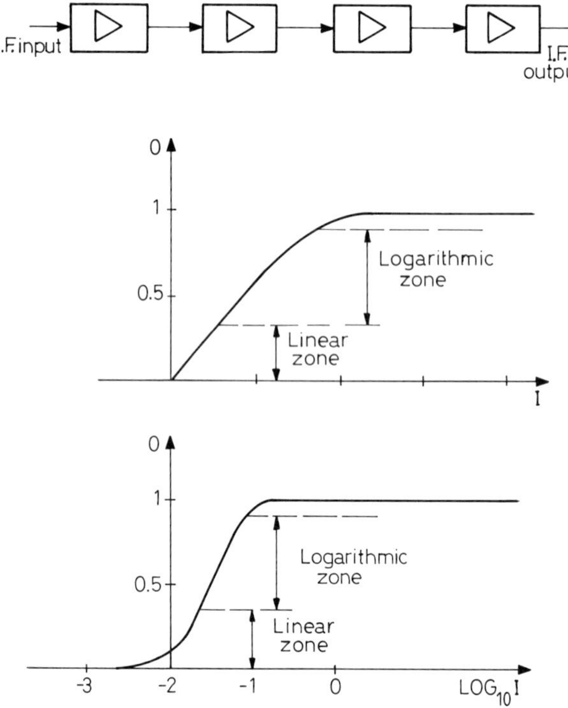

Fig. 4.10 IF amplifier.

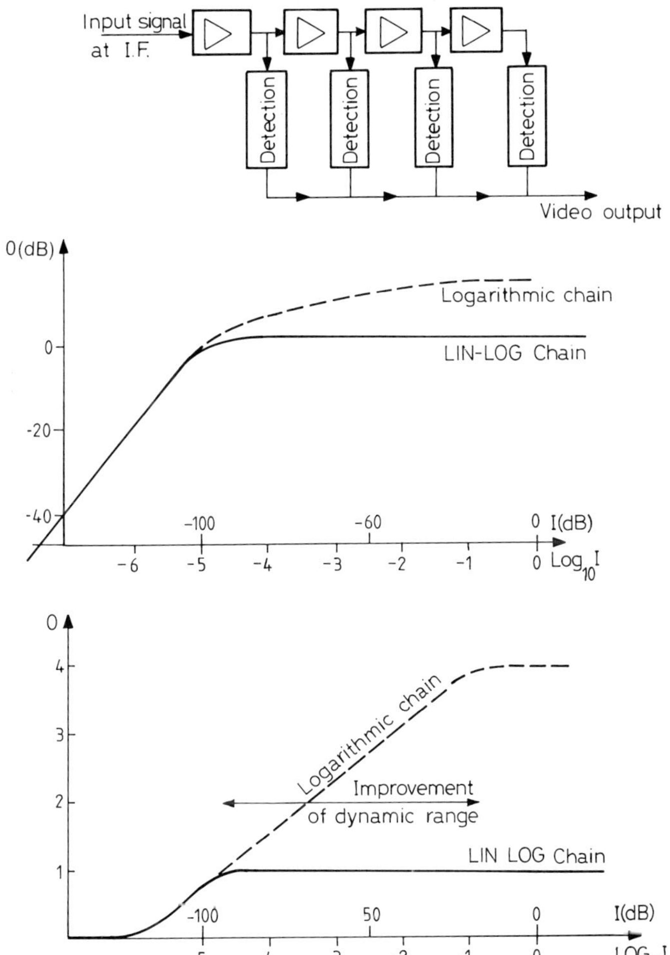

Fig. 4.11 Logarithmic chain.

exists. This simplest IF chain without any processing has the big advantage of presenting to the user, 'in raw video', all the problems coming from clutter, jamming, etc., since it is not protected, in contrast with more sophisticated chains protected against troubles and which could leave the user in total confidence (no trouble present on the 'scope'—the CRT—while in fact the situation is difficult and radar performance degraded).

Logarithmic chain

In order to increase the dynamic range of the receiver and also to provide in analogue the direct computation of the logarithm of the received signal (such as used in some monopulse arrangements—for instance in stacked-beam radars—where the difference of the logarithms of signal amplitude can be used to give angular measurements) logarithmic chains are used as represented in Fig. 4.11 together with the relevant graphs (in the dashed line).

'Log-PLD chain

That is the old name of a chain in use in many radars to cope against sea returns (and possibly cloud returns)—'pulse length discriminator' was a means of achieving a high-pass band filter, which was the reason for the log-PLD name of the chain represented in Fig. 4.12.

A classical situation is indicated where a radar on board a ship receives similar returns from planes A, B and C, but while at the same range as A and B a large return is coming from the sea-clutter giving a continuous detection between α and β which doesn't allow detection of planes A and B.

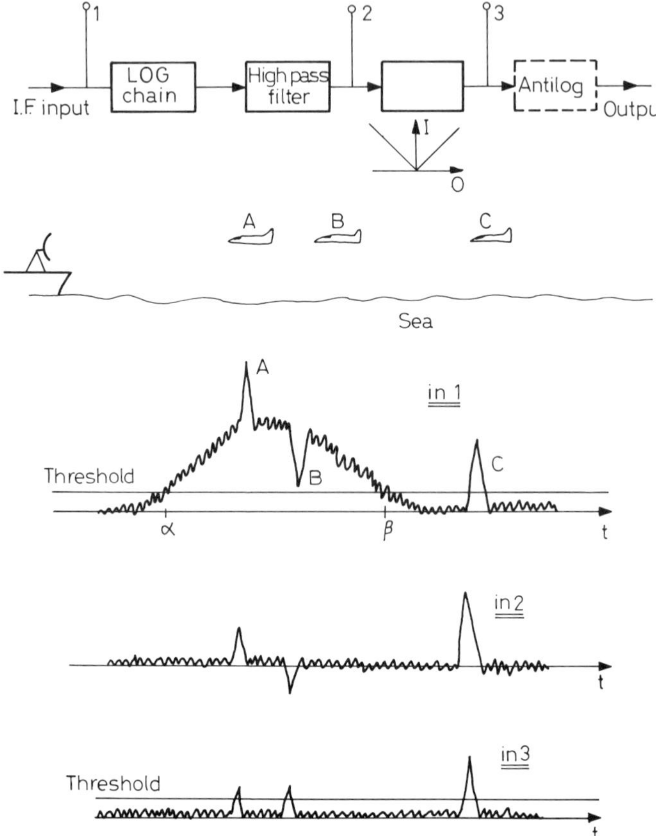

Fig. 4.12 'Log-PLD' chain.

The received signal from sea-clutter can be represented by $A(t) \times p(t)$ where $p(t)$ is a rapidly fluctuating (parasitic) signal and $A(t)$ is a very slowly varying coefficient coming from the variation of reflection index on the sea. Signals for A and B are assumed to be, by chance, in phase or in opposition with the sea-clutter return at the same distance.

A log amplifier transforms $A(t) \times p(t)$ into $\log A(t) + \log p(t)$ and $\log A(t)$ is eliminated by the high-pass band filter. The differential gain of a logarithmic amplifier is smaller on strong returns, which explains why in 2, returns A and B are smaller than return C.

In 3 clutter return is not gaussian, which disturbs the habits of (military navy) users. This is why very often an 'antilog' device is used in the end (with an exponential characteristic) to compensate.

CFAR chain (Carpentier, 1988)

Even if the former receiver is in fact a 'constant false alarm receiver' against sea-clutter returns or equivalent, the name of CFAR is generally limited to systems whose architecture is according to Fig. 4.13.

Before the device, bandwidth is $M\Delta f (M \geqslant 1)$: the device begins with a hard-clipping followed by the normal matched filter (a normal radar IF receiver in case of a conventional—not coded—radar). As a result any noise (coming mainly from jamming but also possibly from clutter returns) is, in 2, a noise definitively not gaussian but with a well defined power.

In 3, the matched filter has made the addition (integration) of $(T/1/M\Delta f)$ independent samples of that noise which gives a gaussian distribution if $MT\Delta f$ is large and a well defined power. In 3 the noise has a well-known distribution: a fixed threshold at the output will then give a constant false alarm rate. This device introduces a loss of 1 dB in the case when $MT\Delta f$ is large (> 100) against a gaussian (jamming) noise.

In case the radar transmits a non-coded signal, the device is known as the Dicke Fix and $M \gg 1$.

Automatic gain control (AGC)

The dynamic range of receivers being limited, it is better to work (regarding for instance monopulse angular measurements) in the 'centre' of the dynamic range,

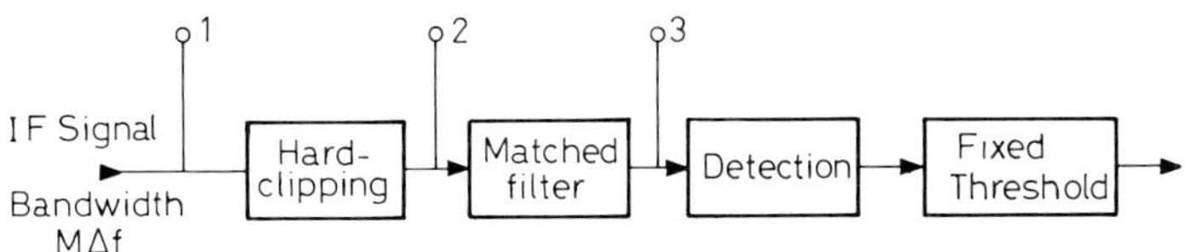

Fig. 4.13 CFAR chain.

which implies reducing the receiver gain in case of a too strong return. This is possible in tracking radars where as many receivers as targets are used, and where automatic gain control is used in each of the receiving 'chains'.

Side-lobe blanking (SLB)

If a target contains a transponder, reproducing consecutively several replicas of the useful signal, at a high level, those replicas will enter the radar not only via the main lobe but also via the (parasitic) sidelobes. In order to avoid that, an auxilliary antenna is used whose gain is above the gain of the parasitic sidelobes, associated with a receiver identical to the main one. When the signal received by this auxiliary channel is higher than that received by the main channel, the output is shut, thus avoiding reception of responses in all directions.

Side-lobe cancellation (SLC)

In this case, means are used in order to automatically change (self-adaptation) the amplitude and the phase of the auxiliary channel in order to be able to subtract the auxiliary signal from the main one in such a way that the jamming signal from a jammer is reduced to a minimum.

4.2.7 Choice of wavelengths—basic examples of radar parameters

A particular case for radars is the case of the over-the-horizon radars (working between 2 and 20 MHz basically), which use reflection on the ionosphere (see chapter on propagation in volume 2). Their frequency is defined by the necessity of being reflected on the ionosphere. For higher and more usual frequencies, propagation is more or less, on average, in a quasi-straight line and the choice of central frequency depends on other considerations.

Starting from the radar equation given in the chapter on propagation in Volume 2 it can be shown (Carpentier, 1988) that for a 3D radar which has to ensure an omniazimuth detection according to a given radar coverage such as represented in Fig. 4.14, in a given context, which means—for a given noise temperature if the main noise is the natural (internal) one, or in a given context of jamming

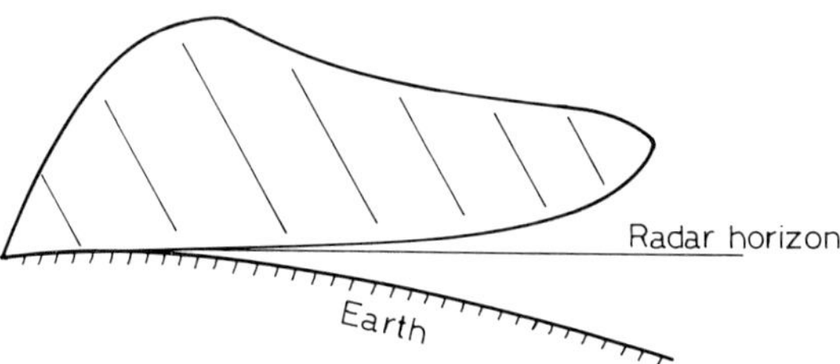

Fig. 4.14 Typical radar coverage of a surveillance radar.

and a given quality of parasitic antenna radiation if limitation comes from jamming – when you give the surface Σ of the coverage, you define the product of the surface of the antenna A by the energy E_T to transmit during one revolution of the antenna (around the vertical axis) by the radar cross-section σ of the target.

In the absence of jamming, $A \times E_T \times \sigma$ varies like the square of Σ and like the noise temperature. The price of a given surface A of antenna increases when the central transmitted frequency increases because decreasing the wavelength requires manufacturing the antenna with more accuracy.

It is generally easier to obtain high mean transmitted power at relatively low frequencies. On the other hand, angular resolution is proportional to the ratio between central wavelength and antenna size, leading to the use of short wavelengths when very good angular resolution is requested (at generally short ranges). As a result, detection at long range is normally achieved in L or S-band (respectively; 0.23 m, 0.1 m). Accurate angular measurement (and tracking) is normally achieved in C, X, K, or even W bands (respectively, 0.05 m, 0.03 m, 0.02 m, 0.003 m) and optical wavelengths are considered.

Some typical radar configurations are given below.

Long range detection of ballistic missiles (10 000 km)

Central frequency:	VHF
mean transmitted power:	2 M$\bar{W}$
peak power:	50 M$\hat{W}$
pulse length:	2 ms
p.r.f.:	20 Hz
duty cycle:	0.04
antenna surface:	2000 m^2
A.E_T	around 2×10^{11} Jm^2

Medium range detection of ballistic missiles (3000 km)

Central frequency:	C-band (around 5 GHz)
mean transmitted power:	1 M$\bar{W}$
peak power:	100 M$\hat{W}$
pulse length:	100 μs
p.r.f.:	100 Hz
duty cycle:	0.01
antenna surface:	300 m^2
A.E_T	around 2×10^{10} Jm^2

3D air defence radar (range around 400 km)

Central frequency:	S-band (around 3 GHz)
mean transmitted power:	30 k$\bar{W}$
peak power:	30 M$\hat{W}$

pulse length:	5 μs
p.r.f.:	200 Hz
antenna surface:	100 m^2
duration of a revolution:	8 s
$A.E_T$	$= 2.4 \times 10^7$ Jm^2

Gapfiller for low altitude detection (range around 100 km)

Central frequency:	L-band (around 1 GHz)
mean transmitted power:	100 $\bar{W}$
peak power:	1.3 k$\hat{W}$
pulse length:	10 μs
p.r.f.:	1 kHz
antenna surface:	5 m^2
duration of a revolution:	4 s
$A.E_T$	= 20 000 Jm^2

An example of a multimode radar for a surface-to-air weapon system in X-band will be given in the problem described in section 4.2.9. Combat aircraft multimode radars work in X or K-band.

4.2.8 Radar cross-section – target fluctuation – stealth targets

The definition of a radar cross-section (see chapter on propagation in Volume 2) is more convenient for mathematical computation than immediately and obviously physically related to the general size and aspect of the target.

On complex targets, which means on targets composed, within the range evolution, of several zones reflecting the radar signal (or reflecting it significantly more than the other ones), slight movements of the target can greatly change the power reflected by the target, because reflected signals have variations in their respective phases and thus are summed or subtracted to give respective returned signals. In general the fluctuation coming from target movements is relatively low, the amplitude of the radar returns changing at a rate of a few times per second. This means that during time measurements of a few tens of milliseconds it does not change, but after a few seconds it has completely changed. In front of a surveillance panoramic radar, this generally gives a fluctuation from scan to scan—cases 1 or 3 of Swerling (1960)—and not from pulse to pulse.

Also some relatively minor changes in the radar frequency may lead to a complete change in the amplitude of the radar return. To have an idea of the frequency change required to completely change the amplitude of a target return, it is, on a very complex target, around $c/5d$ (where c is the speed of the light and d the 'size' of the target), which gives 5 MHz for $d = 10$ m (Carpentier, 1988). That means that for a surveillance radar using frequency agility from pulse to pulse during the dwelling time of the antenna beam in direction of the target, fluctuation is from pulse to pulse and not from scan to scan (cases 2 and 4 of Swerling (1960)).

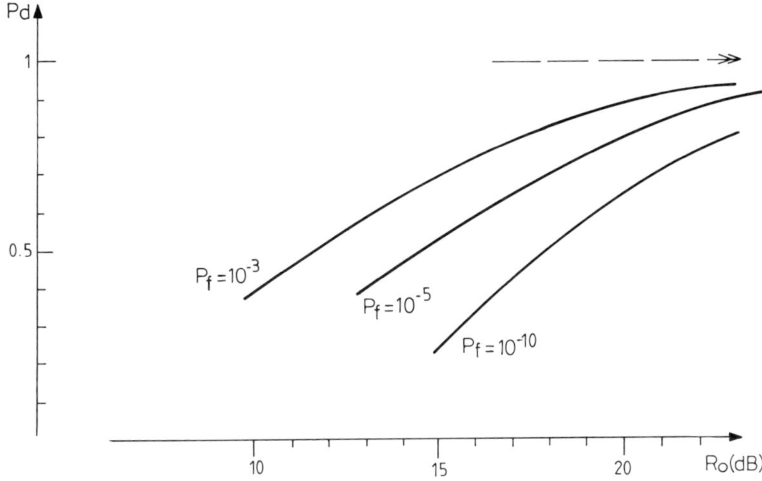

Fig. 4.15 Detection probability versus R for fluctuating target.

One result of target fluctuation is that if the average cross-section is used in the radar equation, and if you are not lucky, the actual cross-section will be much smaller than the average, while if you are lucky it will be higher. A consequence is that it is necessary to replace graphs given in Fig. 4.1 by other graphs giving—for a given false alarm probability P_f—the detection probability P_d versus the average value R_o of R.

Curves to be used are such as indicated in Fig. 4.15 for a 'very complex target' to be compared with graphs of Fig. 4.1 valid for non-fluctuating targets (cases 1 and 2 of Swerling), fluctuation called, in France, Rayleigh fluctuation, target composed of a very large number of reflective zones of all sizes. (Cases 3 and 4 of Swerling correspond to a less unfavourable fluctuation.) It has to be noted that some targets are still worse than cases 1 and 2, while some of them are not as bad. To give an example of a figure, the difference for $P_f = 10^{-8}$, $P_d = 0.9$ between a nonfluctuating target and a target fluctuating with a Rayleigh distribution is around 9 dB, which is quite large.

In order to reduce the effect of target fluctuation on detection, it is possible to use several frequencies during the detection, either consecutively (frequency agility from pulse to pulse) or at the same time (frequency diversity). When enough different frequencies are used, everything is as if the target was no longer fluctuating, with a constant cross-section reduced compared to its average value (by 1 dB in case of Swerling case 1/Rayleigh fluctuation).

Apart from trouble in detection, target fluctuation also introduces problems in measuring target direction (Carpentier, 1988). In fact a normal radar (panoramic radar determining target direction by the direction where received signal is maximum, or a monopulse radar) measures the direction of the target as being the direction of the equiphase surface around it (the surface where the electromagnetic field has the same phase). This direction is that of the target in the case of a non-fluctuating target (a 'point' target), but this direction could be very different in the case of a complex fluctuating target when the radar return is minimum.

That explains why, in the presence of a target flying low above the sea and of its image in the sea, the elevation angle measured by the radar could be very wrong when the return from the target and the return from the image are opposite in phase. The simplest way to deal with that is to eliminate all angular measurements obtained when the received signal is too small.

Some specific devices are built in such a way that for radar they present a large cross-section compared to their size: they are called 'cataphotes', Luneberg lens, corners or trihedrons being examples of cataphotes which reflect back to the radar most of the power they receive. In the same manner a dipole matched to the wavelength could have a good cross-section. On the other side, planes could be designed in order first to have no resonant structure (dipole type) and second to have nothing more or less similar to a cataphote, and they could have, in this way, a very reduced radar cross-section ('stealth' planes).

(Un)fortunately, stealth shapes are generally not very compatible with high performance combat aircraft, and structures of planes with a lot of armaments hanging below the wings (guns, bombs, rockets, missiles, pods) are not compatible with stealth structures either. This is why stealth planes are basically reconnaissance planes or bombers where all equipment is imbedded within the airframe.

Anti-radar paint must be thick enough to absorb the transmitted signal (typically one tenth of the wavelength, taking into account the relative dielectric constant ε_R of the paint, i.e. a tenth of $\lambda \times \varepsilon_R$ if λ is the radar wavelength). This is a very limiting constraint.

4.2.9 Problems analysis of a multifunction radar

1. The radar transmitter utilizes a grid-controlled TWT with the following characteristics: bandwidth 9 to 10.35 GHz gain 40 dB and perveance (peak current/$U_L^{3/2}$)1.65×10^{-6}, where U_L is the voltage between the delay line and the cathode. (V_c, the voltage between the collector and the cathode, is here 70% of U_L-depressed collector-in order to improve the efficiency, here equal to 30%. This efficiency does not take account of the power needed to heat the cathode.)

   Knowing that $V_c = 38\,800$ V, give the relevant peak power P_c. Maximum duty cycle being 0.04, give the maximum acceptable mean power. For a pulse repetition frequency of 4 kHz, which is the pulse length corresponding to that duty cycle? What is the relevant transmitted energy per pulse?
2. That energy is, at each pulse transmission, taken from a capacitor whose energy is recharged between each pulse transmission. Assuming that the relative voltage variation during the transmission of a pulse should not exceed 0.5%, how much energy has to be in the capacitor? Knowing that it is only

possible to store 80 J per litre in the capacitor, what is the minimum volume of the capacitor?

3. The antenna is a square one (1 m × 1 m) with the same illuminating law (in cosine-square) in vertical and horizontal. That introduces a loss of 3.6 dB compared to the maximum gain of an uniformly illuminated antenna, but reduces down to −32 dB the relative gain of the worst sidelobes (the first) and widens the 3 dB beamwidth (in bearing and elevation) compared to its value (in degrees) of $51\lambda/D$ for an uniformly illuminated antenna, increasing it up to $83\lambda/D$.

   Give the antenna gain at 10 GHz, together with the 3 dB beamwidth (in azimuth or elevation).

4. The antenna rotates at 1 turn per minute. How many pulses will be received when the 3 dB beam is in direction of a target, for a prf of 4 kHz?

   Assuming that utilization (integration) of all those 'hits' (received pulses) gives an improvement by 11 dB, that losses on transmission are 3.5 dB and 4 dB on reception (including losses in the phase shifters of the antenna, which is a phased-array antenna), give the energy E received from a target at 30 km, with a radar cross-section of 0.1 m^2 (i.e. the energy received from one pulse at maximum gain, increased by 11 dB).

   Give the spectral density N_o of the thermal noise (kT_B) assuming that the noise temperature T_B of the receiving system is 300 K. Give the value of $R = 2E/N_o$.

5. Compute the Doppler frequency corresponding to the velocity V_A of the edge of the antenna at a frequency of 9 GHz and at a frequency of 10.35 GHz.

   A filter rejecting fixed echoes is utilized which is assumed to let pass (without attenuation) Doppler frequencies between 220 Hz and 3780 Hz (modulo prf) and to completely attenuate Doppler frequencies between + and −220 Hz (modulo prf).

   Give for a transmitted frequency of 9 GHz and a prf of 4 kHz the radial velocities which are suppressed. Same question for a transmitted frequency of 10.35 GHz. What are the radial velocities suppressed at 9 GHz and at 10.35 GHz (assuming the velocity is below 640 m/s). Assuming that a target with V_R radial velocity has, for the radar, all velocities between $V_R - V_A$ and $V_R + V_A$, what are the radial velocities of targets suppressed at 9 GHz and 10.35 GHz?

6. Give the graphs of the gain versus frequency and the gain in decibels versus frequency of a digital filter whose transfer function in Z is as follows:

$$\frac{(1 - Z^{-1})(1 - 1.97022Z^{-1} + Z^{-2})(1 - 1.88176Z^{-1} + Z^{-2})}{(1 - 1.164Z^{-1} + 0.7Z^{-2})(1 - 1.736Z^{-1} + 0.95Z^{-2})}.$$

7. It is assumed that a 'stand-off' jammer is on board a big plane transmitting towards the radar an e.e.r.p. of 125 kW (everything is as if the jammer was tramsmitting 125 kW omnidirectionally).

Assuming that jamming is uniformly spread over all the radar bandwidth and enters the radar via the far-out sidelobes (far from the main beam) assumed to be at -5 dB below the isotropic, give the value of R in conditions of question 4. Same question assuming that the jamming enters via the first sidelobes ('strong' jamming conditions).

8. The antenna provides electronic scanning in elevation and azimuth: so the antenna beam could be in any direction within a cone ($\pm 45°$ angle) around the antenna axis. In 'surveillance' mode, the beam is vertically scanned in the symmetry plan of the antenna, dwelling (prf of 4 kHz)

   2.25 ms at elevation 1.25°
   2.25 ms at elevation 3.75°
   2.25 ms at elevation 6.25°
   2.25 ms at elevation 1.25 and repeat again.

   The number of pulses received from a target such as defined in question 3 decreases, which gives a reduction of R by 4 dB compared to question 4.

   Give the value of R obtained on a target at 30 km, with $0.2\,\text{m}^2$ radar cross-section (RCS), at 7.5° elevation angle (or a zero elevation angle) in case of 'strong jamming'. Same question if RCS is 15 times smaller ($0.013\,\text{m}^2$).

   Give, always under the same assumption, for $P_f = 10^{-5}$, the detection probability of the target, knowing that it is fluctuating, with a probability of 0.50 to have an RCS of $0.2\,\text{m}^2$ and a probability 0.50 to have a RCS of $0.013\,\text{m}^2$ (which gives an average RCS of $0.1 + 0.006 = 0.106\,\text{m}^2$).
9. At what distance does this fluctuating target have to be (always at the same elevation angle: 7.5° or zero, and in the strong jamming situation) detected with a 0.9 detection probability (P_f still of 10^{-5}).
10. In the surveillance mode, transmitted frequency is changed when the beam elevation is changed and then the RCS (which depends on that frequency) changes or not, with for each beam position a probability of 50% to be of $0.2\,\text{m}^2$ and a probability of 50% to be $0.013\,\text{m}^2$. At elevation angle of 2.5°, clearly one measurement in achieved when the beam is aimed at 1.25° and another one when the beam is aimed at 3.75°, which gives, under assumptions of question 8, two possibilities of detecting the target when it is at 30 km.

    At which distance will a detection probability of 0.9 be obtained?
11. Now the jamming power is assumed to enter the radar via the main beam (at the maximum gain beam aimed at the jammer)—very strong jamming—what is the present value of R against $0.1\,\text{m}^2$?

    At what distance does the fluctuating target have to be (at elevation angles of 7.5°, 2.5° and 1.25°) to be detected with a detection probability of 0.9?
12. This range is not enough. In order to have a better detection in the direction of the stand-off jammer, a 'burn-through' mode is used, in which, while the antenna is rotating, the beam is electronically maintained in the direction of the jammer for one second: the 1000 pulses being received in the direction of maximum gain.

Give the value of R per received pulse when $\sigma = 0.2\ \mathrm{m}^2$ and when $\sigma = 0.013\ \mathrm{m}^2$ for a target at 17 km.

The 100 received pulses are grouped in 110 bursts of 9 pulses, the frequency randomly changing from burst to burst. Use of banks of Doppler filters, with 400 Hz bandwidth for each, allows convenient integration of received pulses (in a 'coherent' way) during 2.27 ms, i.e. the 9 pulses of each burst. What is the value of R for each burst (over $0.2\ \mathrm{m}^2$ called R_1 and over $0.013\ \mathrm{m}^2$ called R_2)?

13. It is then assumed, to simplify, that the 110 bursts are equally shared with 55 bursts corresponding to R_1, and 55 bursts corresponding to R_2. The relevant signals are integrated in a non-coherent manner (after detection): 55 being a large number, the theorem of central limit could be applied to addition of various parasitic signals after detection. It is then proven (Carpentier, 1988) that everything is as if, at the place of the useful signal, there are: a useful signal with amplitude given by $(R_1 + R_2) \times 110^{1/2} \times \frac{1}{4}$; and a parasitic random gaussian term of zero mean value and a standard deviation equal to

$$\left(\frac{R_1 + R_2}{2} + 1\right)^{1/2}$$

while in other places there is a parasitic gaussian term of zero mean value and standard deviation equal to 1.

What detection probability is obtained (for $P_f = 10^{-5}$) after non-coherent integration of 110 signals?

Solutions

1.

$$V_c = 38\,800\ \mathrm{V}$$

$$U_L = \frac{38\,800}{0.7} = 55\,400\ \mathrm{V}$$

$$I_{peak} = 1.65 \times 10^{-6} \times (55\,400)^{3/2} = 21.5\ \mathrm{A}$$

$$V_c I_{peak} = 835\,000\ \mathrm{W}$$

$$P_c = 835\,000 \times 0.3$$

$$P_c = 250\ \mathrm{k}\hat{\mathrm{W}}$$

$$P_m = 10\ \mathrm{k}\bar{\mathrm{W}}$$

$$4000\,T_p = 0.04$$

$$T_p = 10\ \mu\mathrm{s}$$

$$E_T = 10^{-5} \times 250\,000 = 2.5\ \mathrm{J}.$$

2. $E = \frac{1}{2} C V^2$

$$\frac{\Delta V}{V} = \frac{1}{2}\frac{\Delta E}{E}$$

$$\frac{\Delta E}{E} = 0.01$$

$$\Delta E = \frac{2.5}{0.3} = 8.35$$

$$E = \frac{8.35}{0.01} = 835\,\text{J}$$

$$\text{Volume of capacitor} = \frac{835}{80} = 10.5\,\text{litres.}$$

3. $f_o = 10\,\text{GHz}$
$\lambda = 0.03\,\text{m}$

$$G_{\text{MAX}} = \frac{4\pi A}{\lambda^2} = 140\,000 \Rightarrow 41.4\,\text{dB}$$

$$G = 41.4 - 3.6 = 37.8\,\text{dB}$$

$$\alpha = \frac{83 \times 0.03}{1} = 2.5^\circ$$

4. $\dfrac{2.5 \times 4000}{360} = 28$ pulses

$$E = \frac{E_T G^2 \sigma \lambda^2 I}{(4\pi D^2)^2 4\pi L_T L_R} = \frac{E_T G^2 \sigma \lambda^2 I}{(4\pi)^3 D^4 L_T L_R}$$

L	$+4\,\text{dB/J}$	
G	$+75.6$	
σ		$-10\,\text{dB/J}$
λ^2	$+9.5$	-40
I	$+11$	
$(4\pi)^3$		-33
D^4		-179
$L_T L_R$		-7.5
	100.1	$-269.5 = -169.4\,\text{dB/J}$

$$E = 1.16 \times 10^{-17}\,\text{J}$$
$$N_o = kT_B = 1.38 \times 10^{-23} \times 300 = 4.14 \times 10^{-21}\,\text{J}$$
$$R = \frac{2E}{N_o} = 5600 \Rightarrow 37.5\,\text{dB}$$

5. $V_A = 0.5 \times 2\pi = 3.14\,\text{m/s} \rightarrow f_D = \dfrac{2V_A}{\lambda}$

188.5 Hz at 9 GHz 216.8 Hz at 10.35 GHz

At $f = 9$ GHz, zones of cancelled velocities correspond to Doppler frequencies between -220 and $+220$ Hz modulo prf.

0 – 3.67 m/s	63 – 70.33 m/s	129.67 – 137 m/s
196.33 – 203.67 m/s	263 – 270.33 m/s	329.67 – 337 m/s
396.33 – 403.67 m/s	463 – 470.33 m/s	529.67 – 537 m/s
596.33 – 603.67 m/s	663 – 670.33 m/s	729.67 – 737 m/s

At $f = 10.35$ GHz they correspond to

0 – 3.19 m/s	54.78 – 61.16 m/s	112.75 – 119.13 m/s
170.72 – 177.10 m/s	228.7 – 235.07 m/s	286.67 – 293.04 m/s
344.64 – 351.01 m/s	402.61 – 408.99 m/s	460.58 – 466.96 m/s
518.55 – 524.93 m/s	576.52 – 589.90 m/s	634.39 – 640.87 m/s

Zones valid for 9 GHz and 10.35 GHz are (in absolute value):

0 – 3.19 m/s   402.61 – 403.67 m/s   463 – 466.96 m/s.

But only targets whose radial velocity is between 0.05 m/s and −0.05 m/s will be completely cancelled at 9 GHz and at 10.35 GHz.

6. See Fig. 4.16 and 4.17.
7. $E = -169.4$ dB/J

$$N_o = \frac{125\,000}{1350 \times 10^6} \times \frac{1}{4\pi 10^{10}} \times \frac{g\lambda^2}{4\pi L_R}$$

with $\lambda = 0.03$ m, $g = -5$ dB, L_R will be taken as equal to 2 dB

$N_o = 125\,000$	+51 dB/W	
1350×10^6		−91.3 dB/W
4π		−11
10^{10}		−100
g		−5
λ^2	+9.5	−40
L_R		−2
4π		−11
	60.5	−261.3 = −200.8 dB/W

$$R = \frac{2E}{N_o} = -169.4 + 3 + 200.8 = 34.4 \text{ dB}.$$

If the jamming enters via the first sidelobes (−32 dB below the main beam) that means that g is now 37.8 − 32 = 5.8 dB instead of −5 dB, i.e. a loss of 10.8 dB. $R = 23.6$ dB.

8. The number of received pulses is only 9 instead of 28 which explains the 4 dB loss. At elevation angle of 7.5° there is an additional loss of 2 × 3 dB on the antenna gain (in transmission *and* in reception). Total loss is 10 dB which gives 13.6 dB on 0.1 m^2 and 16.6 dB on 0.2 m^2 as well as 16.6 − 11.8 = 4.8 dB on 0.013 m^2 ($10 \times \log_{10} 15 = 11.8$).

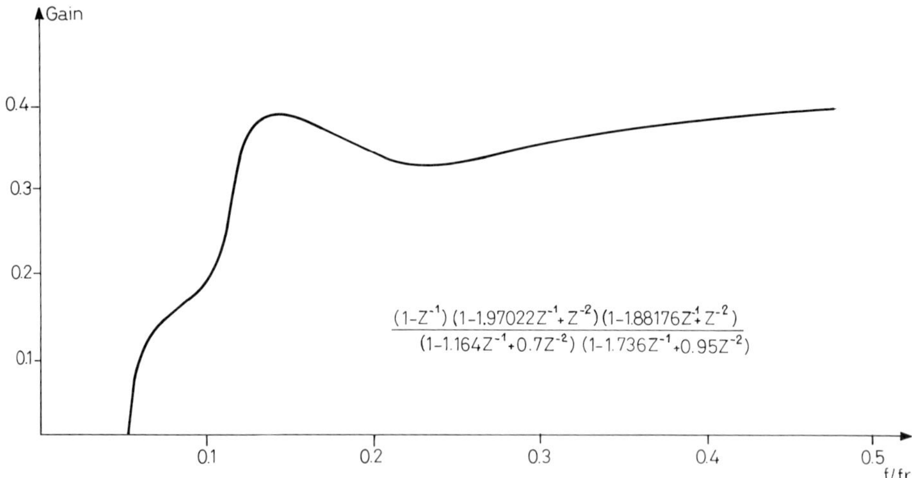

Fig. 4.16 Gain of an anticlutter digital filter versus frequency.

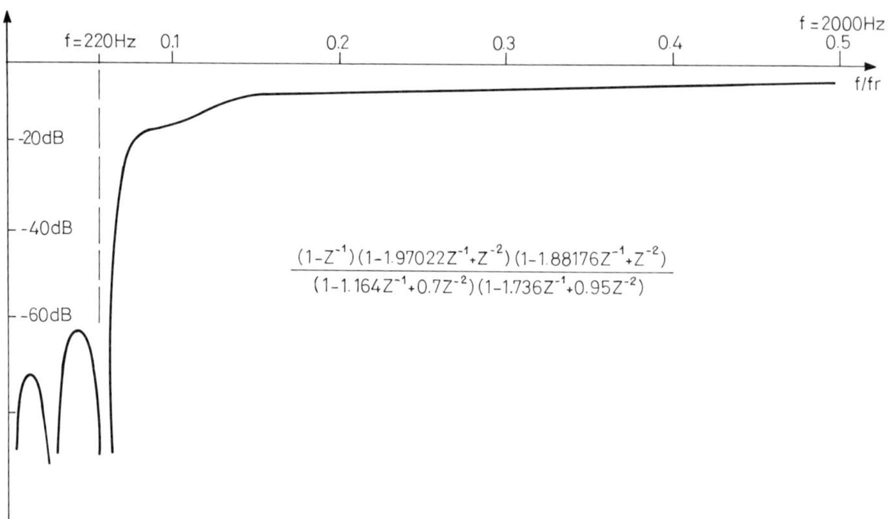

Fig. 4.17 Gain of an anticlutter digital filter versus frequency.

For $P_f = 10^{-5}$ $R = 16.6\,\text{dB}$ corresponds to $P_d = 0.992$ and $R = 4.8\,\text{dB}$ corresponds to P_d negligible.
Detection probability is $0.5 \times 0.992 + 0.5 \times \text{negligible} = 0.5$

9. Practically detection probability on $0.013\,\text{m}^2$ has to be 0.8 (since $0.5 \times 1 + 0.5 \times 0.8 = 0.9$) which corresponds to 14.3 dB. 9.5 dB have to be gained on D^4 and 2.375 on D which has to be multiplied by $10^{-0.2375} = 0.58$ and becomes $30 \times 0.58 = 17.5\,\text{km}$.

10. P_f is increased and becomes 2×10^{-5} instead of 10^{-5}. We need to have at each elevation a detection probability of 0.68 since $1-(1-0.68)^2 = 0.9$. Which means $(0.68 - 5) \times 2 = 0.36$ on $0.013\,\text{m}^2$ corresponding to 11.6 dB to be compared to 4.8 dB, a difference of 6.8 dB, corresponding to 1.7 dB on D which has to be $30\,\text{km} \times 10^{-0.17} = 20.5\,\text{km}$.
11. 32 dB are lost on R, so on $0.1\,\text{m}^2$ it is now 13.6 dB − 32 dB = − 18.4 dB. Those 32 dB have to be compensated (8 dB on range). Range at 7.5° elevation angle becomes $17.5\,\text{km} \times 10^{-0.8} = 2.8\,\text{km}$.

   At 2.5° $20.5\,\text{km} \times 10^{-0.8} = 3.2\,\text{km}$.

   At 1.25° (relative benefit of 6 dB), only 26 dB have to be recovered and the range becomes $17.5 \times 10^{-0.65} = 3.9\,\text{km}$.

12. $$E = \frac{E_T G^2 \sigma \lambda^2}{(4\pi D^2)^2 4\pi L_T}$$

$$N_o = \frac{125\,000}{1350 \times 10^6} \times \frac{1}{4\pi \times 10^{10}} \times \frac{G\lambda^2}{4\pi}$$

$$R = \frac{2E}{N_o} = \frac{2E_T G\sigma \times 1350 \times 10^6 \times 10^{10}}{4\pi D^4 L_T \times 125\,000}$$

$R = 2$	+ 3 dB		
E_T	+ 4		
G	+ 37.8		
σ	+ 3	− 10 dB	for $0.2\,\text{m}^2$
1350×10^6	+ 91.3		
10^{10}	+ 100		
4π		− 11	
D^4		169.2	
L_T		− 3.5	
125 000		− 51	
	+ 239.1	− 244.7 =	− 5.6 dB

and − 5.6 dB − 11.8 dB = − 17.4 dB on $0.013\,\text{m}^2$.

After coherent integration of 9 pulses (gain of $10 \log_{10} 9 = 9.5$ dB) one obtains $R_T = 3.9$ dB (2.48), $R_2 = -7.9$ dB (0.165).

13. Everything is as if, at the position of the useful signal, there was an useful amplitude of 6.94 plus a gaussian signal of zero mean value and of standard deviation 1.524.

    Detection probability for a false alarm probability of 10^{-5} is the probability that the sum of the two above signals be above 4.28 since

$$\frac{1}{(2\pi)^{1/2}} \int_{4.28}^{+\infty} \exp\left(-\frac{v^2}{2}\right) dv = 10^{-5}$$

So it is also the probability that a gaussian value with zero mean value and standard deviation equal to unity by above

$$\frac{4.28 - 6.94}{1.524} = -1.75$$

i.e.

$$\frac{1}{(2\pi)^{1/2}} \int_{-1.75}^{+\infty} \exp\left(-\frac{v^2}{2}\right) dv = 0.96.$$

4.2.10 Pulse compression

General

The name pulse compression is given to organizations in which the transmitted signal is frequency or phase modulated and where, on reception, use is made of a matched filter.

If $\phi(f)$ is the Fourier transform of the transmitted signal, the receiver matched filter has a transmittance equal to the conjugate of $\phi(f)$, say $\phi^*(f)$. In most cases, pulse compression is achieved at low frequency, i.e. at intermediate frequency or possibly around zero frequency. That means that $\phi(f)$ is the Fourier transform of the useful signal at IF.

In that case the Fourier transform of an useful signal crossing the pulse compression device is then multiplied by $\phi^*(f)$, delivering at the output a signal whose Fourier transform is

$$\phi(f) \times \phi^*(f) = |\phi(f)|^2$$

which is a symmetrical signal without frequency or phase modulation, which is the autocorrelation function of the useful signal. It should be recalled that this signal is composed by a significant useful signal at the centre (duration about $1/\Delta f, \Delta f$ being the width of the spectrum of the useful signal) surrounded by smaller signals called sidelobes, with total duration equal to twice the time T_p of the transmitted pulse.

The essential quality of a pulse compression device is to totally remove the frequency (or phase) modulation of an useful signal crossing it: the necessary phase characteristic of the matched filter is obtained if that condition is fulfilled. On the other hand, if a Dirac impulse (a very short one) enters the pulse compression device, the relevant output has a Fourier transform equal to $\phi^*(f)$; it is identical to an useful signal but time-reversed.

Chirp system

Historically the first pulse compression radars used transmission of signals of constant amplitude, linearly frequency modulated between $f_c - (\Delta f/2)$ and $f_c + (\Delta f/2)$ (or vice versa) over the duration T_p of the pulse. In that case, it could be said—on first approximation—that any frequency lasts the same time (linear modulation) with the same power (no amplitude modulation) and, as a consequence, that all the frequencies between $f_c - (\Delta f/2)$ and $f_c + (\Delta f/2)$ have the same energy, which means that $|\phi(f)|^2$ is constant in that zone and zero outside. Since using the 'stationary phase' ('instantaneous frequency') in this way is only valid for a high modulation index, it is valid only for $\Delta f/(1/T_p) = T_p\Delta f$ large enough. If it is not large enough the spectrum is not completely rectangular (Fig. 4.18 gives the real shape of $|\phi(f)|$ for $T_p\Delta f = 10$ and 60).

Matched filtering is then obtained regarding the amplitude by using a rectangular filter on reception (with a loss of 1 dB if $T_p\Delta f$ is around 10, and negligible if $T_p\Delta f$ is above 50). Matched filtering regarding phase is obtained by using a purely dispersive device removing the frequency modulation of a useful signal, which is achieved by using a dispersive device whose (group) propagation time decreases linearly from $T_o + T_p$ to T_o for frequency varying between $f_c - (\Delta f/2)$ and $f_c + (\Delta f/2)$. The complete matched filter is then made of the cascade of the rectangular filter achieving amplitude matching and the dispersive device achieving phase matching (see Fig. 4.19).

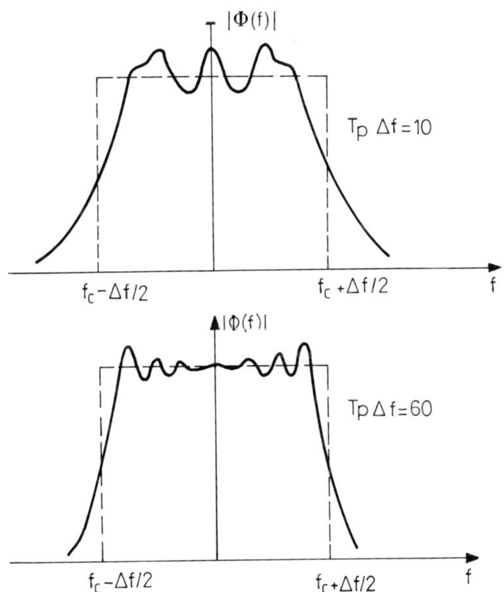

Fig. 4.18 Spectrum of a 'chirp' signal.

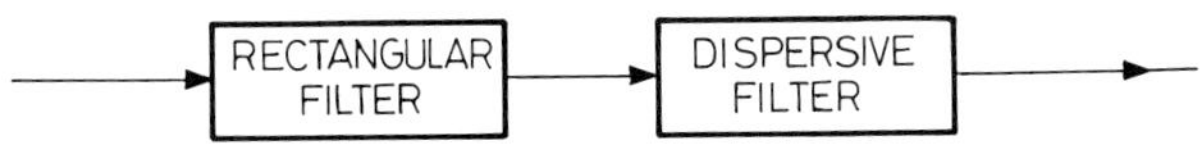

Fig. 4.19 Matched filter in pulse compression systems.

Then if we consider a useful signal crossing that matched filter, the amplitude of the Fourier transform (already rectangular) will not change, while the phase modulation is completely removed. That means that a useful signal after crossing the matched filter has an envelope according to:

$$\frac{\sin(\pi\Delta f(t-t_{o1}))}{\pi\Delta f(t-t_{o1})}.$$

Such a signal is represented in Fig. 4.20: the central signal is surrounded by sidelobes, the main ones being at 13.2 dB below the central signal. In fact since $T_p\Delta f$ is not infinite, the real signal is only close to the former one (limited to a length of $2T_p$). For instance the real shape of the envelope is given in Fig. 4.21 for $T_p\Delta f = 20$.

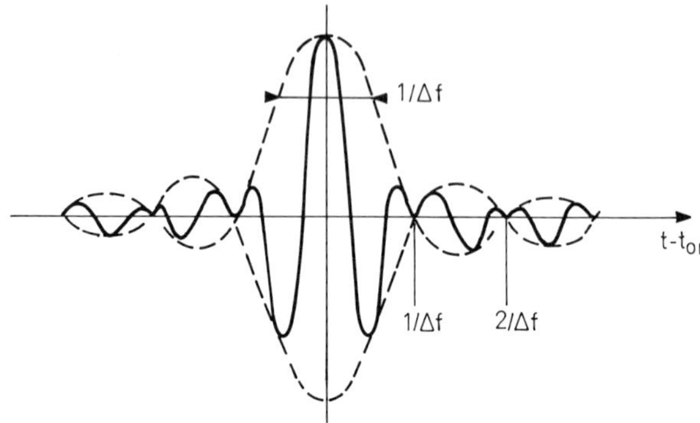

Fig. 4.20 Compressed pulse in chirp systems (very large pulse compression ratio).

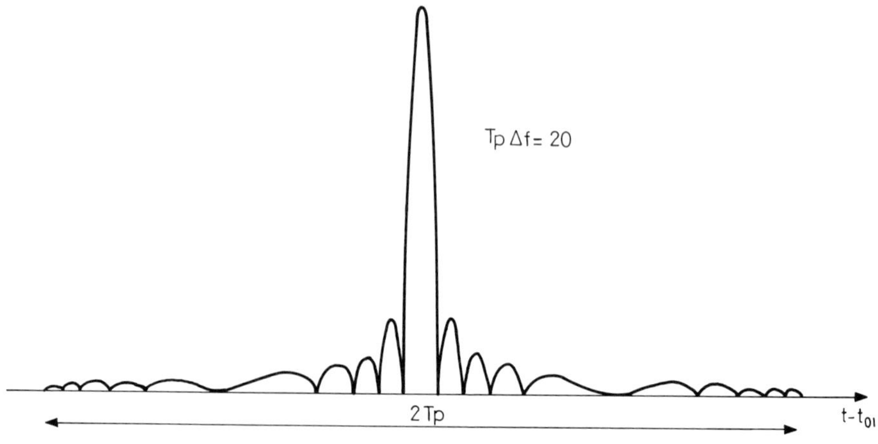

Fig. 4.21 Compressed pulse envelope (pulse compression ratio of 20) in chirp system.

Doppler effect in chirp systems

A Doppler shift f_D of the received signal produces the following consequences.

First, after crossing the dispersive devices, all frequencies contained in the useful signal remain at the output with the same phase, giving an output signal not frequency modulated, but shifted in time by

$$\frac{T_p}{\Delta f} \times f_D = T_p f_D \times \left(\frac{1}{\Delta f}\right)$$

which is $T_p f_D$ times the 'duration' of the output compressed signal.

Second, the central signal is slightly widened (negligible if $f_D \ll \Delta f$) and reduced in amplitude (also negligible if $f_D \ll \Delta f$).

For example

3D Air defence radar $\lambda = 0.1$ m, $T_p = 5\,\mu$s, $\Delta f = 10$ MHz.

$$\frac{1}{\Delta f} = 0.1\,\mu\text{s (15 m)}$$

$$V_R = 600\text{ m/s}, \qquad f_D = \frac{2V_R}{\lambda} = 12\text{ kHz}$$

$$T_p f_D = 0.06 \qquad T_p f_D \times \left(\frac{1}{\Delta f}\right) = 0.006\,\mu\text{s (0.9 m)}.$$

Trajectography accurate tracking radar $\lambda = 0.05$ m, $T_p = 100\,\mu$s, $\Delta f = 500$ MHz.

$$\frac{1}{\Delta f} = 0.002\,\mu\text{s (0.3 m)}$$

$$V_R = 5000\text{ m/s} \qquad f_D = 200\,000\text{ Hz} \qquad T_p f_D = 20$$

$$T_p f_D \times \left(\frac{1}{\Delta f}\right) = 0.4\,\mu\text{s} \qquad \text{(6 m)}$$

Practical implementation of dispersive filters

In the past, dispersive filters have been made using lumped-constant filters or bulk acoustic waves (BAW) devices (thin metallic films) providing a quasi-linear variation of the propagation time versus frequency. In modern equipment surface acoustic waves (SAW) devices and digital transverse filters are used.

In 1967 the possibility of propagating surface acoustic waves along the surface of a piezoelectric substrate was used (Dieulesaint and Hartman, Thomson-CSF), as represented in Fig. 4.22. Used in such devices are an input comb and an output comb engraved on the substrate in such a way that the high frequencies propagate during a short time T_1 and the low frequencies propagate during a longer time

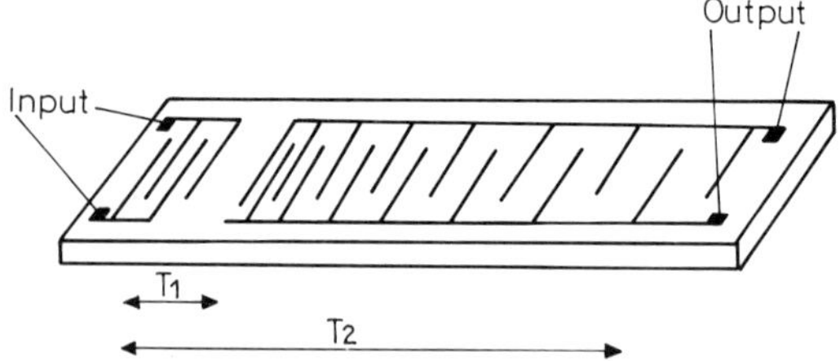

Fig. 4.22 Surface acoustic wave device for pulse compression.

T_2. Repartition of the teeth along the output comb allow for controlling the dispersive curve describing (group) propagation time versus frequency (other organizations are used where the input and output are on the same side of the device, but they basically are of the same nature). The accuracy of the positioning of the teeth directly impacts the accuracy of the curve of phase versus frequency. For instance, if the central frequency of the acoustic wave used for the processing (the IF) is 50 MHz, assuming an acoustic velocity around 5000 m/s will give a (central) wavelength of 100 microns. If we want to obtain the phase with an accuracy of 20 milliradians (which is pretty good—see below) the positioning of the teeth has to have an accuracy of about:

$$\frac{20 \times 10^{-3}}{2\pi} \times 100 = 0.32 \text{ microns}$$

easily achieved by using electron beam machines.

The use of transverse filters is easy to understand when simply considering the useful signal represented in Fig. 4.23 (frequency linearly increasing from 0.5 to 2.5 MHz over a total duration of $T_p = 5\,\mu s$: $T_p \Delta f = 10$) sampled at 6 MHz.

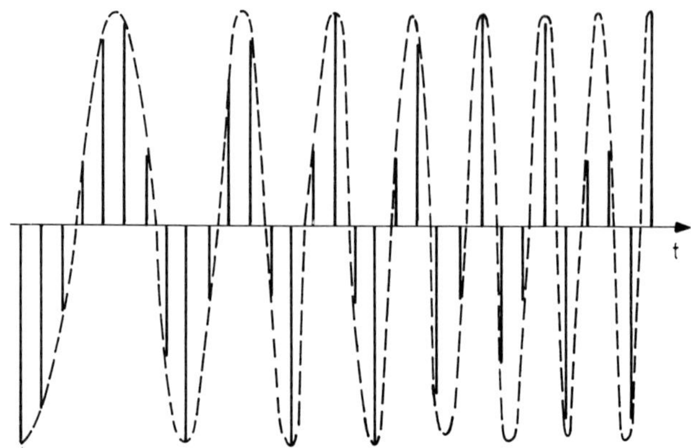

Fig. 4.23 Use of transverse filters.

A filter whose response to a Dirac impulse is that (sampled) signal but time-reversed will be adequate as a matched filter: that will be achieved by using a delay line with successive outputs every 0.167 μs, every output feeding an amplifier, the gain of the successive 31 amplifiers being respectively

$$1, \ -0.85, \ 0.37, \ 0.31, \ -0.88, \ 0.94, \ -0.31, \ -0.62, \ 0.99, \ -0.31, \ \cdots -0.85, \ -1$$

Clearly the device will be realized by using digital technology, the gains of the successive amplifiers being respectively

2^{0}	1 0 0 0 0 0 0 0
$-(2^{-1}+2^{-2}+2^{-4}+2^{-5}+2^{-7})$	−0 0 1 1 0 0 0 0
$2^{-2}+2^{-3}$	0 0 1 1 0 0 0 0
$2^{-2}+2^{-4}$	0 0 1 0 1 0 0 0
etc.	in binary

It is clear that if the curve of phase versus frequency (or time) is not accurate enough, pulse compression will not be perfectly achieved, the main result being the existence of parasitic supplementary sidelobes accompanying the useful signal. As an order of magnitude, it could be recalled that 10 milliradians phase error produces parasitic sidelobes at a level around −46 dB. 100 milliradians phase error produces parasitic sidelobes at a level around −26 dB.

Sidelobe reduction in chirp systems

The existence of large sidelobes accompanying the useful signal is an important drawback in chirp systems in most cases, since it means that a parasitic target close to a useful one with a radar cross-section 14 dB above the useful one will produce a parasitic sidelobe which could cover the useful return. It is often mandatory to obtain pulse compression with reduced sidelobes. Using a 'weighting filter' instead of a rectangular filter, whose gain is smaller on the two edges of the spectrum compared to the centre, is the equivalent of what is achieved on an antenna when the illumination is not uniform along it, being reduced on the edges. In this case, it has long been known that 'weighting' the illumination was the way to reduce the sidelobes of the antenna pattern but with a widening of the beam and a loss in the antenna gain. For the same reasons, replacing the rectangular filter by a weighting filter reduces the pulse compression sidelobes but with a widening of the central signal and a loss in sensitivity, the amplitude matching not being obtained.

In order to avoid the relevant loss, it is necessary to use a transmitted signal whose spectrum is not rectangular, but also reduced on the edges compared to the centre. In the general case of a transmitted signal, which is not amplitude modulated, the energy of the central frequency must be superior to that of the frequencies on the edges of the spectrum, which is obtained by using a non-linear frequency modulation in which the central frequencies last a longer time, as

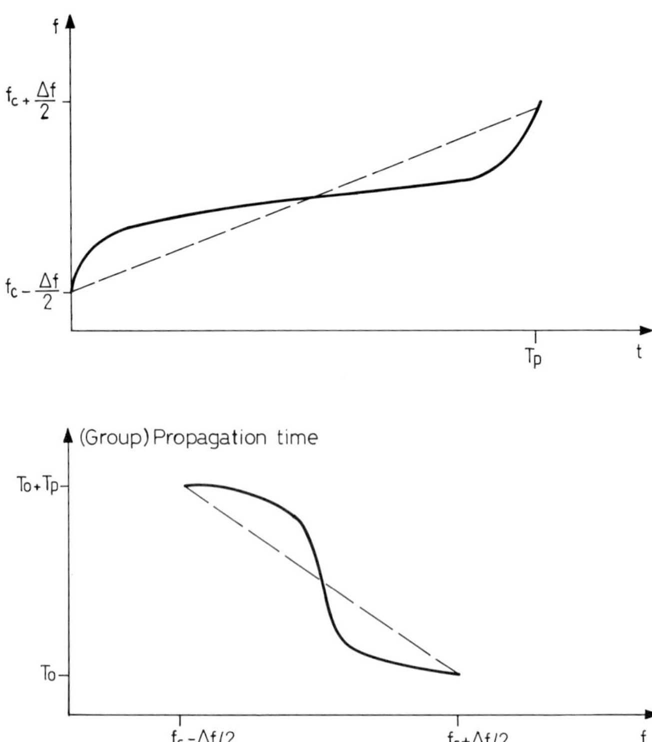

Fig. 4.24 Intrinsic weighting.

represented in Fig. 4.24. Of course to obtain the phase matching on reception, the dispersive device has to provide a curve of propagation time versus frequency which is modified compared to the simple chirp as represented in the same figure (dotted lines represent the case of the 'simple chirp'), and a non-rectangular filter has to be used to get the amplitude matching. Such an 'intrinsic weighting' could provide for instance a reduction of the worst sidelobes (which are close to the central peak) down to -32 dB (instead of -13 dB), together with a widening by a ratio of 1.6 (3 dB) of the central signal length.

Generation of the transmitted signal

In most cases, the convenient frequency modulation of the transmitted signal is obtained by using a dispersive filter on transmission analogous to that used on reception, which is fed by a very small pulse (not frequency modulated) at its input. Frequency modulation is then obtained at the output, where the signal is hard-limited to delete amplitude modulation and gated to obtain the required duration at the good time position. Of course dispersive devices used on transmission and on reception must be complementary.

Use of Fourier transformers on reception

Another way to get pulse compression (and generally matched filtering) is to perform a computation according to the formula given in section 4.2.1 of

$$C(t) = \int Y(f) \times \phi^*(f) \times \exp(2\pi jft)\,df$$

where $Y(f)$ is the Fourier transform of the received signal $y(t)$ and $\phi^*(f)$ the conjugate of the Fourier transform of a useful signal.

That has been achieved in the past by using optical computation which is naturally good as a Fourier transformer. This is now achieved by digital computation using a fast Fourier transform (FFT) (or equivalent). Fourier transform of the received signal is performed, the result digitally multiplied by $\phi^*(f)$, and the new result is digitally Fourier transformed by a second FFT (or equivalent).

Effect of pulse compression on clutter

Let us recall that clutter is the general name given to parasitic echoes existing on the surface of the earth ('ground clutter') or in the atmosphere (rain, snow, birds, atmospheric clutter). Gound clutter could be detected directly or after reflection on the irregularities of the atmosphere in the case of 'abnormal propagation' in which case it could appear as moving at a speed twice the speed of the wind (angels). The clutter could also be artificial ('chaff' dropped by the enemy to disturb radar detection).

If we compare the clutter obtained with two radars A and B using the same pulse length on transmission, the radar A being 'classical' radar without pulse compression and the radar B using pulse compression to improve the resolution range, both radars, A and B, transmitting the same energy per pulse (via the same antenna system), the effect of the pulse compression depends on the nature of the clutter.

If there is a very large number of clutter objects in the 'resolution cell' of the radar B (roughly defined in the case of a panoramic radar by the product of the horizontal beamwidth times the distance and the range resolution, and correctly modified in the case of the volumic clutter such as rain or chaff) the pulse compression, because it has divided the range resolution by the pulse compression ratio, has also divided by the same ratio the number of clutter objects and thus the power of the clutter.

But finally, this is not a frequent situation, since in most cases, the number of clutter objects in the radar B is not large enough to consider that the effect of pulse compression is only to reduce the average power of the clutter by the pulse compression ratio, but also to change the amplitude distribution of the clutter. Two examples will be given to illustrate that situation.

The first one is purely theoretical: it is the case of a pulse compression by 100 in radar B, in a situation in which the clutter is made of very powerful clutter

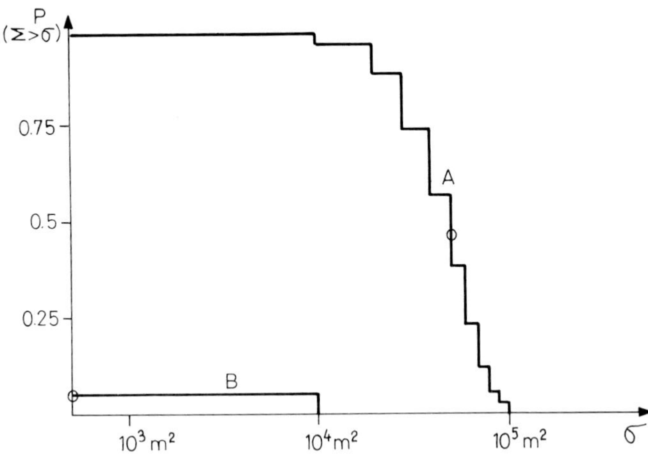

Fig. 4.25 Clutter amplitude distribution with and without very good range resolution.

objects each having a RCS (radar cross-section) of 10 000 m^2, and a probability of 0.05 to have one clutter object in the resolution cell of radar B. At the same time the probability of having several clutter objects in the resolution cell of radar A (100 times larger) is close to one. The curve of Fig. 4.25 indicates the distribution of clutter RCS in radar A and radar B: in radar A, it is a 'normal' distribution, not at all in radar B (the graphs give the probability that the RCS of the clutter exceeds σ). The average value of the clutter (indicated by O) is reduced by 100 (the pulse compression ratio) between A and B, but that has no physical meaning with such different distributions. Practically, since in radar B the probability that a useful target is disturbed by the clutter is only 5%, in many cases we could accept not to detect where the clutter is, keeping 95% detection probability, while the clutter situation is critical for the radar A and requires a very good Doppler filtering.

The second example is a real one represented in Fig. 4.26 which shows the clutter distribution experimentally obtained in a country without buildings and mountains but only farms. The relevant clutter is shown for three identical radars but with different range resolution (50 μs, 5 μs and 0.5 μs). If we consider that required detection probability is 75%, it could be considered that 17 dB gain has, in practice, been obtained by compressing from 50 μs down to 5 μs and 24 dB more by compressing from 5 μs down to 0.5 μs.

In case of 'rocky mountains clutter' or 'buildings clutter', the situation is more similar to that represented in Fig. 4.25.

Digitally phase-coded signals

Instead of using continuous frequency modulation for pulse compression, signals could be used in which the frequency remains constant but where a rapid phase

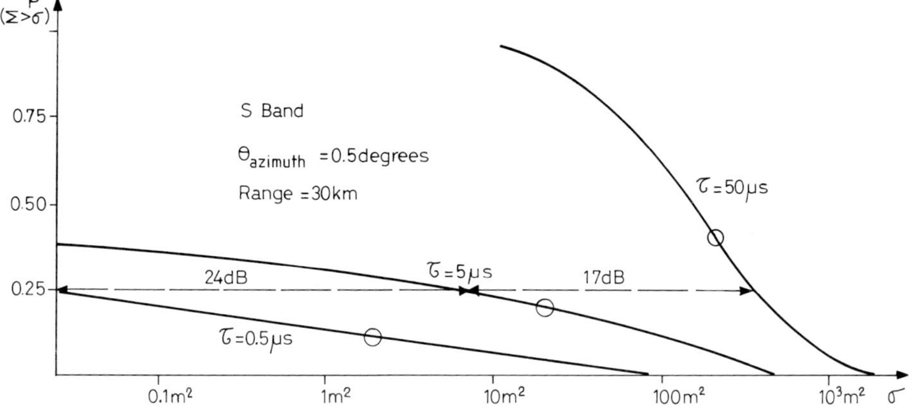

Fig. 4.26 Clutter amplitude distribution with various range resolutions.

change could occur at regular intervals (given by a clock). That could be convenient since the phase modulation could be obtained directly on the microwave signal at the transmitted frequency by using microwave phase shifters. In that case, values of phase are digital, which means that the phase could be 0 or π (binary coded) or it could be 0, $\pi/2$, π, $3\pi/2$, etc. One drawback of such a signal is that it covers a frequency band larger than in the former cases.

4.2.11 About digital processing

Digital processing obviously requires, as indicated in section 4.2.3, that the received signal be transformed from analogue-to-digital at a rate rapid enough. The relevant sampling could be obtained in two different ways.

In the first way, the received signal is shifted down in frequency, the last intermediate frequency f_c being slightly above $\Delta f/2$ (Δf is the spectrum width of the transmitted signal) and the sampling is performed at a frequency slightly above $2f_c + \Delta f$ (Fig. 4.27).

In the second way, the received signal is shifted down to zero frequency but I and Q demodulation is achieved which requires two samplings (one on each component) at a frequency slightly above Δf.

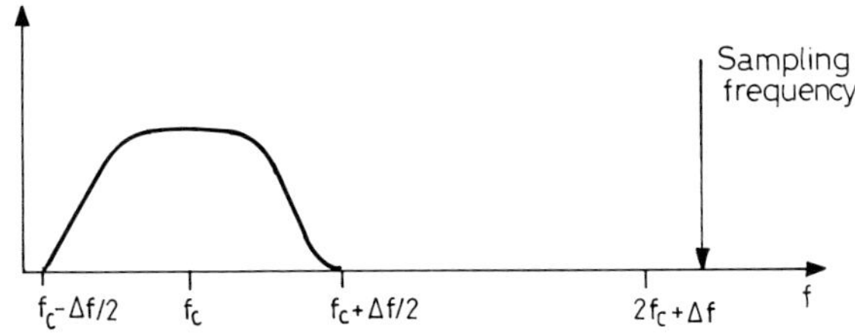

Fig. 4.27 Digital encoding.

Sampling and analogue-to-digital encoding are, the practice, performed together in the same device (A/D encoder) at a rate of $2f_c + \Delta f$, close to $2\Delta f$ in the first case, twice at a rate Δf in the second case (an alternate encoding on I and on Q could be used, at a rate $2\Delta f$ of course).

Feasible characteristics of A/D encoding (regarding accuracy and rapidity), which are constantly improving, are a first limit to the implementation of digital processing in radar systems. The graph of Fig. 4.28 gives the number of bits (accuracy) of digital encoding versus the sampling frequency, indicating the evolution which could be expected by the end of this century.

For instance if a clutter rejection (reduction of clutter compared to useful returns) of 54 dB is requested, the accuracy of computation must correspond to a minimum of $54/6 = 9$ bits let us say 10 bits plus one bit for the sign, which gives 11 bits. 11 bits do not allow a sampling frequency to exceed 150 MHz, which limits Δf to that value in the best case. Evolution of the technology of semiconductors (if it is as predicted in Fig. 4.23) will offer the possibility of a Δf of 2 GHz at the end of this century.

The second limit to the implementation of all types of possible digital signal processing is the performance possible, at a given time, using the existing digital circuitry, taking into account the volume which is available (and the relevant price), in terms of the number of possible operations per second (OPS) at requested format (18 bits −24 bits, etc.). If today Doppler filtering is digital in most cases, pulse compression is often still analogue and other digital processing (digital beam forming, multistatic arrangements, etc.) such as described later in this

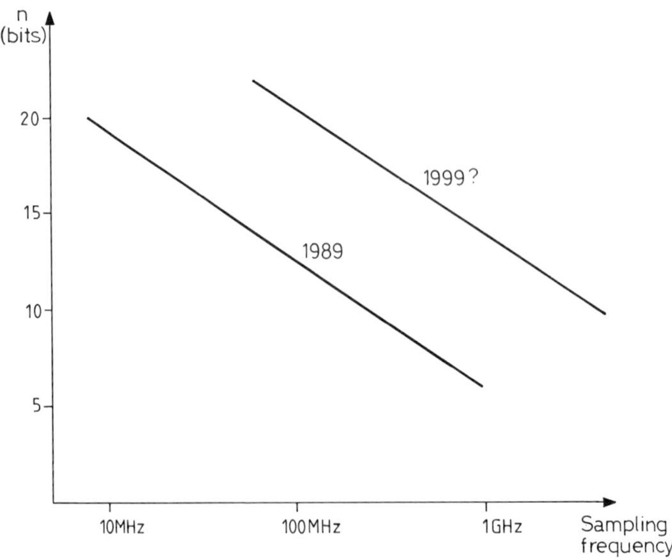

Fig. 4.28 Performance of analogue to-digital converters (ADC).

chapter must wait until acceptable performance in terms of OPS and format is offered.

4.2.12 Action against clutter

In most cases, the best thing to do is to try to limit the quantity of clutter entering the radar receiver. This can be done in several ways.

First, using circular polarization is efficient against rain clutter since against a spherical target, clockwise circular polarization is reflected as counterclockwise and thus does not enter the radar. Practically, imperfection of circular polarization on one hand, and the non-sphericity of rain drops on the other hand, limit the reduction of the rain clutter by this method to 20/25 dB, which is very good.

Second, reducing the parasitic radiation of the antenna limits the quantity of clutter entering the radar, which is always excellent and sometimes absolutely necessary.

Third, reducing the resolution cell of the radar is excellent because first it reduces the average clutter power and second it favourably changes the clutter amplitude distribution in many cases.

When the clutter has been reduced by those means (which are, unfortunately, not always possible to use), further reduction or cancellation is obtained by Doppler processing.

4.2.13 Pulse Doppler radars

Analysis of a surveillance pulse Doppler radar – fixed station – low repetition frequency LRF (no range ambiguity)

The panoramic surveillance radar under consideration operates in S-band ($\lambda = 0.1$ m) with an antenna span of 2 m rotating at 60 revolutions per minute (60 rpm). The horizontal antenna (3 dB) beamwidth is about 3.7° (65 milliradians). The pulse repetition frequency is 7.5 kHz. The maximum range of the radar is 15 km. A prf of 7.5 kHz introduces a range ambiguity of 20 km, since the sequence of pulses returned from a target at 20 km + D is very similar to the same sequence returned by a target at D. Range ambiguity of 20 km is far above the maximum range: except in the case of very powerful targets above 20 km there is no risk of range confusion. (No range ambiguity or low repetition frequency LRF). Transmitted pulses are 10 μs in length, but pulse compression is used in such a way (Δf around 3 MHz) that the compressed pulse is around 0.4 μs (equivalent to 60 m).

Let us examine the nature of the clutter accompanying a useful target at 5000 m range: the resolution cell is $0.065 \times 5000 \times 60 = 19\,500\ \text{m}^2$ equivalent to that of the 0.5 μs radar of Fig. 4.26. In an equivalent country the average value of the RCS of the clutter entering via the main beam will be 2 m^2 with 10% probability of exceeding 4 m^2. Clutter entering via the antenna sidelobes corresponds to a resolution cell of $2\pi \times 5000 \times 60 = 1\,900\,000\,\mathrm{m}^2$ which is relevant to the curve ($\tau = 50\,\mu$s) given in Fig. 4.26, and gives a clutter with an average power of $200\,\mathrm{m}^2$ exceeding $800\,\mathrm{m}^2$ 10% of the time. If the average sidelobes level of the parasitic radiation of the radar antenna is -30 dB, that means that the clutter entering via the sidelobes will be $200 \times 10^{-6}\,\mathrm{m}^2$ on the average ($0.0002\,\mathrm{m}^2$).

Refering to the problem of section 4.2.1, the spectrum of the transmitted signal as well as the return from a fixed echo (clutter) is represented in Fig. 4.29. The second graph of Fig. 4.29 represents (only between f_o and $f_o + 7.5$ kHz) the clutter spectrum, using logarithmic scales on the vertical. In fact, only the average values are used for simplicity.

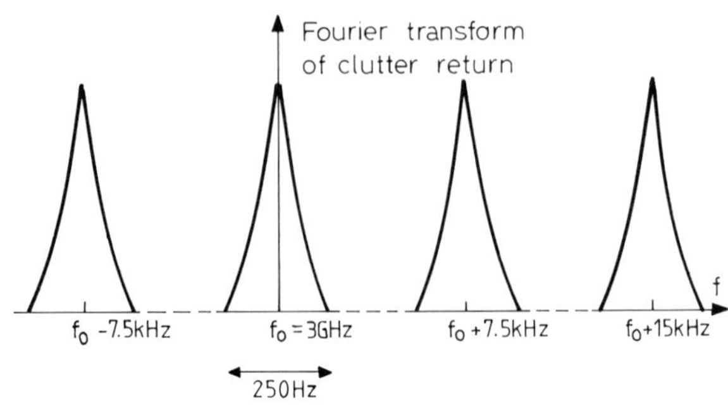

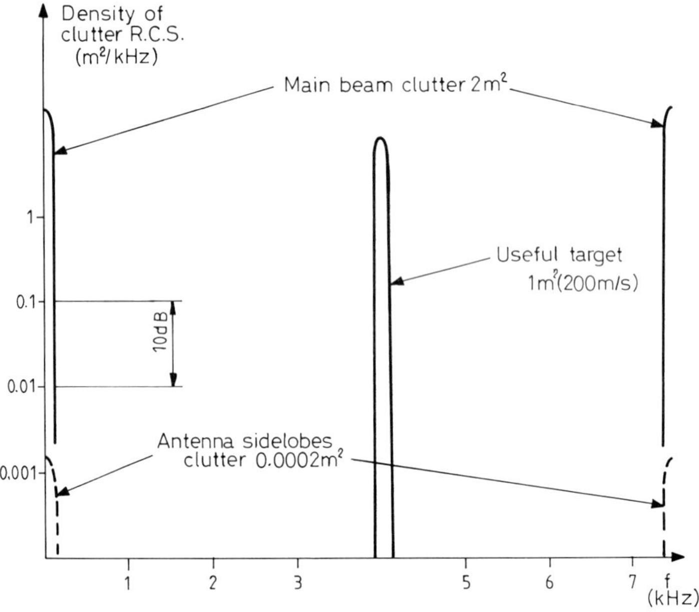

Fig. 4.29 Clutter frequency distribution (fixed surveillance radar).

On the same graph is represented the return from a useful target with an average RCS of 1 m^2 (it should be recalled that if that target is fluctuating according to Swerling then the RCS is below 0.1 m^2 10% of the time) and a radial velocity of 200 m/s. Clearly the graph would also be valid for a target with a radial velocity of $200+375=575$ m/s or with a radial velocity of $200-375=-175$ m/s, because of the stroboscopic effect coming from the periodicity (at 7.5 kHz: $7500=2\times 375$ m/s/$\lambda=0.1$ m).

Figure 4.30 represents the block diagram of the radar. Stable local oscillator (STALO) (1) provides a very stable frequency (e.g. at 2970 MHz). Oscillator (2) and dispersive device (2 bis) provide, at a very stable central frequency, a frequency modulated signal (modulated between 28.5 and 31.5 MHz). Mixing in (3) provides a signal around 3000 MHz which is transmitted after amplification (in 4). On reception, after mixing with local oscillator in (6)—possibly after RF amplification in (5)—the received signal is amplified and compressed in (7). (In case of digital pulse compression, this will be performed later). Compressed signal provided by (7) is in (8) demodulated in phase and in quadrature by frequency coming from (2) and by the same one shifted by 90° (or is demodulated by an equivalent single side band demodulation if Doppler processing is analogue), in order not to have the 'blind phase' problem which existed in the past when only a single demodulation was used (blind phase exists when a Doppler sine wave which is finally obtained is zero around the time when the maximum antenna gain is in direction of the useful target). In Fig. 4.30, the remaining block-diagram is only given for one of the two outputs of the I and Q demodulation. The output(s) of (8) are then range gated which means that the first wire is connected only during the period from 10 to 10.4 μs following the beginning of transmission and then

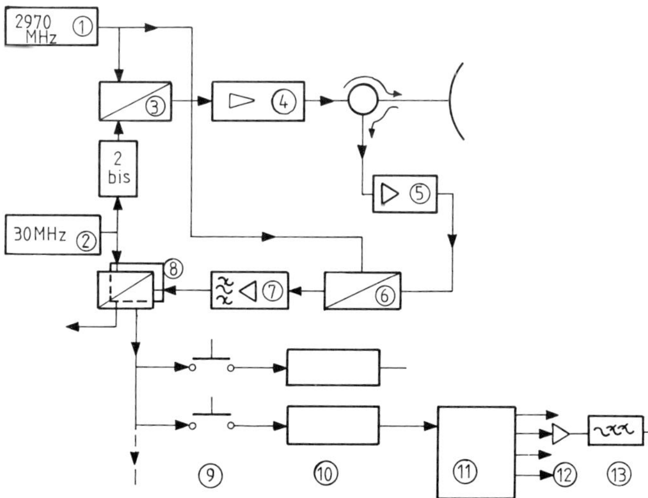

Fig. 4.30 General block diagram of a pulse-Doppler radar equipment.

corresponds to possible returns from targets at 1500 m. The next wire is connected only during the following period of 0.4 μs (corresponding to targets at 1560 m) etc.

Filters (10), all identical, are used to reject the clutter returns: their ideal transmittancy is represented in Fig. 4.31. Each filter 10 is followed by as many Doppler filter banks as there are range gates: a Doppler filter bank is composed of adjacent filters, all identical except for their central frequency. The dwell time of the antenna on a given target is (at −6 dB) around 10 ms (60 rpm is 360° per second: rotating by 3.7° takes about 10 ms) leading to use of Doppler filters 100 Hz wide (at 3 dB) every 100 Hz (see representation of their transmittancy in Fig. 4.32). Each output of a Doppler filter is followed by detection (12). (In case of digital I and Q configurations, detection recovers the vector whose coordinates are filtered I and Q).

In practice, if 200 Hz bandwidth Doppler filters are used instead of 100 Hz, and (12) is followed by (non-coherent) integration during 10 ms, Fig. 4.33 indicates (4) the relevant loss in sensitivity (0.5 dB): the 'price' of Doppler filtering is divided by 2, and paid for by the 0.5 dB loss in sensitivity. Using 500 Hz filters introduces a loss of 1.5 dB and divides the 'price' of Doppler filtering by 5, and that could be a good compromise.

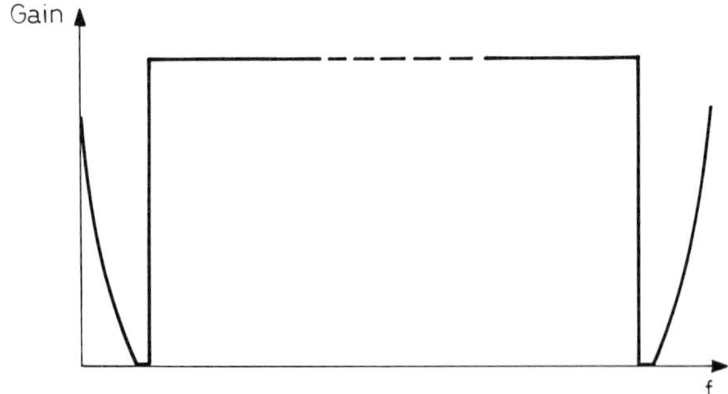

Fig. 4.31 Anticlutter filter.

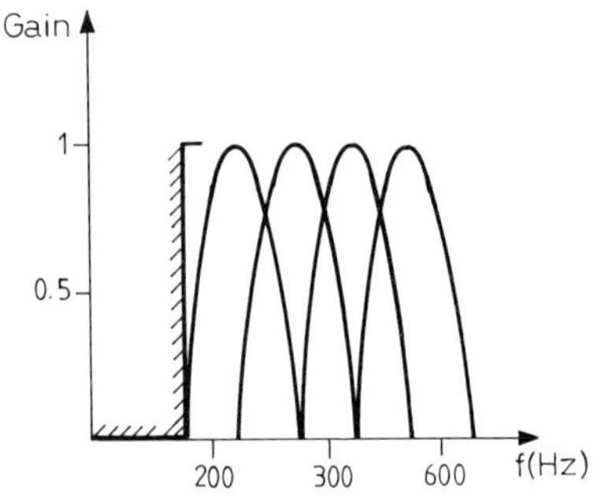

Fig. 4.32 Doppler filters.

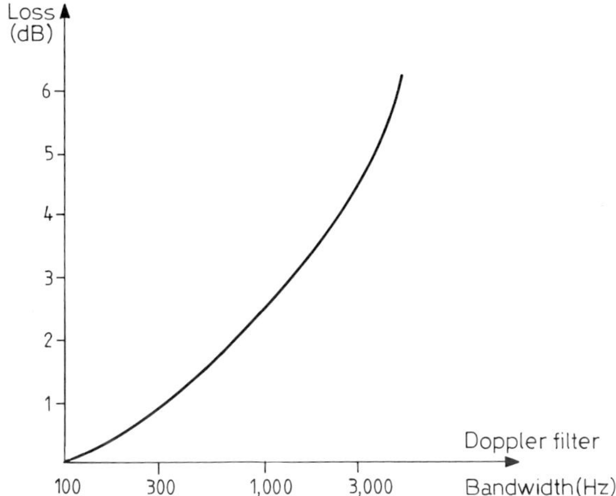

Fig. 4.33 Non-coherent integration losses.

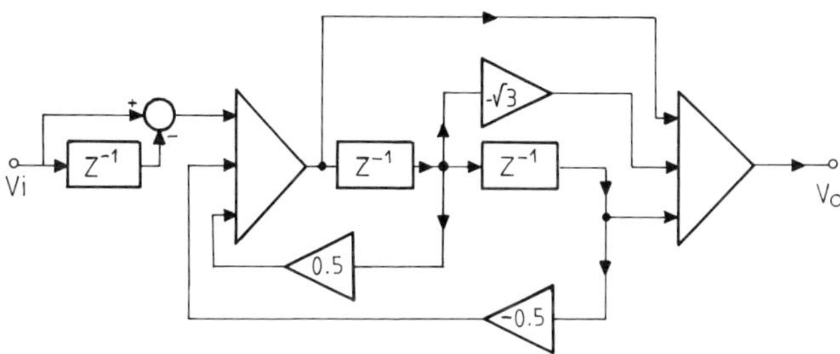

Fig. 4.34 Block diagram of a digital MTI filter.

Many radars have been produced in the past where boxes (9) (10) (11) (12) and (13) were analogue devices, representing a physically large number of Doppler filters. Using digital technology has completely changed the nature of the physical implementation. The range gates (9) and the rejecting filters (10) are now realized in the same physical unit, the 'digital MTI'. Fig. 4.34 gives a practical example of a simple realization with a transfer function in z ($z^{-1} = \exp(-2\pi jft/f_r)$) where f_r is the pulse repetition frequency), given by

$$T(z^{-1}) = (1 - z^{-1})\frac{1 - z^{-1}\sqrt{3} + z^{-2}}{1 - 0.5z^{-1} + 0.5z^{-2}} = \frac{V_o}{V_i}$$

(the relevant curve giving $|T|$ versus f/f_r is at Fig. 4.35).

But other algorithms of computations of V_o from V_i could be used: for instance the one indicated below (with introduction of two parameters u and w) which

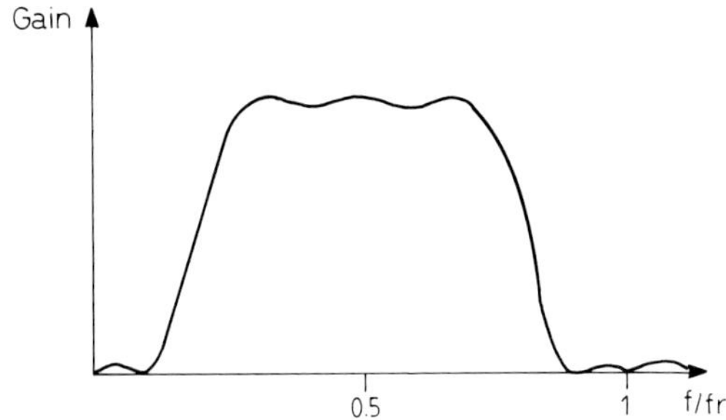

Fig. 4.35 Gain versus frequency of a digital MTI filter.

requires less accuracy in computation:

$$u(1-z^{-1})+w(1.5-0.5z^{-1})=V_i$$
$$w(1-z^{-1})=uz^{-1}$$
$$V_o=u(1-z^{-1})+w(2-\sqrt{3})(1-z^{-1}).$$

Regarding the banks of Doppler filters, they are presently obtained by fast Fourier transform.

Such a radar is compatible with frequency agility from burst to burst but the frequency has ovbiously to remain remarkably constant during the time when clutter rejection and Doppler filtering are achieved. Obviously targets at radial velocity equal or close to 375 m/s are eliminated (blind velocities). That generally being unacceptable, several different prf may be used consecutively to cope with that problem. Anticlutter performance of such a radar is limited (possibly by the number of bits available for A/D encoding and) by the phase instability of the carrier frequency during transmission, and of the frequencies delivered by the various oscillators.

Surveillance pulse Doppler radar on board a ship

Let us now consider a very similar case but where the radar is on board a fast modern ship (100 km/h or 27.8 m/s corresponding to a Doppler shift of 556 Hz), which is attacked by a (small) missile (RCS of 0.1 m^2) flying at 200 m/s in direction of the fast ship (see Fig. 4.36) at a distance of 5 km (see Fig. 4.36). In this radar, no pulse compression is used (transmitted pulse length is 10 μs) and the quality of the antenna pattern is bad (parasitic radiation at −18 dB).

Spectral distribution of the clutter is as indicated in Fig. 4.37. A rejection filter (300 Hz width) around the main beam clutter return will cancel it but if the useful target is at the same place as the clutter received via the antenna sidelobes, it will be impossible to detect it (that is the case if the ground velocity of the enemy is between 312 and 382 m/s). Good parasitic radiation (−30 dB) and pulse compression down to 0.4 μs would have reduced the average clutter entering via the antenna sidelobes by 38 dB, giving an easy situation (on such light clutter).

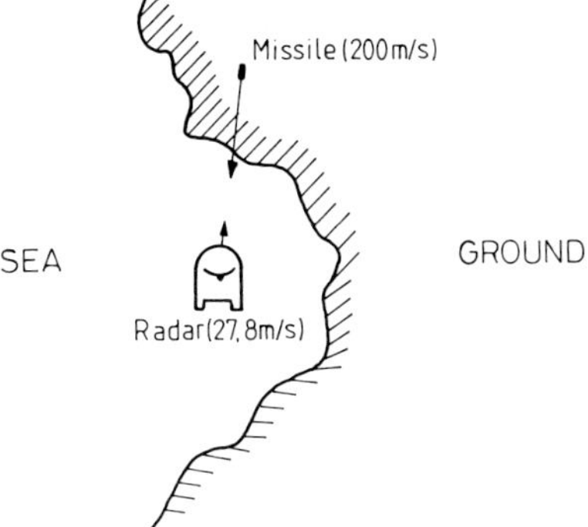

Fig. 4.36 A mobile surveillance radar.

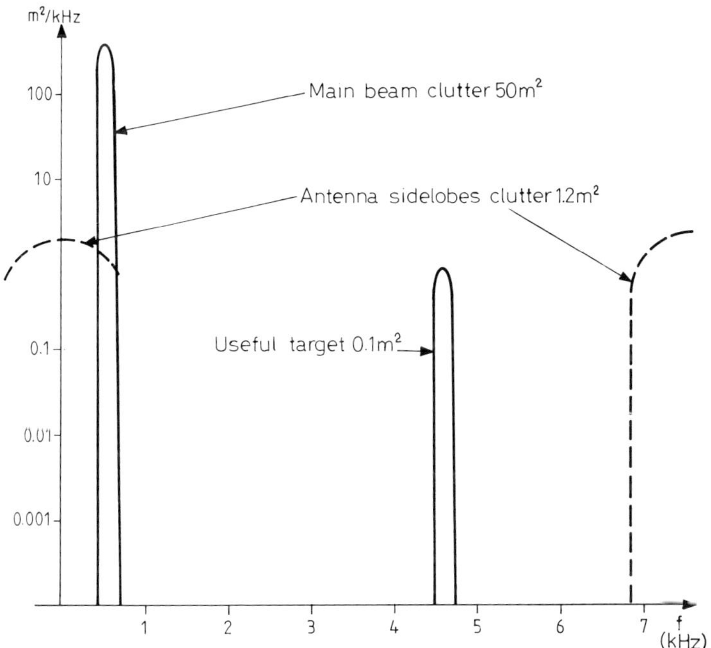

Fig. 4.37 Clutter frequency distribution for a mobile surveillance radar (low velocity).

It is interesting to note that if the same (not very good) radar is on a carrier flying at 600 km/h, the spectral distribution of the clutter is as represented in Fig. 4.38, which leaves no hope for detection of a 0.1 m^2 target if the antenna is not improved and/or pulse compression used (5 dB margin is insufficient taking into account the clutter fluctuation and the target fluctuation).

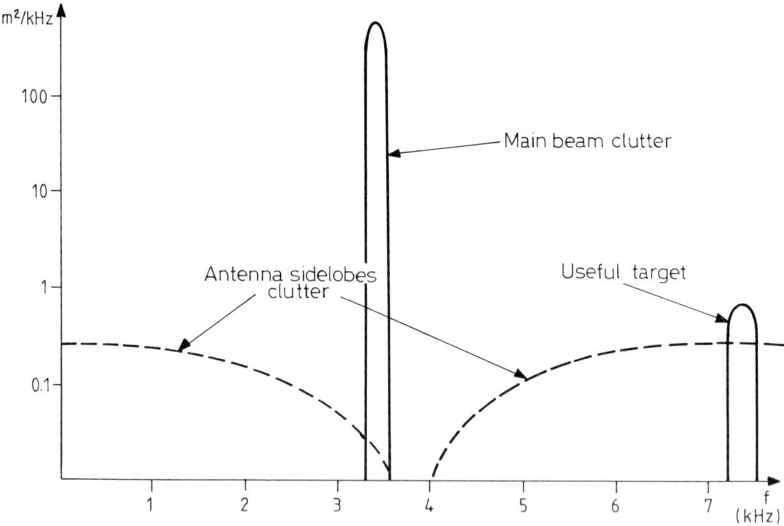

Fig. 4.38 Clutter frequency distribution for a mobile surveillance radar (high velocity).

Airborne surveillance pulse-Doppler radar—high repetition frequence (HRF)—no velocity ambiguity

Let us consider an airborne early warning (AEW) radar installed on board a jet aircraft flying at 222 m/s at 10 000 m above the earth level. S-band is used ($\lambda = 1$ m). The antenna has a span of 9 m and a height of 1.5 m and rotates at 6 r.p.m. Horizontal beamwidth (at 3 dB) is around 14 milliradians, vertical (elevation) beamwidth is around 80 milliradians. It is possible to tilt the beam in elevation at negative angles.

Pulse compression allows us to obtain a range resolution of 0.2 μs. Parasitic antenna radiation is at -30 dB. Pulse repetition frequency is 30 000 Hz: the range is directly measured with an ambiguity of 5 km (modulo 5 km) and blind distances (no reception during transmission) are obtained, both problems justifying the use of several repetition frequencies.

Let us consider the detection of an aeroplane flying at low altitude at a vertical distance of 100 km from the AEW above a clutter such as the one corresponding to Fig. 4.26. The target is flying in front of and towards the AEW. Figure 4.39 represents the spectral distribution of the relevant clutter. It is similar to that of Fig. 4.38 but since the prf is much higher, there remains a wide zone of frequencies free of clutter, where it is possible to have good detection, provided enough stability can be obtained in all frequencies produced.

Detection of targets with ground velocities between zero and 100 m/s is achieved without restriction (except regarding stability of frequencies in use). Detection of targets with ground velocities between -440 m/s and zero depends on the level

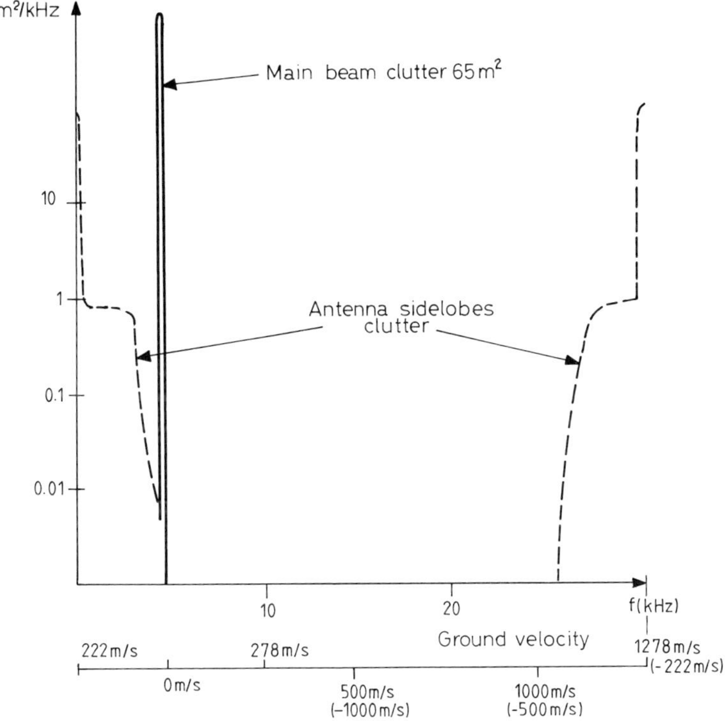

Fig. 4.39 Clutter frequency distribution for an HRF airborne early warning system.

of clutter entering the radar via the antenna sidelobes and on the quality of the antenna pattern and on the range resolution.

4.3 MAIN APPLICATIONS OF RADAR SYSTEMS

4.3.1 Surveillance radars

Production of surveillance radars probably corresponds to the most important activity in the radar field. The problem is to detect and to give coordinates of all interesting targets situated in half a sphere of which the radar is the centre (panoramic radars) or of a portion of that sphere, either delivering only the horizontal coordinates (2D radars) or also measuring their height (3D or volumetric radars). For that purpose most of these radar systems utilize an antenna system mechanically rotating around a vertical axis: in that case the radiated beam may be fixed with respect to the antenna and rotates with it. This beam may be wide in elevation (fan beam), or may be made with a stacked-beam arrangement.

In many modern systems, increasingly used for military air-defence organizations, the beam remains in the vertical symmetry plane of the rotating antenna system, but may change its shape as well as its direction in elevation, by using electronic scanning. More modern solutions (higher performance but more expensive) use a complete electronic scanning (in azimuth as well as in elevation), providing as requested various beam shapes (narrow, fan, monopulse) in any direction compared to the antenna structure, (but generally differing from the axis of the antenna by less than 50). It is clear that, in such a solution, while the antenna rotates (if it rotates) the beam may rotate in the other direction compared to the antenna and remain fixed compared to the carrier ('despun' mode).

In a rotating system, the radar system acquires information on every target position at each antenna revolution: the point representing the target position is a 'plot'. A (digital) computer takes account of successive positions of plots to produce a 'track'. Generally if a plot is seen only once, the computer will consider it as a false alarm, but if successive plots are obtained at three successive revolutions the computer will initiate a track and evaluate some elements of the target trajectory (velocity and number of gs). Those elements are compared with the elements of the known flights (using a possible target identification from the responses obtained to the interrogation of the target transponder by a system called secondary radar or IFF—identification of friend and foe—or SIF—selective identification feature) and the track will be 'dressed' by a number and various characteristics). The computer extrapolates and predicts the position of the next plot and if the actual one is close to the predicted one, the computer will assume that it is the same (same track number) but possibly that some characteristics of the track have changed.

Information on plots or tracks are often presented to the radar operator on a circular cathode-ray tube (panoramic plane indicator (PPI)), where the radar position is at the centre or not (offset picture).

In military systems, users may want to 'intercept' a 'doubtful' or 'hostile' target with a fighter aircraft either to force it to land or to destroy it. In that case the computer will determine the trajectory to be adopted by the interceptor immediately after take-off depending on the characteristics of the track to be intercepted and on the interceptor characteristics, and the relevant information is sent to the pilot by a ground-to-air radiolink, until the interceptor becomes capable of locking onto the target with its own radar and determines by himself what he has to do (for example to guide an air-to-air missile).

Using a radar with complete possibilities of electronic scanning is similar but with at least three additional possibilities:

1. if during the first detection a plot is doubtful, beam could be maintained in the same direction to get several more detections and get a 'confirmation';
2. if a zone of the space is very strongly jammed, it is possible to dwell in the same direction a time long enough to send enough energy to overcome the jamming (section 4.2.9);

3. if a target is particularly interesting, more accurate information may be obtained by doing more measurements on it (or more frequent ones).

4.3.2 Fire control radar systems

In many cases, the weapon to be used against the target will not be a fighter aircraft but possibly a surface-to-air missile or a gun, in which case the track created from the information provided by the surveillance ('acquisition') radar will give to a guidance radar (a 'fire control radar') the indications from which it will first find the 'designated' target, and then spend most of its time in performing measurements on it.

The fire control radar is then a 'tracking' radar, permanently directed (or directed most of the time) towards the target and designed to provide the best angular measurements on a target situated on its axis, and very good distance measurements thanks to the quasi-continuity of detection.

If the weapon to be used is a gun, the tracking radar will permanently evaluate the future behaviour of the target by extrapolation of the past and present (Kalman filtering) to ensure firing as efficient as possible.

If the weapon is a missile, the tracking radar will use two receivers (double chain radar), one tracking the hostile target, and the other one tracking the missile, in order to send to it information to be used for the best interception (at relatively short distances). The tracking radar might 'light' the target with a signal known by the missile ('illumination' radar) leaving the missile, equipped only with a radar receiver, the task of obtaining by itself the information required for the final guidance (interception at longer distances).

Radars are now designed (using electronic scanning) capable of achieving (in one equipment) surveillance, designation, tracking and possibly illumination. They are called multifunctional radars.

4.3.3 Radar systems on-board aircraft

Radars may be used on board aircraft to ensure surveillance. They are not limited by the earth's curvature, but problems coming from the ground clutter become more difficult because of the velocity of the radar carrier (section 4.2).

Radars on board a fighter may also be fire control radars (section 4.3.2). But radars on board aircraft may have other purposes: for instance they may deliver information allowing the aircraft to fly at low altitude ('terrain avoidance' or 'terrain clearance'), in which case, for instance, zones will be presented whose altitude is above the aircraft altitude (with some margin). They may deliver a photograph of the ground to facilitate navigation or bombing, as in the case of synthetic aperture radars (SAR).

Every users' dream is to have a radar capable of all the above functions equally well and if possible (nearly) at the same time. Such a multifunctional radar does not really exist, but some radars do exist which provide pretty well one or two main functions, and also some other functions not too poorly.

4.3.4 Instrumentation radars

They are used in missile firing installations or in missile launching sites. They are tracking radars, dealing with a very limited number of targets and delivering very accurate information (on position or radial speed) on targets to ensure a very accurate guidance or in order to allow reconstitution of trajectories a posteriori. Generally they are built in small quantities and have a high quality mechanisms.

4.3.5 Other applications

There are many other applications for radar systems, such as:

1. antipersonnel radars for troop detection;
2. proximity fuses of missile;
3. systems used to measure the speed of cars;
4. systems used to detect intrusion;
5. systems to facilitate the landing of aircraft;
6. radars to detect clouds (which is the main function of the radars installed on board commercial planes).

Antipersonnel radars may be light radars mounted on a pedestal, rotating around a vertical axis. The beam will be 10° wide in azimuth, compatible in X-band with 20cm antenna span. The radar may be of the pulse-Doppler type; 1 m/s velocity being associated with a Doppler frequency of 60 Hz, that means that the analysis of the Doppler by human ears is very efficient.

Many missiles are equipped with a radar proximity fuse which produces the explosion when a target is for instance between 0 and 15 m from the missile, if the Doppler shift is reasonable. Such radars may use a phase-modulated signal (continuous wave or pulsed) and use correlators on reception.

Precision approach radars (PAR) are still used to facilitate the landing of planes; they localize the landing plane compared to an ideal curve for final approach and send indications to the pilot to place him on that ideal curve.

4.4 EXPECTED EVOLUTION OF RADAR SYSTEMS

4.4.1 Multifunction and multimode in radar systems

A first obvious orientation of the radar evolution is the design of more and more radar systems capable of providing various functions within the same equipment and more or less at the same time. This has been explained about ground-to-air missile systems where radars capable of low altitude surveillance (in clutter), high elevation surveillance, or tracking, possibly of illumination or telemetry or telecommand, are requested by the system designer. This requires the use of electronic scanning, together with use of several radar modes in terms of pulse repetition frequency, pulse length, and phase modulation within the pulse. This has also been evoked for the radars on board combat aircraft for which users require fulfillment of all possible missions with the same radar system: surveillance, combat, navigation, terrain avoidance, missile guidance, etc. This also implies use of electronic scanning together with the use of several radar modes (low repetition frequency, high repetition frequency, medium repetition frequency).

Multimode is becoming ever more possible today for several reasons, the main one being that it is easier to achieve programmable digital signal processing which allows use of the same physical hardware (of reasonable size) to achieve various different filtering laws. Some modes introduce unwanted ambiguities (such as range ambiguities) which can only be resolved by powerful enough data processing. Evolution of silicon digital VLSI (CMOS or bipolar), improvement of performance in analogue to-digital encoding, improvements in conditioning for thermal evacuation are offering, year after year, more OPS* for signal processing and more MIPs (millions of instructions per second) for data processing in reasonable size. This will allow the introduction of more and more digital processing within the radar receivers and will allow the practical implementation of many algorithms and structures which have been known for a long time but which were not feasible until now.

But multimode transmission possibilities also imply the ability to greatly vary the nature of the transmitted signal which is obviously facilitated by the use of travelling-wave tubes with a control grid, as well as by the reduction of the size of capacitor (tank-capacitor) required to store a certain amount of joules in the radar modulator. (While only 20J per litre were feasible in the 1970s, figures of 100J per litre are now possible thanks to the use of new materials such as polyaramides, and it is reasonable to expect figures of 300J per litre in the future.)

4.4.2 Present and future implementation of ancient ideas—active antennas

It is interesting to realize that basic patents on sidelobe cancellation (section 4.2.6) were taken in the middle of the 1950s while the first practical implementation was only achieved at the end of the 1960s, because only analogue solutions were possible in the beginning, not allowing enough accuracy in the control of the amplitude and the phase of the ancillary channel.

In the beginning of the 1960s, digital pulse compression was considered as a solution to be accurate and flexible since one could change the nature of the phase modulation within the pulse, but it was practically impossible since the size of the relevant equipment was enormous and the rapidity of digital circuitry not high enough to compete with the analogue solutions.

*Or FLOPs floating point operations (per second).

Prototype realization of antenna beam forming was achieved in Thomson-CSF in the beginning of the 1960s, where a linear array of low gain receiving feeds was used, and where flexible antenna beam forming was obtained in intermediate frequency by analogue circuitry, but in fact, here again the analogue processing was too large in size and not accurate enough.

Many people remember polystatic systems of jammer location made in the beginning of the 1960s. Although the principle of those systems was perfectly valid, the digital circuitry available at that time led to enormous and expensive systems. In the same manner, models of active array antennas were presented in the beginning of the 1960s using silicon technology and were more or less abandoned, while the present emergence of gallium arsenide MMICs totally opens the way to making active array antennas.

This type of antenna contains any 'transmitting–receiving' modules every half a wavelength approximately, each TR module having two functions: transmission and reception. The transmission part is fed by a very low level frequency/phase reference, and contains a digitally controlled phase (delay) shifter, followed by a power amplifier immediately feeding (via a duplexer) an elementary antenna. The reception part (after microwave duplexing), contains a microwave low-noise amplifier, followed by a protecting limiter and a phase (delay) shifter.

Many advantages will come from the use of active antennas:

1. advantages resulting from electronic scanning possibilities;
2. an increase of reliability resulting from natural redundancy (thousands of T/R modules) and resulting from low voltage;
3. a reduction of the overall volume of the radar equipment;
4. a reduction to nearly zero of losses on transmission (recovering 1.5 to 4 dB) and the same thing on reception;
5. the possibility of transmitting more sophisticated signals (natural amplitude modulation obtained by addition of a large number of different signals). Such signals would be very difficult to reproduce by a transponder which would try to spoof the radar system;
6. the possibility of more flexible installation, opening the possibility of a better implementation on the radar carrier, with associated improvement of angular accuracy (increased antenna sizes);
7. the possibility of sampling, every half a wavelength approximately, the electromagnetic field all along the receiving antenna, allowing digital beam forming and a better angular resolution.

Digital beam forming consists of multiplying the matrix of electromagnetic field samples by another matrix in order to provide the equivalence of any pattern arrangement. This leads to, for instance, the possibility of a stacked-beam arrangement obtained by computation (easy to build and modify) instead of being obtained by a physical waveguide plumbing arrangement.

Regarding the angular resolution ('supradirectivity'), many people have explained for a long time that there was no limit in angular resolution, provided that: the signal-to-noise ratio be high enough; and the relevant (complex) computation be achieved.

In fact the signal-to-noise ratio is very high in two very interesting situations: when the 'signal' is a signal received from a jammer; or when the signal is a 'skin-return' from a target (long time after the first detection) at a relatively small distance from the radar, which is generally the case when an accurate angular measurement is requested. If the signal-to-noise ratio is very large, it is possible, from the analysis of the variation of the electromagnetic field all along the receiving antenna (in phase and in amplitude), to know many things about the 'targets' which are producing that field.

Measurements which were until now performed on most of the (analogue) classical antennas were very simple: addition of all fields ('sum' signal), addition of fields of the right part to be subtracted from the similar addition from the left one, or slightly more sophisticated measurements (such as the 'deviation' channel suggested by Drabowitch for instance, where the addition of central fields is subtracted from the addition of lateral ones). In fact those measurements were very convenient and simple, but a little insufficient since it was in fact assumed that the (electromagnetic) field was zero outside the antenna, which is the only impossible assumption.

If a complex sampling of the electromagnetic field is achieved all along the structure of an active antenna (not too perturbed by noise), realistic assumptions may be used in order to derive realistic information on the angular target situation. It may be assumed that useful targets are point-targets in limited number and/or it may be assumed that targets (jammers) are transmitting signals which are independent from each other, and independent from an useful signal, and/or that the electromagnetic field outside of the antenna aperture is more likely to be according to 'maximum entropy' than to be zero, etc. All those assumptions, which may be combined and which depend on the actual situation, have led to known algorithms (already used in operational sonars) capable of delivering reasonable angular information on targets from the analysis of the electromagnetic field all along an active antenna. Those algorithms, for their use in radar systems, require that a complex computation be performed (in real time), which is increasingly feasible thanks to the rapid increase in the 'power' of digital programmable signal processing.

4.4.3 High resolution in distance

The evolution of components on transmission and on reception will open the possibility of generalizing in radar systems the transmission of signals with a very wide instantaneous bandwidth Δf (a few hundreds of megahertz). Two main advantages are expected from the very good associated range resolution (better than 1 m).

1. Some identification of the targets: accounting for the number of echoes in a given small zone, evaluation of the size of the target, classification of the target, etc.

2. Some possibility of association of Doppler returns with distances: presently many targets deliver several Doppler responses for the same target (coming from various mobile parts of the target), which gives saturation in signal processing and/or in data processing. A very good range resolution will allow the association of each Doppler return to a particular part of the target and will then facilitate the relevant 'data handling'.

4.4.4 New wavelengths

Evolution of technologies is beginning to fill the gap between centimetre wavelengths and infrared. Even if atmospheric attenuation is troublesome when using millimetre wavelengths, they facilitate angular measurements, while atmospheric attenuation it not significant outside the atmosphere or at short ranges. 90 GHz (W-band) radar systems are more and more in use, thanks to the semiconductor technological evolution. 140 GHz is beginning to follow the pace. What about the use of powerful gyrotrons at very high frequencies in the future?

An important advantage of using higher frequencies is clear when thinking of using micrometre (infrared) wavelengths in radars (Lidars). The use of such very short wavelengths will allow very accurate measurement of the radial velocity (if a 'coherent' micrometre lidar is used), and allow even the detection of the motion of the ailerons of a plane, indicating that the plane will turn, while today most of the radars are only able to a posteriori verify that the plane has turned, a few seconds later.

On the other hand, there would be some clear advantages in using very low frequencies between HF and UHF, since the radar cross-section of targets is really difficult to reduce at those frequencies (stealth techniques are not very efficient in metric wavelengths). Unfortunately, lack of angular resolution presently prevents the use of those frequencies. The situation could change in that matter when much more progress in signal processing will be achieved (0.1 to 1 teraflops – 48 bits), because at that time it will be possible to use fully polystatic arrangements of transmitter–receiver antennas, spread all over a territory of thousands of square kilometres, with the capability of processing in a coherent manner all signals received by all antennas, which requires a high dynamic range and then a complex computation.

REFERENCES

Carpentier, M. H. (1988) *Principles of Modern Radar Systems*, Artech House, New York.

Megaw, E. C. S. (1946) The high-power pulsed magnetron: a review of early developments, *The Journal of I.E.E.*, **93, IIIA**, **(5)**, 977–984.

Phillips, V. J. (1978) The telemobiloscope, *Wireless World*, 68–70.

Electrical Magazine—London—(1904) vol. 2 p. 388.

Woodward, P. M. (1950) *Probability and information theory with applications to radar*, Pergamon Press, London.

Swerling, P. (1960) Probability of detection for fluctuating targets, *IRE Trans*, **IT6**, 269–308.

5

Electric confrontation

François Naville

5.1 INTRODUCTION

Whether for armaments or functions in the fields of communication, observations, surveillance, standby or support, electronics play a major part in defence systems.

What is referred to as 'electronic warfare' actually groups together a set of radio-electrical techniques which, in themselves, do not correspond to any specific military end purpose. It is more appropriate to talk in terms of 'electronic confrontation', the aim of which is simply to gain control of radio waves so as to guarantee the use of the radio spectrum to good advantage and to prohibit its use by the enemy. In this game, knowledge of the extent of the threat is an essential element, in particular to defining and designing new equipment.

The stakes in this game are so important that it has justified, and indeed still justifies, specific hardware and concept programs which are considered to be non-traditional, generally bundled together under the name electronic warfare. Much has already been said and written about the notion of electronic warfare. What does the term really mean?

Electronic warfare can be defined as a form of military action aimed at:

1. drawing information from enemy emissions in order to take counter action;
2. reducing or preventing the use of electromagnetic radiations by the enemy, or modifying their use to our benefit;
3. ensuring the efficient use of electromagnetic radiations while depriving the enemy of the advantages of using them.

However, it would be a fundamental error to take all this to mean, as some have written, that electronic warfare is the panacea of military development. In reality, electronic warfare should be considered as a set of techniques which can efficiently lead to satisfying military missions and which are incorporated into systems in which more often than not, they are associated with other more conventional methods. So what are the different points of application of electronic warfare?

The first service rendered to the military by radioelectricity was communication: this resulted in Morse code, followed by voice communication then digital data transmissions with their wide spectrum of sophistication (protection by coding, development of the spectrum used, microwave links, satellite links, etc.).

With the advent of radioelectric techniques in communication, techniques to intercept and exploit enemy communications soon appeared: the first step was called communication intelligence (COMINT). Today, COMINT systems involve a wide variety of techniques to collect and analyse signals: direction finding and transmitter locating, the technical analysis of transmission and finally, when possible, exploitation of the information content of the message.

The second major class of services rendered by radioelectricity is in the area of armaments, with the introduction of radar in the early 1940s. These began with aerial surveillance (early warning) radars, then radars carried on ships, airborne radars, fire control radars and finally missile seeker radars. At the same time, techniques designed to foil their use were developed. The first involved radar signal interception techniques, then came the means of locating the radars and analysing their emissions; these different techniques led to electronic information in the 'non-communication' field (without any conversational content) called electronic intelligence (ELINT). Used at a tactical level, ELINT is also identified with electronic support measures (ESM).

Because of the simplicity of the modulations used and the technical level reached in the bands currently used, it soon became possible to build broadband transmitters which jammed to prohibit the use of the radio spectrum for communication. The jammers became gradually more sophisticated to keep in step with the progress made in frequency management. This progress ranged through fast frequency hopping and the spreading of the spectrum by pseudorandom encoding (modulation using almost white noise) for contemporary systems. The efficiency of jammers against radars and seekers is quite another problem. Jamming techniques used against non-communication systems were only developed at a late stage, when high speed circuits, programmable logic and very advanced technology microwave components appeared.

Jammers, decoys and active deception devices (such as intrusion, range or velocity pull-off, etc.) are what are referred to as electronic countermeasures (ECM).

Two additional areas in which the radio spectrum is used and where electronic warfare is developing should also be mentioned:

1. radio-navigation in which electronic warfare is used to jam signals or to modify them (a technique that specialists refer to as transplexion);
2. optronics (imagery, guidance, designation, range finding using infrared, lasers, ultraviolet, low light level television, etc.) where electronic warfare is making its first steps, beginning with the detection and location of transmission sources and detectors in order to change, jam, neutralize or destroy them, depending on the capabilities of the technique.

In the conventional duel between shell and battleship, each action in a more or less short period of time triggers a protective reaction. This is how many techniques have been developed, aimed at protecting electronic eqipment to make it more efficient. Among them are included: techniques for modifying the electromagnetic signature (absorbant, reflector, etc.); techniques of masking or camouflage extended to the radar electromagnetic spectrum; pulse coding, phase, frequency and polarization coding; the techniques of multi-sensors, multifrequency, multimode and multi-figuration. All these techniques are taken into account by the equipment to be protected and generally result in increased complexity and cost but do not necessarily change the functional characteristics. They are designated under the phrase electronic counter-countermeasures (ECCM).

5.2 ELECTRONIC SUPPORT MEASURES (ESM)

5.2.1 General

To ensure strategic and tactical reconnaissance missions, the facilities monitoring the radiations transmitted by enemy radars have to handle:

1. interception of these signals to detect their presence;
2. analysis of waveforms transmitted by these radars while carrying out a number of measurements such as that of the pulse width or spectrum, the transmission frequency, the received level, the pulse repetition frequency, the illumination period, etc., parameters which will make it possible to establish the signature of a radar and therefore work back to determine its mode of operation;
3. measurements of the signal direction of arrival which, when associated with other parameters, will make it possible to evaluate the location of the radars;
4. identification of the signals received by comparison of the signature obtained by analysis and a priori knowledge of the characteristics of the radars being searched for.

To accomplish all these functions within a radar signal environment which can be very dense in some of the areas observed, the electronic reconnaissance facilities must have the following main performance capabilities:

1. a very broad operation frequency band to intercept surveillance radars generally at low frequency (L-band or below) as well as tracking or fire control radars capable of operating in the X or Ku-bands or even in the millimetric bands (Ka-band);
2. high sensitivity for detecting radars at long distances, taking account of the fact that such radars may not illuminate the monitoring facilities (detection on the side or scattered lobes of the radar) or radar with very low radiated peak power levels (LPI low probability of intercept radars);
3. an excellent dynamic in order to be capable of detecting both powerful near signals and remote weak signals without any distortion in the information gleaned from measurements made on such signals;
4. excellent selectivity and suitable means of signal processing to allow sorting and identification of a determined radar signal in a very dense electromagnetic environment leading to a great number of simultaneous signals;

5. measurements to detect the direction of arrival (DOA) of the signals to a high level of precision thus governing the correct location of radar transmitters;
6. a very wide angular coverage adapted to the mission in order to capture the different signals whatever their direction of arrival.

These different performance characteristics, aimed at ensuring a good probability of interception and good measurements on radar signals are partly contradictory and require a trade-off regarding the means of monitoring. For instance, very high sensitivity can only be obtained by reducing the instantaneous frequency passband of a receiver in order to reduce its thermal noise level but this runs counter to the method of detecting weak radar signals in a very broad frequency band with a high probability of interceptions. The same applies for the angular or time related aspect of the received signals.

This chapter outlines the different techniques of reception, measurement of direction of arrival and the means of location used to obtain the best trade-off regarding performance. It should be noted that these electronic intelligence (ELINT) receivers, unlike radar receivers, do not in theory know the signals they are designed to receive leading to a difference in structures between the two types of receivers.

5.2.2 Reception techniques

The main reception techniques used are summed up, with their main characteristics, advantages and drawbacks, in Table 5.1 and are discussed below.

Direct detection receivers (crystal video detectors) One of the first used for electronic intelligence, this type of receiver used with a very broad instantaneous frequency band antenna (covering one or several octaves) carries out video detection on the received microwave signal.

The advantages are a very broad instantaneous frequency band, therefore good probability of intercepting the signals and low cost. The drawbacks are low sensitivity because of the type of detection and the broad band covered in terms of frequency (high detector noise level) as well as its non-selectivity in terms of frequency.

To remedy these two major faults—the limited sensitivity and non-selectivity for frequencies—the following facilities were developed:

1. with the improvement of the technology, setting-up of very broad band amplifiers in order to compensate for antenna-detector link losses thereby reducing the noise figure and leading to improved sensitivity;
2. use of microwave filter benches by which detection in several instantaneous frequency sub-bands is possible in order to improve frequency selectivity.

Measurements on such a detector concern the envelope of the signals received i.e. the levels, pulse widths, pulse repetition period, illumination period, etc.

The frequency, a major parameter of the radar signature, is not measured, leading to the development of devices such as IFM (instantaneous frequency measurement) receivers. This technique is based on the principle of instantaneous phase measurement between two channels, one delayed by τ from the other. For a measurement of the phase difference $\Delta\phi$, the frequency F of the incident signal is then given in the equation:

$$F = \frac{\Delta\phi}{2\pi\tau}.$$

Phase measurement is nevertheless ambiguous (measurement to within 2π) so that several measurement channels have to be used, presenting different delays to clear up any such ambiguity leading to a relatively high cost.

The association of an IFM receiver with a direct detection receiver thus makes it possible to measure the received level and instantaneous frequency in a wide frequency band.

Superheterodyne detection receivers The limited sensitivity and the lack of frequency selectivity of direct detection receivers have led to the use of superheterodyne receivers. The frequency transposition of the received microwave signal by beating using a local oscillator, and filtering in a limited band around the lower intermediate frequency obtained lead to good frequency selectivity (because of filtration) and good detection sensitivity (because of the limited band), the main advantages of this type of receiver.

The main drawback of this is a low probability of intercepting radar signals spread out through a very wide frequency band (several octaves) whereas the instantaneous band of a superheterodyne receiver extends merely from a few MHz to a few dozens or hundreds of MHz. This means that a receiver like this must be confined to a frequency scan offering a reduced probability of interception (probability of the superheterodyne receiver being at the radar frequency when the radar is transmitting at that frequency).

To overcome this drawback, several superheterodyne receivers have been connected in parallel (multichannel receiver) to increase the instantaneous band coverage and thereby the probability of intercepting radar signals. But this involves complex processing and high cost. The cost of a superheterodyne receiver is already more than that of a direct detecting receiver because of the components it employs, (frequency-controlled local oscillator, filters, intermediate-frequency amplification, etc.).

Spectrum analysers To improve the instantaneous broadband, high sensitivity, frequency selectivity, high intercept probability compromise, techniques based on instantaneous spectrum analysis have been developed in two main areas.

First, the use of frequency dispersive electro-acoustic lines in compressive receivers or SAW (surface acoustic wave) receivers. If the incident microwave

Table 5.1 Reception techniques

Type	Basic structure	Main characteristics		Advantages	Drawbacks
		Instantaneous band	Simultaneous signal processing		
Direct detection	 A, FILTER	> octave	No	Interception probability	Sensitivity Non-selective
Instantaneous frequency meter (I.F.M.)	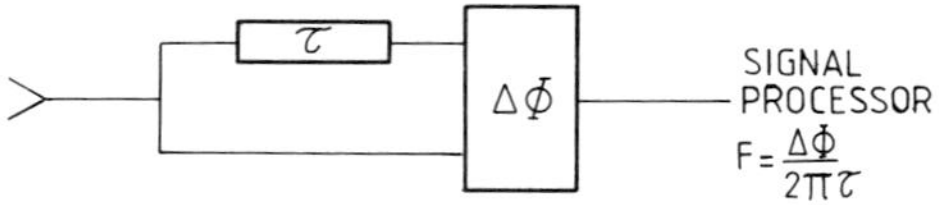 τ, $\Delta\Phi$, SIGNAL PROCESSOR $F = \frac{\Delta\Phi}{2\pi\tau}$	> octave	No	Low cost Interception probability	Cost Non-selective
Superheterodyne	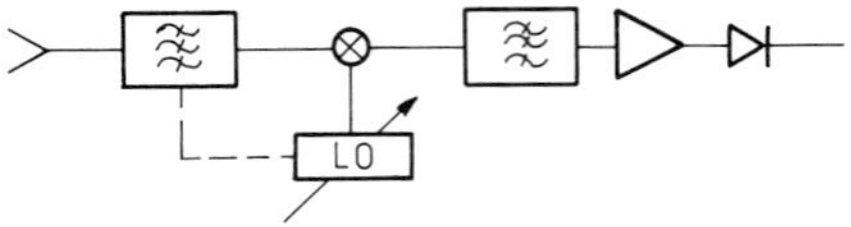 LO	Some MHz or tenths of MHz	Yes (outside IF band)	Sensitivity Selectivity	Very low intercept probability
Multi-channel	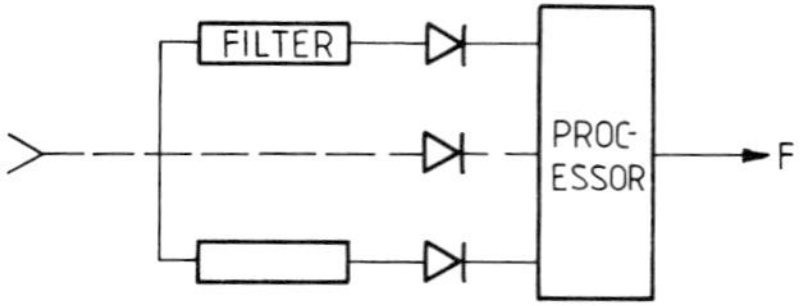 FILTER, PROCESSOR, F	Number of filters	Yes	Sensitivity Selectivity Average interception probability	Cost Processing

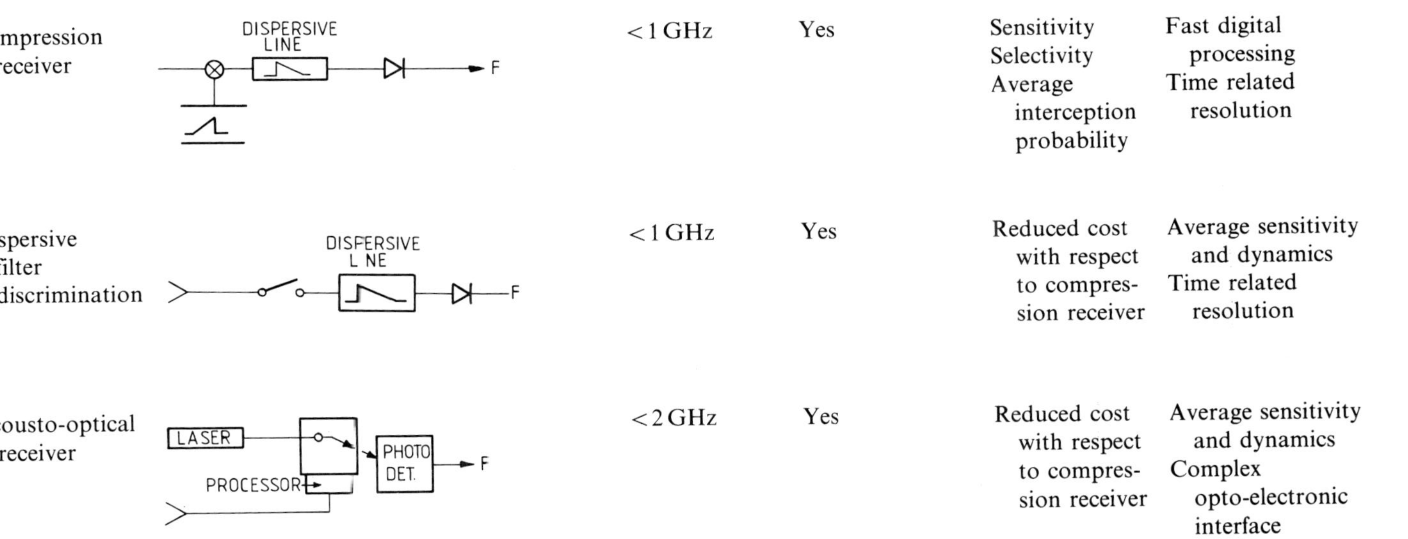

Compression receiver	<1 GHz	Yes	Sensitivity Selectivity Average interception probability	Fast digital processing Time related resolution
Dispersive filter discrimination	<1 GHz	Yes	Reduced cost with respect to compression receiver	Average sensitivity and dynamics Time related resolution
Acousto-optical receiver	<2 GHz	Yes	Reduced cost with respect to compression receiver	Average sensitivity and dynamics Complex opto-electronic interface

signal is modulated by a frequency sawtooth signal (variable frequency LO), itself generated by the excitation of a dispersive line and feeding the intermediate frequency beat obtained into a dispersive line whose slope, delay versus frequency is the inverse of that of the sawtooth, we obtain a time/frequency transposition of the signal received during the sawtooth signal.

In this way, an instantaneous spectrum analyser is produced capable of separating two or several simultaneous signals at different frequencies in the compressive receiver passband. This process gives an instantaneous passband of around 500 MHz to 1 GHz while providing excellent selectivity and high sensitivity (e.g. equivalent band of 1 MHz or less) through the use of spectrum analysis. Conversely, a compressive receiver like this costs more because of the components it uses and the very fast signal processing means it demands, for instance, instantaneous 500 MHz band analysis in 1 μs, which means processing signal samples every 2 ns to obtain 1 MHz resolution.

Simpler receivers are also available, derived from the above type, referred to as dispersive delay line discriminators, directly using the properties of dispersive lines: signal delay at line output proportional to the incident frequency. The cost is less with respect to compressive receivers but the sensitivity and selectivity are not as good.

Second, the use of acoustic-optical processes based upon the principle of Bragg cells working as follows: a coherent beam of light (laser) is directed into a medium which can be electrically excited in order to generate acoustic waves. The laser beam is then deviated in proportion to the excitation beam frequency.

An optoelectronic interface comprising photo-detectors measures the deviation of the laser beam in order to work back to the frequency of the incident electromagnetic signal. This type of receiver, which requires the development of integrated optical circuits, makes it possible to widen the instantaneous frequency band up to values of around 2 GHz but leads to an instantaneous dynamic and lower sensitivity than those of a compressive receiver as well as an output optoelectronic interface which calls for complex digital processing techniques.

5.2.3 Direction-finding techniques

The main methods of direction measurement can be based upon amplitude, phase or time of arrival measurement on the incident signal. Table 5.2 sums up a number of these methods with their main characteristics.

Direction-finding by amplitude measurement Directional measurement by amplitude measurement can be made conventionally.

1. Either with a high-gain revolving antenna whereby the measurement of the signal direction is given by the position of the antenna at the time when the highest levels are received. This solution has the advantage of presenting good

angular selectivity but the instantaneous coverage sector is very small, and these devices are very large and slow due to the rotation mechanisms. This means these systems are hardly compatible with installation in aircraft.

2. Or by the use of several antennas with angular aiming-off so that directional measurements are made on the basis of the differences in the received levels on each of them. In this way, there can be solutions comprising twin-lobe antennas comparable to the 'monopulse' technique used in different radars or multiple antenna solutions capable of covering 360° in azimuth. This principle applies well to direct detection receivers (crystal video detectors) with each antenna having its associated reception channel leading to an instantaneous direction measurement.

More complex solutions using a microwave lens (such as Rotmann lens) make it possible to measure the direction accurately but at the cost of high complexity.

Finally, some time of arrival compensation techniques, by transforming the phase difference between two received signals on two different antennas into amplitude modulation, make it possible to measure the direction by locating a maximum (or minimum) level.

Direction-finding by phase measurement Two main methods can be used.

1. A plane interferometric network for which, at the output of two (or several) antennas, the phase difference is measured. The direction of arrival is reconstituted by calculation from these phase measurements. The accuracy of direction-finding increases in proportion to the size of the base made by the two antennas. However, the phase is only measured to within 2°. Directional ambiguities may appear as a function of the wavelength of the incident signal relative to the dimensions of the measurement base. To remove such ambiguities, more than two antennas are used in order to cover broad frequency bands. The angular coverage of a system like this is nevertheless limited, making it necessary to use several networks to obtain a omnidirectional coverage.
2. An antenna array using microwave circuits arranged in a Butler matrix makes it possible to carry out direction measurement by relative phase measurement and provides omnidirectional coverage. Such a technique leads to high complexity of microwave circuits and makes it difficult to install antennas on board aircraft.

Direction-finding by time of arrival measurement In principle, this method is very similar to the previous method. At the output of two antennas receiving the same incident signal, we no longer scan the phase differential between two received signals but the time of arrival difference. It is however necessary to have large bases (several hundreds of meters) in order to ensure good angular accuracy.

Table 5.2 DF techniques

	Type		*Advantages*	*Drawbacks*	*Limitations for airborne utilization*
Amplitude	Monolobe rotating antenna		Simplicity Selectivity	Slowness Mechanism	Slowness incompatible with evolution of an aircraft
	Twin lobe rotating antenna		Simplicity Selectivity	Slowness Mechanism	Slowness incompatible with fast evolutions
	Multiple beams	Aimed off antennas	Monopulse Low cost	Sensitivity Selectivity	
		Rotmann's matrix	High gain Accuracy Monopulse	Complexity	
Phase	Compensation for time of arrival		Simplicity	Non-instantaneous	
	Interferometric network		Monopulse	Processing	Antenna clearance
	Butler's matrix network		Omnidirectional Monopulse	Microwave circuit complexity	Antenna installation on aircraft

5.2.4 Location measurements

To use the information detected by monitoring receivers in order to determine the location of radar transmitters, a variety of methods can be used. Basically, these include the following.

1. Methods based on measurement of the received level, either by directly estimating the transmitter range as a function of the received level or by dynamic estimation of the range on the basis of a variation of this level with time. However, these are only approximate methods because the received level depends on parameters which are not all controlled, such as the attenuation of radiations during their propagation, fluctuation of antenna pattern gains (by ground effect) and, above all, the possibility of radar transmitters being able to modify their transmission level.
2. Methods based on geometrical calculations such as:
   (a) analysis of direction-finding bearing straight lines from a single carrier moving with respect to the radar to be localized.
   (b) analysis of direction-finding bearing straight lines by triangulation from one or several monitoring sites (on a mobile or other carrier);
3. Analysis of the time of arrival difference by hyperbolic location from two (or several) monitoring sites.

The joint or separate use of these different methods depends upon the relative situation of the monitoring facilities with respect to the radars to be located. For monitoring facilities carrying out angular measurements from a mobile carrier vehicle, the direction-finding bearing methods during the movement of the carrier vehicle will supply acceptable accuracy if the angular movement of the radar to be located with respect to the carrier vehicle is sufficient. For instance, with a direction finder having 1° of accuracy, range location accuracy of 5% requires angular movements between carrier vehicle and radar to be located to approximately 30°. This introduces constraints regarding the monitoring missions accomplished.

For monitoring means which make location measurements from several sites, the precision of location is linked directly to the dimensions of the measurement base formed by these sites whether they use direction-finding or time of arrival measurements for this purpose. For instance, angular measurements to within 1° from a 1 km base will supply a locating relative accuracy of around 50% only for a radar transmitter located at 50 km perpendicular to the measurement base.

Different approaches have been developed to minimize the locating errors as a function of the number of elementary measurements made (direction finding or time of arrival) in time. But it should be noted that these methods apply effectively when the radar transmitters to be located are totally separated from the remainder of the environment. In the case of a dense environment where the radars to be located may be very close to each other, general sorting-localizing methods have been developed. These take account of all the measured parameters whether they are radioelectrical, or whether they relate to direction or time of arrival. The elementary data which may then be used for radar transmitter sorting-location control then have to be grouped together and classified in a multidimensional observation space ('clustering' methods).

5.2.5 Evolution of the systems

Because of the evolution of the signals transmitted by the radars, the monitoring facilities will evolve in the following directions:

1. expansion of the frequency band covered in order to take account of modern radars;
2. the joint use of a direct detection receiver and a superheterodyne receiver with spectrum analysers to optimize performance levels in terms of sensitivity, dynamics, selectivities, and intercept probabilities;
3. the use of new technologies (MMIC, VHSIC) enabling higher integration within the receiver units;
4. a search for greater selectivity by developing spatial, frequential and time-related filtration facilities for the incident signals;
5. the development of high performance processing facilities for the sorting and locating of radar signals in dense environments (for instance, clustering methods) and for signal identification (artificial intelligence techniques).

5.3 ELECTRONIC COUNTERMEASURES (ECM)

5.3.1 Introduction

The term ECM generally applies to all of the facilities used for implementing jamming or decoy application procedures against radars of enemy setups. To trigger countermeasure actions, ECMs must include a receiver for analysing the environment and working out the best suited action to neutralize enemy radars.

The general description given in section 5.2 refers to reception and directional measurements of radar signals using electronic reconnaissance devices and also applies to this type of receiver. However, these devices must be carefully designed to allow real-time triggering in the automatic mode of jamming and decoy launching operations during a mission while informing the operator or the pilot of the environment in order to provide him with time for reactions. In this chapter we will attempt to describe more particularly the different types of jammers.

5.3.2 Main operational uses of jammers

The classification of the different types of jammers can be made according to the position of the jammer with respect to the targets to be protected against enemy radars. In this way, we encounter the following.

1. Stand-off jammers (SOJ), so called because they are located outside the danger area created by enemy weapons systems. Their initial role is to prevent enemy radars from detecting friendly air strikes. Therefore, they are designed first and foremost to jam early warning surveillance radars within the territory of the enemy forces. The long distance between jammer and radar, combined with the need to be able to jam such radars outside their main lobe means that high radiated power jammers are necessary, installable only upon long haul aircraft, or installed on the ground.
2. Escort jammers, accompanying a raid to prevent detection and evaluation of the threat by enemy radar. The smaller distance between jammers and targets to be protected allows the radiated power levels to be lower.
3. Self-protection jammers aimed at protecting the carrier vehicle on which they are mounted. The initial objective of this type of jammer is to perturb the operation of weapon systems which constitute an imminent threat, i.e. fire control radar and missile seekers.
4. Stand forward jammers, i.e. jammers located nearer to the radars to be perturbed than the target to be protected. For instance, drones or rockets can be used for this purpose.
5. Devices which can be towed, dropped or ejected by the targets to be protected in order to create a decoy effect with respect to enemy weapon systems.

5.3.3 Jamming techniques

Jamming techniques are aimed at two main effects.

1. Masking, by transmitting a jamming signal resembling noise, producing a desensitization in the victim radar receiver which, if it is sufficiently strong, prevents the detection of the targets being searched. The distance and the target speed can no longer be measured by the radar.
2. Decoy launching obtained by transmitting likely false echoes either in large numbers (saturation effect) to prevent the victim radar from acquiring the true target, or in limited numbers but attracting the victim radar to one of the false target forms so as to perturb more particularly its fire control facilities (generating measurement errors on target speed or distance).

Noise jammers

The generation of a noise jamming signal by a jammer is a way of protecting the target with respect to a given radar up to a minimum distance (referred to as the burn-through range). This self-protection range depends upon the radar characteristics (radiated power, receiver processing gain, instantaneous frequency band), the characteristics of the jammer (radiated power and transmitted spectrum width) and the target characteristics (equivalent radar cross-section).

To minimize this self-protection range, the transmitted jamming spectrum widths are adapted to the type of radars being processed. In this way, the following are used:

1. a barrage jammer (barrage noise, broad spectrum) against frequency agility or diversity radars, or in the case of several radars close to one another in terms of frequency;
2. narrow spectrum jamming (spot noise) against radars at a fixed or slowly varying frequency;
3. very narrow spectrum jamming against CW or pulse-Doppler radars.

The schematic diagrams currently used for this type of jammer are given in Fig. 5.1.

Jammer generating false echoes

The creation of false echoes by a jammer is obtained by generating echoes offset in time (range) and/or in Doppler frequency (radial velocity) with respect to the radar signal. This offset can vary gradually in time.

The time offset can be obtained conventionally in two ways.

1. Either by transmitting synchronously or not with radar signals, jamming signals the frequency and duration of which are similar to those expected by the radar receiver. This type of jamming is however not effective against coherent radars or when pulse compression is used.
2. Or by using delay lines. The signal received after amplification is delayed in the line and retransmitted (transponder type jammer, see schematic diagram Fig. 5.2). These lines may consist of:

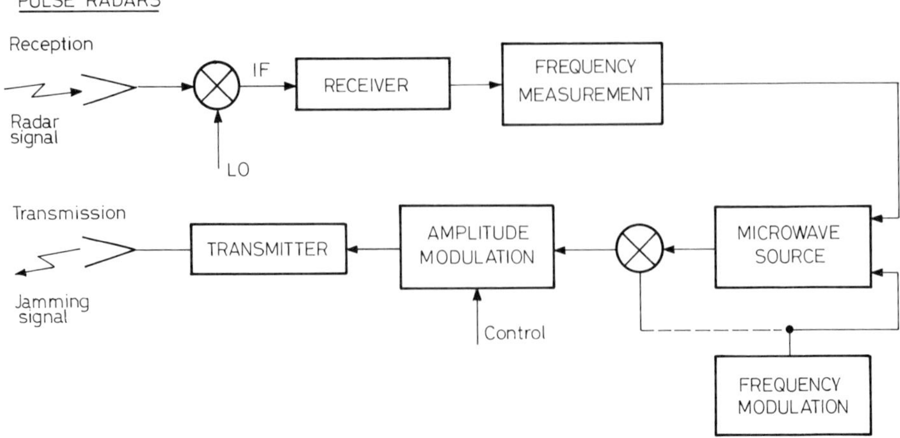

Fig. 5.1 Schematic diagram of noise jammers.

(a) microwave lines (coaxials) which, however, will only create short delays (10 m of coaxial will create a delay of 40 to 50 ns); an increased delay can be obtained by the recirculation of the signal in the line but the part of the signal stored is at the most equal to the delay duration of the line leading to limited efficiency in particular in the case of a compressive and/or coherent radar;
(b) acoustic lines which are capable of creating greater delays (several microseconds) by using looped lines or several switched lines.

The Doppler frequency offset cannot be obtained simply other than by using a transponder type jammer (see schematic diagram Fig. 5.2) which ensures the coherence of the signals retransmitted by the jammer with respect to coherent processing radars which measure the Doppler speed. In this way, it is possible to create the following effects.

1. Creation of many false plots in surveillance radars or tracking radars computers and displays. The retransmission of large numbers of echoes by the jammer, illuminated or not by the jamming radar, makes it possible to generate false plots which are spread out in range and angle so that extraction and tracking of these plots becomes very difficult (saturation effect) preventing the identification of the plot from to a real target in this context.
2. With respect to fire control radars or missile seekers, it is possible to create an effective deception on tracking computers onboard systems. For this

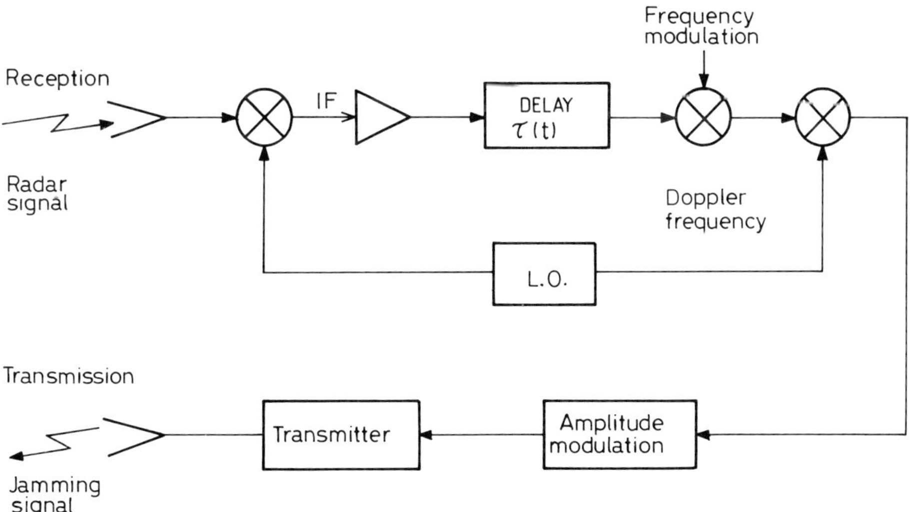

Fig. 5.2 Schematic diagram of transponder type jammer.

purpose, false echoes are created which are first overlaid in range and/or speed upon the target echo, then shifted gradually in time and Doppler frequency so as to draw the radar tracking circuits toward the false echo. These techniques of jamming are referred to as range gate pull off (RGPO) or range gate stealer, and velocity gate pull off (VGPO) or velocity gate stealer.

5.3.4 Main effects of jammers

The jamming devices are designed to disturb enemy radar receivers or missile seekers so as to prevent them reaching their targets. These perturbations can be obtained by action on radar measurement devices (range, speed, angle), or specific circuits of the receivers. A description of the main effects is given below. An exhaustive description does not form part of this book.

Action on measurement devices

The perturbation of radar or speed measurements These can be made by the following:

1. by noise jamming while masking target echoes using the jamming signal in the radar receiver (desensitization effect);
2. by decoy launching, while creating false echoes in range and/or speed thus generating either a saturating effect against surveillance radars or a deception effect against tracking systems.

These principles serve as a basis for the two major types of jamming and the association of these two devices makes it possible to increase the effects of each of them.

The perturbation of angular measurements The devices used to perturb or prevent angular measurements are adapted to the measurement devices used by the radars. For radars which measure direction on the basis of an amplitude measurement of the signals reflected by the targets, the jammers attempt to modulate the transmitted jamming signal (noise or false echoes) in such a way that the measurement made by the radar is wrong. This is the case, for instance, in inverse gain jamming techniques used against scanning radars or for the application of amplitude modulations against conical scanning radars.

For mono-pulse radars, which are in the theory, insensitive to amplitude modulations, an endeavour will be made to create: either a distortion of the phase plan they are measuring (cross eye or artificial glint techniques) using several phase-controlled jamming sources, or by creating equivalent bright spots spatially separated from the targets, for instance the use of jammers on several aircraft flying in formation (buddy mode), the use of electromagnetic decoys (chaff and active decoys).

Action upon the specific radar receiver circuits

Onto the jamming signal transmitted by the modulation jammer, several types of signals can be superimposed in order to perturb some specific radar circuits, for instance, the following.

1. amplitude and/or duty cycle modulations around the cut frequencies which can be applied to perturb circuits such as the automatic gain control (AGC), antenna servo loops, or some counter-countermeasure circuits (home on jam or jamming detector). For instance, these are jamming techniques such as audio noise, count down, cover pulse, etc.;
2. the use of frequency modulation at a more or less great speed in order to perturb the information received by radar receivers (generating false plots, AGC perturbations, etc.);
3. the use of several simultaneous jamming signals which by beating, perturb the radar intermediate frequency circuits, etc.

This description is far from exhaustive but does give an oversight of different jamming techniques used.

Possible protections

Confronted by these different jamming techniques, radars have developed electronic counter-countermeasure (ECCM) circuits as described in section 5.5.

Logically, some forms of jamming therefore have a level of efficiency which is greatly reduced or which is disappearing. On the other hand, jamming modulations have been created to perturb some particular counter-countermeasure circuits. Therefore, there is a constant evolution in countermeasure facilities and in radar facilities so that each one preserves its efficiency, hence the need to offer excellent adaptivity of countermeasures to handle a wide variety of radars.

5.3.5 Evolution of jamming facilities

Confronted by the evolution of radars, the advent of new techniques and technologies is improving and increasing the performance of countermeasures facilities. Principally, this involves digital memories (DRFM digital radio frequency memories), electronically scanned transmitters, and expendable jammers.

Digital memories Digital memories (see schematic diagram, Fig. 5.3) are used for sampling, encoding in terms of amplitude and phase, and storing in a fast digital memory the received radar signals. Not only are they used for reception, they are also used at any time by reading the memory and decoding the stored signals to regenerate a replica signal of the received radar signal. This makes it possible to create delays which are as long as necessary, to generate a high number of successive echoes, and to store long duration signals (depending on the memory

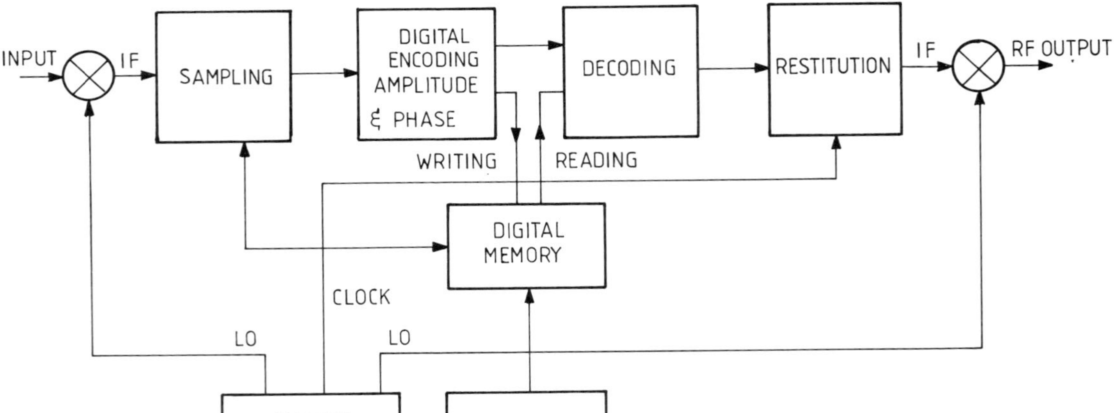

Fig. 5.3 Schematic diagram of digital memory used for jamming.

size established), while creating the desired Doppler offsets. The fact that the phase of the receiver signal is preserved makes it possible to create echoes which are totally coherent so that not only can conventional radars be processed but Doppler or pulse compression radars can be dealt with too. In addition, the fact that it is constantly possible to retransmit the same signal means that by superimposing noise on the retransmitted signal, it is possible to operate as a noise jammer. Finally, by injecting data into the memory, DRFM makes it possible to synthesize the desired signals.

The instantaneous band of these memories can now be up to one GHz or more with the possibility of storing information for several hundreds of microseconds.

Electronically scanned transmitters The jamming power levels to be radiated are linked with the transmitted jamming forms, with the respective radar–jammer-target distances, with the equivalent radar cross-section of the targets to be protected and with the characteristics of the radars to be jammed. In addition, the evolution of radar counter-countermeasure circuits is generally leading to an increased amount of jamming power to be radiated in order to maintain good effectiveness. However, it should be noted that the stealth techniques at the level of the targets to be protected should lead to the reduction of the radiated jamming power levels for a given level of efficiency.

The general requirements of wide angular coverage to simultaneously counter radars centred in different directions had led to the use of high power TWT transmitters associated with broad beam low gain antennas. Electronic scan techniques will either make it possible, by fast beam switching in space (less than one microsecond) or by the simultaneous generation of two or several beams in different directions, to concentrate the radiated power in the direction of the radars to be jammed (multi-threat jamming). In this case, the antenna gain can be higher than for conventional transmitters leading to the possibility of generating high transmitter power levels or, for a given radiated power, a lower transmitter power, and therefore consumed energy levels facilitating their installation on a carrier vehicle (aircraft).

Particularly, amongst the various solutions for electronic scanning, the evolutions of technology for high power and broad band solid-state amplifiers, are leading to the design of active antenna arrays.

Expendable jammers To handle the various counter-countermeasure circuits generated by the radars, one possible action is to create artificial targets resembling real targets in the form of the retransmitted signals and by their trajectory with respect to the radars. The capacities of circuit integration (MMIC for microwave, VHSIC for analogue or digital techniques) will make it possible to obtain small sized jammers which can either be dropped by the target to be protected or towed or propelled in order to generate likely echoes. These techniques, associated with more powerful jammers are one of the facilities by which the radars of the future can be countered.

5.4 ECCM APPLIED TO RADIO FREQUENCY LINKS

5.4.1 General

Telecommunications cover a far wider frequency range than radar, extending from ELF to millimetric. Within such a wide range of frequencies, there are many conceivable electronic warfare threats, and many conceivable electronic countermeasures. Conversely, whatever type of link has to be protected, the techniques used will necessarily take into account the specific characteristics of military telecommunications.

It is rarely possible to use high gain directive antenas. Except in the case of fixed satellite telecommunications stations, it is very unusual to be able to depend on an antenna pattern to separate the useful signals from the jammers. Communications with mobile elements use frequency ranges which do not allow a high level of directivity to be attained. At radio frequencies, the practical directivity of the antenna is limited by propagation. The direct path is generally masked and the link is by diffraction. The various diffusers in the side lobe diffract the signal from any jammers directed at the receiver. This results in an apparent increase in the side lobe level, reducing protection against jamming which would be obtained from a directive antenna.

In tropospheric radio frequencies, the received energy comes from a diffusion volume delimited by the intersection of transmit and antenna beams. It is as if the energy were obtained from a bright spot fluctuating in terms of direction and amplitude. In addition, the received field is very weak.

Emissions are continuous or quasi-continuous. This is an advantage for electronic warfare systems which thus have plenty of time to analyse transmissions, locate them or jam them.

Propagation attenuation varies extensively. In free space, the received power is proportional to the inverse of the square of the distance. Beyond optical range, the level decreases very fast. Thus between 30 MHz and 1 GHz it is considered that for a ground-to-ground link, attenuation increases to the power of four of the range. It is also affected, at the same range, by static deviation fluctuations of around 10 dB. For a desired range, it is therefore very difficult to guarantee both high operational probability of the link and low interception probability, even by a far more remote electronic warfare system.

Similarly, the efficiency of a jammer will depend far more upon propagation conditions than upon the power of the transmitter. Thus, it will be a relatively easy matter to provide a link with some protection against electronic warfare but it will be almost impossible to affirm that this protection will be enough in a given operational context.

5.4.2 Jamming protection techniques

Jammers intended for telecommunications have particular characteristics which also govern the appropriate counter-countermeasures. A radio link is jammed when it drops below a given quality threshold, beneath which transmitted signals become unusuable. For present-day digital transmissions, this limit can be expressed by the acceptable error rate. This rate varies from 10% for telephone transmission, to less than 10^{-6} for inter-computer exchanges.

Packet errors are generally less of a hindrance. A delta modulation speech link will tolerate jamming for 30% of the time or even half of the time if the perturbed periods can be identified by the receiver.

A jammer must assure that during the transmission of any message the error ratio remains higher than a limit value corresponding to the type of link attacked. If the modulation used is binary, the error rate (ER) can be expressed as:

$$\mathrm{ER} = \tfrac{1}{2}p(J > S)$$

where $p(J > S)$ is the probability of the jamming signal (J) present in the receiver being greater than the useful signals. This jamming signal consists of the sum of the receiver thermal noise, atmospheric noise and a fraction of the jamming signal which has passed through the receiver filters. The adapted jammer will concentrate its energy and time on exceeding the power of the useful signal for a maximum period of time.

The use of error corrector codes

Initially invented for protection against accidental errors caused by thermal noise or industrial interference, the error corrector codes provide effective protection against jamming (See Fig. 5.4).

To limit the complexity of decoding, two cascaded codes are often used. The internal code is simple and rugged and detects the jammed parts. The external code, for instance a Reed–Salomon code, corrects any erasures determined in the decoding of the internal code and any undetected errors. The coding strategy depends on the nature of the expected interference and therefore on the type of jamming for which protection is needed.

This protection method is highly attractive. However, remember that its use cuts down on the effective flow of the channel in a ratio which increases rapidly in proportion to the level of protection obtained.

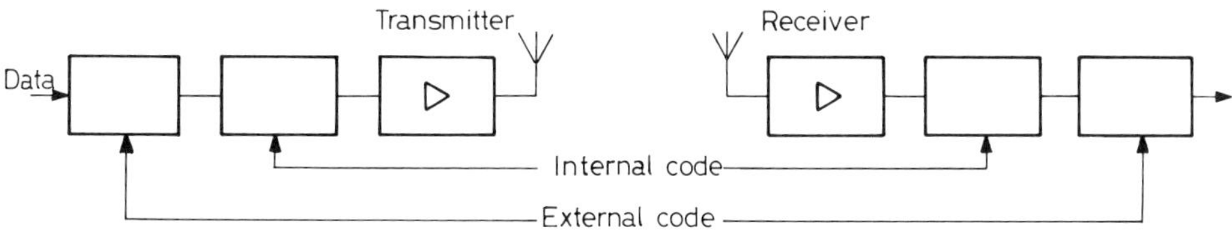

Fig. 5.4 Use of error corrector codes for ECCM jamming protection.

Spectrum spreading by encoding

For transmission and reception alike, this method of protection against electronic warfare consists of the addition to the transmission chain of a signal of a multiplication by a code derived from a pseudo-random generator whose clock frequency F is far faster than the maximum transmission frequency.

It is acceptable that during a code bit, the signal to be transmitted is almost stationary. The transmitted spectrum is therefore that of a phase modulated signal at a rate F. Therefore, it is far wider. Because it is assumed that the radio channel is a broadband device, cascading two multiplications during transmission and reception will not modify the signal transmitted as long as the transmit and receive clocks are perfectly synchronized.

The thermal noise power received by the receiver is also unchanged. However, only a fraction of the jammer power contained in the frequency band within which the signal was spread is transmitted to the receiver. This makes the link less sensitive to narrow band jamming but is not an improvement with respect to broadband jammers. In practice, the width of the spread signal band is limited by a filter, as is the receiver band, but this does not modify the principle.

The only practical difficulty in implementing this technique is due to the fact that it is necessary to synchronize the receiver code with the emitted code. The simplest method consists in using short codes and slowly sliding the relative clock phase until it is possible to demodulate the useful signal. When a digital link is used, the signal to be transmitted and the spreading code can be combined by an exclusive OR function whereby the resulting signal will modulate the carrier phase. Since the functioning of a transmission coder is a known process, the code or the few codes corresponding to the transmission of a symbol can be determined. They need only be computed by correlation. This is by far the best method but requires very high computation power, often excessive. The part played by system engineers is thus to cut down on the computation volume needed without excessively decreasing synchronization performance.

Combined use of spectrum spread by encoding and error corrector encoding

In its digital form, a receiver with a spectrum spread out by coding correlates the received signal with the spreading sequence. This operation is simply linear decoding (soft decoding) of an error detector code working on isolated information bits. De-spreading is then, in principle, the simple first step of error corrector decoding. Accordingly, this process includes the combined use of two or three codes which simultaneously reduce the flow of information and the error rate. Highly varied code combinations can thus be chosen. In particular, when several correlators are used, it becomes possible to translate some information bits simultaneously using orthogonal codes. This makes more efficient use of the channel and brings matters closer to the Shannon limit of energy per transmitted bit.

Frequency jump

This is the oldest technique. It corresponds to the application of simple common sense. If the reaction time of an electronic warfare system is *T*, it is a matter of never using a frequency for a greater period of time to avoid jamming. This gave birth to frequency hopping which consists of changing the carrier frequency in a pseudo-random manner within a lot of predetermined frequencies. In many ways, performance is similar to that obtained by spectrum spreading by encoding. It becomes necessary to synchronize the receiver with the transmitter. Protection against broadband jammers is identical to that obtained with conventional links or by spreading with encoding. The existence of occupied channels, self-jamming by links of the same type, or of jamming parts of the frequency band used produces an irreducible error rate. Therefore, an error corrector code must be used for data transmission. As in spectrum spreading links, it is often worth using two codes in succession, an internal code for evaluating the quality of the different frequencies and an external code whose bits belong to different frequencies corrects for errors or deletions resulting from the decoding of the internal code.

If we observe the spectrum of a frequency jump transmission over a sufficiently long period of time, we observe that it is practically white within the band used when the channels of the attributed lot are adjacent. In this case, we can accept that the frequency sequence used is a code spread modulated by the signal to be transmitted. The frequency jump then appears as a particular spread-out spectrum transmission in which the instantaneous spectrum of the spreading code is very narrow. Accordingly, it is easier to synchronize and is more resistant to jammers in the narrow band appearing in the outspread band. Conversely, it requires the development of fast tuning synthesizers which are relatively complex.

Adaptive antennas

It is also possible to eliminate jamming by antenna processing. By combining the signals received from *n* different antennas, it is theoretically possible to eliminate *n*-1 jammers. In theory, this technique seems to apply to any transmission. The rejection of jamming should go hand in hand with spectrum spreading or frequency jump techniques. In reality, the problems posed by adaptive antennas are relatively complex.

The first approach consists of attempting to protect digital or analogue fixed frequency links. Its advantage is to provide protection against jamming for conventional equipment. The actual performances of the many systems built today are relatively disappointing. This is mainly due to the following difficulties.

1. The useful signals are transmitted continuously. Therefore, there is no way of adjusting the antenna when there is no signal in such a way as to eliminate jamming. The useful signal could thus be mistaken for a particular jammer and eliminated by processing.
2. It is rarely possible to take into account the direction of the correspondents

using the frequency links. This is often a little known fact. In any case, the frequent presence of multipath interference partially destroys the relations between the direction of arrival and the local form of the wave front.

3. Jamming cancellation is only perfectly effective for one frequency. It is rarely sufficient in the entire useful band.

Finally, producing an adaptive antenna for telecommunications requires the resolution of one of the most complex problems in signal processing: the separation of the sources from unknown linear combinations.

The combination of adaptive antennas with other anti-jamming methods

Initially, the combination of signals from the different antennas was provided in analogue form. This limited the use of adaptive antennas to fixed frequency links. Today it is increasingly possible to digitize broadband signals and to process them in digital form. This allows the use of algorithms, the implementation of which would otherwise be impossible, and in particular makes the most of the natural characteristics of the broadened spectrum signals to improve the adjustment accuracy of adaptive antennas.

For frequency jump possibilities, a single jammer can be listened to before and after the appearance of the useful signals. For spectrum spreading by encoding, it is also possible to make the most of *a priori* knowledge of the signal in order to extend the performance limits of the adaptive antenna. It can thus be estimated that the receiver is attempting to ensure optimum processing of the signals from several antennas. In such a case, there is no need to attempt to identify individual functions within the overall signal processing of the receiver but instead to consider them as a whole.

5.4.3 Signal interception protection techniques

For the above reasons, it is very difficult to guarantee that a signal will not be intercepted by ESM receivers. Only extensive absorption with distance, such as that of the millimetric waves in the 60 GHz band, can provide any serious guarantee.

The only propagation losses for $1/D^2$ signals are generally insignificant. The signal can be intercepted at a level well below that needed for the correct reception of the transmitted message. In this case, we speak in terms of wave forms with a low probability of interception (LPI). Modulations improving resistance to jamming will simultaneously diminish the probability of interception.

Frequency jump

In a conventional interceptor, the level detected in a channel is averaged out over a certain period of time or over several successive measurements in order to increase sensitivity at a given false alarm rate. If the transmission time for a frequency is very short, the sensitivity of the interception receiver will decrease. The accuracy of direction finders does the same thing. Frequency jumping therefore offers some protection against the interception and location processes of present-day radio direction finders.

Spectrum spreading by encoding

The spectrum spreading by encoding widens the band of the transmitted signal without modifying the power needed for transmission. The signal-to-noise ratio in a narrow band interceptor is therefore decreased within the spread ratio. An interceptor with an adapted band, by time related integration, will give the best possible results as long as several broadened band signals do not occupy the same spectral band. It is possible to design specific interceptors for a particular form of modulation but their performance is generally inferior to that of interception receivers with respect to narrow band signals. In addition, present-day direction finders are totally incapable of locating signals which are spread out by encoding. Therefore, for the time being, these are the best protected signals against ESM systems.

5.4.4 Conclusion

For telecommunications, counter-countermeasures are difficult to implement and their performance is limited. There is a theoretical limit to anti-jamming performance obtained by spectrum spreading. The use of adaptive antennas, where possible, can provide complementary protection when the number of jammers present at the same time is very small. Estimating protection against interception is far more difficult. If it is possible to protect links against present-day interception systems, it is difficult to ensure that it will not be possible to design equally effective, or even more effective measures to counter such new modulations. Only signal processing complexity appears to be slowing developments in this area. Therefore, we must be extremely vigilant and always provide for several operating modes, including a conventional fixed frequency mode which will be the best when coping with certain threats.

5.5 ECCM APPLIED TO RADARS

5.5.1 General

For radar techniques, the name ECCM relates to all electronic means used to counter the ECMs described in a previous chapter. The days are past when radar engineers simply made a few modifications to an existing radar, particularly in signal reception and processing, to protect it from ECMs.

To be truly efficient—that is, to enable a radar to ensure its acceptable level of operation in the presence of an active or passive jammer—ECCM functions have to be brought into the radar design. ECCM functions are integral to the choice and definition of all the subassemblies: antenna, transmitter, generation–reception, signal processing, data processing and display, as well as radar management.

5.5.2 Radar range in the presence of jamming

In the more general case of a jammer transmitting continuous noise, the jamming power density received by the radar is defined by:

$$J = \frac{ERP \times G_r \times \lambda^2}{(4\pi)^2 R_j^2 B_j L_j}$$

where $\text{ERP} = P_j \times G_j$ = effective radiated power, P_j is the jammer power, G_j is the jammer antenna gain, B_j is the jammer bandwidth spectrum, G_r is the radar antenna gain in jammer direction, λ is the wavelength, R_j is the jammer-to-radar range, and L_j is the atmospheric loss (one way).

The total spectral noise density becomes:

$$N_j = J + N_0$$

where $N_0 = F \cdot k \cdot T_0$, F is the receiving system noise factor, k is the Boltzmann's constant, $T_0 = 290\,\text{K}$, the standard noise temperature and N is the radar noise power spectral density.

If R_0 represents the range of the radar in clear conditions, its range R in the presence of jamming is defined by relation:

$$\frac{R}{R_0} = \left(\frac{N_j}{N_0}\right)^{-1/4} = \left(1 + \frac{J}{N_0}\right)^{-1/4} \simeq \left(\frac{J}{N_0}\right)^{-1/4}$$

(the approximation is valid if the jammer is effective, $J \gg N_0$).

This expression shows that the radar would be better protected, all other conditions being equal, if: the jammer must transmit throughout a broadband (B_j), and the antenna gain is low during reception (G_r).

An example can be found in the case of a surveillance radar in the S-band with 150 km range, facing a stand-off jammer (SOJ), with the following parameters:

$$\text{ERP} = 15\,\text{kW} \qquad \text{R}_0 = 150\,\text{km}$$

$$\lambda = 0.1\,\text{m} \qquad \text{F} = 4.5\,\text{dB}$$

$$\text{D}_j = 100\,\text{km} \qquad \text{L}_j = 0.5\,\text{dB}$$

$$J = \frac{15 \times 10^3 \times (10^{-1})^2}{1.58 \times 10^2 \times (10^5)^2 \times 1.12} \cdot \frac{G_r}{B_j} = 8.5 \times 10^{11} \frac{G_r}{B_j} (\text{W/Hz})$$

$$N_0 = 2.82 \times 1.38 \times 10^{-23} \times 290 = 1.13 \times 10^{-20} (\mathrm{W/Hz})$$

$$N_j = 7.5 \times 10^9 \frac{G_r}{B_j}.$$

There are two extreme cases. First the jammer is received in the main lobe, with $G_r = 30\,\mathrm{dB}$, and concentrates its power in a narrow band whereby $B_j =$ 30 MHz.

$$N_j = 7.5 \times 10^9 \frac{10^3}{20 \times 10^6} = 3.75 \times 10^5.$$

The jamming power is approximately 56 dB above the thermal noise and the radar range is reduced to $R = 6\,\mathrm{km}$.

Second the jammer is received by the side lobes with $G_r = -5\,\mathrm{dB}$ and distributes its power in a broadband, $B_j = 400\,\mathrm{MHz}$

$$N_j = 7.5 \times 10^9 \frac{0.316}{400 \times 10^6} = 5.93.$$

The jammer power is approximately 8 dB above the thermal noise and the radar range is still $R = 92\,\mathrm{km}$.

This shows the advantage there is for the radar manufacturer:

1. to reduce the probability that an SOJ will enter the main lobe;
2. to construct an antenna with low side lobes;
3. to prevent the jamming from concentrating its power in a narrow band.

5.5.3 General principles used against jamming

A radar which has to operate in an intense jamming environment must be conceived according to three basic principles.

1. Minimizing the power of the jamming which penetrates the space of the useful signal. This is achieved by spatial, frequential or time-related filtering.
2. Preventing the jammer from determining the characteristics of the direction, frequency and time of the signal transmitted as this is the best way to increase its effectiveness.
3. Regulating the remaining false alarm in order to prevent the radar operating system from becoming saturated, at the cost of a sensitivity loss, which should be kept as low as possible.

5.5.4 Main ECCM techniques

In the same way that jammers adapt their signature to the type of radar aimed at, ECCM techniques can be classified into two categories: those which apply to surveillance radars and those which apply to tracking radars. Indeed, most of the techniques developed to protect surveillance radars apply to cases of tracking radars. In addition, the advent of multifunctional electronic scan radars has more or less eliminated this distinction. Therefore, in succession, we will examine the various ECCM techniques which also apply to surveillance and tracking functions while limiting ourselves to the most efficient. We will finish with a description of a few methods which are especially adapted for the protection of tracking radars. To aid the presentation, these techniques will be described according to the conventional breakdown of a radar into sub-units, from antenna through to data processing.

ECCM with antenna

The example presented in the calculation of the radar range in the presence of jamming clearly demonstrates the important role played by the antenna in the protection of the radar.

Antenna beamwidth To aid the detection of silent targets near those directions occupied by jammers, the antenna lobe during reception must have the smallest possible aperture in azimuth and in elevation.

For example Fig. 5.5 illustrates the superiority of a 3D surveillance radar compared with a 2D surveillance radar. It preserves the possibility of detecting a silent high altitude target at a short range, protected by a long range stand-off jammer at the same azimuth. In addition, nearby side lobes have an angular scope proportional to the aperture of the beam angle of around ten times the width of the beam at 3 dB. This domain will be smaller in proportion to the narrowness of the aperture.

The beam aperture in elevation during transmission is not preponderant. It is possible to choose transmission with a wide beam and to simultaneously receive on several narrow lobes spread out in elevation, or to transmit and receive with a narrow lobe which moves electronically. Each of these solutions has its advantages and its drawbacks.

The first solution requires several receivers, imposes the same wave form whatever the elevation and increases the illumination time on the target, which is favourable to the implementation of the passive antijamming technique (MTI). The second solution requires a more complex antenna but increases the energy on the target while adapting the transmitted wave form as needed.

Near and far side lobes As seen in Volume 2, the antenna pattern follows the Fourier transformation of illumination and the level of the nearby side lobes is defined in theory by the illumination weighting law. Therefore, very low side lobes are possible as long as:

1. the antenna dimensions are enlarged to compensate for the widening of the beam due to weighting;

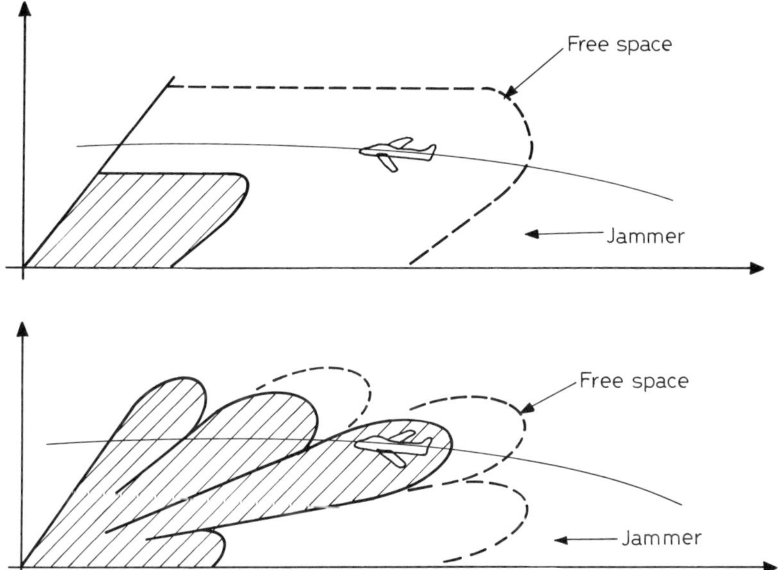

Fig. 5.5 Comparison of 2D (top) and 3D (bottom) surveillance radar coverage in presence of long range stand-off jammer.

2. the construction technology for antennas is totally mastered from the elctrical and mechanical points of view in order to respect the theoretical illumination law as accurately as possible.

For this reason, today most radars which have to ensure active antijamming techniques use phased array antennas. For instance, a high gain phased array antenna can achieve 35 dB attenuation on first side lobes and −10 to −20dB below isotropic for far side lobes.

Side lobe cancellation (SLC) In the presence of very powerful SOJs or medium range jammers penetrating nearby side lobes, it is possible that the previous techniques are insufficient. In such a case, an attempt may be made to improve the antenna pattern while keeping to the directions occupied by the jammers.

Briefly, let us return to this principle by considering the simplest case of an SLC with only one auxiliary channel. The jamming signal picked up by the auxiliary omnidirectional antenna, after weighting, is subtracted from signal S_m received on the main channel (Fig. 2). Expressed mathematically:

$$S_m = G_m J(t) + N_m$$

$$S_a = G_a J(t) + N_a$$

where G_m, G_a are the gain of the main channel and auxiliary channel on the jammer, and N_m, N_a are the thermal noise of the receivers of the two channels where N_m and N_a are independent and have the same power P_n. $J(t)$ is the jamming signal with power P_j in the Δf radar signal band. Stating that $J(t)$ is the same on both channels expresses the spatial coherence of the jammer.

To eliminate the jammer, attempt to carry out weighting W which mimimizes the jamming power on the main channel after opposition, W such that

$$\varepsilon = E[|V_m - W \times V_a|^2]_{min}$$

This is conventional least squares calculation.

$$W_{op} = \frac{E[V_m V_a^*]}{E[|V_a|^2]}$$

and

$$\varepsilon_{min} = E[|V_m|^2] - \frac{|E[V_m V_a^*]|^2}{E[|V_a|^2].}$$

Therefore

$$\varepsilon_{min} = P_n + \frac{|G_m|^2 P_j}{|G_a|^2 P_j + P_n} \cdot P_n < P_n + |G_m|^2 P_j.$$

If the jamming power on the auxiliary channel is high compared with the thermal noise then:

$$\varepsilon_{min} = P_n + \frac{|G_m|^2}{|G_a|^2} P_n.$$

The first term of this expression corresponds to the natural noise of the main channel and the second to the noise of the auxiliary channel brought onto the main channel after weighting and opposition.

The jammer is eliminated but performance is limited by the antenna gain ratio. However, note that the power after opposition is, at least theoretically, always less than the input power whatever the gain.

The SLC does not make an a priori hypothesis regarding the nature of the jamming signal. Conversely, it is essential to preserve spatial coherence i.e. the identity at each moment of signals on the two channels through to the point of cancellation; more particularly this implies a differential delay between the two channels which must be totally controlled.

Conditions for good performance Conditions for good performance include the following.

1. The auxiliary channel must overlap the strongest side lobes of the principal

channel. For instance, with a margin of 3 dB, the sensitivity loss will be limited to $10 \log(1.5) = 1.8$ dB.

2. The SLC is a linear filter. For the signals observed on both channels to preserve their coherence, it is essential to avoid saturation in the reception channels which precede cancellation, and to adapt the characteristics of the reception channels (delay time, gain according to frequency, etc.) from the antennas through to the point of cancellation. Antenna separation helps to limit SLC performance because it creates a more or less considerable delay time, according to the direction of the jammer.
3. Any signals other than those of the jammer will bias the estimation of W. It is therefore worth making this estimation on samples not contaminated by clutter or by the useful signal.

SLC processing applies in the case of several jammers. In this case, we must have as many auxiliary channels as there are jammers. Three to four auxiliary channels are the maximum allowed with this type of technique. To efficiently process a high number of jammers, the concept of the antenna must be checked in order to obtain digital-beam-forming solutions.

Integrated SLC Some antenna architectures are more than suitable for the layout of an SLC function without an auxiliary antenna having to be added. The most elaborate anti-jamming form integrated in the antenna corresponds to beam forming systems by computation. The signal received by each elementary feed is amplified, transposed into intermediate frequency then coded in digital form on two channels in phase quadrature. The elementary signals are combined linearly to form one or several simultaneously pencil beams with hollows in the directions of the jammers.

At the present time, the best cost-performance trade-off is obtained by grouping the feeds together in the conventional manner in high frequency to form sub-networks. The output of each sub-network is transposed into intermediate frequency then coded into digital. The main beam is reconstituted by a linear combination of the suitably weighted sub-networks.

The number of jammers which can be eliminated is theoretically equal to or less then the number of sub-networks. The active antennas, in which cach radiating source is preceeded by a solid state transmit-receive module, is particularly suitable for the installation of an integrated SLC of this type.

Side lobe blanking (SLB) Side lobe blanking eliminates pulse, transponder or relatively slow scanning jammers penetrating through the side lobes. SLB works on the principle of comparing the signals received by the main channel with the signal received by an auxiliary channel which overlaps the side lobes of the main channel (Figs 5.6 and 5.7). If the signal received by the secondary channel is higher, with a margin of X dB, than the signal received by the main channel, the

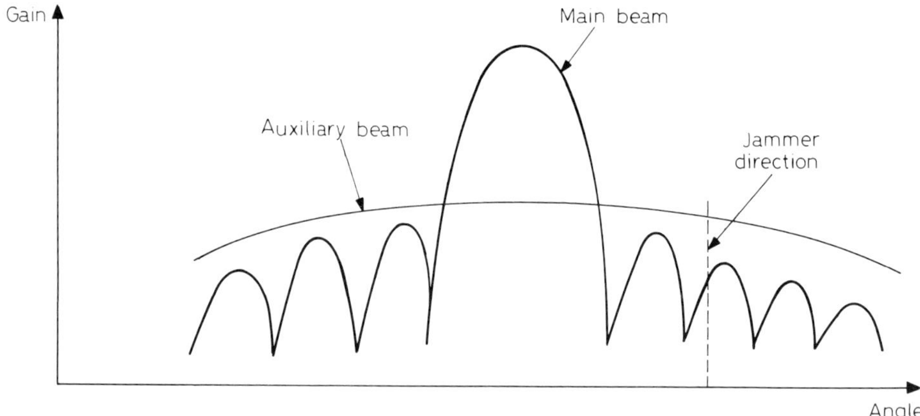

Fig. 5.6 Main beam and auxiliary beam signals used for side lobe blanking (SLB).

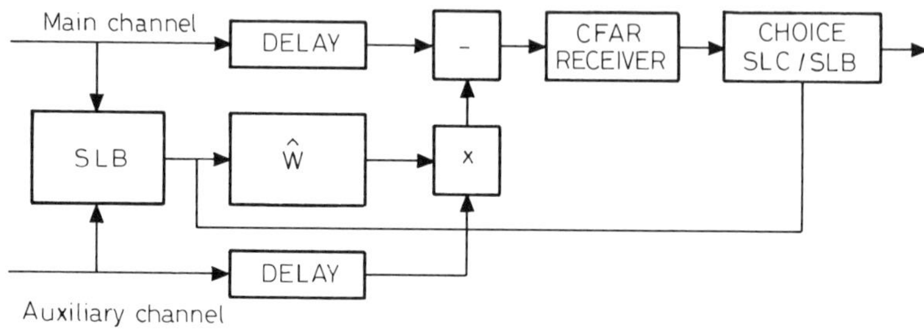

Fig. 5.7 Block diagram of SLB implementation.

main channel will be blocked. The value of X is chosen to ensure that the main channel operates normally during periods when there is no jamming.

Figure 5.8 shows the azimuth coverage of a 2D search radar in the presence of a single stand off jammer (solid curve) and without jammer (dashed circle).

ECCM with transmitter

The development of very stable high power amplifier tubes has considerably improved the resistance of radars to jamming.

Frequency agility The best means for the jammer to improve efficiency is to concentrate its available power into the radar signal band. The optimal way of preventing this from happening is to transmit with frequency agility, i.e. to change

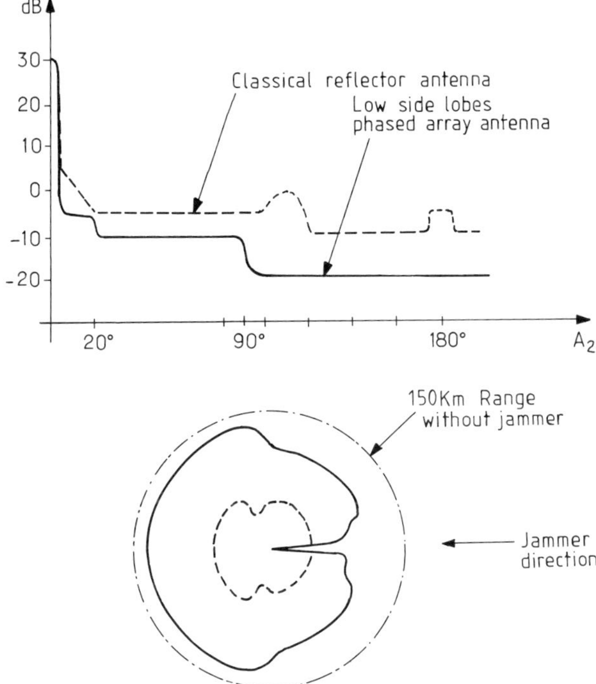

Fig. 5.8 Azimuth coverage of a 2D search radar in the presence of one stand-off jammer (SOJ).

transmission frequencies from one pulse to another so that the jammer has to spread its energy throughout the entire agility band.

Table 5.3 indicates the reduction in jamming density caused by a frequency agility using as a reference a 20 MHz narrow band and a wide band which is 10% of the transmitted frequency.

Pulse compression. Pulse compression consists of transmitting a long frequency— or phase – modulated signal which occupies a broad spectrum. From the ECCM standpoint, compression offers three advantages.

1. For the same amount of emitted energy, the signal peak power is less. For the jammer, detection and measurements of received signal characteristics are more difficult (quiet radar).

Table 5.3 Jamming density reduction with frequency agility in a 10% bandwidth

Radar band	L: 12 GHz	S: 3 GHz	C: 5 GHz	X: 10 GHz
ERP reduction	8 dB	12 dB	14 dB	17 dB

2. While transmitting a long pulse, the radar preserves range resolution equal to $1/\Delta f$.
3. The pulse transmitted by a jammer, or by another radar, and which do not have the transmitted signal modification law, are extended at the output of the adapted filter and their peak power is attenuated by the compression ratio $T \cdot \Delta f$. They are easily eliminated at the outset by the range CFAR with a reduced loss in terms of sensitivity.

The widening of the signal band by modulation is not a drawback when a broadband jammer is involved. At the output of the adapted filter, the transmitted energy to noise power spectral density ratio is what counts, and not the bandwidth.

Let us consider the case of a radar having 50 kW peak power, emitting a 20 μs long pulse modulated in a 10 MHz band, and that of a 1 MW peak power radar emitting a 1 μs pulse ($\Delta f = 1$ MHz). A compression radar, all other things being equal: has the same range because it radiates the same energy ($50 \times 10^3 \times 20 \times 10^{-6} = 10^6 \times 10^{-6} = 1$ J); is more discrete because the emitted signal is 13 dB weaker; and has a range resolution which is 10 times better.

ECCM with receiver

To detect small signals in the presence of very powerful clutter, modern radars feature improvements which are extremely useful against jamming: double frequency change, gain control and high dynamic.

Double frequency changes (see Fig. 5.9) The input band of an agile radar is very wide and the intermediate frequency (IF) preceding transposition into video,

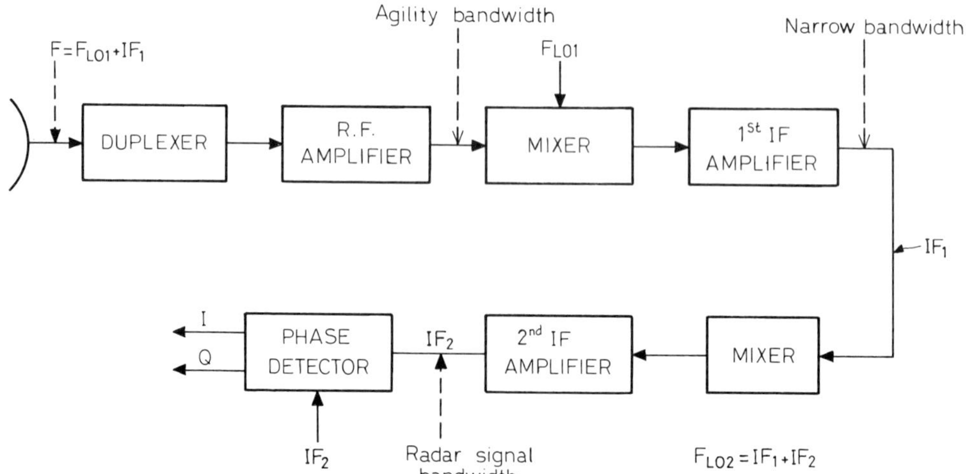

Fig. 5.9 Block diagram of double frequency change ECCM technique.

is necessarily low, around several tens of megahertz. With a single frequency change we obtain, at IF frequency at the mixer output:

1. the useful signal corresponding to transmission frequency f_t;
2. interference centred on the image frequency ($f_i = f_{LO} - \text{IF}$, if $f_t = f_{LO} + \text{IF}$);
3. the spurious signals from a jammer which is using the non-linear feature of the mixer by emitting two frequencies whatsoever, separated by the intermediate frequency (IF).

The use of a double frequency change with the first IF higher than the radar agility band eliminates these spurious signals by filtering at the receiver input.

Gain control and high dynamic A high dynamic associated with gain control is essential for avoiding, in most cases, the saturation of the receiver, and for ensuring the efficiency of the linear filters located downstream and thus conserving the sensitivity of the receiver.

Least jammed frequency (LJF) It is not always easy for a jammer, when confronted by a radar using frequency agility transmission, to emit at each moment a very broadband signal with a properly uniform power spectral density. Besides, the amplitude of the side lobes of an antenna varies with frequency: in the radar agility band there can be a frequency for which the antenna pattern has a hollow or even a zero in the direction of the jammer. So, it is possible that, at every moment of transmission, there has to be one frequency less jammed than all of the others and that the radar has to choose it to maximize sensitivity.

In the LJF mode, before each change of frequency, the radar measures the noise power received on n frequencies distributed throughout the agility band. The frequency for which the noise power in the Δf band of the radar signal is minimum is chosen as the transmission frequency. (The radar can also use the least jammed frequency to make measurements concerning the jammer.) A search for the least jammed frequency can be carried out without taking too much time.

A high-speed synthesizer can switch in 2 μs. We have to count on approximately 20 or so independent samples (separated by $1/\Delta f$) to correctly estimate the noise power received on a given frequency. Thus, $2 + 20 \times 0.1 = 4\,\mu\text{s}$ for $\Delta f = 10\,\text{MHz}$. Thus, in 40 μs the radar can test ten frequencies.

We can estimated the LJF gain on a jammer by using a simplifying hypothesis that the jamming power J^2 received on the side lobes varies as a function of the frequency according to an exponential law. The attenuation of the jamming power is given, in this case by

$$\frac{\text{E}[\text{MIN}_1^n(J_i)^2]}{\text{E}[J^2]} = \frac{1}{n}$$

i.e. $-10\,\text{dB}$ for $n = 10$ frequencies. Using the same hypothesis, it can be demonstrated that the attenuation decreases as a function of the number of jammers.

ECCM with signal processing

Doppler filtering and MTI Conventional techniques developed to ensure the visibility of target echoes hidden in ground or rain clutter apply in the case of chaff. In an intense jamming environment characterized by the simultaneous presence of SOJ, chaff and ground clutter, the most efficient processing for the radar consists of emitting bursts of ten or so pulses with a change of transmission frequencies and bursts or repetition frequency to eliminate blind speed zones.

Pulses relative to the same burst go through Doppler processing carried out using a narrow band filter bank. The output of each speed channel is detected and followed by range CFAR in order to eliminate mobile clutter and continuous jammers (Fig. 5.10). This type of processing combines the advantages of Doppler processing with those of frequency agility.

Confronted by a jammer which measures the radar frequency on the first received pulse, it may become necessary to use very short bursts with MTI-type adaptive processing. The system incorporates two filters of the three pulse canceller type, separated by an operator measuring the average velocity of the chaff after ground clutter elimination and centres the zero of the second filter on the chaff

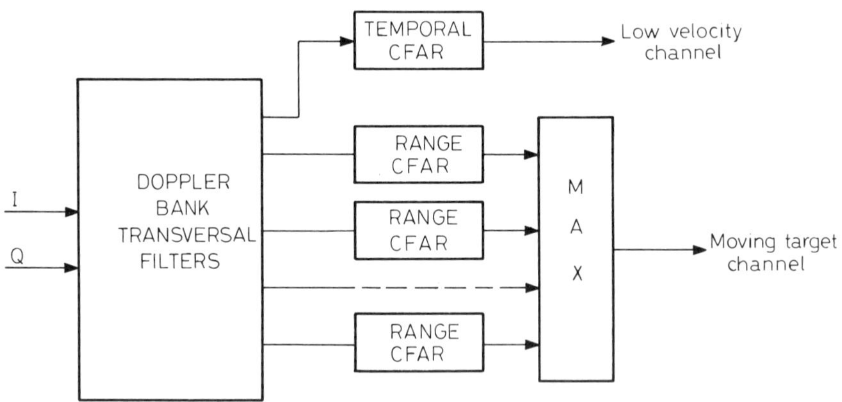

Fig. 5.10 Doppler signal processing.

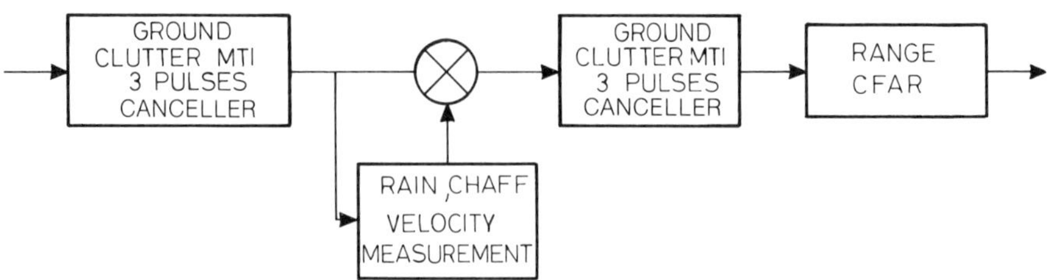

Fig. 5.11 Adaptive MTI processing.

Doppler frequency (Fig. 5.11). The visibility of targets whose speed is a multiple of the ambiguity speed is ensured by modifying the period of pulse-to-pulse repetition (staggering).

CFAR In spite of the previously-described measures, there is no way of totally eliminating jamming and clutter which penetrate the radar up to the detection threshold. To prevent the saturation of the operating system and of the operator, it is essential to keep a constant and low false alarm rate at all times. The different CFAR procedures can be grouped together in two main classes: CFAR with a single sample in which false alarm regulation is by a non-linear limiting operation and CFAR with two samples using homogeneity tests.

Single sample (or hard-limiting) CFAR This class contains the Dicke-fix, limited pulse compression and limited speed filtering. Its principle can be modelled by a normalization operation in which the amplitude information is eliminated from the input signal to preserve only the phase information, followed by integration for time T corresponding to the transmitted signal duration (matched filter) (see Fig. 5.12 for schematic).

The basic hypothesis is that the phase of the spurious signals is equally distributed. To obtain sufficient contrast at the output between the noise and the echo of the target, the product of B times T, the input signal band times the integration time, must be sufficient ($BT > 10$). This condition is obtained by pulse compression. In the well-known case of the Dicke fix, the BT product is obtained by a broadband IF amplification before limiting at a low threshold within the noise.

After integration, the spurious signal is more or less gaussian with standardized power ensuring that the false alarm rate is constant whatever the probability of detection of the input signal (non-parametric character). This type of CFAR gives excellent results against slow scan jammers and broadband noise jammers. On the other hand, this limitation perturbs the detection of a small target near a strong target with pulse compression and trends to muffle target echoes in the presence of strong clutter.

Two-sample CFAR To find out whether a sample X taken from an elementary cell represents the useful signal or the spurious signal, take a sample $\{Y_i\}$, size n, near the sample to be tested, in a certain space representative of the spurious signals (range, frequency, time). See Figs 5.13 and 5.14. Sample X is compared with the group of Y_i by a test procedure to decide whether X is of the same nature (spurious signal) or of a different nature (useful signal).

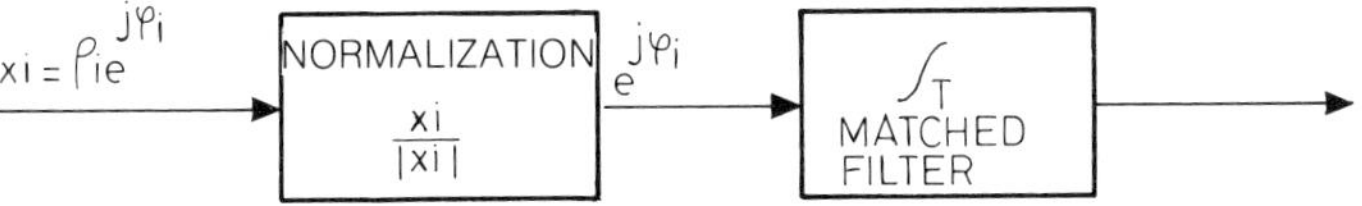

Fig. 5.12 Hard limiting CFAR principle.

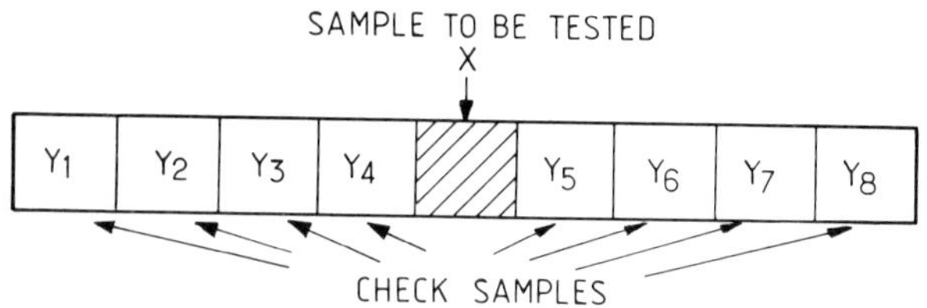

Fig. 5.13 Homogeneity test.

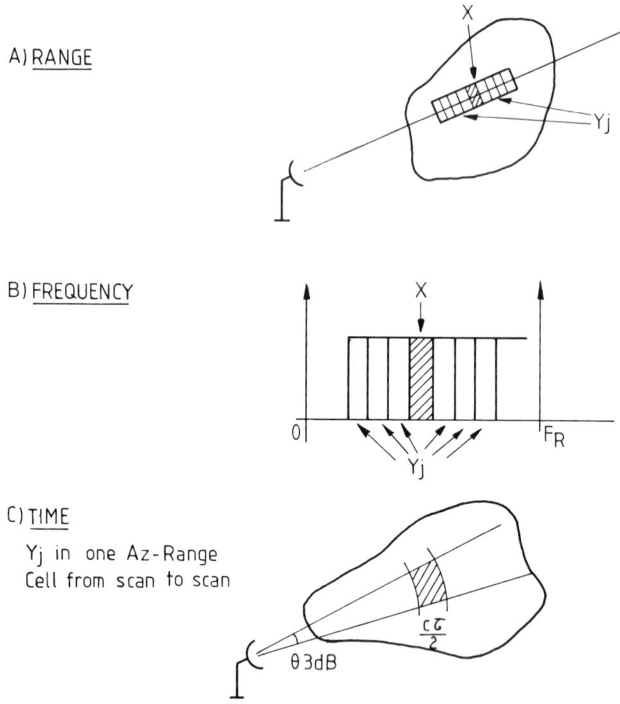

Fig. 5.14 Sampling fields for (a) range, (b) frequency, (c) time.

Test procedures Different test procedures are used in the radars. Very often, they are designed to normalize the clutter power or the jamming. This is particulaly the case with mathematical and geometrical normalizations.

Mathematical:

$$z = \frac{X}{\frac{1}{n}\sum_i y_i}.$$

Geometrical:

$$z = \frac{x}{(y_i)^{1/n}}$$

(or FTC log). These procedures are very effective in the presence of noise jammers.

Sampling domain The sampling domain Y_i must be chosen according to both the characteristics of the spurious signal to be eliminated and those of the useful signal, so as to contain only the spurious and none (or very little) of the useful signal.

In order to cut down on atmospheric clutter (rain, clouds, etc.), passive jamming (chaff) or continuous jammers, the samples should be taken within the radar repetition period in the range cells surrounding the cells to be tested (range CFAR, Fig. 5.14(a)). When the useful signal is present, it has a small extent and will therefore not be eliminated on its own.

To regulate the false alarm rate with respect to radar instabilities (fixed echo residue), against continuous or pulsed jammers, sample the frequency (velocity CFAR), (Fig. 5.14(b)): the useful signal only occupies a narrow band.

Against ground clutter resulting from a mixture of diffused echoes and local echoes, sampling will be carried out in time, i.e. on the basis of signals derived from the same azimuth-range cell (Fig. 5.14(c)). (The useful signal changes cells from one antenna revolution to the next.)

Special features for tracking radar (or tracking functions)

This section addresses jammers located in the main lobe for which the radar has to make precise angular range and velocity measurements to ensure weapon control. The range equation (see section 5.5.2) demonstrates that it is an easy matter for a jammer to prevent range measurement by emitting continuous noise.

The key to radar protection is its capacity in such cases to make good angular measurements and obtain target range, if necessary by triangulation, correlation or by association with another type of sensor. The radar will therefore be equipped with a mono-pulse receiver practically insensitive to continuous wave jammers modulated in amplitude, and which also resists cross-polarization jammers if it has a good antenna. The receiver will have a high dynamic (>60 dB), and a fast automatic gain control will allow framing onto the maximum echo of the range gate at the beginning of each transmission burst.

To counter the transponder jammers which break off range and velocity locking (RGPO, VGPO), the radar will use leading edge tracking and a variable repetition period so as not to have any false echoes in front of the target. These measurements will be completed by advanced gate openings to provide fast re-acquisition when a spurious echo is locked on.

ECCM with jamming analysis and radar management

The set of procedures presented here are an efficient way of dealing with the various forms of jamming. But not all these procedures are compatible with one another. For instance, pulse-to-pulse frequency agility and Doppler processing conflict; some are effective against continuous jammers, others against pulsed jammers; finally, their use generally leads to a loss of sensitivity, consumes radar time to improve the signal-to-jamming ratio and downgrades the improvement factor of the signal on clutter.

Therefore, the procedures must be used in a coherent manner, in order to fashion an effective response and it is preferable to use them only in the observed zones where they are essential. This presupposes a sufficiently accurate knowledge of the environment in which the radar is employed, as well as adaptive management in time and space of the transmitted wave forms and of the reception chain configuration.

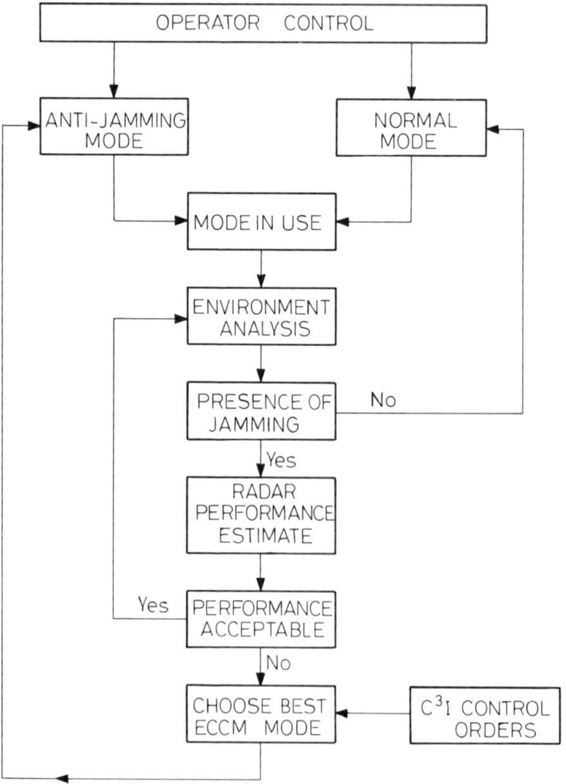

Fig. 5.15 Adaptive radar management block diagram.

Knowledge of the environment is provided by a specialized analysis processor which is mainly used to alert the operator to the presence of jamming, and to inform him of the performance in terms of radar range in the mode being used while measuring the received jamming power. The second role of the analysis processor is to gather the information needed to choose the best response to the jamming. Analysis must be performed in the space, time, and frequency domains.

The desired parameters are the type of jamming (active or passive), the number of jammers, their location, the form of the transmitted signals (continuous, pulsed), their spectrum (broadband or narrow band), etc. Some measurements will necessarily be made by devices located along the radar reception line (search for least jammed frequency, evaluation of performance degradation). Conversely, it may be worth using a specialized receiver, employing the SLB channel for instance, to characterize certain jammer types (for instance, frequency scanning jammers). This arrangement has the advantage of being able to operate constantly at all the frequencies in the agility band without requiring any of the radar's operational time.

The diagram Fig. 5.15 is a simplified outline of the linking together of tasks by which adaptive radar management can be obtained under the control of the operator while taking into account the information and instructions supplied by the C^3I system in which the radar is incorporated.

5.6 SYSTEM DESIGN METHODOLOGY

5.6.1 New tools to be incorporated into operational systems

Electronic warfare encompasses a group of techniques whose setup, implementation and efficiency can only be analysed on the basis of a mission in which these particular techniques are harmoniously combined with other techniques, forming part of the system and aiming towards a final military goal. These are tools of a new type, made available to the military to improve its chances of accomplishing its missions, to be combined with other, more conventional, tools already in use. Today, many areas bring electronic warfare techniques into play, in particular: strategic and tactical intelligence, self protection, air strike, counter mobility, radioelectric superiority, anti-C3 and elimination of enemy AA defence.

5.6.2 Strategic intelligence: an in-depth analysis

Strategic intelligence embraces all activities involving the research, gathering, analysis and use of intelligence for strategic purposes. Electronic and communication intelligence contribute to this activity in the same way as documentary research, the analysis of the fruits of foreign visits or other means relating to the field of espionage. The sources are correlated and information is analysed in detail without time constrainsts. To meet these requirements, the electronic intelligence measuring equipments (COMINT and ELINT) must give priority to fine detailed technical analysis, even over the probability of an enemy message being intercepted, as well as the speed of reaction.

The techniques used are slow, searching techniques which provide for intelligence, in-depth characterization and sophisticated off-line processing. All measures will be taken to correlate the results obtained with those drawn from other sources so as to compile reliable data banks. Environmental, maintenance and operational constraints will generally be minimal, benefiting technical analysis performance.

5.6.3 Tactical intelligence: quick analysis of situations and priorities

Tactical intelligence is involved directly in analysis of the immediate threat, in triggering warnings and in designating targets. Fast integration of situations and the priorities is essential as is an effective presentation of the results of the analysis to the people in charge of decision-making and command.

The characterization of the messages to be analysed cannot be allowed to override interception probabilities and considerable room is left for correlation tasks between sources, or with previously established databases.

Electronic intelligence equipment (COMINT and ELINT) can participate directly in air defence surveillance as long as the knowledge is there of how to integrate them into other equipment (radar, observation, electro-optic observation) within the system.

When correlated by a radar, electronic warfare techniques are capable of characterizing a target or even identifying it by its electromagnetic signature. (This is the case, for instance, for ships at sea, for an air defence system, or an active missile.) The result of this is to present a very complete tactical situation in almost real time. In some situations the techniques can even ensure enemy detection faster than the radar can.

5.6.4 Self-protection of weapon systems: a highly complex function

The first contribution of electronic warfare support measures (ESM) to self-protection is warning detection: warnings of air defence batteries, or of missiles, for aircraft; warnings of missiles and pointed radars for ships; warning of laser illumination systems for tanks and helicopters. This function is characterized by the fact that the warning is immediate and the danger is deadly. The reaction must be immediate. The reaction may sometimes involve other electronic warfare techniques (chaff, jammers, etc.) jointly with other means such as avoidance manoeuvers, guns, missile, etc.

Self-protection is also one of the goals of electronic countermeasures (ECM) which interfere with enemy means intended for observation, characterization of targets and weapon guidance (modification of signatures, masking, etc.). If we refer to the example of self-protection of aircraft and ships, extremely complex systems are required involving electronic counter-countermeasures adapted to production and utilization conditions (for instance, low observable techniques), as well as electronic warfare support measures (ESM) used as warning detectors and electronic countermeasures (ECM) associated with conventional threat neutralization measures.

The ESM equipment used as warning detectors has specific characteristics: the possibility of intercepting signals characteristic of the threats at maximum range is close to 1; the false alarm probability is very low to permit automatic reaction modes and to give credible results to the user; reliability must be tailored to its 'vital function' condition; integration is more and more thorough with conventional or electronic countermeasure systems in order to cut down on the reaction time.

In reality, this reception equipment bears little resemblance to the equipment used for strategic or tactical intelligence. The reaction means are nevertheless as specific and well-integrated as the previous ones.

On aircraft and ships alike, the ejection of chaff cartridges or infrared decoys is harmonized with the use of jammers and with the evolutions of the moving target. All this takes place in a highly integrated and automated system and with reaction times of less than one second.

5.6.5 Air strike: active support at several levels

The strike of an airspace by piloted aircraft is supported at several levels by electronic warfare techniques. To help air strikes, accompanying aircraft open a corridor through the air defence means using copious electronic countermeasures (ECM): the airborne jammers create zones or blind sectors in the air surveillance system; the electronic support measures (ESM) permanently establish an electronic battle order for air defence weapons and their activation levels; the chaff generates false echoes, dummy raids and creates zones of total confusion for air surveillance radars.

Here again, the techniques used are specific to the assignment. Thus, the volume of chaff carried has nothing in common with the means needed for self-protection. The jamming modes and powers are halfway between self-protection jammers and electronic superiority jammers.

5.6.6 Counter-mobility: neutralization by jamming

The mobility of the adverse means, principally aircraft, is backed by a radioelectric support for navigation, friend–foe identification and traffic regulation. Counter-mobility aimed at hindering, slowing down or even inhibiting this mobility uses electronic warfare techniques jointly with other actions like the neutralization of air bases or the breaking off of supply chains for fuel or spare parts.

These electronic warfare techniques are also very specific and refer to a variety of radio navigation systems such as beacons, transponders, VOR-DME, ILS, etc. in order to neutralize them by jamming or even better, to bias them so that they give false information while the user remains unaware of the fact. They are also suitable for neutralizing friend or foe identification systems and can be used to provide a sufficiently high level of jamming of communications so as to inhibit traffic regulation and to paralyse it.

5.6.7 Radioelectric superiority: preventing the use of the spectrum by the enemy

The radioelectric superiority concept covers all the actions guaranteeing use of the radio spectrum by friendly forces and which, in so doing, inhibits its use by enemy forces (similar to the concept of 'airborne superiority').

Considering the position taken by communication networks, data routes and command processes in the use of weapon systems, the anti-C3 (control, command and communications) is a very up-to-date subject. Electronic warfare techniques too are contributing extensively to anti-C3 but they are far from monopolizing it. Their first contribution is in the restitution of the enemy 'electronic battle order' so that any critical points can be identified.

A variety of electronic intelligence measures are used: COMINT restores the radioelectric communications networks and ELINT restores the radioelectric sensors used and, in particular, characterizes the 'radio-electric sites'.

The second application of anti-C3 electronic warfare techniques is that of neutralization by countermeasures (jamming, chaff, decoys, etc.).

5.6.8 Elimination of the enemy anti-aircraft defence: extensive use of ESM and ECM

Another present-day concept involving electronic warfare techniques and equipment especially tailored for the task involves the suppression of enemy air defence (SEAD). This concept, aimed at the physical destruction of these sites, is nevertheless widely based upon strategic intelligence and therefore on electronic support measures to analyse the situation; it provides for the massive use of electronic countermeasures to temporarily knock out the enemy air defence systems before destroying them physically.

To understand better the scope and the issues involved in electronic confrontation, it is convenient to get an idea of 'the other side' of its current situation, of its operations intentions and of its outlook. In short, it has to do with knowing how to analyse the threat. The purpose of threat analysis is to understand today's electronic threat and that of tomorrow (in 15 to 20 year's time) as far as possible, in technical and operational terms (deployment, conditions of use, coordination with other means available, etc.).

To clarify our thinking and work, in general we distinguish between three levels of threat: direct threats, front support and rear support. At these three levels there are three corresponding types of electronics:

1. direct threat electronics contributing directly to an immediate risk of destruction and whose neutralization brings this risk down to an acceptable level;

2. front support electronics, directly participating in the enemy offensive action but whose neutralization would not stop that action;
3. rear support electronics, not participating in the offensive enemy action.

How do we analyse this threat? By drawing up an inventory of the sections to be examined. Each threat category corresponds to a certain level of priority and a certain type of confrontation. Examining these sections is necessary if we are to obtain an idea of the environment to be provided against and what is needed in equipment design so as not to be neutralized in the field in the event of confrontation.

6

Infrared

Jean Dansac, Yves Cojan and Jean-Louis Meyzonnette

6.1 INTRODUCTION

6.1.1 General definition

For the general public, the word infrared generally evokes an image of mysterious powers, sometimes wonderfully beneficial to mankind, sometimes evil and terrifying. For some years now, certain television programmes have been helping to magnify these misconceptions. It is true that the ability to see in the dark or in poor visibility is a wonderful thing, whereas missiles which automatically home onto their target can be frightening. On the other hand, nobody is surprised at the way in which infrared has become part of our daily life, thanks to burglar alarm detectors, process control, machining, or diagnostic and medicare equipment. The ability to see in the dark, to detect intruders, to guide missiles, measure distances, detect tumours, control temperature, machine complex parts, prevent tooth decay, etc., these are just a few examples of the possible applications of IR among a host of others.

However, infrared terminology only covers a fraction of the wavelength spectrum in the electromagnetic wave field. We should, then, start by defining the limits of this spectrum and this is precisely where things become a little complicated. In practice, we normally locate infrared between the wavelengths in the visible spectrum (field of sensitivity of the human eye) and the so-called millimetric wavelengths.

The wavelength limits of the visible spectrum based on the standard eye are somewhere in the neighbourhood of 0.4 μm (or 400 nm) in the violet part of the spectrum, and 0.75 μm (or 750 nm) in the red part of the spectrum.

The lower limit of the IR spectrum is therefore around 0.75 μm which corresponds to a frequency of approximately 4×10^{14} Hz. The upper limit of the IR spectrum is somewhat ill-defined. There is a certain amount of interpenetration between infrared and millimetric waves. Although there is no clearly defined limit, state-of-the-art technology (particularly in the military field) points to a limit wavelength of 20 μm (or 20 000 nm) giving a frequency of 1.5×10^{13} Hz.

6.1.2 Spectral bands

Having thus set the limits of the spectral domain of infrared at between 0.75 μm and 20 μm, we can—without pre-empting the detailed description of these aspects in the paragraphs which follow—mention the subdivisions which we normally make within the infrared domain.

These subdivisions or 'spectral bands', have been defined principally in relation to the environmental conditions and the technological limitations. Environmental conditions differ greatly depending on whether we are looking at military, industrial or consumer applications. Generally speaking however, the most stringent conditions are those dictated by military applications where operation is almost always 'in the field'. This means working in extremely variable climatic conditions (sunlight or darkness, humidity, visibility, temperature, altitude, etc.). In addition we are normally looking for long range, good contrast and signal-to-noise radio, good angular resolution, etc.

Under these circumstances, three parameters predominate:

1. atmospheric transmission in relation to wavelength;
2. spectral distribution of the sun's energy scattered by the environment (including that of the targets);
3. spectral distribution of the energy radiated by the environment and the targets, which is linked to the inherent temperature of the objects.

A typical atmospheric transmission curve in shown in Fig. 6.1, in which we recognize the following transmission bands:

1. 0.7 to 1.1 μm band;
2. 1.2 to 1.8 μm band;
3. 2 to 2.4 μm band;
4. 3 to 4.2 μm band;
5. 4.5 to 5 μm band;
6. 8 to 13 μm band.

When designing an infrared system the choice of spectral band (or of a combination of several bands) will also depend on:

1. solar radiation, which is particularly sensitive in the short wavelengths and reflectance factors of the background and of the targets;
2. background radiation which is found mainly at high wavelengths;
3. radiation inherent to the targets to be detected, tracked or displayed which depends essentially on their temperature.

Of course, the choice of wavelengths may also be limited by technological considerations and in particular, the availability of optical materials, the sensitivity of sensors, and the possibility of using laser transmitters. These various aspects are developed in the pages which follow.

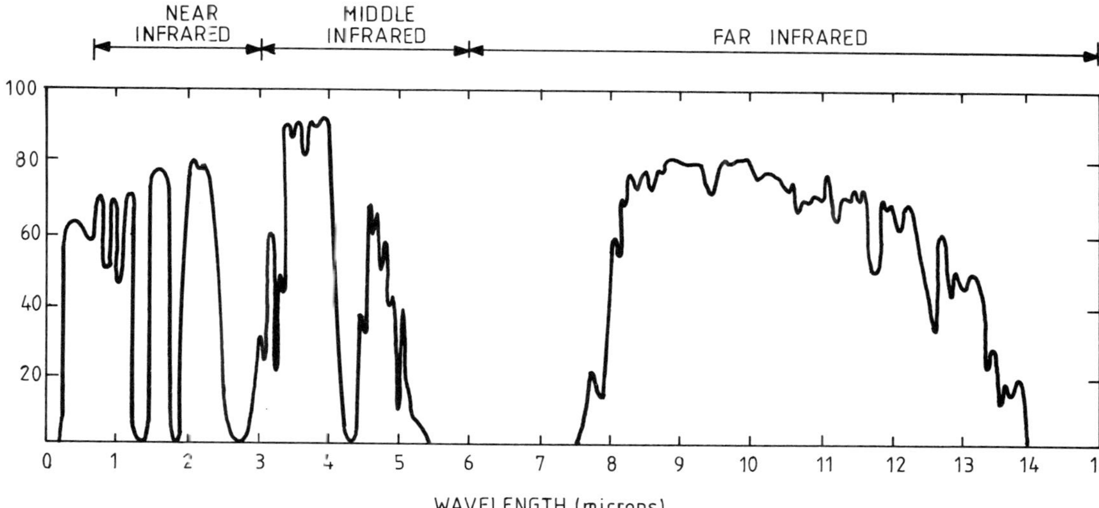

Fig. 6.1 Typical atmospheric transmission.

6.1.3 Infrared system classification

Infrared (IR) systems may be broken down into four major categories:

1. point-to-point IR transmission systems;
2. active IR systems;
3. semi-active IR systems;
4. passive IR systems.

Point-to-point IR transmission systems

A typical system is shown in Fig. 6.2 in which E represents an IR radiation emitter: a few decades ago such emitters were composed of mechanically modulated tubes, whereas today they mostly contain lasers (semiconductor, solid or gas type), combined with electrical or electro-optical modulation. R represents a receiver which picks up the signal sent by emitter E along atmospheric path A and then transmits it to a processing unit P and then onto a display unit D.

Such point-to-point transmission systems can be illustrated by means of the following examples:

1. the path E to R may be a passage which needs to be protected; in this case any interruption of the link by an intruder will trigger an alarm;
2. path E to R is used to transmit audio or video data;
3. link E to R can also send commands, as is the case in particular for missile guidance.

Active infrared systems

The functions and principles of active IR systems are fairly similar to those of active radars. A flow diagram is given in Fig. 6.3. In this case, emitter E and receiver R are co-located (monostatic system). The emitter therefore sends an IR signal to the target (or targets) T and to the background. The reflected signal is received by receiver R and therefore crosses the atmospheric path A twice.

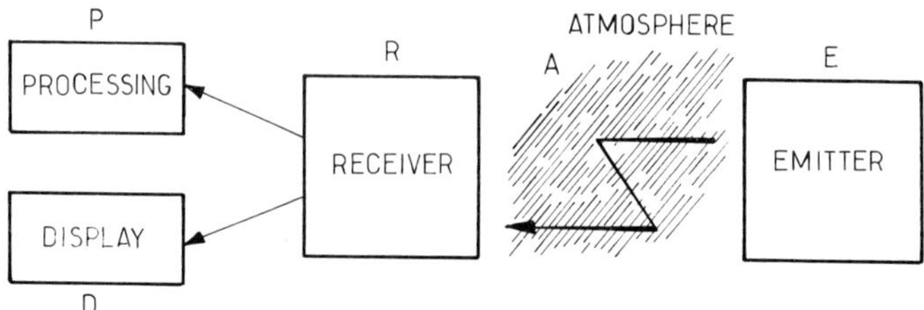

Fig. 6.2 Point-to-point transmission.

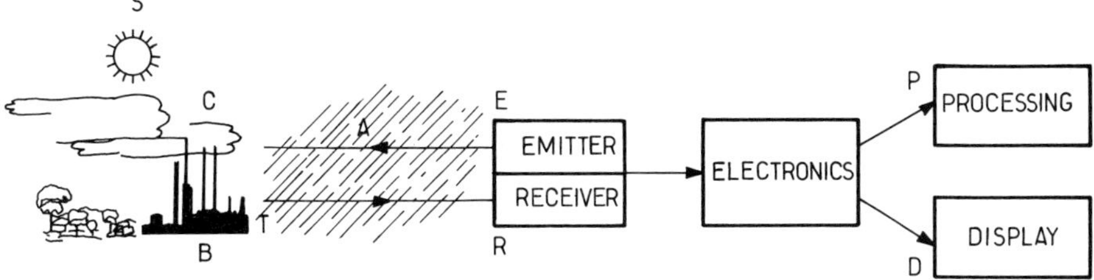

Fig. 6.3 Active IR system.

As examples, such system structures are used:

1. in laser range finding: surface-to-surface, air-to-surface, surface-to-air and air-to-air;
2. in active imaging, whether it be the light amplification active imagers of the 1950s or the laser imagers of the 1980s.

Semi-active IR systems

This third type of IR system is shown in Fig. 6.4. In this kind of configuration, emitter E illuminates the target T from a geographical point. Backscattered radiation is picked up by receiver R located at another point which may be a considerable distance from the emitter (bistatic system). Usually a processing unit P provides target angular deviation measurement and if necessary automatic target tracking.

This type of system architecture is frequently used in the military field.

1. Emitter E is a ground-based IR laser which illuminates target T for an aircraft or helicopter carrying receiver R. The aircraft or helicopter is thus provided with automatic unambiguous target acquisition.
2. Let us now suppose that receiver R is carried by a missile, rocket or even a shell. In this case, it could act as a homing head guiding the weapon to the point illuminated.
3. In another configuration, emitter E is carried by an aircraft and receiver R by a missile or another aircraft.

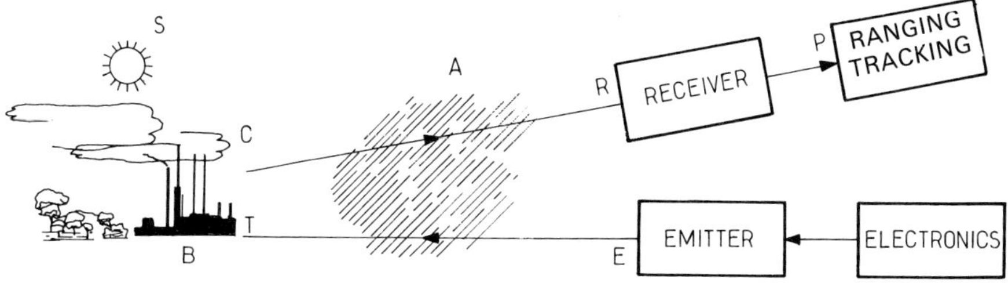

Fig. 6.4 Semi-active IR system.

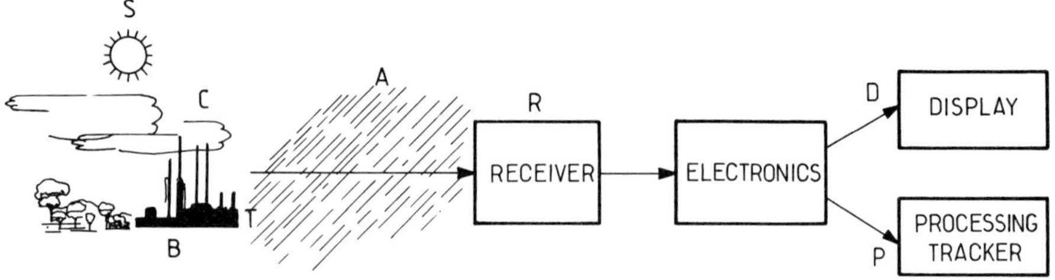

Fig. 6.5 Passive IR system.

Passive IR systems

These systems do not really possess a proper emitter (Fig. 6.5). Receiver R picks up the IR radiation linked to the nature and temperature of the objects composing the landscape and solar scattered radiation. These objects may include targets and background. Receiver R analyses the landscape and by means of a series of suitable processing operations: either automatically detects, locates or tracks targets; or produces thermal imaging of the scene observed.

This, then, is the broad outline of the subjects which will be developed in this chapter.

As we shall see, the environment has an important role: atmospheric transmission, relative humidity, presence of aerosols, etc. These are also important factors: insulation, scattered or specular solar reflections may be either spurious or useful elements depending on system configuration and applications.

In addition, infrared system design is generally complex as it requires multiple knowledge of the surrounding environment, the targets (thermal signature, reflectance), as well as: optics (materials, combinations, processing...), electromechanics (stabilization, servos, gyroscopics), electronics (analogue, digital) and display (presentation, eye interface).

If is hardly surprising then that the path to present-day system performance has been a long one, even now we are still only at the beginning of operational infrared applications.

6.2 SHORT HISTORICAL BACKGROUND

The history of infrared dates back over nearly two centuries; it was in 1800 that Herschel first demonstrated the presence of radiation outside the visible spectrum. He observed the solar spectrum obtained using a prism and, placing a thermometer outside the visible wavelengths (above the red part of the spectrum), noted the presence of radiation creating a temperature rise. During the 19th century a number of experiments confirmed the existence of this radiation thanks to the discovery of new more sensitive sensors such as thermopiles and bolometers. However it was not until the start of the 20th century that the first quantum detectors appeared which were to form the basis of present day infrared receiver systems.

It took a number of decades to produce operational equipment. As a matter of fact there were some particular difficult technological problems to be overcome:

1. the manufacture of sensors sensitive in the useful spectral bands:
   (a) performance reproducibility;
   (b) long-term reliability;
   (c) production efficiency;
   (d) environment withstanding capability;
2. production of optical systems suited to the spectral bands in question:
   (a) transparent materials,
   (b) surface coatings;
   (c) stability of materials;
   (d) environment withstanding capability;
   (e) cost;
3. production of sensor cooling devices:
   (a) operational lifetime;
   (b) autonomy;
   (c) life cycle;
   (d) cool-down time;
4. design and production of spectrally suitable laser emitters:
   (a) CW power;
   (b) peak power;
   (c) short-term stability;
   (d) pulse repetition rate,
   (e) influence of the environment.

These technological aspects produced a development curve, which more or less followed an order of increasing wavelength. Thus in the military field, the first passive IR equipment which appeared during the Second World War, used bands in the near infrared located between 1.5 μm and 3 μm, because the traditional fused quartz or fused silica based optical materials possess the appropriate characteristics, and the detectors (and in particular lead sulphide based detectors) required no cooling.

Similarly, the first night vision devices using active infrared (with illumination by means of search lights fitted with a filter cutting off the visible spectrum) used the band of 0.75 to 1 μm in which certain tube photocathodes (type S_1) are sensitive.

It wasn't until the 1950s that the first passive infrared systems sensitive in the 3 to 5 μm band were developed, with the advent of cooled or uncooled lead selenide sensors (PbSe) and especially indium antimonide sensors (InSb), cooled to the temperature of liquid nitrogen (77 K) using open circuit systems. Whereas the 1.5 to 3 μm band is well suited to the detection of high temperature objects (1000 to 2000 K, i.e. missile propellers or jet engines), the 3 to 5 μm band can be used for detecting lower temperature inherent radiation (profile sighted aircraft, helicopters and tanks).

The third infrared band (8 to 12 μm) presented the hardest problems both from the point of view of: optical materials and their surface coatings; and sensors and their cooling devices. Here again, a considerable technological effort was required in order to exploit all the possibilities offered by this spectral domain. This is the domain in which most objects produce radiation at the ambient temperature (around 300 K). This explains the interest in passive detection and especially in what is called thermal imagery.

The first field-usable thermal cameras appeared in the 1960s. However, the sensors used at the time—gold doped germanium (Ge:Au), copper doped germanium (Ge:Cu) and mercury doped germanium (Ge:Hg)—required cooling at very low temperatures: 60 K for Ge:Au, 28 K for Ge:Hg, and 4 K for Ge:Cu; which resulted in complex coolers using cascaded circuits of nitrogen with hydrogen or helium.

In spite of this handicap, the first airborne thermal cameras were developed in the USA at the start of the 1970s. These systems were known as FLIR systems, for *f*orward *l*ooking *i*nfra*r*ed. At the same time in the Europe. Ge:Hg sensors were also being used for thermal cameras. However, in the USA, as in Europe, the real surge in thermal imaging occurred around the end of the 1970s, with the development of mercury cadmium telluride sensors (HgCdTe) often known as MCT.

We can see then that the development of sensor technology and the associated optical systems required several decades of effort which continues today in the USA, Europe and Japan in order to extend the capabilities of existing sensors (high density arrays, surface matrices, etc.), in order to create new types of sensors and to produce more resistant optical materials.

Now let us return to the history of active or semi-active infrared. Immediately after the Second World War the armed forces procured active night vision equipment using bulbs and S_1 photocathode tubes. These systems have several disadvantages, for example: poor visibility factor, size of the lights, and beam easily detected. They were soon replaced by light amplifier tubes giving low light level vision and then by FLIRs. Active infrared applications did not come into their own until the development of lasers however.

The theoretical description of stimulated emission of radiation was presented for the first time by Einstein in 1917. In France, Professor Kastler developed the 'optical pumping' technique in 1950, which was later used in the production of solid lasers (ruby, etc.) It wasn't until 1958 that the first announcement of the laser (Light Amplification by Stimulation Emission of Radiation) appeared in an article by Schalow and Townes. Eighteen months later, Maiman demonstrated the operation of the first laser. Since then development has been rapid and the number of different types of lasers have been invented for a variety of applications, including: scientific, industrial, medical, artistic, and military; and using different technologies: crystal and glass lasers, gas lasers, dye lasers, semiconductor lasers, etc.

Faced with this great diversity of applications and techniques, we shall limit ourselves to retracing the main stages of the military applications of the laser because they are leading to mass production programs.

Maiman's first laser was a solid ruby laser (Al_2O_3/Cr^{3+}) with an emission wavelength around 0.69 μm, capable of emitting high power in short pulses. This type of laser was militarized in the 1960s for range finding, in particular in artillery or tank fire control systems. This laser did however have one major handicap. As it emitted in the visible part of the spectrum (red), it was 'non-stealthy' and as a result easily detectable and locatable. Ruby lasers for range finding were therefore rapidly dethroned when the first neodymium lasers made their appearance at the end of the 1960s. Neodymium range finders emitting at 1.06 μm were developed for combat tanks and some air-to-ground fire control systems at the start of the 1970s.

However, the YAG lasers (YAG—yttrium aluminium garnet) ($Y_3Al_5O_{12}/Nd^{3+}$) with a main emission line also around 1.06 μm were to open up new application horizons in the military field (as in the civil field). Range finders, target location illuminators and illuminators for laser spot guided weapons were developed around the YAG principle throughout the 1970s and 1980s.

Still in the field of solid lasers, work has been carried out on 'eye-safe' lasers since the middle of the 1980s. This work is producing erbium lasers (ER^{3+}) and YAG-Raman lasers with emission lines around 1.54 μm.

Also during the 1970s, studies were carried out on certain gas lasers in the civil and military fields. As far as infrared is concerned, the CO_2 laser is particularly interesting as its main emission line at 10.6 μm is the 8 to 12 μm atmospheric window used for thermal imaging. In addition, its characteristics enable its use for both direct and heterodyne detection. So the first CO_2 range finders were developed in the 1980s and experimental work is in progress on heterodyne lidars using techniques comparable to radar, on guidance systems, on communications and active imaging systems, etc., but the military fields of application for CO_2 lasers have not yet been completely explored.

At the same time, important work has been carried out in the last two decades in the field of semiconductor lasers (GaAs – AlGaAs) with emission wavelengths normally between 0.8 and 0.9 μm. This explains their use as pulsed illuminators for communications for example, or proximity fuses.

As with passive infrared, the applications of active infrared lasers are only just being discovered and considerable research is underway in the USA, Europe and Japan, which should lead shortly to the development of new systems.

6.3 THEORY NOTES

Characteristics of optical instruments

It would take too long and here be inappropriate to study all the characteristics of optical instruments. We shall limit ourselves to four fundamental values for the user: magnification, optical power, magnifying power and field of view.

Linear magnification Magnification is the ratio between a linear dimension y' of the image and the corresponding dimension y of the object.

$$\gamma = \frac{y'}{y}.$$

Optical power Optical power is defined in the case of instruments designed for the observation of close objects (magnifying glasses, eyeglasses, etc.). The optical power of an ocular instrument is the ratio between the angular dimension θ' of an object, as seen through the instrument, and its corresponding linear dimension y.

$$P = \frac{\theta'}{y}$$

where θ' is in radians, y in m, and P in diopters.

Magnifying power Magnifying power is the ratio between the angular dimension θ' of the object as seen through the instrument and its angular dimension θ as seen with the naked eye.

$$G = \frac{\theta'}{\theta}.$$

Field of view The 'total field of view' of an instrument is the area of space visible through the instrument, or in the image plane for a lens. The aperture is the diameter of the physical element which limits the passage of the rays through the optical system, and the focal plane is the plane of convergence for rays from a single point at infinity.

6.3.1 Optical quantities and relationships

Abbe's law (or Lagrange's constant) In an optical system the notions of field of view and aperture are independent of each other. Let us consider, for example, a system working to infinity, composed of three lenses (O_1), (O_2), (O_3) (Fig. 6.6). If n is the index of the object medium R the radius of the entrance pupil ($= D/2$), θ the angular radius of the object y'_1, y'_2, y'_3 are the intermediate images, α'_1, α'_2,

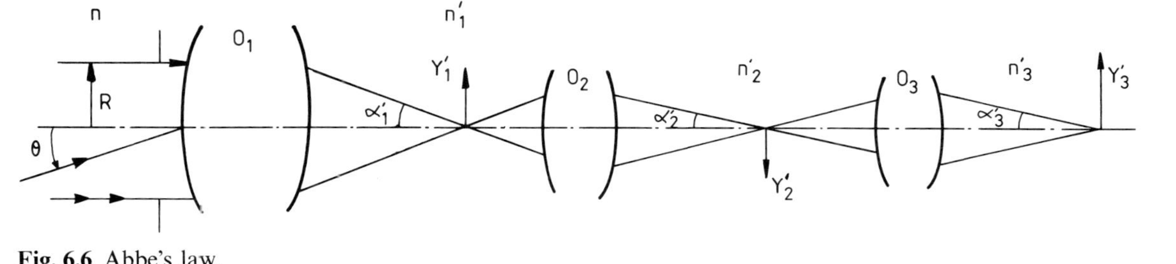

Fig. 6.6 Abbe's law.

α'_3 are the intermediate aperture angles and n'_1, n'_2, n'_3 are the intermediate medium indices; the constant product 'field times aperture diaphragm' is written as follows:

$$nR\theta = n'_1 y'_1 \sin\alpha'_1 = n'_2 y'_2 \sin\alpha'_2 = n'_3 y'_3 \sin\alpha'_3 = \cdots$$

For a unique system:

$$nR\theta = n'y' \sin\alpha$$

or

$$R\theta = y' \sin\alpha'$$

if first and last media are identical. This notion is fundamental. It merely expresses the conservation of energy.

Sine law In a corrected optical system, Lagrange's constant dictates, for small values of the field angle θ:

$$\frac{D}{2}\theta = F'\theta \sin\alpha'.$$

F' being the image focal length in which:

$$\frac{F'}{D} = N = \frac{1}{2\sin\alpha'}$$

and in which N represents the geometrical aperture number, thus in the limit case, $N = 0.5$ when $\alpha' = 90°$.

Photometric parameters

Spectral transmittance The ratio between the energy E' transmitted through a lens and the incident energy E is defined as the lens transmittance τ:

$$\tau = \frac{E'}{E}.$$

Spectral transmittance $\tau(\lambda)$ is the transmittance at each wavelength λ.

Vignetting Vignetting by an optical system is the ratio between its transmittance for an off-axis source and its value on axis, taken as a reference.

Image of point source.

DIFFRACTION

Let us consider a perfect instrument. The image of a point source is a spot whose shape depends on the shape of the aperture diaphragm. If the diaphragm is circular, the image spot exhibits axial symmetry and is called Airy's diffraction spot. The image is practically reduced to the central spot which contains 83% of the total energy.

For a perfect lens of focal length F and aperture diameter D, the radius of this spot is equal to:

$$\frac{1.22\lambda F}{D}$$

where λ is the wavelength.

DIFFRACTION LIMITED RESOLUTION

It is generally admitted that two point sources can be detected if the centres of their diffraction spots are separated by a distance equal to the first Airy's ring. This corresponds to an angle in the object space equal to $1.22\lambda/D$. This angle is the 'angular resolution' of a diffraction limited instrument of diameter D.

IMPERFECT INSTRUMENT

For a great many applications, the image quality of a lens is specified by the encircled energy, i.e. by the percentage of energy of the image spot contained inside a disk of given diameter: for example, one may specify that 80% of energy be inside a disk 50 microns in diameter.

Image of extended objects – modulation transfer function The overall performance of any instrument can be evaluated by studying its response to sinusoidal signals. That general rule applies to the field of optics, where the luminance of a two-dimensional object may be represented by the superimposition of an infinite number of sinusoidal signals with amplitudes and phases depending on the period p (varying from 0 to infinity).

A lens may be characterized by its ability to transmit these various components, i.e. by its transfer factor as a function of the period p, or rather of its opposite

$$v = \frac{1}{p}$$

called the spatial frequency.

Let us consider the typical case in which the luminance of the object is the sum of (Fig. 6.7) a uniform value L_o and a sinusoidal component of amplitude l and period p (frequency v).

Contrast $C(v)$ of the object at this frequency is defined by:

$$C(v) = \frac{l}{L_o}.$$

The image is represented by an average luminance L'_o and a sinusoidal component of amplitude l' and period p (frequency: v) the magnification is assumed to be

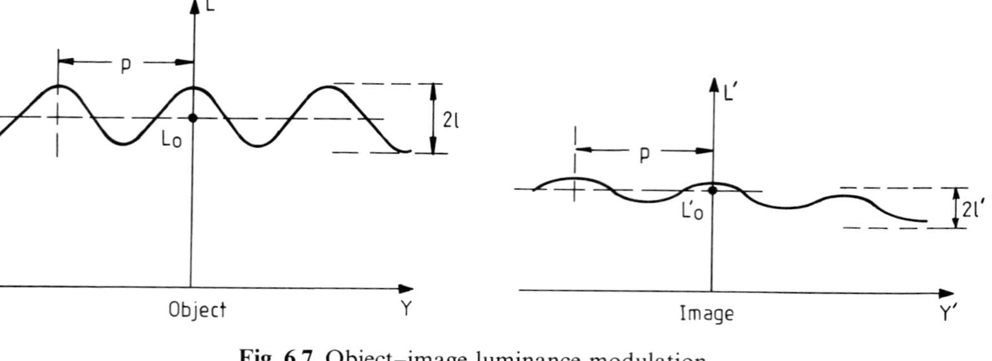

Fig. 6.7 Object–image luminance modulation.

equal to 1) or a contrast:

$$C'(\nu) = \frac{l'}{L'_o}.$$

In fact L'_o is different from L_o owing to the transmittance τ of the lens:

$$L'_o = L_o \tau.$$

We can therefore write:

$$C'(\nu) = d(\nu)C(\nu)$$

in which $d(\nu)$ is the 'contrast transfer' of the lens for a sinusoidal object at spatial frequency ν.

In order to achieve a function standardized at 1 for $l = 0$, we say that $L'_o = L_o$. In this case this function is called the modulation transfer function (MTF). MTF characterizes the quality of a lens and should be given for each point in the optical field and for each wavelength. This law is shown in Fig. 6.8 for a perfect, diffraction limited lens with a circular aperture.

The higher the object spatial frequency, the lower the contrast transfer of the lens, up to a cut-off frequency ν_c, for which the MTF is down to zero. A lens is therefore equivalent to a low-pass filter.

The cut-off frequency of a perfect lens with a circular aperture is the following:

$$\nu_c = \frac{2 \sin \alpha'}{\lambda}$$

where $\sin \alpha'$ is the image numerical aperture of the lens, and λ is the wavelength.

No information at spatial frequencies above ν_c is transmitted through the lens. Being the inverse of a length, spatial frequencies are expressed in mm^{-1} or as a number of pairs of lines per mm (lp/mm). The dotted curve in Fig. 6.8 represents

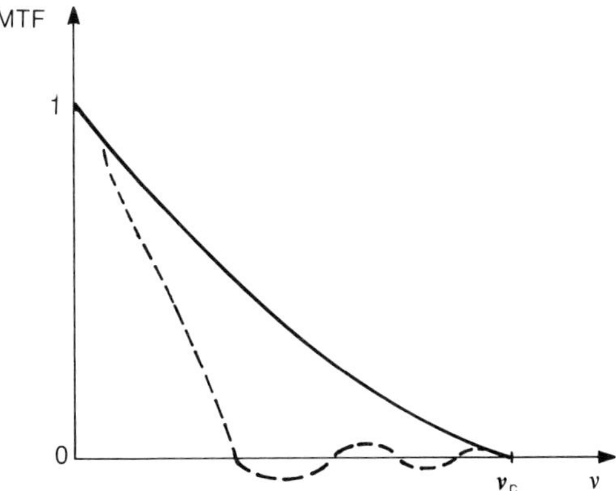

Fig. 6.8 Modulation transfer function.

the example of an MTF for a lens with aberrations; negative values of the curve represent contrast inversions.

Aberrations, causes of faults in instruments. In the best case, the quality of a lens is limited by diffraction, but one must also consider lens defects, due to several causes, including aberrations, and design errors, for example.

ABERRATIONS

There are two types of aberrations in optical systems, chromatic aberrations and geometrical aberrations.

Chromatic aberrations are due to the fact that indices of refraction in optical materials indices are wavelength dependent. As a result, the characteristics of a lens will depend upon the wavelength. Chromatic aberrations may be reduced by the proper choice of different materials that will balance each other out.

Geometrical aberrations are due to the fact that optical systems are not geometrically perfect. Rays do not converge exactly at a given point whatever the conditions of aperture or field. The study of aberrations and their correction are very complex. Aberrations are normally broken down into five types:

1. spherical aberration which degrades image quality on-axis;
2. coma which degrades image quality off-axis;
3. astigmatism (off-axis degradation);
4. field curvature (the surface image is not flat);
5. and distortion which does not affect image quality, but simply modifies image geometry.

MANUFACTURING DEFECTS

The defects may be due to:

1. non-homogeneity of the glass;
2. poor optical surfacing;
3. off-centering of the different lenses or mirrors;
4. tilting of the optical elements;
5. mechanical or thermal stresses, that create surface deformation;
6. stray light, that reduces image contrast.

6.3.2 Photometry and radiometry

Energetic units

POWER RADIATED BY A SOURCE

To begin with, let us consider the case of an isotropic point source, which radiates equally in all directions in space. The power which it radiates is expressed in joules per second, i.e. in watts (ϕ).

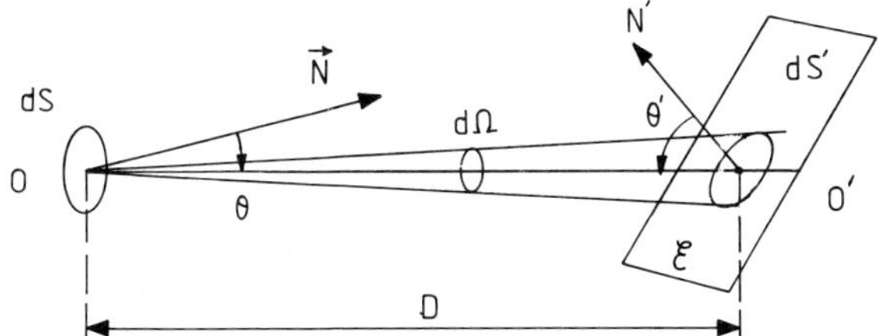

Fig. 6.9 Luminance of a source.

INTENSITY I OF A SOURCE

Intensity I of this isotropic point source is defined by

$$I = \mathrm{d}\phi/\mathrm{d}\Omega$$

($\mathrm{d}\Omega$ = solid angle), i.e. the flux it radiates per unit solid angle. The units are watt per steradian ($\mathrm{W \cdot sr^{-1}}$).

LUMINANCE OF A SOURCE

Let us consider a source made up by a small area dS, whose normal N makes an angle θ with the observation line 00′ (Fig. 6.9). We define luminance L of the source by means of the equation:

$$\mathrm{d}^2\phi = L \cdot \mathrm{d}S \cdot \cos\theta \, \mathrm{d}\Omega$$

in which $L = \mathrm{d}^2\phi/\mathrm{d}S \cdot \cos\theta \cdot \mathrm{d}\Omega$ the units of luminance are watts per square metre per steradian ($\mathrm{W \cdot m^{-2} \cdot sr^{-1}}$).

IRRADIANCE

Irradiance, E, of the surface dS' is defined by:

$$E = \frac{\mathrm{d}\phi}{\mathrm{d}S'}$$

the units are watts per square metre ($\mathrm{W \cdot m^{-2}}$).

Radiance, or emittance, or exitance, R, of a radiating surface dS, is the total power emitted per unit area in all possible directions is:

$$R = \int_\Omega L \cos\theta \, \mathrm{d}\Omega = \frac{\mathrm{d}\phi}{\mathrm{d}S}$$

the units are watt per square metre ($\mathrm{W \cdot m^{-2}}$).

SPECTRAL UNITS

Spectral flux (in $\mathrm{W \cdot \mu m^{-1}}$):

$$\phi_\lambda = \frac{\mathrm{d}\phi}{\mathrm{d}\lambda}(\lambda)$$

spectral intensity (in $W \cdot sr.^{-1} \cdot \mu m^{-1}$):

$$I_\lambda = \frac{dI}{d\lambda}(\lambda)$$

spectral radiance (in $W \cdot m^{-2} \cdot sr.^{-1} \cdot \mu m^{-1}$):

$$L_\lambda = \frac{dL}{d\lambda}(\lambda)$$

spectral irradiance (in $W \cdot m^{-2} \cdot \mu m^{-1}$):

$$E_\lambda = \frac{dE}{d\lambda}(\lambda)$$

VISUAL AND ENERGETIC UNITS

These are shown in Table 6.1.

IRRADIANCE FROM A POINT SOURCE

The flux $d\phi$ radiated by a source of intensity I into the solid angle $d\Omega$ is the following:

$$d\phi = I \cdot d\Omega$$

The irradiance E of a surface dS' situated at range D from the source and whose perpendicular makes the angle θ' with the direction of the radiation, is given by (Fig. 6.9):

$$E = \frac{d\phi}{ds'} = \frac{I\,d\Omega}{ds'} = \frac{I \cos\theta'}{D^2}$$

LAMBERT'S LAW

In many cases, the luminance $L(\theta)$ of a point or extended source is independent of the emission direction θ of its radiation. In this case the source is said to satisfy Lambert's Law:

$$L(\theta) = L.$$

The total flux emitted by a source radiating in a complete hemisphere, is given by:

$$d\phi = \pi \cdot L \cdot dS$$

Table 6.1 Different systems of units and their relations

Nature	*Visual units*	*Symbol*	*Dimensional relations*	*Energetic units*
Flux	Lumen	ϕ		W.
Intensity	Candela	I	$1\ cd = 1\ lm\ sr^{-1}$	$W.sr^{-1}$
Luminance	Candela m^{-2}	L	$1\ cdm^{-2} = 1\ lm\ m^{-2}\ sr^{-2}$	$W.m^{-2}.sr^{-1}$
Irradiance	Lux	E	$1\ lx = 1\ lm\ m^{-2}$	$W.m^{-2}$

and its radiance:

$$R = \pi L.$$

REFLECTANCE OF A DIFFUSE SURFACE

If one considers a surface dS' which receives a flux $d\phi$, the surface scatters a flux $d\phi'$, which is a fraction of the incident flux. The reflectance of the surface, ρ, is defined as follows

$$\rho = \frac{d\phi'}{d\phi}.$$

CONSERVATION OF LUMINANCE IN AN OPTICAL SYSTEM

Figure 6.10 shows a source of luminance L and of area $S = \pi y^2$, which is imaged through the optical system ○ onto a surface of area $S' = \pi y'^2$. The half aperture angle being α, the flux ϕ which enters the system (○) is:

$$\phi = \pi L S \sin^2 \alpha = \pi^2 L y^2 \sin^2 \alpha.$$

The transmitted flux ϕ' is the following:

$$\phi' = \pi L' S' \sin^2 \alpha' = \pi^2 L' y'^2 \sin^2 \alpha'.$$

Input and output fluxes are related to each other by the optical transmittance τ of the system:

$$\phi' = \tau \phi.$$

On the other hand, the sizes of object S and its image S' are related to each other through Abbe's law which may be expressed as the following, when object and image have the same index of refraction:

$$y \sin \alpha = y' \sin \alpha'.$$

We therefore draw the two following important conclusions:

$$L' = \tau L$$

the luminance of the beam is conserved within the limits of the transmission factor; and in an optical system, the 'etendue' of the beam is conserved.

6.3.3 Atmosphere

The propagation of infrared radiation is dealt with in Volume 2 in the chapter 'Propagation of electromagnetic waves'. A part of the chapter is particularly concerned with atmospheric transmission in infrared.

6.3.4 Sources

In order to design an optronic system, we must first characterize the environment and the medium in which it is intended to be used in the most reliable way

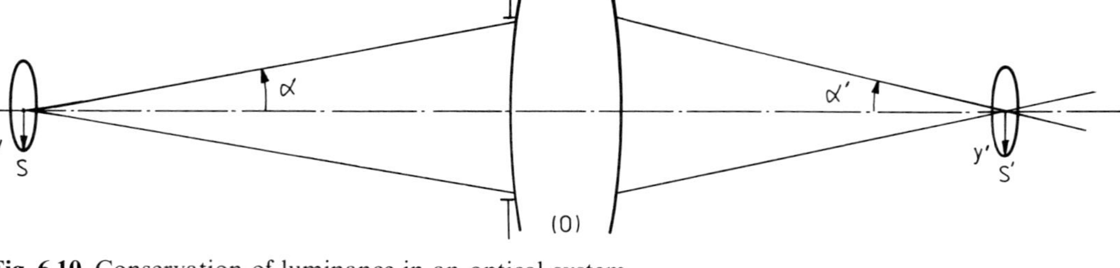

Fig. 6.10 Conservation of luminance in an optical system.

possible. This means identifying (and quantifying) all the sources formed by the landscape which is observed by the system. Whether natural or artificial, they all represent input data which will be 'seen' by the optronic system.

The landscape is composed of different sources of radiation which are normally;

1. the object itself;
2. the natural background against which it is presented;
3. the propagation medium, normally the atmosphere;
4. all the other artificial sources which may or may not form part of the system.

Very often, infrared sources are described as 'heat sources'. This is because radiation from most natural sources is related to their temperature according to black-body radiation laws. We shall therefore consider these sources at first, before dealing with non-thermal sources that radiate on specific spectral lines, such as lasers for example.

Black-body

Definition of a black-body Any body, which is at a temperature different from 0 K, behaves like an emitter. Generally speaking, any body that is being illuminated absorbs, transmits and reflects some percentages of the incoming light (respectively a, t and r), in such a way that:

$$a + t + r = 1$$

A black-body is such that $a = 1$ for any λ, which means that a black-body completely absorbs all radiations incident upon it.

Properties of black-bodies

1. The luminance of a black-body is independent of the direction of emission (Lambert's law).
2. Total radiance only depends upon absolute temperature (Stefan–Boltzmann's law).
3. Spectral luminance depends on temperature and wavelength (Planck's law, 1906).

Spectral luminance (Planck's formula) At each wavelength, λ, the spectral luminance of a black-body at temperature T is given by Planck's formula:

$$L_\lambda^{BB}(\lambda, T) = \frac{C_1 \lambda^{-5}}{\exp \dfrac{C_2}{\lambda T} - 1}$$

where $C_1 = 1.1905 \times 10^{+8}\ \mathrm{W \cdot m^{-2} \cdot sr^{-1} \cdot \mu m^{+4}}$ and $C_2 = 1.4385 \times 10^{+4}\ \mu\mathrm{m \cdot K}$. In this formula, the wavelength λ should be expressed in micrometers (μm) and the temperature in Kelvin (K). The spectral luminance $L_\lambda^{BB}(\lambda, T)$ is then evaluated in $W \cdot m^{-2}$, sr^{-1} μm^{-1}.

Total luminance – Stefan–Boltzmann's law The total luminance of a black-body at a temperature T depends only on this temperature;

$$L^{BB}(T) = \int_0^\infty L_\lambda^{BB}(\lambda, T)\, d\lambda = K_3 T^4$$

with T in K, and $K_3 = 1.806 \times 10^{-8}\, W \cdot m^{-2} \cdot sr^{-1} \cdot K^{-4}$.

Maximum spectral luminance – Wien's displacement law The spectral luminance of a black-body goes through a maximum at a wavelength λ_m defined by Wien's displacement law:

$$\lambda_m T = K_1 = 2898$$

λ_m is evaluated in μm and T in Kelvin.

Maximum value of the spectral luminance is given by:

$$L^{BB}(\lambda_m, T) = K_2 T^5$$

with T in K and $K_2 = 4.095 \times 10^{-12}\, W \cdot m^{-2} \cdot sr^{-1} \cdot K^{-5} \cdot \mu m^{-1}$.

Special properties of black-bodies The shape of black-body spectral distribution at different temperatures is shown on Fig. 6.11. It should be noted that the wavelength corresponding to the maximum in spectral radiance is inversely proportional to the black-body temperature ($\lambda_m = K_1/T$), the curve of a black-body at a given temperature is below the curves of other black-bodies at higher temperatures and therefore, for a given wavelength λ, the hotter the black-body the higher its spectral luminance.

Emissivity of materials

SPECTRAL EMISSIVITY

The spectral luminance $L_\lambda(\lambda, T)$ of any given material at temperature T can be written as follows:

$$L_\lambda(\lambda, T) = \varepsilon(\lambda, T) \times L_\lambda^{BB}(\lambda, T)$$

where $L_\lambda^{BB}(\lambda, T)$ is the spectral luminance of the black-body at same temperature T, and $\varepsilon(\lambda, T)$ is the spectral emissivity of the material.

Three different types of natural sources may be defined, depending upon their characterized spectral emissivity curves. If $\varepsilon(\lambda, T) = 1$, the source is said to be a black-body; if $\varepsilon(\lambda, T)$ is a constant different from 1, it is a grey-body; and if $\varepsilon(\lambda, T)$ varies over the spectrum, the source is a selective emitter.

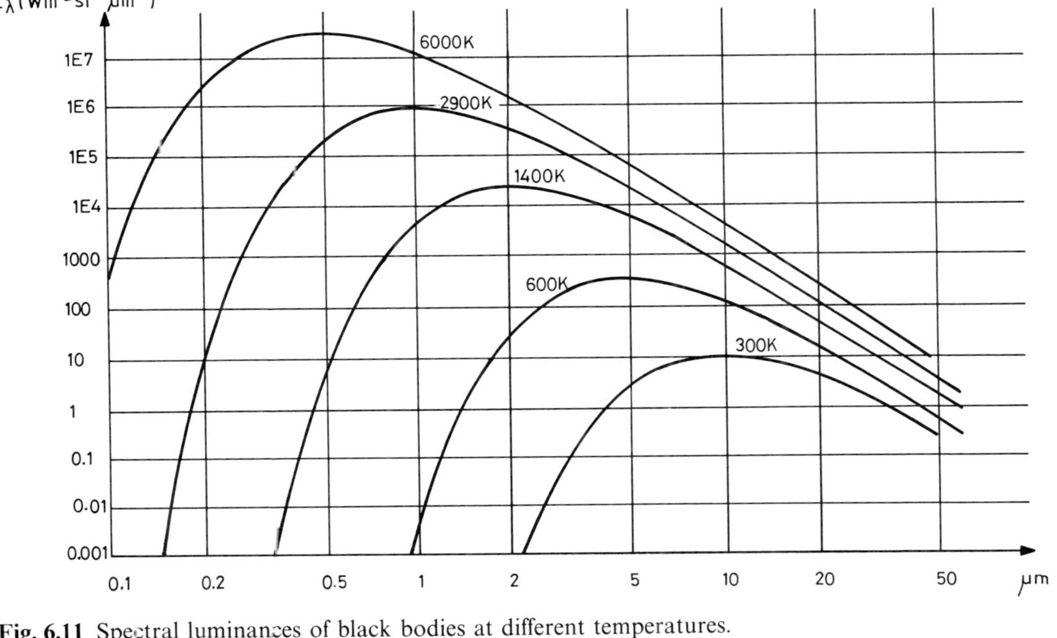

Fig. 6.11 Spectral luminances of black bodies at different temperatures.

AVERAGE EMISSIVITY

Average emissivity of a source is the ratio between the energy W' emitted by that source and the energy W emitted by a black-body at same temperature:

$$\bar{\varepsilon}(\lambda, T) = \frac{\int_0^\infty L_\lambda^{BB}(\lambda, T) \times \varepsilon(\lambda, T)\,d\lambda}{\int_0^\infty L_\lambda^{BB}(\lambda, T)\cdot d\lambda} = \frac{\int_0^\infty L_\lambda(\lambda, T)\,d\lambda}{K_3 T^4}.$$

KIRCHHOFF'S LAW

This law states that for any body at a given temperature T, the ratio between its spectral luminance and its spectral absorption is equal to the spectral luminance of the black-body at same temperature:

$$\frac{L_\lambda(\lambda, T)}{a(\lambda, T)} = L_\lambda^{BB}(\lambda, T) \Rightarrow \frac{\varepsilon(\lambda, T)\cdot L_\lambda^{BB}(\lambda, T)}{a(\lambda, T)} = L_\lambda^{BB}(\lambda, T).$$

One result is that $\varepsilon(\lambda, T) = a(\lambda, T)$. In other words, at each wavelength spectral absorption and spectral emissivity of a given body are identical.

'Natural' sources

Any given object will emit light, because it is either a 'primary source' or a 'secondary' one, or both.

Primary source A 'primary source' radiates light by its own means, by converting different types of energy (electrical, chemical, nuclear) into light. In fact, as we have seen any material at temperature T emits radiation.

If the detector is not sensitive to this thermal radiation, the object does not appear as a primary source. The detection of an object as a primary source is therefore determined by the choice of the detector used for its observation.

It is recalled that the spectral luminance of an object of emissivity $\varepsilon(\lambda, T)$ and at temperature T is the following:

$$L\lambda(\lambda, T) = \varepsilon(\lambda, T) \times L_\lambda^{BB}(\lambda, T).$$

Secondary source An object is defined as a 'secondary source' if it re-emits light incident from other sources. That is the case for most of the objects surrounding us, and which we observe with the naked eye. They are visible because they reflect part of the light received from primary sources (sun, lamps, etc.).

If the object is a diffuse, Lambertian surface, its luminance is:

$$L = \frac{\rho E}{\pi}$$

where E is the illumination received and ρ is the reflectance of the object.

In practice, most objects only scatter light in a much more complex way which depends on both illumination and observation, angles θ, as well as wavelength λ. If spectral irradiance of the object E_λ, its apparent spectral luminance in direction θ will be:

$$L_\lambda(\theta) = \frac{\rho(\theta)}{\pi} E_\lambda$$

In most passive infrared applications, the 'objects' or targets are illuminated by natural (primary) sources. In active or semi-active infrared applications, this illumination is produced by artificial sources, lasers for example.

Emission characteristics of natural 'backgrounds' The background against which the object is observed can vary a great deal: sky, sea, earth, forest, etc. Like the object, the background is characterized essentially by its spectral luminance. One must therefore also consider it either as a primary or a secondary source, or even as both, depending on its nature and the spectral band envisaged.

Defining the 'background' therefore implies knowing its emissivity, and its reflectance or reflection coefficient.

Their fluctuations in time often are of utmost importance at the design level since they may represent sources of noise for the detection system. For example, one might mention fluctuations in the luminance of the sea surface in the case of thermal observations. Depending on the application, the background is either a surface background or a sky background.

SURFACE BACKGROUND

Made up of either earth or sea background, this type of background is characterized by a total spectral luminance L_λ such that:

$$L_\lambda = \varepsilon(\lambda, t) \times L_\lambda^{\mathrm{BB}}(\lambda, t) + \frac{\rho_\lambda(\theta, \varphi)}{\pi} E_\lambda.$$

It is therefore composed of a primary, or self-emitting, source normally similar to a grey body at ambient temperature and a secondary or reflecting source characteristic of its scattering properties.

Figure 6.12 shows the typical shape of such surface background for daylight applications.

SKY BACKGROUND

This paragraph deals with the characteristics of radiation from the sky itself, i.e. apart from the known primary sources such as the moon, planets and stars, or the sun. Radiation from the sky essentially comes from:

1. the atmosphere itself;
2. the 'zodiac' light (scattering of light by cosmic dust);
3. emissions produced by the upper ionized layers of the atmosphere, known as the luminescent layers.

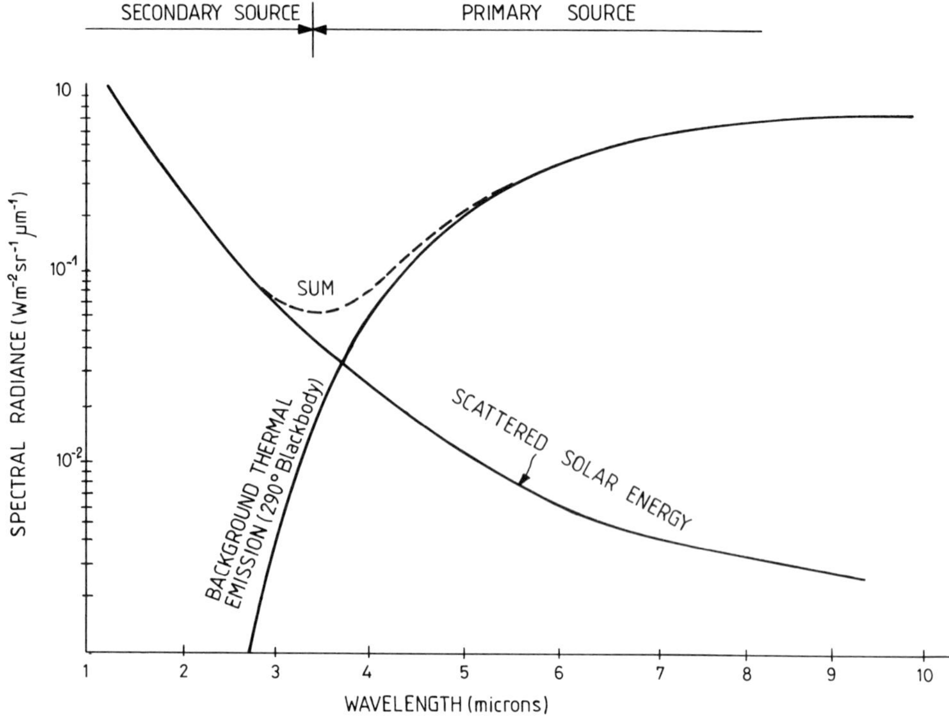

Fig. 6.12 Theoretical surface background (daylight).

The luminances of these sources are extremely variable and cover a considerable dynamic range. They should always be analysed spectrally.

In the 7 to 15 μm (infrared) spectral band the luminance of the sky as observed from the earth arises from thermal emission by the atmosphere. Figure 6.13 shows its spectral characteristics, which vary with the elevation angle of observation above the horizon.

In the 2 to 5 μm spectral band, the sky behaves as a grey body at ambient temperature, whose spectral emissivity is equal to the spectral transmittance of the atmosphere, and, in case of day-time observation, a secondary source from the scattering of solar illumination by the atmosphere. Figure 6.14 shows these components in the case of observation from the earth near the horizon.

For wavelengths below 2 μm, the sky background luminance is provided during day-time, by the scattering of solar illumination by the atmosphere (Rayleigh and Mie scattering, respectively on molecules and aerosols), and at night, by thermal radiation from the zodiacal light and from the ionized layers of the atmosphere at high altitudes. Figure 6.15 shows spectral luminance of the night sky in this domain of the infrared spectrum.

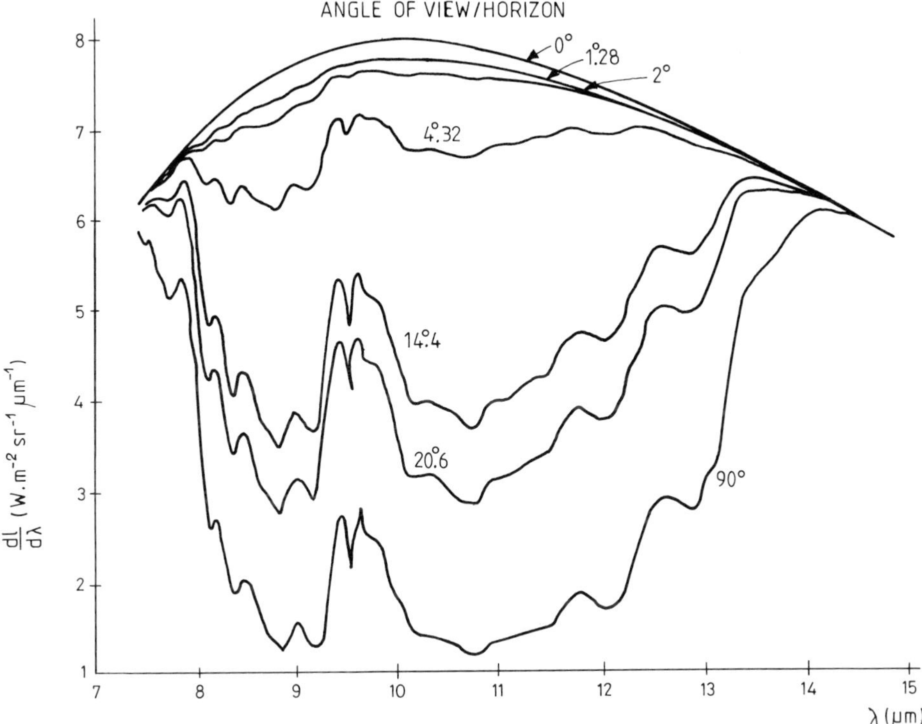

Fig. 6.13 Spectral luminance of sky background (7 to 15 μm band) as a function of the elevation angle above the horizon.

Natural primary sources The main natural primary sources are the sun, the moon, the stars and planets.

THE SUN

Outside the atmosphere, solar radiation can be compared to that of a 5900 K black-body. With an average angular diameter of 30′, the sun provides an irradiance of the order of 1400 W·m^{-2}, and around 10^5 lux. At ground level, the spectral luminance of the sun is modified by the transmittance of the atmosphere, whose influence increases as the sun approaches the horizon.

THE MOON

The moon is in fact a secondary source in the visible and near IR parts of the spectrum, where it only reflects solar light. Ground illumination of the earth from the moon is variable and depends upon several parameters:

1. the phase of the moon;

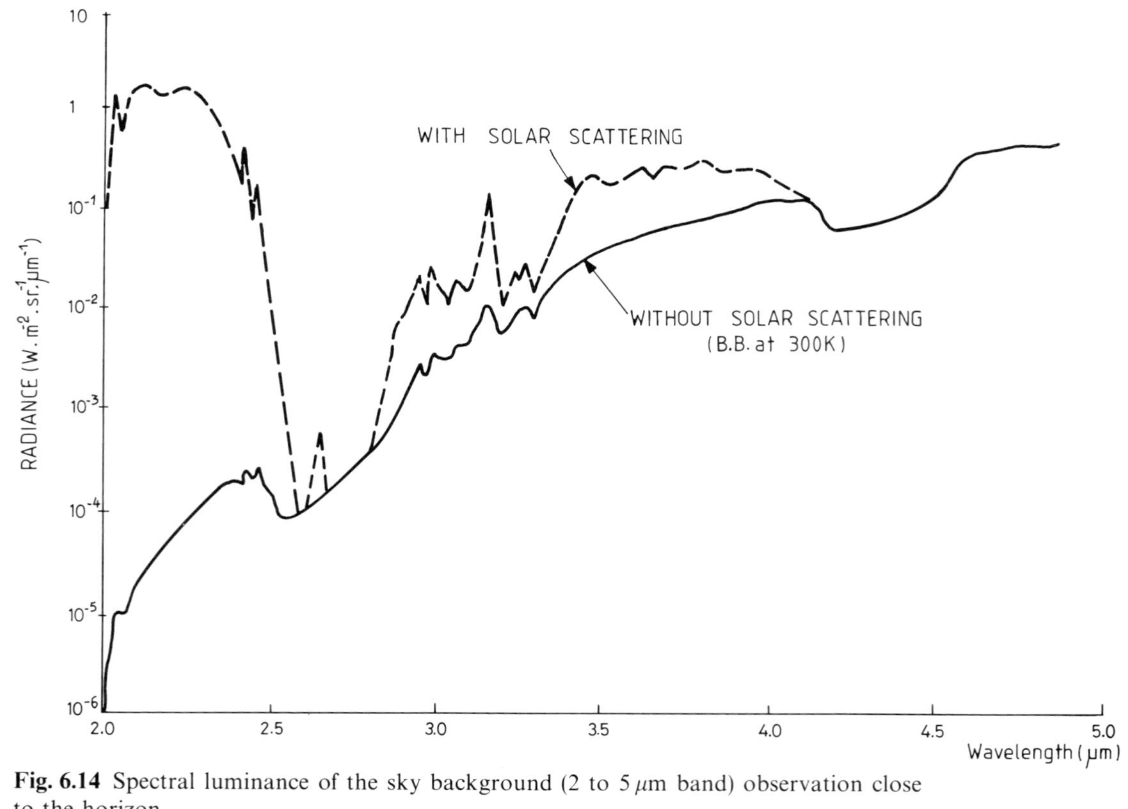

Fig. 6.14 Spectral luminance of the sky background (2 to 5 μm band) observation close to the horizon.

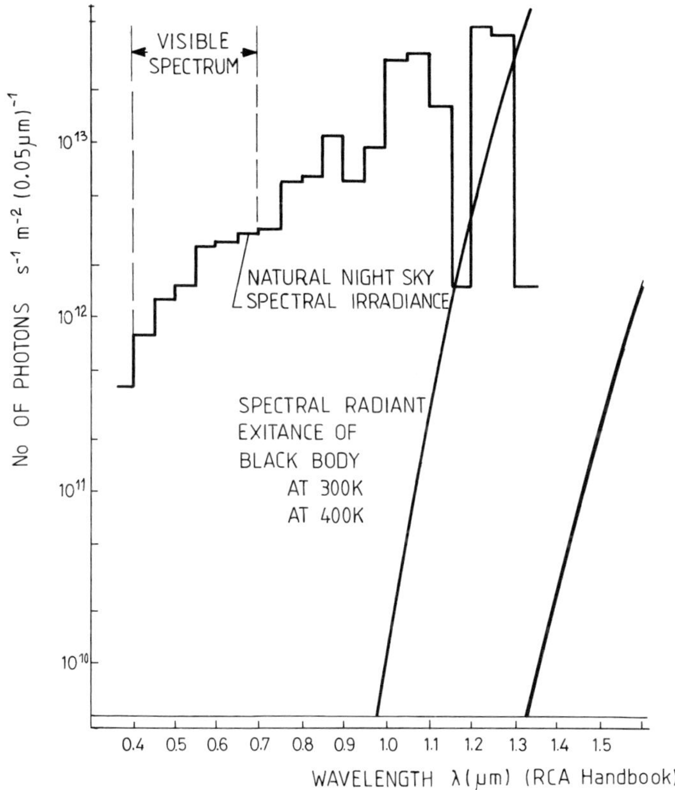

Fig. 6.15 Natural night-sky spectral irradiance on horizontal earth's surface and the spectral radiant existence of black-bodies at 300 K and 400 K.

2. variations in the distance between the earth and the moon during the lunar cycle, which can be as much as 26%;
3. local variations in reflectance on the lunar surface;
4. the elevation angle of the moon above the horizon.

On the average, the visual luminance of the moon is 10^{-1} lux, and its solid angle from the earth is between 5.8 and 7.2×10^{-5} sr.

Artificial sources

Black-body sources Theoretical studies on black-bodies have demonstrated the interest of such types of sources for assessing the performances of infrared systems, since their spectral luminance is dependent solely on their temperature.

The design of a black-body or at least of a realistic simulator is beset by two difficulties: temperature stability and uniformity; and emissivity as close as possible to 1.

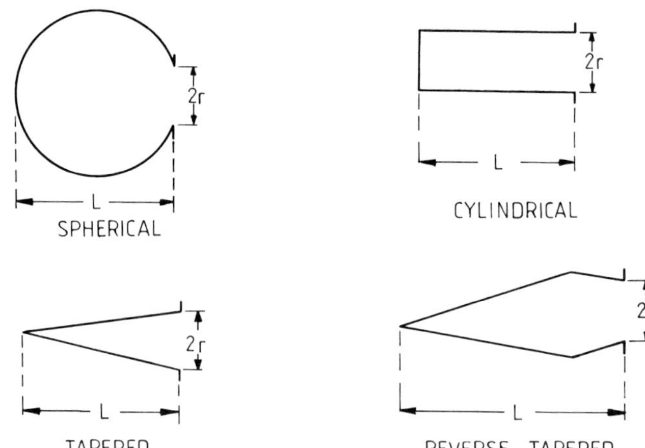

Fig. 6.16 Typical cavity shapes.

Flat surfaces with emissivity close to 1 are difficult to make. One way to simulate black-body radiators is to design cavities whose apparent emissivity is larger than that of their walls. Most frequently used cavities are of four types, as shown in Fig. 6.16 (characterized by depth L and aperture diameter $2r$).

Lamps Several sources are available for use in system checkout and alignment, spectrometers, communication devices, and solar simulators. In general, the characteristics of these sources are not as well known as those of a black-body, but their intended applications do not require such detailed knowledge. In addition, such sources are often relatively inexpensive, readily portable, and simple to use.

THE NERNST GLOWER

A Nernst glower is often found in infrared spectrometers that are used to measure the transmittance, reflectance, or absorptance of various materials. It consists of a relatively fragile cylinder made by sintering a mixture of zirconium, yttrium, thorium, and certain oxides. When cold, it does not conduct, but when it is heated to 400 °C by a flame or self-contained tungsten filament, it becomes conductive and can be further heated by passing an electrical current through it. For an average glower, that is, one about 3 cm long and 0.15 cm in diameter, an input of 0.5 A at 20 V is required after the initial heating. Under these conditions the effective temperature* of the glower is about 2100 K. Because of the large negative temperature coefficient of resistance, a current-limiting ballast is needed. The emissivity of the glower varies somewhat with wavelength and has an average value of about 0.67 from 2 to 15 μ.

*The effective temperature of a non-black-body source is defined in terms of the temperature of a black-body having the same spectral luminance, at a particular wavelength, as that of the source.

THE GLOBAR

Another source often used with infrared spectrometers is the globar. It is a rod of silicon carbide, typically 5 to 10 cm long and 0.5 cm in diameter. It is heated to an operating temperature of about 1500 K by an input of 3 to 5 A at 50 V. Unlike the Nernst glower, the globar needs no separate heater since the heating current is passed directly through the silicon carbide rod. The emissivity varies somewhat with wavelength and has an average value of about 0.8 from 2 to 16 μ.

THE CARBON ARC

A low-intensity carbon arc has been used as a spectrometer source when a greater luminance than that of the globar or Nernst glower was needed. A source temperature of about 3900 K is reached. A five-fold decrease in emissivity occurs as the wavelength increases from 2 to 10 μ. The high-intensity carbon arc, which operates at 5800 to 6000 K, is used in solar simulators. The arc current is three to four times greater than that of the low-intensity arc and the operating life of the electrodes is proportionately less.

THE TUNGSTEN LAMP

Tungsten lamps are used as sources, but only for the near infrared since their glass envelopes do not transmit radiant energy beyond 4μ. Filament temperatures as high as 3300 K can be obtained. The average emissivity of a tungsten filament at 2800 K is about 0.23 from 2 to 3μ. Tungsten lamps provide a solution, although not always a satisfactory one, to the problem of finding a suitable source for field calibration of near-infrared equipment. Since the radiant emittance changes rapidly with changes in filament current, it is imperative that this current be closely monitored during measurements.

Tungsten lamps are surprisingly inefficient sources of visible light. Ten per cent of the input power to a typical 100 W household lamp is radiated beyond the bulb as visible light. Seventy per cent is radiated in the near infrared, and 20% is absorbed by the gas in the lamp and by its glass envelope. The glass envelope can readily reach a temperature of 150 °C. As a result, equipment operating in the intermediate- and the far-infrared may receive strong signals from tungsten lamps. It is important to note that the signals are from the heated envelope and not the filament, since the spectral distribution is quite different for the two.

THE XENON ARC LAMP

The xenon arc lamp has been used in near-infrared communication systems. Its particular advantage is the ease with which the output can be modulated by varying the current supplied to the lamp. Most of the energy from the xenon arc is radiated in the visible and ultraviolet, but there is a useful output in the near infrared, extending to a wavelength of about 1.5 μ.

Laser sources Laser sources have already been described in Volume 1.

6.3.5 Optical materials

The characterization of optical materials, whether infrared or not, requires a knowledge of all their optical, physical and mechanical properties. This is a very important point as all optronic systems use optical materials which, depending on the duty and environmental constraints, will dictate to a great extent the operational performance of the optronic system. This means that one does not go about choosing the optical material for a dome on a missile homing head in the same way as one would have the dome on the electronic equipment installed on a tank, or again, for material designed for a laboratory application as against that for space-type applications.

In order to optimize the choice of an optical material, one must therefore know all its characteristics, including:

1. spectral transmittance;
2. refraction index;
3. hardness;
4. resistance to abrasion;
5. density;
6. thermo-conductivity;
7. expansion coefficient;
8. specific heat;
9. modulus of elasticity;
10. melting point or softening temperature;
11. radio wave transmittance.

This being the case, the final choice of optical materials for use in optronic systems will always be a trade-off between:

1. the environmental constraints of the system;
2. the required optical performance as well as optical surface coating performance;
3. acceptable price for the system (raw materials, construction, coating, transport, etc.);
4. and finally, duty requirements (construction, toxicity, size, procurement lead times).

Table 6.2 summarizes the optical and mechanical properties required for normal infrared materials.

6.3.6 Detectors

General

Infrared radiation detectors have partly been described in Volume 1, in particular focal plane arrays. We shall limit ourselves here to mentioning the more common detectors for infrared applications from the systems aspect. It should

Table 6.2 Optical and mechanical characteristics of the main infrared materials

Material	Refraction index n at 25 °C for 1 μm	3 μm	5 μm	8 μm	12 μm	dn/dt at 25 °C ($10^{-6} K^{-1}$)	Fall-off in transmission as temperature increases	Density at 25 °C (g/cm^3)	Hardness knoop	Coefficient of linear expansion $\frac{dl}{ldt}$ at 25 °C ($10^{-6} K^{-1}$)	Solubility at 25 °C for 100 g of water (g)
Ge		4.0449	4.0151	4.0054	4.0019	280–400	Very high	5.33	700–880	5.5–6.1	<0.005 insoluble
GeAsSe	2.6055	2.5187	2.5109	2.5034	2.4904	50–100	Low	4.40	170	13	Insoluble
GeSbSe	2.7235	2.6263	2.6[illegible]5	2.6076	2.5921	79	Low	4.67	150	16	Insoluble
Si		3.4324	2.4221	3.4184		80–168	Very high	2.33	1150	2.3–4.7	<0.005 insoluble
Hot pressed ZnS	2.2907	2.2558	2.2447	2.2213	2.1689			4.09	325–350	6.6–7.9	Insoluble
CVD ZnS	2.2923	2.2572	2.2461	2.2228	2.170	43–50	Appreciable for >9.5 μm	4.09	200–250	7.9	6.5×10^{-5} insoluble
Cleartran ZnS	2.270	2.254	2.243	2.221	2.170	39–50		4.09	160	7.9	Insoluble
ZnSe	2.4888	2.4375	2.4295	2.4173	2.3928	53–60	Very low	5.27	100–120	7.1–8.5	<0.001 insoluble
AsGa	3.455 (=1.127 μm)			3.34	2.97 (=13 μm)	150–200	Low	5.31	720–750	5.7–6.8	<0.005 insoluble
M_gF_2	1.3778	1.3640	1.3374	1.2634			Appreciable for $\lambda > 6 \mu m$	3.18	576	9–12	Insoluble
C_aF_2	1.4289	1.4179	1.3990	1.3498		(9–13)		3.18	mono 158 poly 200	24	0.0017
Sapphire	1.7557	1.7122	1.6239			13–15	Low: cut at 4 μm for 500 K	3.98	1525–2200	5.8–6.7	9.8×10^{-5} insoluble
Spinelle	1.704	1.698	1.659				Low	3.58	1300–1820	5.6–6.5	Insoluble

be remembered first that the detectors in question are either image detectors or thermal detectors.

IMAGE DETECTORS

The best way of defining an image detector is to make an analogy with a photographic film whose essential characteristics are time integration and two-dimensional representation of the illumination of the scene.

It is also a wavelength integrator:

$$E(y, z, t) = \int_0^\infty E(\lambda, y, z, t) S(\lambda) \, d\lambda$$

where $S(\lambda)$ is the relative spectral sensitivity of the detector.

FLUX DETECTORS

This type of detector performs spatial integration of the energy across its whole surface. As in the case of the image detector, this type of detector also performs spectral integration. The physical principle of the conversion of photons to electrons has two possible origins which result in the following two types of detectors: thermal detectors, and quantum or photonic detectors.

THERMAL DETECTORS

The temperature rise caused by infrared radiation produces modifications in the electrical balance. These detectors have properties which are not related to wavelength but only to energy received.

QUANTUM OR PHOTONIC DETECTORS

In this case we are dealing with photon–electron interaction. Each incident photon creates electrons. The energy characteristics of these detectors are essentially dependent on wavelength.

Factors influencing the design of a flux detector

RESPONSIVITY, SENSITIVITY

Detector responsivity or voltage sensitivity is defined as the ratio between the output voltage and the flux incident upon the detector.

$$R = \frac{V_S}{\phi}$$

where V_S is the detector output voltage and ϕ is the flux incident upon the detector. The voltage response is expressed in volts per watt (V/W).

NOISE EQUIVALENT POWER (NEP)

NEP is the incident optical flux onto the detector that produces an output voltage identical to the r.m.s. noise voltage, V_N.

If an effective flux ϕ generates an effective output voltage V_S on the detector, the NEP (in W) can be written as:

$$\text{NEP} = \frac{\phi}{V_S/V_N}.$$

DETECTIVITY

Detectivity is defined as the inverse of NEP (and is expressed in W^{-1}):

$$D = \frac{1}{\text{NEP}}.$$

SPECIFIC DETECTIVITY

The concept of specific detectivity D^* has been chosen for characterizing a detector because it is independent of the area A_d of the detector and of the measurement method (bandwidth Δf):

$$D^* = \frac{\sqrt{A_d \Delta f}}{\text{NEP}}$$

or

$$D^* = \frac{\sqrt{A_d \Delta f} \times V_S/V_b}{\phi}$$

$(\text{cm} \cdot \text{Hz})^{1/2}\, W^{-1}$.

It is important to note that as with the response factor, detectivity is highly dependent on the spectral composition of the radiation.

SPECTRAL RESPONSE

We have seen that responsivity R and specific detectivity D^* depend upon the wavelength of incident radiation and therefore one defines spectral response as the relative response:

$$S(\lambda) - \frac{D^*_\lambda}{D^*_{\lambda\,\max}} = \frac{R_\lambda}{R_{\lambda\,\max}}.$$

SPATIAL RESPONSE

A detector must be characterized geometrically (shape, size), or more precisely by its spatial response $R(y, z)$. This value defines detector responsivity at each point. If the detector responsivity is uniform, definition of its surface area will be sufficient.

ACCEPTANCE SOLID ANGLE

If a detector is fitted with a screen limiting its field of view, the solid angle inside which the detector is sensitive to infrared radiation must be defined.

ELECTRICAL CHARACTERISTICS

In order to match a detector to the first stage of its amplification system, one must know its resistance and its capacity.

EXTERNAL PARAMETERS

Overall performance of a detector does not only depend upon its inherent characteristics, which have been defined above, but also upon certain external parameters, the most important of which are the following:

1. spectral composition of the source;
2. detector temperature;
3. modulation frequency;
4. detector bias voltage;
5. bandwidth of the measuring instrument.

Main types of high performance infrared detectors

The choice of the infrared detector in an optronic system will have a direct influence on:

1. system sensitivity and therefore performance;
2. field of view/angular resolution;
3. focal length of the lens;
4. maximum diameter of the pupil for a given numerical aperture;
5. complexity of the scanning system where appropriate;
6. complexity of the proximity electronics;
7. complexity of the cryogenic system;
8. cost efficiency of the system.

Table 6.3 · Main types of quantum detectors

Temperature (K)	*Spectral band* 1 to 3 μm	3 to 5 μm	8 to 12 μm
295	1.2 μm:Si 1.8 μm:Ge PbS	PbSe HgCdTe	
200 to 250	PbS	PbSe HgCdTe	
77	PbS	PbS (– >4.2 μm) PbSe PtSi InSb	HgCdTe PbSnTe
<35			GeHg CuHg

Table 6.4 Main types of detector cooling

Temperatures to be achieved (K)	*Possible solutions*
250	Peltier effect 1 to 2 stages
200	Peltier effect 4 stages
90	Joule-Thomson argon expansion Cryogenic machine
77	Joule-Thomson air or nitrogen expansion Liquid nitrogen dewar Cryogenic machine
60	Cryogenic machine
20	Liquid hydrogen dewar
4	Liquid helium dewar

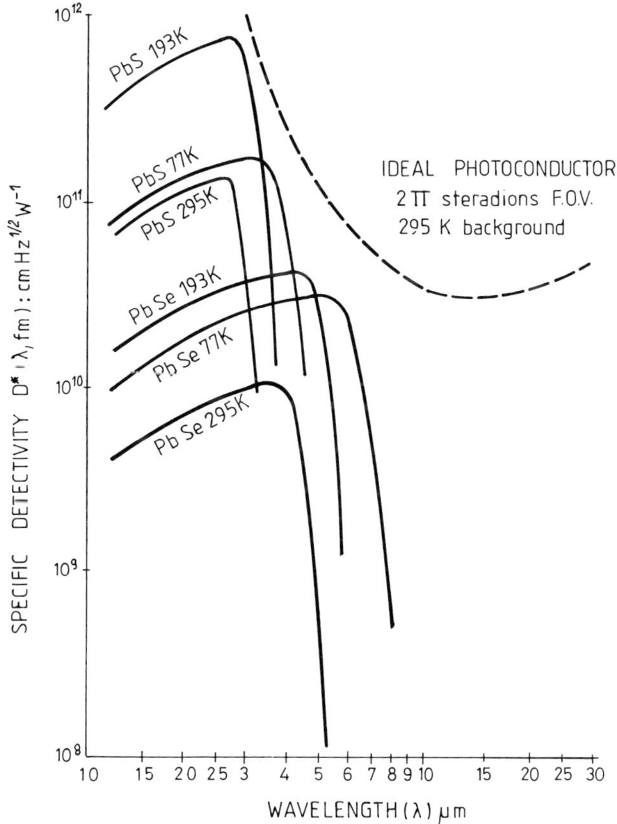

Fig. 6.17 Typical specific spectral detectivity.

The detector is therefore a particularly critical optoelectronic component in the design of an optronic system. There are two major families of detectors: thermal detectors and photonic detectors.

Detection by thermal detectors is based on the absorption of the incoming light flux and measurement of the electric signal generated by the resulting temperature rise. In this category, there are:

1. thermocouples, thermopiles;
2. bolometers;
3. pneumatic detectors;
4. pyro-electric detectors.

These detectors are not very sensitive, slow (response time > 1 ms) and often not cooled. The most efficient detectors are photonic detectors. Based on the principle of the liberation of electrons by interaction between light and matter, they are the most frequently used.

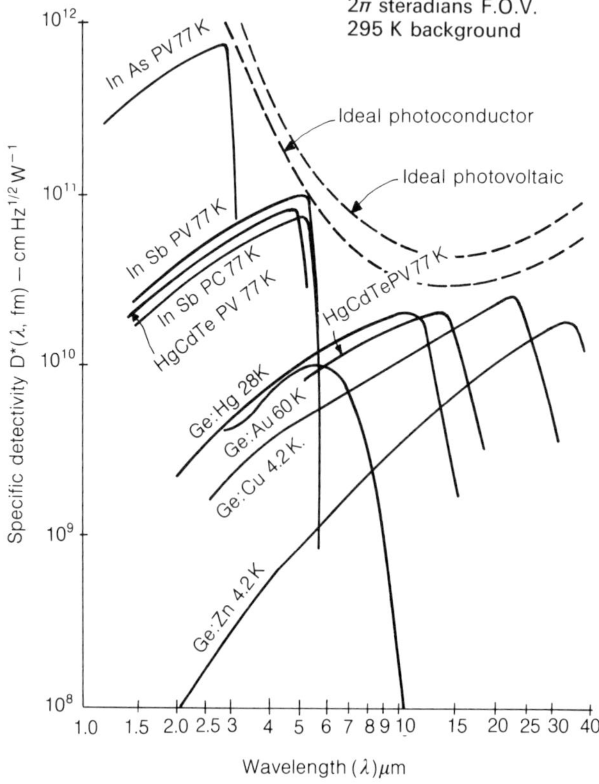

Fig. 6.18 Typical specific spectral detectivity.

The mode of interaction that serves as a basis for detection leads to the classification of photonic (or quantum) detectors into three main categories:

1. photoemissive
2. photoconductive
3. photovoltaic.

The main types of quantum detectors are summarized in Table 6.3. The optimum cooling temperature of the detector is given in this table. Main cooling methods are given in Table 6.4.

Figures 6.17 and 6.18 give the specific spectral detectivity performances of the most commonly used quantum detectors.

6.4 INFRARED TECHNIQUES

6.4.1 Instrument design considerations for passive IR detection

General

Before going into detail on the techniques and technologies involved and although some of these concepts have already been developed in previous chapters, we nonetheless consider it useful to restate a few general considerations.

The concept of signal-to-noise ratio In all systems, useful sources are only operational in the presence of two types of spurious signals: internal noise produced by the equipment (detector, amplification etc.); and signals arising from the background surrounding the object observed.

The strength of these spurious signals compared to the useful signals limits system performance. There are therefore two possibilities.

1. The predominant phenomenon is noise inherent to the system (internal noise), in which case performance is limited by the signal to internal noise ratio. The equipment is considered as operating under optimum conditions.
2. The predominant phenomenon is the signal arising from the background. In this case, system performance is defined by the useful signal-to-background-signal (or noise) ratio.

Flux detector – image detector There are two types of detectors: flux detectors (spatial integration) and image detectors (temporal integration). Many infrared systems have flux detectors, unlike the visible domain where there are a variety of image detectors.

This constraint means we must design detection systems which can distinguish between the various sources seen in the field of view by means of their specific characteristics. This explains the reason for certain spatial filtering techniques using reticles and the use of these reticles to provide indications as to the position of useful sources. We shall also describe a few multiple detector type filtering techniques.

Duty frequency Optimization of the detector and amplification circuit (in particular low frequency noise) often means modulating the signal at a higher frequency than that of the observed source. The techniques most often used in order to optimize the characteristics of infrared systems while making due allowance for these concepts are essentially: spectral filtering, spatial filtering and modulation. In addition, we shall look at the application of modulation to the determination of the angular co-ordinates of a source.

Spectral filtering

The purpose of spectral filtering is to isolate the useful spectral interval in order to:

1. improve contrast by limiting the background signal without attenuating the useful signal;
2. limit spurious light;
3. study a phenomenon which only exists in a spectral band (study of atmospheric gas, pollutants).

Let us take an example in order to explain the principle of this type of reasoning. We assume:

1. a useful source in the form of a black-body at 600 K emerging from a background located at the same distance;
2. a spurious: a background having the spectral distribution of a terrain: at short wavelengths ($< 3\,\mu$m), spectral luminance corresponds to a black-body at 6000 K reflected by the ground.

In order to find the spectral band which gives the best useful signal-to-background-signal ratio between the two bands 3 to 5 μm and 8 to 13 μm we shall compare the apparent radiations of the useful source and of the background in both spectral bands.

The apparent radiation of a source in terms of measurement is the integral (in the spectral band) of the product of the spectral luminance of the source by the spectral transmission for the distance in question. Figure 6.19 shows the various spectral values used in the calculation: relative spectral luminance of the useful source black-body at 600 K, background spectral luminance, and atmospheric transmission for a path of the order of 2 km (20 mm precipitable water).

Table 6.5 gives, for the two spectral bands in question, the luminance of the useful source, the mean transmission of the atmosphere, the apparent luminance of the useful source and these same values for the background. The last column in this table gives the ratio of useful and spurious luminance in the two spectral bands. This table shows the gain achieved (in the order of 20) in signal-to-noise ratio by the use of the band 3 to 5 μm compared to the band 8 to 13 μm, whereas

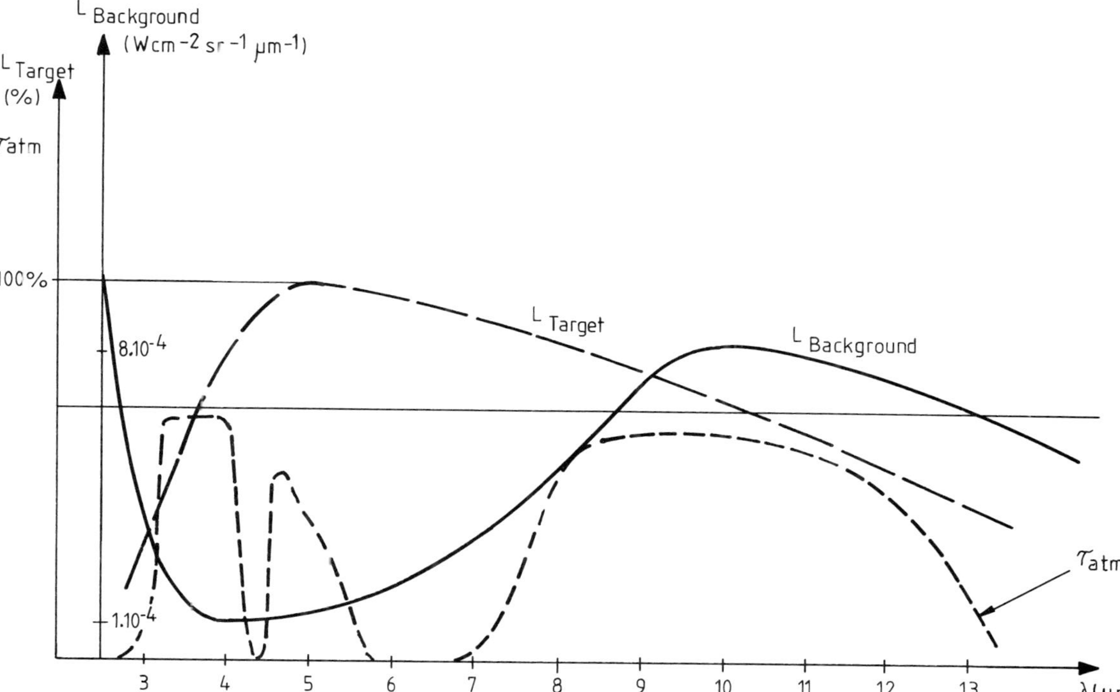

Fig. 6.19 Spectral bandwidth choice.

Table 6.5 Comparison of apparent luminances in the bands 3 to 5 μm and 8 to 13 μm

		3 to 5 μm	8 to 13 μm	ratio $\frac{3 \text{ to } 5\,\mu\text{m}}{8 \text{ to } 13\,\mu\text{m}}$
Useful source	Luminance	0.23	0.24	≃0.96
	Mean transmission	0.40	0.60	0.67
	Apparent luminance	0.092	0.144	0.64
Background	Luminance	2×10^{-4}	40×10^{-4}	0.125
	Mean transmission	0.40	0.60	0.67
	Apparent luminance	0.8×10^{-4}	24×10^{-4}	2.033

if the only useful signal is taken into account these two bands would seem to be equivalent.

This type of argument forms the basis of the choice of spectral bands for an infrared system, but in order to be precise when applied practically, we should take into account all the items which influence the spectral band (the optical system, the detector, etc.).

Spatial filtering

Spatial filtering is based on the following experimental finding:

1. useful sources which are difficult to detect are normally distant and seen under a narrow apparent angle;
2. the background is seen under a wide apparent angle compared to the dimensions of the useful sources and its variations in luminance are slow compared with the dimensions of the useful source.

The aim is to optimize the signal from small sources (adapted to the dimensions of the useful sources) and to reduce, or even to eliminate, the signal emitted by extended sources. Two major types of processes are involved: spatial filtering with a single detector (or a very limited number of detectors), and spatial filtering with multi-detectors.

Spatial filtering with a single 'detector' A diagram of detection with a single detector is given in Fig. 6.20. It essentially comprises:

1. an optical system focusing the scene on a reticle;
2. a reticle and transparent slots in an array, rotating at uniform frequency ν_0 around its axis;
3. an amplification system with a narrow bandwidth centred on the modulation frequency obtained from a small source i.e. $N\nu_0$.

Electrical signals obtained at the output of detector A and at the output of the amplification system B show the gain obtained from a small source (matched

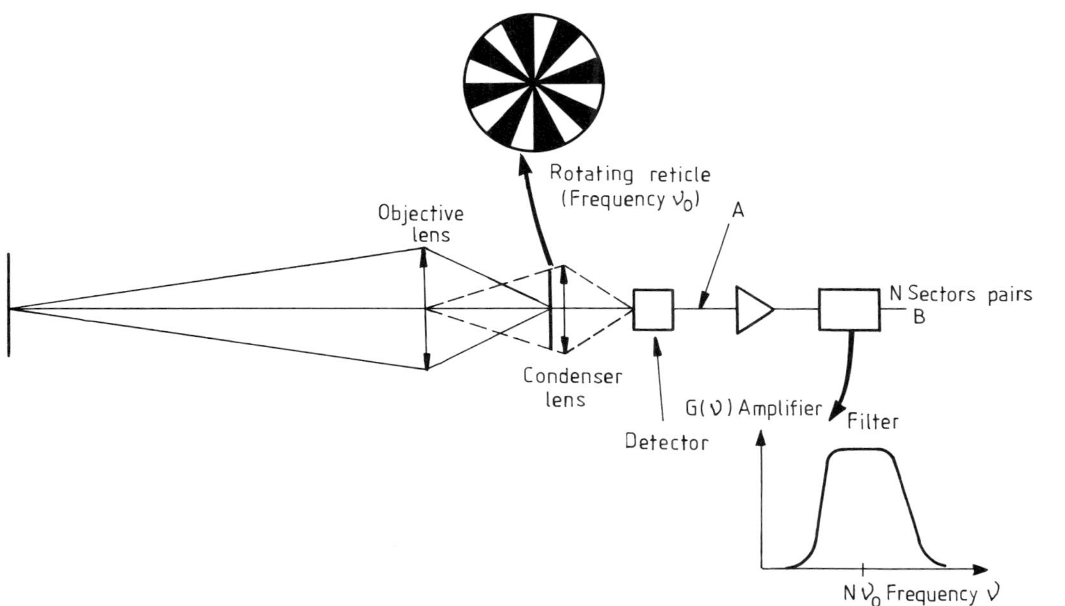

Fig. 6.20 IR detection with rotating reticle.

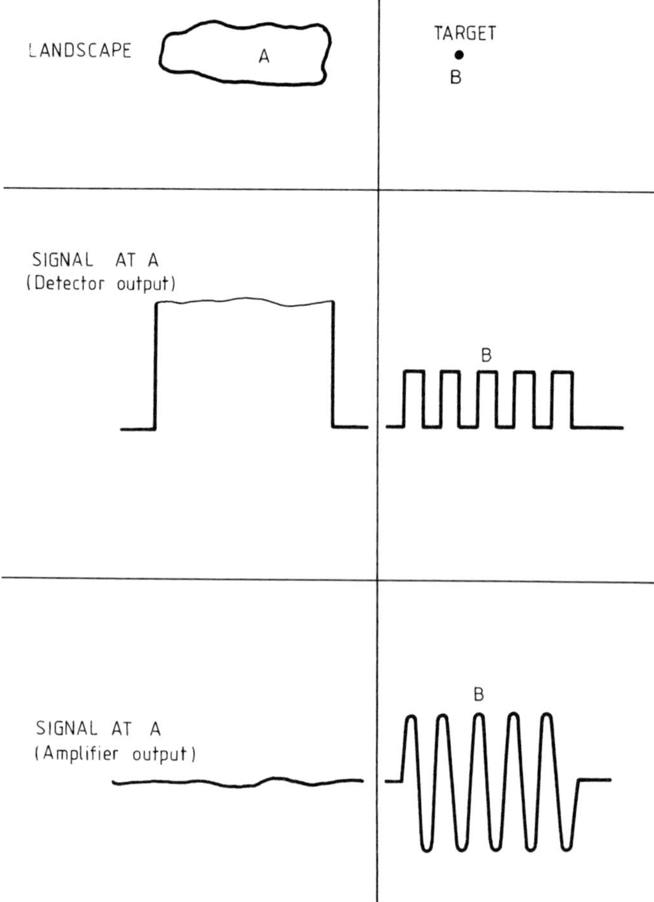

Fig. 6.21 Output signals.

frequency) compared to an extended source in which only the modulation residuals give an amplification output signal (Fig. 6.21). As we can see, the frequency spectra, at the output of the detector, of the background and the source are very different: the background spectrum contains essentially low frequencies and the useful source spectrum is centred on $N\nu_0$. The matching of the frequency band of the filter to the spectrum of the small sources enables us to effect background rejection of several orders of magnitude. The operational limitations of such a device are the insertion of an intense area into the optical field, and the non-uniformity of the background. In order to optimize this technique, we must match the size of the 'slots' in the reticle to the size of the spot produced in the focal plane by the useful sources.

The spatial modulator can also be used to determine the direction of the useful source. In this case, its geometrical configuration must be adapted to its dual func-

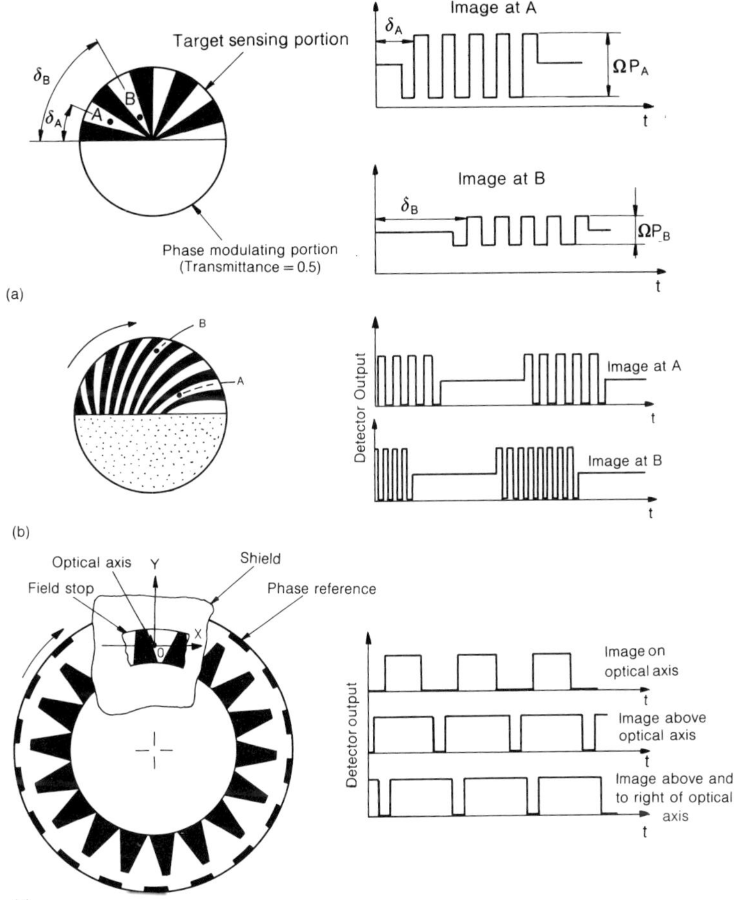

Fig. 6.22 (a) Amplitude modulation; (b) phase and frequency modulation; (c) pulse-width and phase modulation.

tion: spatial filtering and direction measurement. A number of different types of reticles have been invented using: either amplitude modulation, frequency modulation, or pulse modulation.

As an example, Fig. 6.22 shows three types of reticles and the shape of the signal at the output of the detector:

1. amplitude modulation: the amplitude of the signal provides a measurement of the radial distance of the source, whereas the phase gives the polar angle;
2. phase and frequency modulation: the shape of the reticle enables us to measure the polar co-ordinates of the source matched to the creation of the variation in the modulation frequency with radial distance and phase variation polar angle;

3. pulse-width and phase modulation: movement of the source along the axis OY creates a modulation of the signal width whereas movement along OX generates phase variations.

Spatial filtering with multiple detectors Very often multiple detectors are used to perform spatial analysis of the field of view: this process is used in particular in most thermal cameras. This type of analysis also enables spatial filtering by adapting the size of the basic detector to the size of the image of the useful source which we wish to extract from the background and whose position we wish to determine.

Figure 6.23 shows various types of process.

1. Analysis and filtering are carried out by a single small detector which, combined with optomechanical X, Y scanning facilities, covers all points of the field in turn. The crossing instant over the useful source (or target T) determines its position. If we try to achieve a fairly high analysis rate (50 Hz for example), the scanning mechanism becomes complex, the crossing time of the detector over the source gets shorter and therefore the electronic bandwidth increases while equipment sensitivity drops off. In order to avoid these problems, we multiply the number of detectors by n which multiplies sensitivity by $\sqrt{n}$.
2. n detectors are used to carry out parallel scanning in strips. However, we still need two scanning axes.
3. The N detectors cover the whole of the field vertically and one scanning axis is sufficient.
4. Several detectors p are arranged along the X axis of the lines. The p detectors therefore cross the target T which enables us to perform signal integration using TDI (time delay and integration) circuits.
5. $n \times p$ detectors are used to produce combined series parallel system.
6. We have a focal plane array where scanning is carried out electronically as in television techniques.

Analysis and spatial filtering – conclusion It is not possible to say which is the best of the different solutions given above. Each application is a particular case and it is therefore up to the engineer to make his calculations and to carry out simulations and then make the best possible choice on the basis of performance and reliability, size and cost!

6.4.2 Performances of passive infrared optronic systems

Range performance assessment of FLIRs (infrared cameras)

The performances of a camera are evaluated along two criteria: its ability to detect small temperature differences inside a given scene, and its ability to pick out fine details, i.e. high spatial frequencies. Four parameters are commonly used to evaluate a thermal (or infrared) camera:

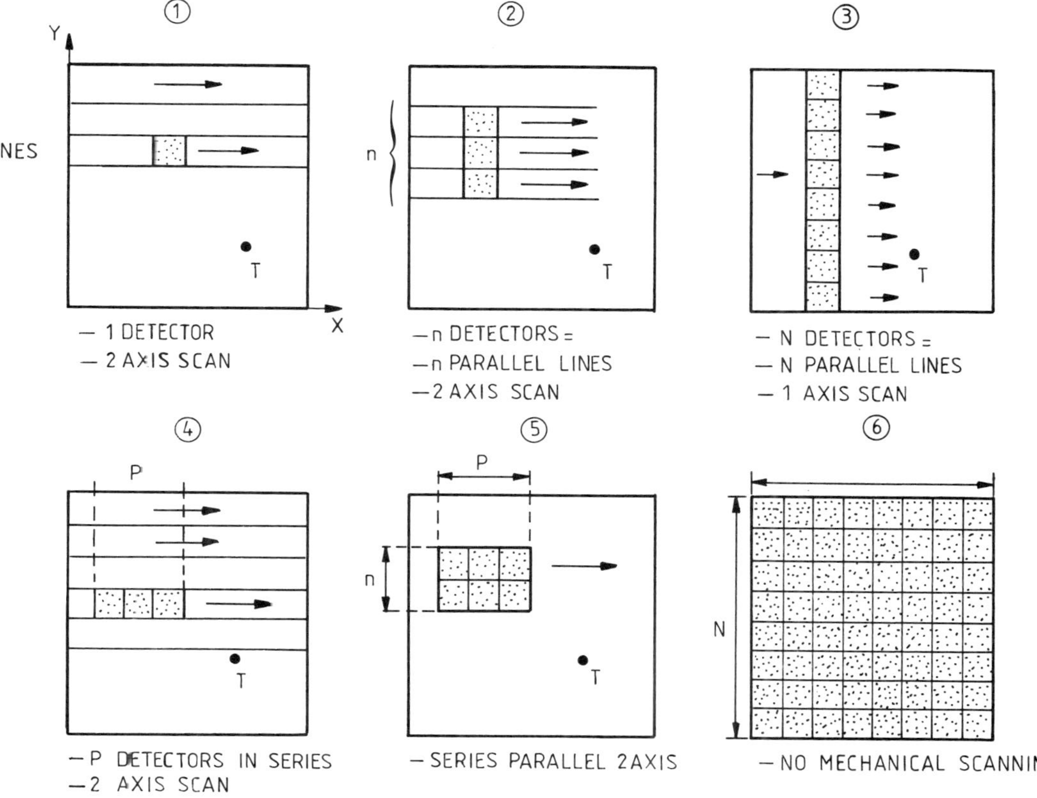

Fig. 6.23 Scanning and spatial filtering with multidetectors.

NETD (noise equivalent temperature difference) This parameter characterizes the ability of a camera to detect small temperature differences. It is defined as the temperature difference for which the output signal-to-noise ratio from the detector is equal to 1. This parameter is defined without taking into account atmospheric absorption (which reduces temperature differences); therefore NETD can only be measured at close range.

NETP (Noise equipment temperature perceptible) The NETP is the equivalent of NETD, but in addition takes into account the characteristics of the human eye (integration time). It therefore represents the temperature difference which can be distinguished by the human eye. It is a lower value than the NETD owing to the averaging of noise by the eye; NETD = 5 NETP at a frequency of 25 Hz.

MTF (modulation transfer function) MTF characterizes the capacity of a camera to distinguish between hot lines and cold lines making up a thermal grid or test pattern at a given temperature difference. Ability falls off as the lines get closer together, i.e. when spatial frequency increases. MTF therefore takes the value 1 for low or null spatial frequencies and then drops to 0 at the so-called cut-off frequency which occurs when the field of view of an individual detector exactly covers a period on the object (i.e. one hot line and one cold line: Fig. 6.24). This transfer function therefore depends on the geometry of the detectors (size, spacing, shape), but also on many other factors: optical, electronic, etc.

MRTD (minimum resolvable temperature difference) (Fig. 6.25) Each of the parameters defined previously quantifies one of the properties of a camera, but

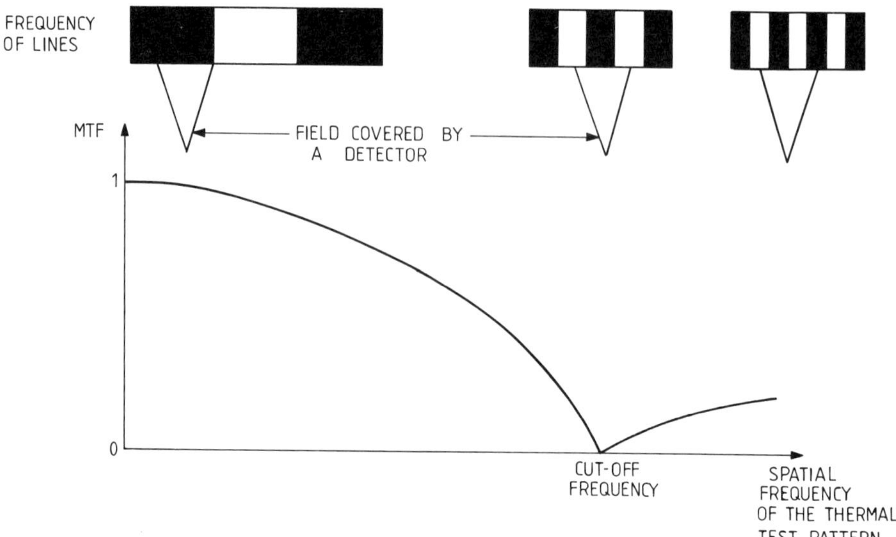

Fig. 6.24 Modulation transfer function (MTF) as a function of test pattern spatial frequency.

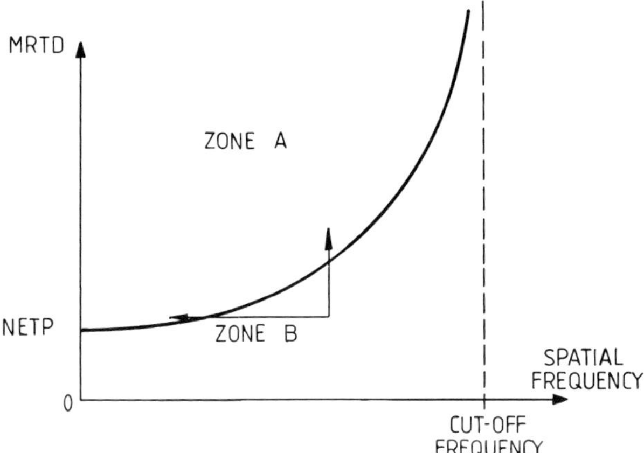

Fig. 6.25 Minimum resoluable temperature difference.

provides no information as to its overall performance. In fact, a thermal camera may well be able to detect very small temperature differences (and therefore have good NETD record) while having a very poor transfer function, and hence be unable to sort out small details. The end results for such a camera will be poor performance during the recognition or identification phase. The role of the minimum resolvable temperature difference is precisely to illustrate the overall performance of a thermal camera on both grounds: thermal resolution and spatial discrimination.

MRTD is proportional to the ratio between NETP and global MTF. The MRTD curve is a graph showing MRTD versus spatial frequency. Two areas, A and B above and below the curve, corresponds to the following working conditions of the camera. In area A, the camera can distinguish a given temperature difference at a given frequency; in area B, the camera cannot distinguish this difference. For it to do so, the temperature difference must be increased or the spatial frequency reduced (or both simultaneously). The curve itself is plotted for an S/N ratio of 2.25, which corresponds to a recognition probability of 50%.

Visual recognition range for an infrared target Most usual recognition criteria on infrared targets suppose that a spatial frequency of 3.5 cycles (7 lines) be placed on the target. Usually a recognition probability of 50% is considered.

The spatial frequency v associated with the observation distance d (at which a square target of side a seen at an apparent angle α) is:

$$v = \frac{2a}{7d}\frac{2\alpha}{7} \quad \text{(cycles per mrd).}$$

If one is given the temperature difference ΔT_0 between the target and its background, as well as the characteristics of the atmospheric path between the scene and the thermal camera, one can plot the graph of the apparent temperature

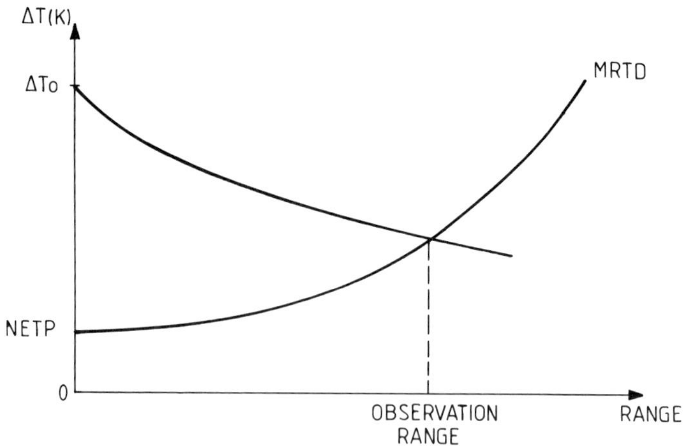

Fig. 6.26 Range performance of a thermal camera for target recognition.

difference as seen from the camera, with respect to the observation range. It we superimpose the MRTD versus range curve onto this graph, the intersection of the two curves will correspond to the range at which the target can be recognized with a probability of 50% (Fig. 6.26).

This basic procedure can be extended to other applications, such as target detection and identification. Different target shapes (rectangular, etc.) can be put into the model, as well as detection probabilities other than 50%.

DETECTION AND IDENTIFICATION

For square shaped targets, the usual detection and identification criteria call for the following number of cycles (hot and cold lines) over the target: 1 cycle (2 lines) for detection and 7 cycles (14 lines) for identification. As far as the MRTD curve is concerned, the conversion of the spatial frequency scale into the range scale must therefore be modified accordingly.

OTHER PROBABILITIES

If a 90% detection probability is asked for, instead of 50%, then one should use a signal-to-noise ratio of 5 instead of 2.25. In order to avoid plotting a new MRTD curve, one may change initial temperature difference between the target and the background by the ratio $2.25/5 = 0.45$, which will give identical results.

Range performance assessment of passive infrared optronic systems for target detection

This section will describe a method generally used to calculate the range performance of optronic systems in which an automatic mode operates once the target has been acquired. These systems do not require any visual interpretation from a human observer since no operator is in the loop. Most of these systems usually compare to the output signal from the detector with an acquisition threshold level defined by the system's false alarm rate and specification target detection probability.

In order to describe the performance of this type of equipment, the following is assumed:

1. the target to be detected as well as the background around it, are at the same range from the system (which corresponds to air-to-surface and surface-to-surface applications);
2. the background immediately around the target is uniform, over an area much larger than the angular resolution of the system;
3. when detected, the target is smaller than the spatial resolution of the optronic system.

Detectors used in passive infrared optronic equipment respond to the flux coming from both the target and the background. The continuous signal from the background is eliminated and the only signal used for detection is the output modulation when the target goes across the instantaneous field of view of the detector.

In passive infrared systems, there is no carrier for the signal and the noise is the difference compared to its mean. Whether accompanied by a signal or not, noise always has a gaussian statistic. The detected signal S then corresponds to modulation on the detected flux $\Delta\Phi$, during the passage of the detector field of view over an area of the background and then over the target. The signal may be positive or negative, depending on whether the target is hotter or colder than the background.

We shall call S the standard deviation of noise, s_o the detection threshold, B the standard deviation of noise from signal, b_o the standard deviation of threshold noise, P_d the detection probability and PFA the false alarm probability. A false alarm occurs whenever gaussian distribution exceeds the agreed threshold in the absence of a signal. The corresponding probability of false alarm is therefore:

$$\text{PFA} = f\left(\frac{s_o}{b_o}\right)$$

with

$$f(x) = \frac{1}{\sqrt{2\pi}} \int_{-\infty}^{x} \exp(t^2/2)\,\mathrm{d}t$$

i.e.

$$\left(\frac{s_o}{b_o}\right) = -g(\text{PFA})$$

where $g(y)$ is the reciprocal function of the function $f(x)$. Table 6.6 shows the reciprocal function values given by Gauss' law.

Table 6.6 Values of the reciprocal function of PFA

(y)	$g(y)$	y	$g(y)$	y	$g(y)$
10^{-12}	−7.034484	2×10^{-3}	−2.878162	0.6	0.253347
10^{-11}	−6.706023	5×10^{-3}	−2.575829	0.7	0.524401
10^{-10}	−6.361341	0.01	−2.326348	0.8	0.841621
10^{-9}	−5.997807	0.02	−2.053749	0.9	1.281552
10^{-8}	−5.612001	0.05	−1.644854	0.95	1.644854
10^{-7}	−5.199338	0.1	−1.281552	0.98	2.053749
10^{-6}	−4.753424	0.2	−0.841621	0.99	2.326348
10^{-5}	−4.264891	0.3	−0.524401	0.995	2.575829
10^{-4}	−3.719016	0.4	−0.253347	0.998	2.878162
10^{-3}	−3.090232	0.5	0	0.999	2.090232

6.4.3 Laser detection techniques

All laser systems comprise two basic sub-assemblies: a laser transmitter and a receiver. In the case of so-called active or semi-active systems, these two sub-assemblies are directed at the same object, (for example, a target) and in point-to-point systems they are directed at each other. In this paragraph, the laser signal to be detected by the receiver is evaluated in the most representative configurations of active and point-to-point systems. Then a comparision is made between the sensitivities of the two detection techniques (direct and heterodyne) most commonly used for the processing of these signals. Finally some typical range performances of such systems are evaluated and commented upon.

Laser signal calculations

Active or semi-active laser systems The overall configuration common to both active and semi-active laser systems is given in Fig. 6.27. In most applications, the transmitter illuminates the landscape, or the target, by means of a narrow laser beam and picks up a laser signal reflected or scattered from the illuminated object, by pointing the receiver unit at the laser point of impact.

Depending on its relative dimensions with respect to the laser illumination spot, the object may be characterized either by the reflectance of its surface at the laser wavelength (more precisely by its bidirectional reflectance), if the object is 'extended', i.e. larger than the laser spot or by its laser cross-section (which takes into account both its reflectance and its surface area) if the illuminated object is small compared to the laser spot. The first case is to be found usually during air-to-surface airborne missions where the laser beam is directed onto large targets on the ground (buildings, dams, bridges, etc.), whereas the second configuration is found in air-to-air or surface-to-air long-range or space missions.

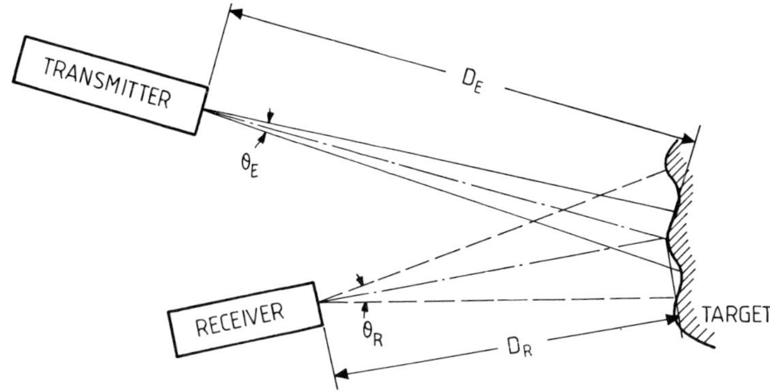

Fig. 6.27 Overall configuration of active and semi-active laser systems.

In the first case, the laser power picked up by the receiver is given by one of the two formulae below (corresponding to the bistatic and monostatic configurations respectively):

$$P_R = \frac{\rho}{D_R^2} P_L \cdot T_E \cdot T_R \cdot S_{op} \cdot T_{atm}(D_E) \cdot T_{atm}(D_R)$$

$$P_R = \frac{\rho}{D_R^2} P_L \cdot T_E \cdot T_R \cdot S_{op} \cdot T_{atm}(2D)$$

where P_R is the laser power picked up by the receiver, ρ is the directional reflectance factor of the target at the laser wavelength, T_E, T_R are optical transmissions of the transmitter and receiver sub-assemblies, D_E, D_R are propagation distances of the laser beam on its way to and from the target ($D_E = D_R = D$ in the case of monostatic systems), P_L is the laser output power, and S_{op} is the area of the receiver collecting optics.

In the above it is assumed that the transmitter and receiver fields of view are matched with each other and in particular that the receiver field does not introduce any geometrical diaphragm reduction of the laser beam backscattered from the target.

If the size of the laser beam at the target is larger than the target itself, the above formulae assume a form similar to that of the radar equation. For example, in the case of monostatic laser systems, the laser signal is given by the following equation:

$$P_R = \frac{\mathrm{LCS} \cdot P_L \cdot T_E T_R \cdot S_{op} \cdot T_{atm}(2D)}{\pi \theta E^2 \cdot D^4}$$

where LCS is the laser cross-section of the target and E is the total divergence of the transmitted laser beam.

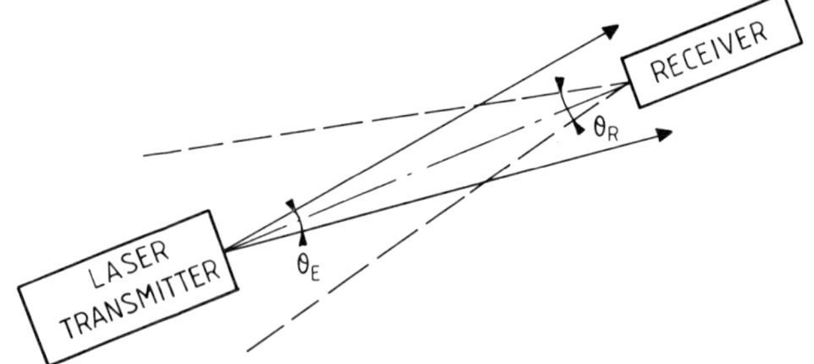

Fig. 6.28 Configuration of a point-to-point laser system.

Point-to-point laser systems (Fig. 6.28) In a point-to-point laser system, the receiver unit, located at a distance D from the transmitter, is aligned to it. The laser signal detected is then given by the following general formula:

$$P_R = \frac{P_L \cdot T_E \cdot T_R \cdot S_{op} \cdot T_{atm}(D)}{\pi \theta E^2}.$$

Detection techniques

General considerations Generally speaking, a laser signal is detected by means of one of the two following basic techniques:

1. incoherent, or direct, detection;
2. coherent, or heterodyne, detection.

The choice of one or the other is obviously dictated by the technical specifications of the equipment, such as the nature of the information sought (target direction, range or radial velocity), data rates, sensitivity (maximum range), available laser sources, etc., but also by other criteria such as price, complexity, etc.

The architecture of the sensor in a direct detection laser system is very similar to that of a passive infrared sensor since the laser flux picked up by the receiver optics is focused onto a quadratic detector. Main components of this type of sensor are described in section 5.4.1 together with their basic characteristics.

The architecture of a heterodyne detection laser system, on the other hand, is similar to that of a coherent radar since the laser signal to be detected is fixed upon the detector area with a reference laser wave or local oscillator.

As is the case with the other infrared systems, various types of photodetectors may be considered for laser systems: although thermal detectors are normally much too slow for operational laser systems, most photonic detectors are worthy of consideration, in particular photoconductors, and photodiodes and avalanche photodiodes. As with radar or passive infrared systems, performance of laser systems is evaluated by computing the signal-to-noise ratio at the output of the receiver, and the following paragraphs describe some of the main results achieved by one or the other of these two techniques (direct and heterodyne detection).

Sensitivity of incoherent laser systems (direct detection) In a direct detection laser system, the electric output from the photodetector is proportional to the laser flux P_R picked up by the receiver optics (detector). The signal current i_s from the detector can then be written:

$$i_s = R_i P_R = \frac{\eta e}{h\nu} P_R$$

in which η and R_i are respectively the photodetector quantum efficiency (number of photoelectrons/incident photon) and current response (in A/W), ν the laser frequency, and e the electron charge.

System sensitivity is defined as the minimum laser flux detectable by the receiver. As in the case of passive infrared systems, quantities such as noise equivalent power (NEP) or noise equivalent illumination (NEI) are used, which define laser power or illumination levels for which the electrical output from the photodetector is equal to the rms noise of the processing system (at input of the preamplifier). Various sources of noise at the output of the detector, may limit system sensitivity, among which one can quote: thermal noise from the load resistor R_L, shot noise due to dark current, and shot noise due to background illumination.

If $\overline{I^2_{N,T}} \overline{i^2_{N,O}}$ and $\overline{i^2_{N,B}}$ are the the respective variances of these noise currents, the detector output noise power can be expressed as follows:

$$P_N = R_L[\overline{i^2_{N,T}} + \overline{i^2_{N,O}} + \overline{i^2_{N,F}}]$$

By substituting the traditional expressions of these noise currents into the above equation and making allowance for the preamplifier noise factor and internal gain G of the detector (for detectors such as avalanche photodiodes, or photomultipliers), one finds that the noise equivalent power of a direct detection IR (NEP_{DIR}) system obeys the following law:

$$NEP_{DIR} = \frac{h\nu}{\eta e}\left[\frac{4kT_o(F - 1 + t_E)}{R_L G} + 2e(i_o + i_B)\right]\Delta f$$

where Δf is the video bandwidth of the signal processing system, k is Boltzmann's constant, T_o is the reference temperature, F is the pre-amp noise figure, t_E is the normalized equivalent noise temperature of the detector load resistance, i_o is the detector dark current, i_B is the average current generated by background radiation, G is the pre-amp gain and R_L is the load resistor.

Sensitivity of coherent laser systems (heterodyne detection) Certain lasers are sufficiently stable in frequency to be used in coherent laser systems or heterodyne/homodyne detection lidars similar to RF radars. Figure 6.29 shows a

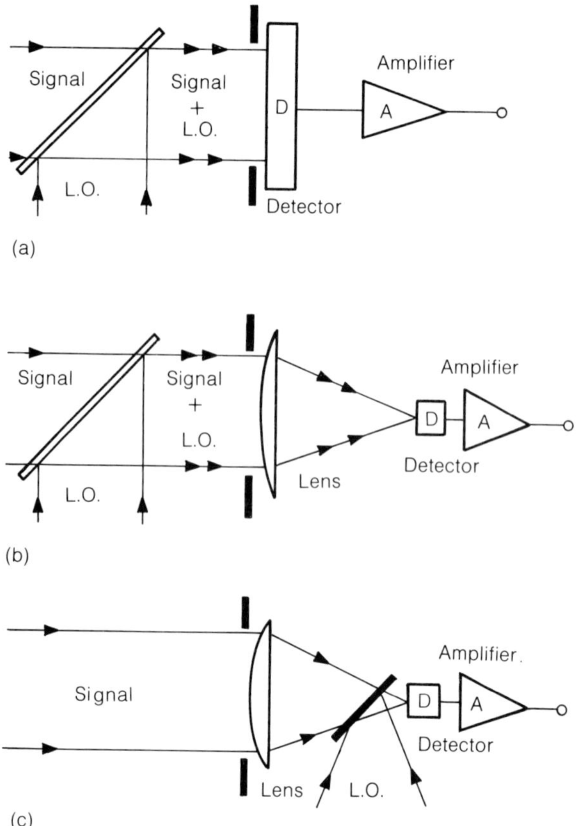

Fig. 6.29 Various configurations of heterodyne laser receivers: (a) without collector lens; (b) and (c) with collector lens.

typical configuration for a heterodyne laser receiver. The laser flux to be detected is focused onto a photodetector along with a much more powerful reference beam coming from a laser source that belongs to the receiver. The photodetector output current in response to this reference (or local oscillator) laser beam sets a noise level (local shot noise) that dwarfs all other sources of noise found in direct detection systems.

If we assume that the two laser beams (signal and local oscillator), at frequencies ν_s and ν_{LO}, are spatially coherent across the whole surface of the quadratic photodetector, the output current of the quadratic detector is proportional to the square of the electric field ε_T resulting from the mixing of these two waves:

$$I_s(t) \sim |\varepsilon_T(t)|^2$$

with

$$\varepsilon_T(t) = \varepsilon_{LO} \exp(-j\omega_{LO}t) + \varepsilon_s \exp[-(j\omega_{st} + \varphi_s)]$$

with $\omega_{LO} = 2\pi \times \nu_{LO}$, $\omega_s = 2\pi\nu_s$ and φ_s is the relative phase shift between signal and local oscillator wavefronts.

This photoelectric current can then be written

$$I_s(t) = \frac{\eta e}{h\nu}[F_{LO} + F_s + 2\sqrt{F_S F_{LO}} \cos[(\omega_{LO} - \omega_s)t + \varphi_s]$$

or since $F_{LO} \gg F_S$:

$$I_s(t) \cong i_{LO} + 2\sqrt{i_{LO} i_S} \cos[(\omega_{LO} - \omega_s)t + \varphi_s].$$

Its first component i_{LO}, is constant to a first approximation, and proportional to the local oscillator power incident upon the detector. The second component, called the heterodyne signal, oscillates at the intermediate frequency, $\nu_{IF} = \nu_{LO} - \nu_S$. This heterodyne signal is the only useful output about the target from the detector. It can be detected with large bandwidth photodetectors (bandwidths of photovoltaic diodes can research up to several gigahertz).

Since the noise equivalent power of the heterodyne detector is set by the shot noise due to the local oscillator current, we can write:

$$\begin{aligned} \overline{I_N^2} &= 2ei_{LO}\Delta f \\ &= 2\eta \frac{e^2}{h\nu_{LO}} F_{LO}\Delta f \end{aligned}$$

which gives the following value for the laser noise power of a heterodyne laser receiver:

$$\text{NEP}_{\text{HET}} = \frac{h\nu}{\eta} L\Delta f.$$

Thus, the theoretical limit of noise spectral density of a coherent laser system is the photon energy at the laser wavelength ($\eta = 1$, $\Delta f - 1$ Hz).

6.4.4 IR laser system performance

Laser system performances are normally assumed on the basis of some or all of the following criteria:

1. maximum range, i.e. the distance beyond which this system no longer fulfils its function:
2. accuracy, particularly on range and radial velocity measurements;
3. angular resolution, field of regard or total illumination field;
4. guidance precision;
5. data rate, etc.

As laser system architectures are very close to those of radar systems, the performance assessment procedures are very similar; the points specific to laser systems, as compared to conventional radar systems, are due to differences in wavelength arising essentially from:

1. atmospheric properties (transmission, scattering, absorption);
2. target reflectance or backscatter;
3. the narrow width of the laser beam, which can either cover the whole target (which is normally the case with radar), or illuminate it locally and analyse it point by point.

The paragraphs below illustrate the assessment procedures for range-finding laser systems by means of range equations with direct and heterodyne detection. Accuracy on range and Doppler velocity will also be evaluated.

Simplified range equations for a laser system

Range equations for a laser system on a given target are based on the evaluation of the signal-to-noise radio delivered by the system when it illuminates the target. For a set of parameters (laser power, atmospheric conditions, target reflection characteristics, size and transmission of the optical system), this signal-to-noise ratio depends on the range of the target. Under these conditions, maximum range of the system is the target range for which the signal-to-noise ratio reaches a limit value (or threshold) imposed by the operational conditions (normally summarized as a target detection probability P_d and a false alarm rate FAR). The values of this threshold as a function of detection probability and false alarm rate have been tabulated in a very detailed fashion: basically they depend on the fluctuation of the signal to be detected and the detection mode (direct or heterodyne).

Useful data for range calculations can be found on charts showing the detection probability of a target as a function of signal-to-noise ratio for a given false alarm rate. Figures summarize some of the main results. The two charts corresponding to heterodyne detection are the same as those used in the field of radar: signal-to-noise ratios are ratios between electric signal and noise powers. In case of direct detection, signal-to-noise ratios are the ratios between signal and noise voltage or current (video signal-to-noise ratios).

Maximum range of direct detection laser systems

Some of the main noise sources in a direct detection active system are essentially detector dark current, background generated current and electronic noise (particularly pre-amp and proximity electronic noise).

Background noise is generated by illumination of the detector by natural sources other than the laser signal. It may be caused by the reflection of sunlight on the earth surface or luminance of the day-time sky if the laser wavelength is in the visibile or near infrared domains. This illumination is due to thermal emission of the landscape if the laser emission is further in the infrared (for example, in the 8 to 12 μm atmospheric window).

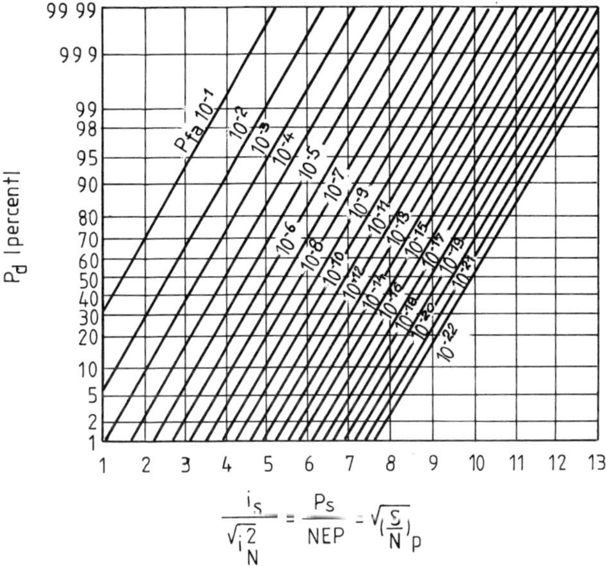

Fig. 6.30 Direct detection (non-fluctuating target).

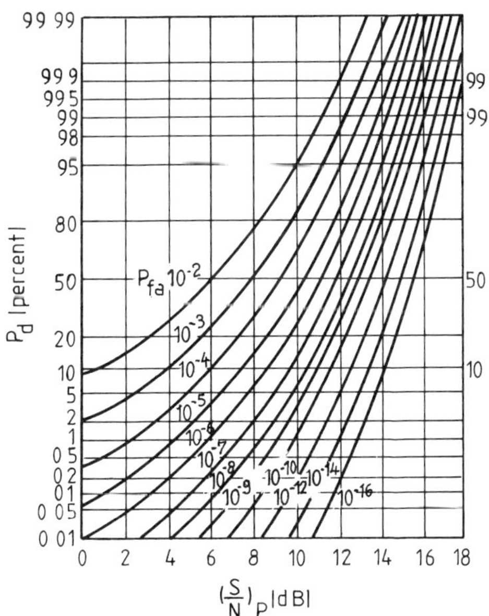

Fig. 6.31 Heterodyne detection (non-fluctuating target).

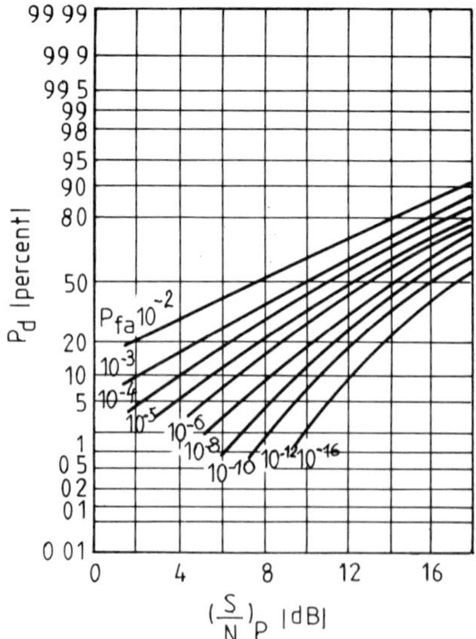

Fig. 6.32 Heterodyne detection (fluctuating target).

The general signal-to-noise ratio (voltage) equation for a direct detection laser range finder can thus be written:

$$[S/N]_v = \left[\frac{RF_s^2 R_L G^2}{2e\Delta f(RF_B + i_o)R_L G^2 + 2FkT\Delta f}\right]^{1/2}$$

where R is the detector response (in A/W), R_L is the load resistance (Ω), G is the internal gain of the photodetector, F_B is the spurious background flux incident on the detector, i_o is the detector dark current (A), F is the preamplifier noise factor and F_s is the return laser signal flux.

The optimization of this signal-to-noise ratio and therefore of the laser system maximum range is conditioned by the following basic points:

1. the choice of laser wavelength is dictated by the optical properties of the atmosphere, the available laser power and the operational conditions of the equipment (climate, ocular hazards, etc.);
2. the divergence of the beam should be matched to the dimensions of the target and the aiming precision of the system;
3. spurious background flux should be reduced to a minimum by narrow spectral filtering centred on the laser emission line.

Maximum range of coherent laser systems

Coherent (or heterodyne detection) laser systems are still at a development stage, and for the moment there exist only mock-ups or prototypes based upon the use of CO_2 lasers. However, heterodyne detection in the near infrared is spreading thanks in particular to research work being carried out on diode pumped solid-state lasers.

The general procedure for performance evaluation of coherent laser systems is similar to that described above: in particular, computation of the laser back-scattered by the target onto the receiver optics is identical in both cases (direct or heterodyne detection). A rough estimate of the maximum range for a coherent laser system can be done by computing the signal-to-noise ratio at the detector output (quantum efficiency η),

$$[S/N]_{HET} = \frac{\eta F_R}{h\nu \Delta f}.$$

Comparison of the theoretical sensitivities of the two types of detection (direct versus heterodyne) shows heterodyne detection is more effective particularly at longer wavelengths since:

$$NEP_{HET}/NEP_{DIR} = \frac{h\nu D^*}{\eta\sqrt{AD}}\sqrt{\Delta f}.$$

For an infrared detector operating at $\lambda = 10.6\,\mu$m, the ratio between the theoretical sensitivities of these two modes of detection, direct and heterodyne, is typically of the order of 10^2 to 10^3 (for a bandwidth of $\Delta f = 10$ MHz). However the theoretically enormous advantage of heterodyne over direct detection is quite often partially upset in most experimental conditions by two main factors.

First, it must be taken into account that thresholds must be set higher in heterodyne than in direct detection (for identical P_d and FAR), because targets are usually more fluctuating under coherent illumination (speckle effect): for example, if the target is an extended diffuse object, only its axial area will backscatter laser waves matched to the reference wave (local oscillator). Other areas of the object, even those inside the laser illuminated spot will backscatter sets of randomly phased wavefronts towards the receiver optics, owing to the object roughness. When it reaches the receiver optics, the return laser echo is composed of laser spots or grains of similar dimensions but with random phases (speckle figure, Fig. 6.33), and wild irradiance fluctuations.

Secondly, if a direct detection laser receiver can be considered as a photon collector (i.e. it responds to the whole incident flux F_R), a heterodyne receiver is an interferometer in which the return laser wavefront quality determines which fraction of the laser signal the receiver will respond to. The wavefront quality of the return laser signal is generally expressed by a quantity known as its degree of coherence (mathematically given by the Zernicke Van Cittert's theorem). It indicates the heterodyne efficiency (i.e. the percentage of the incoming laser power that gives rise to heterodyne signal) of a coherent laser system operating on a given target.

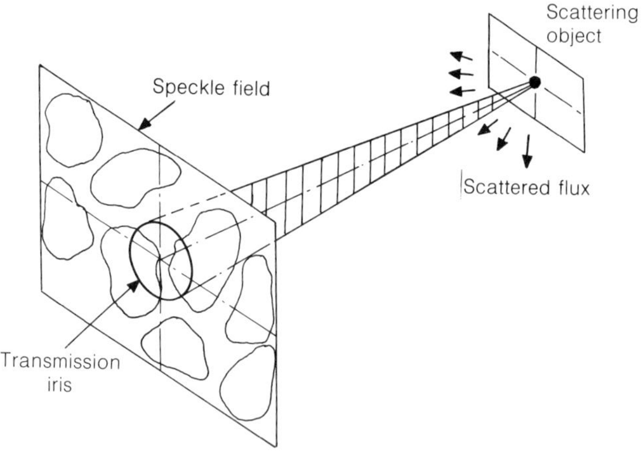

Fig. 6.33 Speckle phenomenon at the entrance pupil of a coherent laser receiver.

In the very usual case of monostatic laser systems with common transceiver optics, application of Zernike Van Cittert's theorem shown that this efficiency on an extended diffuse target is of the order of:

$$\eta_{\text{SPECKLE}} \cong 0.5.$$

Atmospheric turbulence can play a very important role in the experimental performances of optronic systems (active or passive) and in certain cases, its presence can make equipment that would be otherwise highly performing, practically useless. It arises essentially from inhomogeneity in the refractive index of the propagation medium, created by local variations in air temperature. These increase as ambient temperature increases (sunlight areas in summer) and as the aiming axis of the system gets closer to the ground.

Laser systems accuracy

If maximum range is one of the basic performance criteria of any laser system, so is the accuracy of its measurements, particularly on target range, radial velocity and angular coordinates. This section summarizes the theoretical limits of laser systems in these three areas and gives some typical experimental results.

Theoretical accuracies of laser systems

CASE OF HETERODYNE DETECTION

Theoretical accuracy of coherent laser systems is deduced by applying results of modern radar and matched filtering theories into the optical domain. As in the case of radar, a laser system essentially determines target range D and radial velocity V_R by means of two types of measurements:

1. on the one hand, measurement of the flight time of the beam, since $t = 2D/c$;
2. on the other, measurement of the heterodyne signal frequency due to the laser echo, since Doppler frequency shift, f_{Doppler}, due to target radial motion, is equal to:

$$f_{\text{Doppler}} = \frac{2V_{\text{R}}}{\lambda}.$$

As a result, accuracy on target range and radial velocity is limited by the measurement errors on the time of arrival and frequency of the detected laser signal.

In order to simplify the problem we assume that the target has the following characteristics: it is angularly small with respect to the transverse size of the beam, shallow compared to the range accuracy sought, and it moves in a pure translation at a constant radial velocity V_{R} throughout the measurement.

Toward this target, the laser system transmits a single mode wave, which is frequency modulated over a domain ΔF throughout a duration T. It detects the return signal by means of a diffraction limited optical system of diameter ϕ_{op}, and a heterodyne receiver followed by a matched filter (whose transfer function is the conjugate of the modulation law of the transmitted laser wavefront), in order to optimize the signal-to-noise ratio at the end of the chain. The matched filter output for each laser echo from the target is a pulse of duration ('compressed pulse') equal to:

$$\tau \approx 1/\Delta F.$$

Signal theory shows that the chronometry accuracy σ_t on this so-called compressed pulse (its duration τ may be a lot shorter than that of the target illumination by the laser, T) depends upon pulse duration and signal-to-noise ratio:

$$\sigma_t \geqslant \frac{\tau}{(2\text{S/N})^{1/2}} = \frac{1}{\Delta F(2\text{S/N})^{1/2}}.$$

In the same way, the uncertainty σ_{f} of the heterodyne signal frequency is minimized by increasing both the duration T of the measurement (or laser illumination) and the signal-to-noise ratio:

$$\sigma_{\text{f}} \geqslant \frac{1}{T(2\text{S/N})^{1/2}}.$$

The results that theoretical accuracies of coherent laser systems on target range and radial velocity are limited to the following values (Woodward's formulae):

$$\sigma_{\text{R}} \geqslant \frac{c}{2\Delta F(2\text{S/N})^{1/2}} \qquad \text{(Range)}$$

$$\sigma_{\text{D}} \geqslant \frac{\lambda}{2T(2\text{S/N})^{1/2}} \qquad \text{(Doppler).}$$

Similarly, angular accuracy of laser systems is limited by diffraction, to the following value σ_θ:

$$\sigma_\theta \geqslant \frac{\lambda}{\phi_{op} \times (2S/N)^{1/2}}.$$

For a given signal-to-noise ratio (which depends only on laser energy), the theoretical performance of a coherent laser system depends only on:

1. the laser frequency modulation domain (range accuracy);
2. the duration of each measurement (Doppler velocity accuracy);
3. the diameter of the receiver antenna (angular accuracy).

In order to match the specifications on range and Doppler accuracies, the designer of a coherent laser system must optimize the transmitted laser waveform by defining T (duration of illumination) and ΔF (spread of the spectrum) independently from each other.

The following formula:

$$\sigma_D \sigma_{VR} \geqslant \frac{c\lambda}{8T\Delta F(S/N)}$$

derived from the above accuracy formulae for a range and radial velocity shows that the global performance of a given system (at a fixed wavelength λ) over both measurements improves as the product $T\Delta F$, or compression ratio increases.

If no frequency modulation is applied onto the laser beam during emission, the duration of the measurement T and the laser wave spectrum ΔF are inversely proportional to one another, which results in a compression ratio close to unity. This solution leads to an optimization either on range accuracy (using short laser pulses) at the expense of radial speed or vice versa (quasi-continuous laser emission with spectral analysis of the return signal and no account taken of range).

CASE OF DIRECT DETECTION

The limiting accuracy on range measurements by direct detection laser systems is identical to that obtained by heterodyne systems. The bandwidth Δf to be used in its formulation is that of the filter matched to the width T of the laser pulse itself ($\Delta f \sim 1/T$). It therefore follows that in order to optimize range performance and range accuracy of an incoherent laser system in which the mean power (or energy per pulse) of the laser is dictated, one must choose the laser configuration which transmits the shortest pulses.

Insofar as angular measurements are concerned the limit performances of laser systems are theoretically equivalent for both direct and heterodyne detection, and can be evaluated as a fraction of the diffraction limit (λ/ϕ_{op}). It should however be noted that while the angular field of view of a heterodyne system is diffraction limited (antenna theorem) to about λ/ϕ_{op}, that of a direct detection system can be much larger. In this case, the angular accuracy of an incoherent laser system (which is a certain percentage of the field of view) is degraded with respect to the diffraction limit dictated by the diameter of the optics.

Experimental range and Doppler accuracies Experimental accuracies of laser systems may depart more or less from the theoretical values given above, depending on operational conditions. Very often, actual performance limitations are not set by the system itself, but by outside elements such as mechanical environment, carrier vibrations, atmospheric turbulence or target characteristics. Along with atmospheric turbulence, mechanical vibrations and deformations constitute some of the basic limitations of target angular acquisition and tracking by laser systems. They also induce laser frequency variations or fluctuations which are detrimental to Doppler measurement performance of heterodyne systems.

The target itself must also be taken into account when evaluating range, Doppler or angular performance of a laser system: thus the depth of the target likely to be illuminated by the laser beam can be the limiting factor in the range accuracy of the system. Similarly, the angular size of the target and the geometrical distribution of its 'bright spots' (highly reflecting areas) inside the field of view may also have a bearing on the specifications of the system both on angular resolution and accuracy. Finally, the motion of the target itself (rotation, acceleration) and its geometrical deformation over time are the basis for the definition of a 'target coherence time' which imposes an experimental limit on Doppler resolution and accuracy of coherent laser systems.

TYPICAL ACCURACIES OF PULSED LASER SYSTEMS

Table 6.7 summarizes typical accuracies of laser systems, either on range measurement only (direct direction) or range and Doppler velocity (heterodyne detection).

TYPICAL ACCURACIES OF CONTINUOUS WAVE LASER SYSTEMS

Continuous wave (CW) laser transmitters are of interest if one needs to optimize simultaneously both Doppler and range performance of active systems (heterodyne) by comparison with the set of results given in Table 6.7. CW lasers

Table 6.7 Typical accuracies of pulsed laser systems

			Accuracy	
Laser source	*Trigger mode*	*Pulse width*	*Range* (m)	*Doppler* (m/s)
Nd:YAG	Q-switch	15/20 ns	1	20
CO_2	Q-switch	100 ns	5	30
CO_2	Cavity dump	20 ns	1	150
CO_2	TEA	100 ns/1 μs	5	3

can also be used in the design of point-to-point laser systems using direct or heterodyne detection (for example in communications or for missile guidance by means of beam riding techniques).

RANGE MEASUREMENTS

Laser (Doppler) rangefinding by means of laser frequency modulation and heterodyne signal demodulation is based on the same principles and techniques (pulse compression, FMCW) as those used in radars. However, the methods used for laser frequency modulation are specific enough to be mentioned here. From their basic characteristics one can then deduce the performance to be expected from continuous wave laser systems. A distinction is usually made between types of modulators, depending on whether they are situated inside or outside the laser cavity. Table 6.8 below summarizes the characteristics of the main frequency modulation techniques (for CO_2 lasers), and the corresponding range accuracy.

DOPPLER VELOCITY MEASUREMENTS

In case of the CO_2 laser (which is the most frequently used laser for Doppler velocity measurements), it can be said that frequency stability depends essentially on the mechanical environment. Ground installed CO_2 lasers for trajectory calculation systems show very low frequency fluctuations (< 1 kHz) giving Doppler accuracy figures better than 1 cm/s. In airbone applications, short term laser frequency fluctuations (during the flight time of the laser pulse) can reach 50 kHz, which leads to Doppler accuracy of 20 cm/s on each measurement.

Among the components already designed and developed for frequency modulation and continuous wave radars, those which respond best to the overall requirements of laser systems (for both measurement accuracy and range) are the surface acoustic wave devices (SAWD).

Experimental angular precision of laser systems Although the directivity of the laser beam forms the basis of the angular precision of laser systems and is used for example for targetting or weapon guidance (active or semi-active systems) or for alignment between optronic systems (point-to-point transmissions), another factor essential for achieving good precision is the angular measurement technique used. We describe below one of the methods most commonly used in operational

Table 6.8 Range accuracies for different frequency modulation techniques

Technique		*Response time*	*Frequency domain*	*Range accuracy* (m)
Extra cavity	A/O	0.5 μs	<50 MHz	>2
	E/O	1 ns	1 GHz	0.1
Intra cavity	E/O	10 ns	>100 MHz	1
	piezo	100 μs	<100 MHz	1

laser angular deviation measurement devices to illustrate the angular precision capabilities of these systems.

GENERAL OPERATION

The most common laser receiver used for angular measurements comprises a collecting optical system and a multiple element laser detector. The whole assembly is then mounted in such a way that its response to an incident laser illumination is characteristic of the direction of incidence of the beam with the desired precision.

In its simplest configuration, the detector comprises 4 separate elements ('4 quadrant detector') which break the field of analysis down into 4 areas. These quadrants supply 4 signals S_1, S_2, S_3, S_4 used to measure the direction of the incident laser flux. The relative distribution of these signals produces 2 operating modes: linear angular deviation measurement in the central area of the field, and saturation in the peripheral area (Fig. 6.34).

Linear angular deviation measurement: in the case of a laser illumination near the optical axis of the device, the laser spot formed by the collecting lens illuminates a part of each quadrant. The weighted outputs from respectively the left/right and up/down quadrants are then closely related to the azimuth and elevation of the incoming laser beam. The angular deviation signals from the device, s_{az} (azimuth) and S_{el} (elevation) are then defined as follows:

$$S_{az} = [(S_1 + S_4) - (S_2 + S_3)]/\Sigma$$
$$S_{el} = [(S_1 + S_2) - (S_3 + S_4)]/\Sigma$$

in which $\Sigma = S_1 + S_2 + S_3 + S_4$ (sum signal).

The angular deviation measurement device is optimized by designing the optical system (collecting lens, choice of focusing plane, distribution of laser irradiance

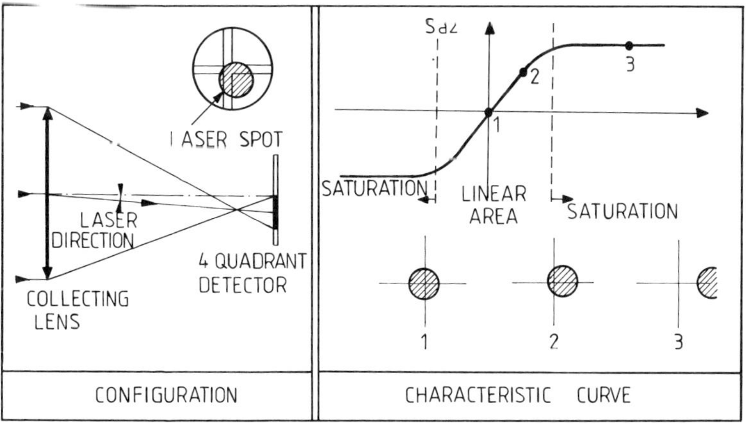

Fig. 6.34 Configuration and characteristic curve of a 4 quadrant laser angular deviation measurement device.

in the detector plane) in such a way that the angular deviation signals S_{az} and S_{el} are proportional to the corresponding angular coordinates of the incoming laser beam.

Saturation: when the laser source is farther away from the optical axis of the device, the laser spot illuminates at most 2 of the 4 quadrants, and hence there appears a saturation of the angular deviation signals (see Fig. 6.34). The angular deviation measurement device can no longer measure accurately the angle of incidence of the laser beam, but it approximately indicates where it comes from.

EXPERIMENTAL RESULTS

In heterodyne detection laser angular deviation measurement devices, the incoming laser beam is focused down onto the detector by means of a diffraction limited collecting optics. The linear field of view is therefore of the order of (λ/ϕ_{op}). This type of angular deviation measuring device is used for hyperfine tracking or angular stabilization and requires prior precise acquisition of the target. In direct detection, the total field of a laser angular deviation measurement device is between 10 and 30° with a linear field of a few degrees.

Angular distance measurement devices exist which operate in the saturation mode only, the laser beam being focused on the 4 quadrant detector. Only one quadrant is illuminated for any direction of the laser beam which explains the term 'on/off' or 'bang/bang' angular deviation measurement. This mode is less precise than the previous one.

6.5 MILITARY APPLICATIONS OF INFRARED

For the purpose of describing military applications, it is useful to locate the question within the global framework of a weapon system and examine the possible benefits which can be supplied by infrared techniques.

The design of a modern weapon system, whether land based, surface vessel or submarine carried, aircraft or helicopter borne or even satellite carried, involves the consideration of a number of essential functions which are dealt with in the following chronological order of events:

1. search and target acquisition;
2. communications (where appropriate);
3. reconnaissance and identification;
4. weapon guidance or self-guidance.

A system is sometimes required not to interrupt a function when another function is triggered, for example identification or guidance should not interfere with search.

Most often, in state-of-the-art systems, for reasons connected with 'all weather' aspects, with covertness and with countermeasures, the same technique is not used for all functions; in preference a combination of techniques (such as radar and infrared) is used. This section, which is angled towards those military applications of infrared which have involved considerable investment compared to those in the civil field, gives a brief description of certain equipment which carries out the above-mentioned functions. We also give examples of equipment with special features other than those quoted previously. Often military investment has repercussions in other sectors of activity, and finally we shall quote a few applications of infrared in the civil field: industrial, medical and other applications.

6.5.1 Military applications of passive infrared

Target search and acquisition

The first function required from our weapon system is to carry out surveillance of air space in order to detect any threats. This task is often carried out by a radar system, but increasingly in recent years discretion and countermeasure resistance considerations have led to a radar being combined with passive discrete infrared search. Passive infrared search systems use a variety of techniques and architectures, depending on the specifications which are imposed. The major considerations are given below.

The vehicles Infrared search systems can be installed in a variety of configurations: ground based, either fixed or on a carrier vehicle; at sea, on board ships or submarines; and airborne, on aircraft or helicopters or even spacecraft.

The targets By their nature and the environment in which they operate, targets figure in equipment design in several different ways. The IR signature of targets (aircraft, missiles, helicopters, tanks, ships, etc.), in fact play a large part in the choice of useful spectral band (or bands). In addition, search systems must have target priority functions (for example low flying aircraft). These functions not only influence the choice of spectral band, but also that of angular coverage and information renewal rate.

The environment Equipment design includes several environmental aspects. First of all, the nature of the 'likely environment' may influence the choices made, for example noise and false alarms will be different depending on whether the target is moving against a sea background, a land background or a sky background.

Secondly, climatic aspects must be taken into account, for example, sea aerosols do not have the same effect as land aerosols which in turn have a different influence depending on whether the system is at high or low altitude.

These kind of operational considerations result in technical specifications and concepts provided in which these are properly expressed.

There are three major categories of search systems: sector infrared search, PPI infrared search, and omnidirectional infrared search.

Sector search is limited to a sector of a few tens of degrees in elevation and azimuth and normally finds airborne applications. PPI search is not limited in azimuth but is restricted to a few degrees in elevation. This configuration is suited to surface weapon systems. Omnidirectional search, as its name indicates, is a search without angular limitation. Its requirements are such that there is no satisfactory operational system in use today.

Regardless of whether the search system is sector, PPI or omnidirectional, there are a number of elements which are common to all:

1. an optomechanical scanning device which analyses the system's field of coverage;
2. an optical system suitable for the spectral band and the field to be analysed;
3. a set of infrared detectors which may assume a variety of different configurations.

There are a number of different detector architectures which simultaneously meet the different requirements of;

1. angular coverage;
2. range;
3. target designation precision (resolution);
4. data rate;
5. false alarm rate;
6. cost, etc.

Among others, two possible solutions are emphasized.

1. The angular field to be covered is analysed by a linear array with a small number of detectors. If the elevation field so requires, scanning must be carried out along two axes (azimuth and elevation).
2. Angular coverage in elevation and azimuth is provided by uniaxial scanning in azimuth of a multi-element array. This solution, which is similar from a mechanical point of view, nevertheless requires total mastery of the production technology required for multi-element arrays.

The choice of architecture is often dictated by false alarm rate considerations which may result in the use of several arrays so as to give several spectral bands. Multi-band numerical processing can then be used to reject spurious targets.

Localization – tracking

We shall remain inside the framework of a major weapon system.

The search process, whether visual, electromagnetic or infrared, enables acquisition of the target with a certain precision in elevation and azimuth. Depending on the method used, angular precision can vary between values of the order of one degree and one milliradian. The target coordinates are sent to a localization and tracking device. The 'localization' and 'tracking' functions can be carried out by a passive infrared system, sometime combined with a laser rangefinder, and a processing unit for automatic target tracking.

As with infrared search, the design of the localization device is based on considerations concerning: the carrier, the targets, the environment, and the required range and precision.

Thus the equipment shown in Fig. 6.35 is composed of two subassemblies: the first (on the left of the photograph) operating in a wavelength band between 3 and 5 μm, and the second (on the right of the photograph) using the band of 0.7 to 0.9 μm. The simultaneous operation of two spectral bands ensures a high level of automatic tracking safety.

The units are mounted on either side of a high precision turret. The 3 to 5 μm subassembly has an InSb detector array cooled to 77 K combined with a scanning mirror. The 0.7 to 0.9 μm subassembly is a camera using a silicon Vidicon tube from which the visible spectral band has been suppressed by optical filtering.

Reconnaissance and identification

Once the target has been located and tracked with precision, the operator can be provided with an image whose quality should enable him to recognize and then identify the target.

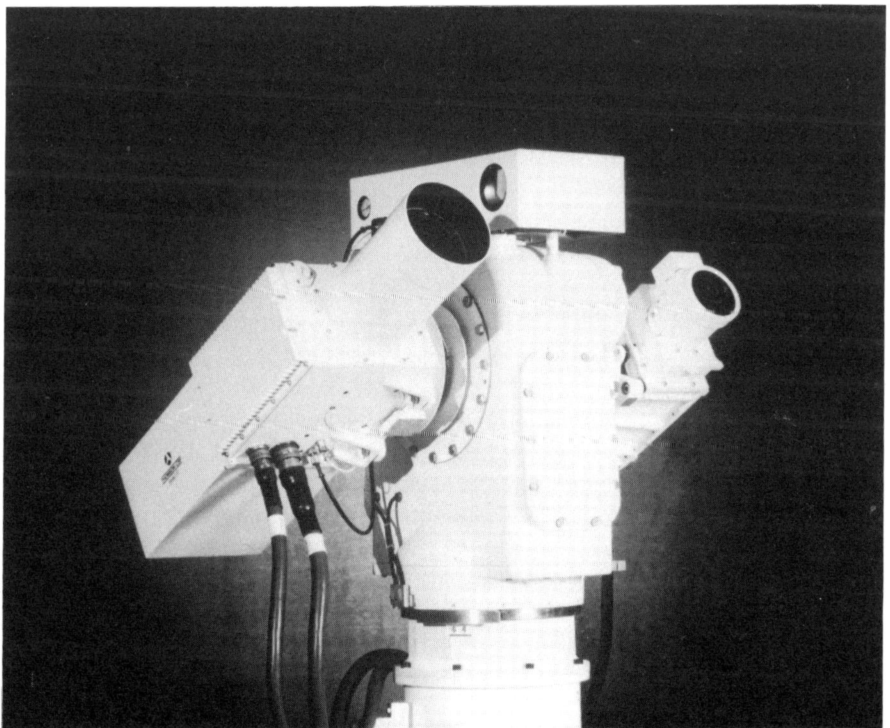

Fig. 6.35 Dual-band infrared tracking device.

The infrared cameras available are as follows:

1. near IR cameras (0.7 to 0.9 μm) with silicon Vidicon tubes;
2. near IR cameras (0.7 to 0.9 μm) with silicon CCD matrices, which are increasingly replacing Vidicon cameras;
3. 8 to 12 μm thermal cameras (or FLIR (forward looking infrared));
4. 3 to 5 μm thermal cameras.

Certain thermal cameras can be used for simultaneous localization, tracking, reconnaissance and identification.

Most of the research effort in the last two decades has been devoted to 8 to 12 μm thermal cameras, as the requirement for high quality day/night images has become manifest in all military sectors and for all types of carriers: tanks, helicopters, aircraft, surface-to-air batteries, ships, submarine, satellites, etc. There has also been a considerable development of modular thermal cameras (FLIR common modules) in the USA as well as in France and Great Britain. However the techniques used are different. In the USA the technique is based on single linear arrays with a large number of HgCdTe elements (60, 120, 180). In Great Britain, the technique is based on arrays with a limited number of HgCdTe SPRITE detectors. In France, the technique is based on the use of several juxtaposed HgCdTe arrays with a limited number of points, but operating in the time delay and integration mode (TDI).

The main modules are as follows:

1. detection
2. scanning
3. cryogenics
4. display
5. electronics.

These thermal cameras are now used in: fire control systems (tank to tank (Fig. 6.36), helicopter to tank, surface to air, air to surface); navigation/control of aircrafts and helicopters; and aerial reconnaissance. Small thermal cameras have even been developed for remotely piloted vehicles and for infantrymen.

The subject of 'aerial reconnaissance' is worth looking at in more detail, as passive infrared is producing profound changes in this area and enables reconnaissance using real-time data transmission. There are two major types of process: 'oblique imagery' using near IR cameras and thermal cameras, and 'vertical imagery' in which a line scan process is often used. Figure 6.37 illustrates the general principle: the aircraft moving in the direction of vector V at velocity V is fitted with an infrared system comprising one or more detector arrays parallel to the vector V. An opto-mechanical scanning device perpendicular to the vector V enables the IR detectors to scan in a direction perpendicular to the aircraft trajectory. The 'aircraft/scanning' combination thus images a strip of terrain underneath the aircraft.

Fig. 6.36 Tank-mounted IR fire control system.

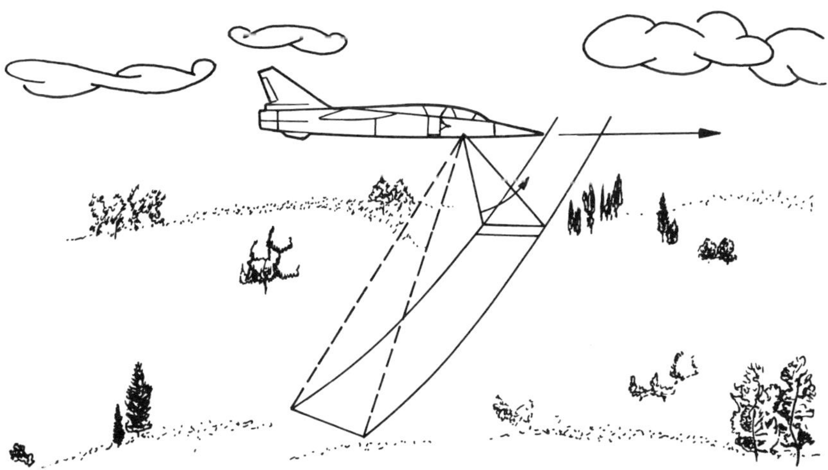

Fig. 6.37 Aerial reconnaissance with IR detector scanning.

Guidance – self-guidance

Weapons guidance was one of the first military applications of passive infrared. The infrared guidance which emerged during the Second World War has been considerably developed since then. There are two major categories: angular deviation measurement and remote control guidance or CLOS (command line of sight), and self-guidance.

Command line of sight (CLOS) The principle of this technique, known as CLOS, is represented in Fig. 6.38. It consists of aiming at the target, tracking the target until weapon impact, and at the same time, detecting and measuring the angular deviation between the weapon with respect to the target, and transmitting orders to the weapon to bring it on the line between firer and target.

Aiming and target tracking can be carried out by the human eye and manual tracking; a television camera or a thermal camera with automatic tracking; or a radar or lidar.

Angular deviation measurement, that is the measurement of the apparent angular deviation (X, Y) between the target and the weapon can be carried out by a passive infrared device. This IR device can use either a TV tube, a CCD, or a specific IR/CLOS equipment can be designed with: one or several IR detectors with a spatial modulator, a linear array of IR detectors, a focal plane array. The spectral bandwidth has to be matched with the spectral emission of the weapon propeller or the weapon flare.

The transmission of guidance orders to the weapon may be transmitted by wires, connecting the firing unit to the weapon; radio channel; or coded laser beam.

This type of CLOS guidance has been developed mainly for surface-to-surface weapons and surface-to-air weapons.

Self-guidance The second type of passive infrared guidance involves IR homing-heads. In this case, the weapon (missile, rocket, shell, bomb, etc.) is fitted with an IR detection head gyrostabilized which detects and tracks the target using its own radiation (Fig. 6.39). This gyrostabilized head ensures not only the aiming

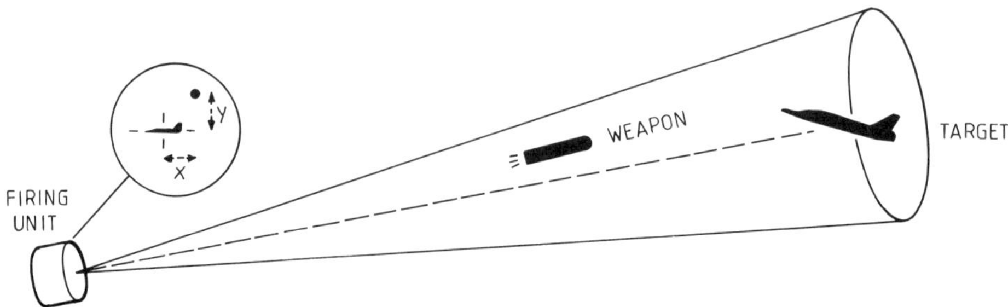

Fig. 6.38 CLOS (command line of sight) fire control.

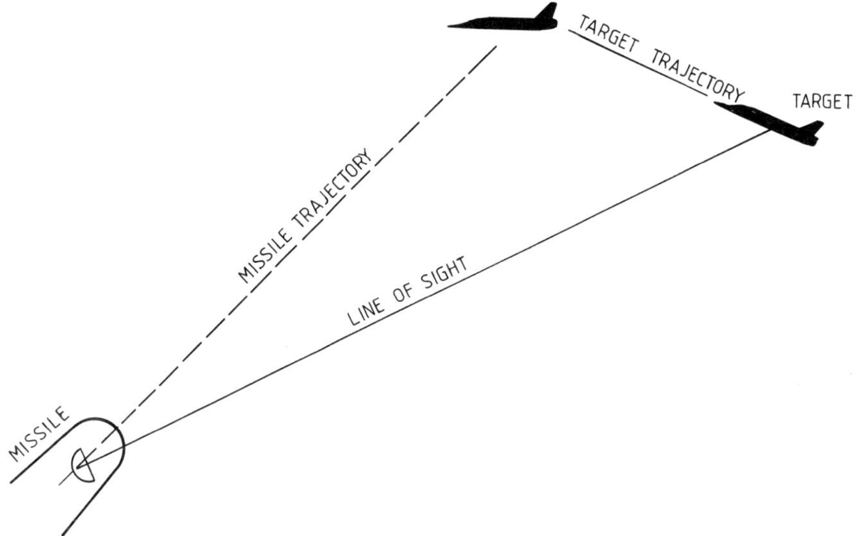

Fig. 6.39 Passive IR self-guided missile homing head.

stability necessary for precise tracking of the target, but also calculation of a proportional navigation law which guarantees optimum missile interception conditions.

The post-Second World War period saw considerable development in the use of IR homing-heads for air-to-air and ground-to-air systems, the reason being that the nature of aerial target signatures—in their back sector—enables the use of simple lead sulphide detectors (PbS), combined with spatial optomechanical modulators. Using this technology a whole range of now familiar homing-heads were developed and mass produced in both the NATO and Warsaw Pact countries. With the advent of indium antimonide detectors (InSb) more sophisticated homing-heads were produced.

Today, a considerable amount of work is being carried out in the field of infrared imaging homing-heads opening up new self-guidance possibilities not only in the air-to-air and surface-to-air fields, but also for surface-to-surface and air-to-surface applications.

From military to civilian

This overview of passive IR military applications is not intended to be exhaustive, we will also need to mention other achievements such as proximity fuses, terminal guidance and trajectory correction detectors, etc. As with other disciplines, civil activities have benefited from military investments, and passive IR applications for non-military ends has multiplied in recent decades.

In the medical field IR thermography (or IR radiometry) is used mainly for the detection of tumours. This process uses thermal cameras combined with measuring instruments which make precise measurement of signals supplied by detectors during spatial analysis of a subject. The spectral band around the 10 μm wavelength is frequently used, however lower wavelengths are sometimes used for the detection of circulatory malfunctions.

In the industrial field Passive infrared is a valuable inspection tool either on production lines or in-service inspection of equipment. There are two main techniques.

The first is the line-scan, the general principle of which is described in section 6.5.1. However, in industrial applications it is not the analyser which moves but the part (or the materials inspected) which travels in a given direction across the analyser whose detectors scan the perpendicular direction. Depending on the temperature range involved, the detectors are selected to suit the emission spectral bands of the material to be tested. The choice of detector can also be dictated by the operating autonomy of the thermographic analyser. Detectors cooled by the Peltier effect (thermoelectric cooling) are sometimes used for this reason. Certain types of equipment use several varieties of detectors in order to cover several bands.

The line-scan process is particularly suited to the following types of inspection:

1. sheet steel on exit from the rolling mill;
2. plastics;
3. glass during the cooling stage;
4. weld quality, etc.

The second industrial inspection process is similar to that used in thermal cameras and can therefore be used for automatic measurement and display of thermal maps of fixed objects, for example:

1. temperature control in furnaces;
2. monitoring thermal operations on materials;
3. in-service temperature measurements of electrical or electronic circuits.

In the field of security and environmental protection Infrared is increasingly used in connection with security and ecology:

1. monitoring of automobile pollution by means of multiband detectors;
2. measurement of atmospheric pollution;
3. meteorological readings;
4. measurements of maritime pollution using airborne line-scan;
5. detection and mapping of crop disease;
6. detection of intruders, etc.

This short list provides ample evidence of how the non-military applications of infrared will continue to develop in the future. We should not, however, lose sight of other highly important applications of infrared.

In the scientific field

1. measuring instruments for astronomy and spectrography,
2. atmospheric investigation or materials analysis devices and above all in the space field, where IR analysis and imagery techniques are increasingly being developed for:
   (a) weather forecasting;
   (b) pollution detection;
   (c) analysis of agricultural production and crop disease, etc.

6.5.2 Military applications of active infrared systems

As the classification given in section 6.5.1 shows, active infrared systems are very similar in principle to active radars. The use of a laser as the transmitter however opens up the field to specific applications which radars cannot fulfil in a satisfactory fashion.

The angular directivity of laser sources and their ability to produce short pulses enable an optical beam to be addressed very precisely both in space and in time. This ability to precisely focus electromagnetic energy, which in a radar system would require the use of enormous antennas, makes lasers the ideal choice when faced with the increasing demands for precision generated by modern weapon systems.

We shall now take a brief look at the basis of active infrared systems and then describe a few typical applications such as laser rangefinding, and active imagery.

Basic principles

The general principle of an active infrared system is based on the emission of a narrow beam of light (normally from a laser source), from the system that also detects and processes the echo backscattered from the object thus illuminated. The receiver is located near the transmitter and in most military applications where system size must be optimized, the transmitter and receiver optics are either common or they share a number of elements (monostatic assemblies). The laser beam which is backscattered towards the illuminator and then picked up by the receiver optics, generates an output signal (from the photodetector) which, depending on the requirements, may be used for: detecting the presence of a target; measuring its angular coordinates, range and Doppler velocity; automatic target tracking.
The basic components of an active infrared system are normally the following (Fig. 6.40):

1. the infrared source (laser with its control electronics);

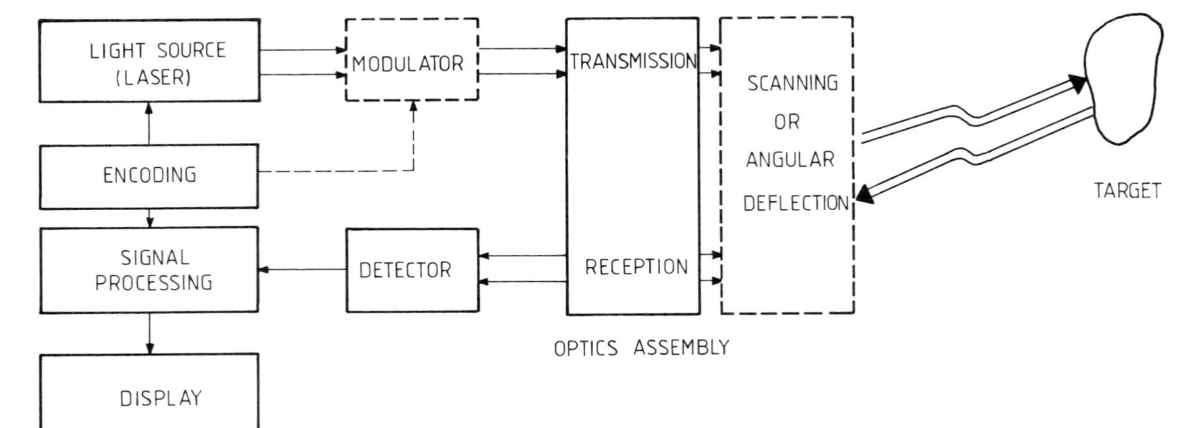

Fig. 6.40 Basic components of an active infrared system.

2. the transmitter and receiver optical systems (common or not);
3. the laser beam steering device (random deflection or systematic scanning);
4. the photodetector with its spatial and spectral filters and preamplifier;
5. the signal processing, synchronization and logic electronics;
6. the image processing and display units.

The laser beam which illuminates an area of space located at a range D from the system must propagate through twice that range of atmosphere (or any other propagation medium) prior to detection. The design of military systems and in particular the choice of the laser source, are therefore not only dictated by the characteristics of available lasers, but also and more importantly by the optical properties of the atmosphere, the propagation distances being large (several kilometres or tens of kilometres).

The designer of an active infrared system must also make allowance for laser stray light backscattered by the atmosphere (as well as by the optical system itself), particularly with continuous wave or quasi-continuous wave laser systems. In some cases the flux backscattered by the atmosphere (and in particular the layers of the atmosphere closest to the system) can be as large or larger than the laser signal from the target, which means that the system may 'blind' itself by its own emission.

The atmospheric scattering and absorption coefficients at various laser wavelengths depend on the ambient meteorological conditions (temperature, pressure, relative humidity, type and concentration of aerosols, dust, smoke, etc.). In practice, they are calculated by means of special codes, among which one may mention FASCOD, developed by AFGL (Air Force Geophysics Laboratory).

In addition to these attenuation and scattering effects, the laser beam may be subjected along its path to deformations that can be detrimental to certain applications (high precision angular measurement systems, for example). This deformation, caused by non-homogeneities in the refractive index of the atmosphere results in:

1. beam widening;
2. angular deviation from its initial axis (beam wander);
3. amplitude and phase fluctuation inside the laser spot (scintillations).

These effects get more and more important as temperature fluctuations along the path become larger and laser wavelength shorter. Thus these perturbations occur during daylight hours, in summer, at ground level, and are more important with visible or near IR systems than at longer wavelengths.

Laser rangefinding

Principle Laser rangefinding is based on the measurement of the flight time of a laser beam to and from the target. Several methods can be used, among which one can quote (Fig. 6.41): flight time measurement of short laser pulses, and phase

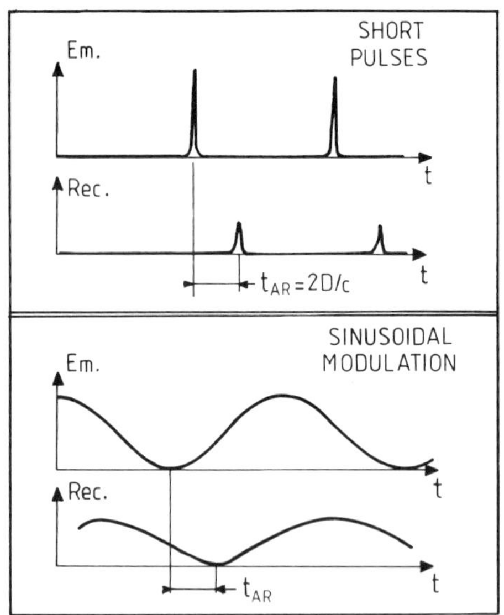

Fig. 6.41 Example of laser rangefinding techniques.

shift measurement between transmitted and received laser signals, the transmitting source being sinusoidally amplitude modulated.

The first method is the most frequently used in the military field, the second being preferred for short range but high precision applications such as military robotics (automatic control of ground vehicles) or satellite rendezvous. We shall therefore only describe a few types of rangefinders to illustrate the so-called 'flight time measurement' method, the schematic of which is shown in Fig. 6.41.

Solid state laser rangefinders.

GENERAL DESCRIPTION

The transmitter in these rangefinders is a solid-state pulsed laser that emits laser bursts (typical width of 20 ns) with rather high peak power outputs (several MW), at rates of several hertz or tens of hertz.

The first pulsed laser used for rangefinding was the ruby laser, which was abandoned because of its lack of discretion (visible emission at 0.68 μm) and because of its output fluctuations with operating temperature. It was replaced by lasers whose active medium is a glass or yttrium and aluminium garnet (YAG) matrix doped neodyme: their output power is more stable (four level lasers) and their wavelength ($\lambda = 1.06\,\mu$m) makes the beam less sensitive to atmospheric scattering.

The active medium of these lasers (cylindrical rod approximately 100 mm long, 6 mm in diameter) is optically pumped by a flash lamp (pulses of several hundred μs). During population inversion by the flash light, these lasers are usually triggered by means of an electro-optic cell (Pockels cell) inside the cavity or by a rotating prism or mirror.

The transmitter optics reduces the divergence of the output beam to make it compatible with the angular dimension of the targets to be processed. In most cases, this divergence does not exceed 1 mrd. The receiver optics collects the flux backscattered by the targets and focuses it on the sensitive area of the photodetector. An interference filter and a limiting aperture optimize system sensitivity by matching the geometry and spectral band of the receiver to those of the laser signal. One the most commonly used receivers is the avalanche silicon photodiode which gives these rangefinders a sensitiviy of the order of 1 nW (for a pulse duration of 20 ns).

Finally, the signal processing circuit typically comprises the following modules:

1. an amplifier, possibly with variable gain to minimize dummy echoes from atmospheric or optical backscattering;
2. a matched filter (i.e. matched to the pulse width);
3. a threshold type extractor;
4. a clock whose rate is matched to the required range accuracy (for example, a rate of 30 MHz corresponds to a range precision of ± 5 m).

USE/PERFORMANCE/DEVELOPMENTS

Solid-state laser rangefinders are at present used both by infantry (binocular rangefinders), artillery (sight glass rangefinders), tanks and helicopters (rangefinders for stabilized aiming devices). They are also found on air-to-ground attack aircraft and naval optronic turrets.

The photograph in Fig. 6.42 shows a Nd-YAG rangefinder for combat aircraft fitted with a Risley prism head (set of two rotating prisms) that steers the laser beam onto the target in response to the target angular coordinates generally delivered by a passive optronic automatic target tracker.

A few typical characteristics of an up-to-date version of this type of rangefinder are given below:

laser peak power ~ 5 MW
pulse width ~ 20 ns
typical maximum range at sea level (good meteorological visibility) ~ 15 km,
range accuracy ~ ± 5 m,
output data rate ~ up to 10 Hz.

The work in progress for the optimization of solid laser rangefinders concentrates typically on two areas. First, improvement of the electrical efficiency of lasers by the use of laser diodes as optical pumps, since the neodymium absorption spectrum coincides with the emission spectrum of certain diodes.

Second, research on eye safe solid-state lasers: their emitting wavelength must be larger than 1.4 μm, upper limit of the spectral transmission of the human eye (aqueous humour). Absorption of this radiation by the pupil of the eye prevents

Fig. 6.42 Nd-YAG rangefinder for combat aircraft.

its focusing down onto the retina. Among the promising lasers in this field, currently at the testing stage, we can quote erbium and Raman shifted Nd-YAG lasers (Nd-YAG laser with Raman frequency shift in a gas such as methane) which emit at $\lambda = 1.54\,\mu m$, or the holmium laser ($\lambda = 20.8\,\mu m$).

CO_2 *laser rangefinders*

OPERATIONAL REQUIREMENTS OF A RANGEFINDER IN THE 8 TO 12 μm BAND

Given their emission spectrum (between 1 and 2 μm), solid-state laser rangefinders are well matched to passive search and track systems operating in the visible or near infrared (such as the human eye or TV cameras), and their range performances are excellent under favourable meteorological visibility. However, their performances depend to a great extent on this visibility. The development of long wavelength passive infrared systems (thermal cameras of FLIRS in the 8 to 12 μm band) often results in a range disparity between passive acquisition systems and solid-state laser rangefinders. There are in fact a great many circumstances in which passive systems are able to detect targets beyond the range of solid-state laser rangefinders (particularly in the presence of smoke, fog, etc.), or vice-versa.

CO_2 gas lasers (which emit inside the thermal band to 12 μm) are increasingly being used in a number of rangefinder projects designed to meet the following requirements:

1. maximum range improvement at low meteorological visibility and particularly in smoke and dense aerosols;
2. compatibility with thermal imaging systems of the FLIR type;
3. 'eye safety.'

Depending on the type of excitation used, CO_2 lasers can emit single mode or multimode, continuous wave or pulsed beams. They are compatible with either direct or heterodyne detection, and the emission line may be selected between 9.2 and 11.2 μm (maximum output at 10.6 μm). This diversity of transmitted waveforms has resulted in several designs of CO_2 laser rangefinders, some of which are described below:

DIRECT DETECTION CO_2 LASER RANGEFINDERS

The schematic for a direct detection CO_2 laser rangefinder is identical to that of a pulsed solid-state laser rangefinder given above. In this configuration, the CO_2 laser is a transversely excited pulsed laser with the gas mixture at atmospheric pressure (TEA laser). Pulsewidths are of the order of 100 ns with peak powers of the order of 1 MW, and output rate up to 10 Hz.

Transceiver optics are made up of the same materials as those used in a FLIR (germanium, zinc selenide, etc.) and the detector is most often a mercury/cadmium telluride photodiode (MCT) cooled to liquid hydrogen temperature (77 K), for example by means of a Joule Thomson cooler. The field of view of these rangefinders is typically less than 1 mrd. The first operational programme concerning CO_2 laser rangefinders is the installation of such systems on tanks.

CO_2 LASER DOPPLER RANGEFINDERS AND HETERODYNE DETECTION

The very high frequency stability of single mode CO_2 lasers makes it possible to produce coherent CO_2 laser rangefinders with the same architecture and signal processing techniques as those of radar systems. Heterodyne detection of the laser echo yields not only target range (as is the case for direct detection rangefinders) but also its Doppler velocity component along the line-of-sight.

As in the case of radars, infrared heterodyne detection has produced two major families of coherent rangefinders which are at present being developed both in the USA and in Europe:

1. pulsed CO_2 laser rangefinders;
2. pulse compression and continuous wave CO_2 laser rangefinders.

The pulsed CO_2 laser coherent rangefinder is based on the use of a single mode TEA CO_2 laser as the illuminating source and a second, much less powerful continuous wve CO_2 laser as a local oscillator.

Generally speaking, the pulse emitted by this type of laser has a spatial temporal shape (Fig. 6.43) with an initial narrow peak (of the order of 50 to 150 ns) and a much longer tail (of the order of 1 to 2 μs), the total energy reaching 100 mJ

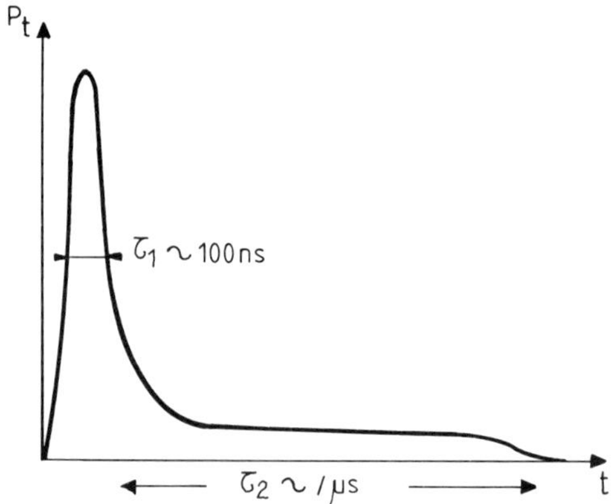

Fig. 6.43 Typical shape of a TEA CO_2 laser pulse.

at rates of less than 10 Hz. Accuracy expected from this type of rangefinder is of the order of ± 10 m on range and ± 5 m/s on Doppler velocity.

The schematic of a pulse compression, continuous wave CO_2 laser heterodyne rangefinder is given in Fig. 6.44. The optical design looks like that of a Mach Zehnder interferometer in which the continuous wave CO_2 laser beam is split into two unbalanced channels outside the cavity. The weaker channel (100 mW) constitutes the reference beam of this interferometer and is focused onto the receiver as the local oscillator. The main channel is frequency modulated (using acoustooptic cells for example) before being transmitted toward the target by the optical system.

The return laser signal, collected by the receiver optics, which carries data on target range and radial motion is also focused on to the detector where it is mixed with the local oscillator. The resulting heterodyne signal is matched filtered

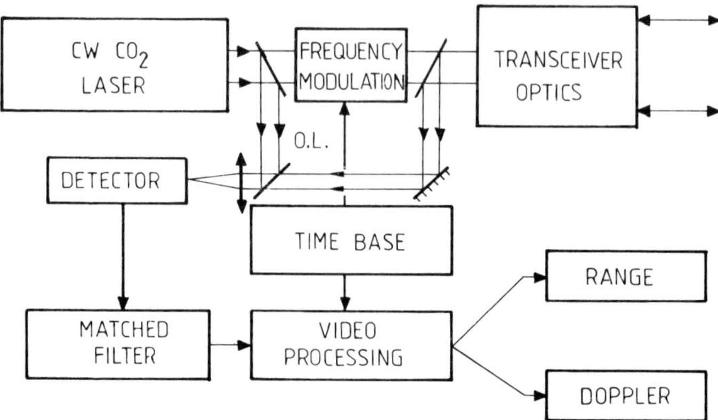

Fig. 6.44 Schematic diagram of a pulse compression CO_2 laser rangefinder.

(electro-acoustic dispersive delay lines) and supplies both target range and radial velocity data through traditional pulse compression radar techniques.

As an example, a 5 W CW laser rangefinder has a range of the order of 5 km with accuracies of ± 5 m for range and ± 0.5 m/s for Doppler velocity.

Lidars

In the same way that radars have evolved from essentially range measuring devices to the present day multifunctional systems, active infrared system design has matured (at least at the research level) to a point where it leads to the definition of systems (described previously) which are more diversified than rangefinders. These are known as lidars or irdars (infrared detection and ranging).

Most of the work on lidars used the CO_2 laser as the basic component owing to the diversity of its waveforms (high or low rate, pulsed/continuous wave) and to its compatibility with heterodyne detection. Lidars are therefore infrared coherent radars, as such, should be capable of carrying out a great number of functions traditionally associated with radars in areas such as target detection and acquisition, rangefinding and Doppler analysis, with the limitations and advantages inherent to the optical spectrum. It should also be noted that the small divergence of laser beams gives lidars additional possibilities in high resolution imagery which are difficult to achieve with radars.

The functions below illustrate a few of the possibilities of lidars:

1. Air to ground targed acquisition:
   (a) 3-dimensional imagery;
   (b) moving target indication;
2. target identification:
   (a) high resolution;
   (b) Doppler signature analysis;
3. automatic target tracking;
4. navigation aids:
   (a) wide angle imagery and map correlation;
   (b) Doppler navigation;
5. terrain following, cable and obstacle avoidance;
6. wind field measurements;
7. telecommunications;
8. missile guidance.

Various lidar programmes are being conducted in particular for airborne applications. One of the most ambitious of these is certainly the infrared airborne radar programme (IRAR) by the MIT Lincoln Laboratory, which is aimed at developing the technology and components necessary for coherent laser radars, and at demonstrating lidar capabilities in moving target detection, high resolution imagery, target tracking and ranging, etc.

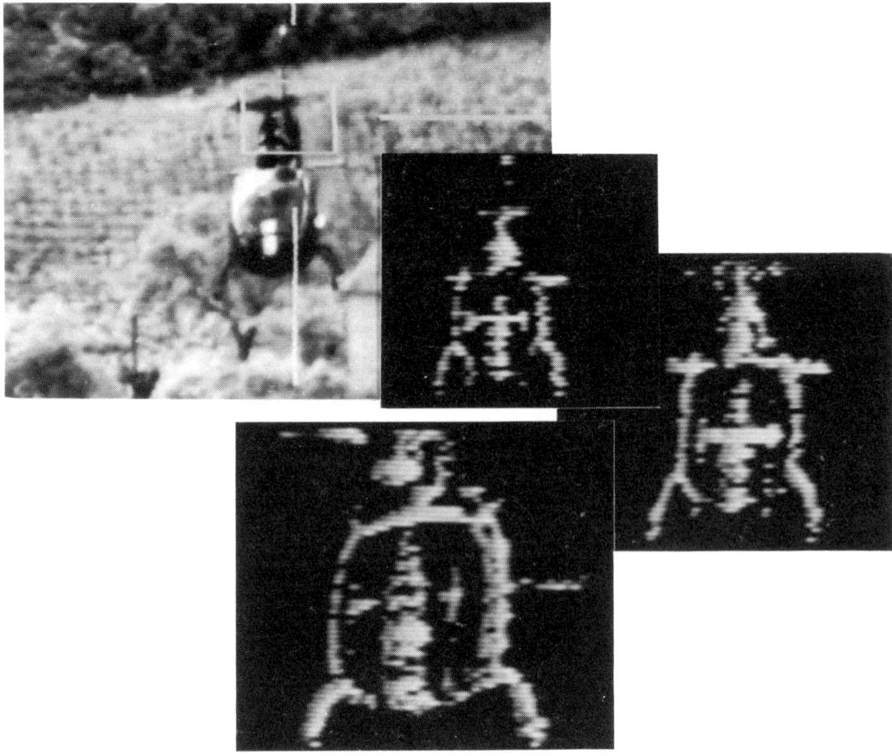

Fig. 6.45 Helicopter Doppler image (CO_2 laser radar).

Other projects in the USA and Europe are concerned more particularly with the use of lidars on helicopters for cable detection (laser obstacle and terrain avoidance warning system), or on cruise missiles for terminal guidance. Finally, one can also mention programmes on tank detection and tracking using laser imagery.

In this area, a pulse compression CO_2 laser imaging system has been designed in France for validating some of the functions listed above, in particular fire control, moving target detection and target recognition and tracking. This equipment, which is still at the prototype stage, confirms the high Doppler sensitivity of 10.6 μm lidars since it easily detects walking pedestrians. Figure 6.45 shows a typical Doppler image of a moving helicopter with automatic background suppression.

6.5.3 Military applications of semi-active infrared

The design of a semi-active infrared system is similar to that of the active systems described in the above section. It is based on the use of a laser transmitter and a laser receiver aimed at the same target but separated from each other.

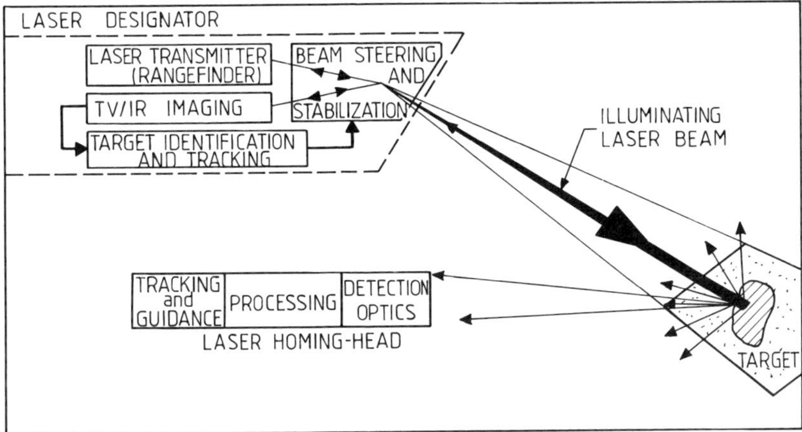

Fig. 6.46 Basic components of a weapon guidance system by laser target designation.

One of the main military applications of semi-active infrared systems is laser target designation for weapon guidance, particularly in air-to-ground attack missions. In this guidance mode shown in Fig. 6.46, the laser transmitter, called 'illuminator' or 'laser target designator', illuminates the target, usually by means of a pulsed laser. This transmitter may be mounted either on the weapon carrier aircraft, on an accompanying aircraft, on the ground, or on a helicopter. As for the laser receiver or 'laser seeker', it is located in the nose of the weapon itself, and points at the target. It directs the weapon onto the target by extracting the target angular coordinates and delivering them to the weapon navigation system.

A typical airborne, air-to-ground, attack mission using laser target designation is described below, for example in the case of a single seater fighter carrying both laser target designator and munitions fitted with homing-heads (See Fig. 6.47 for illustration). Usually, the laser illuminator is mounted inside a pod underneath the aircraft with an automatic search and track imaging system to which it is optically coupled.

In the first phase of the attack, the pilot performs target acquisition using either the onboard radar, the head-up display, the visor display or the aircraft's inertial unit. Following visual target recognition on the monitor (from TV or thermal imagery), the pilot designates it to the tracking system which ensures image stabilization (phase 2). Once the aircraft enters the weapon firing envelope, the weapon is fired in quasi-alignment with the target and begins its run in inertial fashion (phase 3) while the aircraft breaks away to avoid enemy defences (phase 4). Laser illumination of the target is then automatically activated by the aircraft fire control computer (phase 5).

The homing-head or laser seeker used for weapon navigation comprises a flux collecting optics with a four quadrant cell near its focal plane. Laser flux scattered from the target is picked up by the homing-head optics and focused on the four

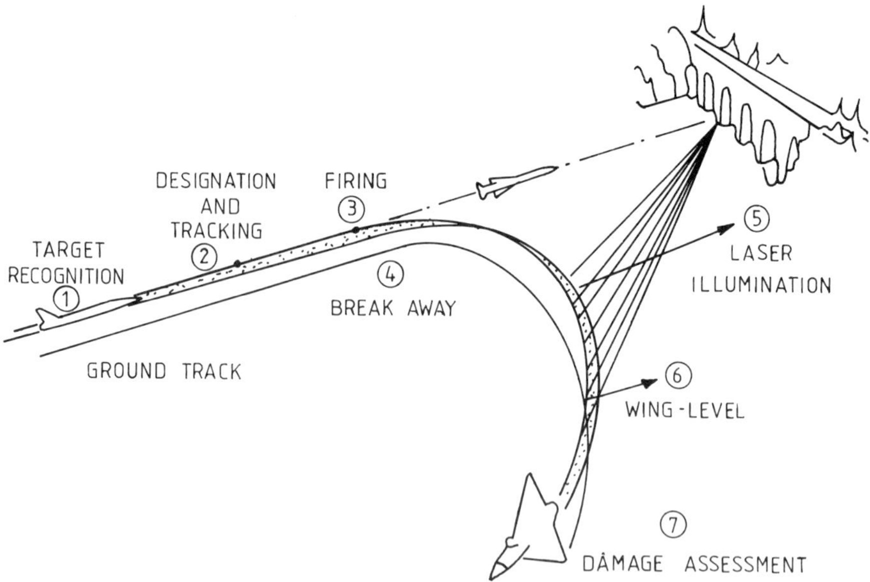

Fig. 6.47 Typical attack sequence with a laser guided weapon system.

Fig. 6.48 Laser designation pod and laser homing-head missiles belly-mounted on a combat aircraft.

quadrant cell (silicon photo diode in the case of Nd:YAG laser illuminators), each of which generates a signal proportional to the fraction of incident flux. The target angular coordinates are then extracted from the relative weighting of each of these signals and fed to the weapon navigation system.

The photograph in Fig. 6.48 shows a laser designation pod for single seater fighter aircraft. It automatically acquires and tracks targets by means of TV imagery from a camera operating in the visible or near infrared and designates them with a Nd:YAG pulsed laser transmitter to laser homing-head missiles or 'smart' bombs.

Semi-active infrared weapons guidance is extremely effective as it considerably improves firing accuracy, while at the same time increasing weapon launch distance and, as a result, mission safety (stand-off launch). This type of system can be used for either day or night operations using either TV or thermal cameras.

Another military application of the semi-active type concerns designation of a target by several operators. As an example, the illuminator is located on the ground and a laser receiver (similar in principle to a homing-head) is installed on the aircraft. After laser signal processing, a marker indicates the target position to the pilot on this head-up display.

6.5.4 Military applications of point-to-point links

In an infrared point-to-point link, the receiver is aimed directly at the laser emitter. Among the applications of such systems we would mention laser communications and beam riding missile guidance. Because of the directivity of a laser beam, laser communications have a definite discretion advantage over conventional free space communication techniques.

As far as beam riding techniques for missile guidance are concerned, a typical system (Fig. 6.49) is comprised of a fire control unit for target tracking and laser illumination and a laser receiver located in the rear of the missile.

The line-of-sight of the fire control unit is maintained onto the target by means of a passive automatic tracking imaging system (for example TV or infrared),

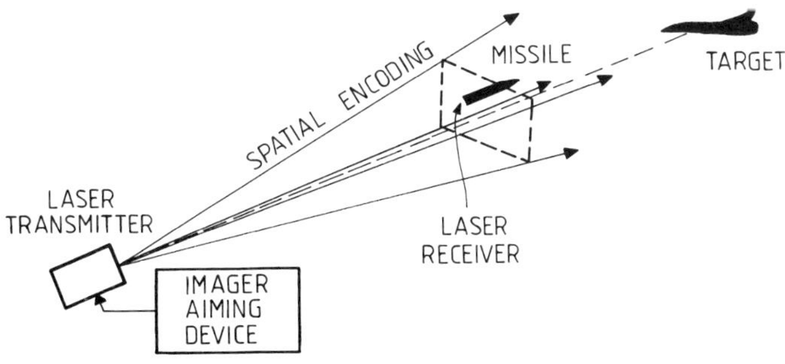

Fig. 6.49 Components of a beam riding missile guidance system.

that also directs the laser beam at the target. The beam itself is spatially coded in such a way that the laser receiver at the rear of the missile computes the missile position inside the laser upon laser signal detection and delivers the appropriate commands to the missile navigation system. Space encoding is carried out by amplitude modulation or beam scanning around the direction of the target. The laser receiver on the missile 'looks' back and therefore receives the space encoding data which enable it to determine its position compared to the axis defined by the firing unit and the target. The missile can thus calculate its position error and make the necessary trajectory corrections.

6.6 DEVELOPMENTS AND TRENDS IN THE INFRARED FIELD

This Chapter has shown the diversity of the parameters involved in the design of infrared systems. This very diversity gives an idea of the technical and technological difficulties which have had to be overcome in order to achieve present day performances. For the future, infrared, which already plays an important part in the field of optronics, holds considerable possibilities of new development.

Looking again at the overall design of an item of IR equipment, we find:

1. optical windows or IR domes;
2. optomechanical stabilization and/or scanning devices;
3. optical design for energy transfer or image formation;
4. radiation detectors;
5. cooling devices;
6. laser emitters;
7. processing units;
8. display units.

6.6.1 Optical windows and IR domes

Military infrared systems on aircraft and missiles in particular are required to operate under increasingly difficult conditions. These spectral bands are increasingly distant (or wide), the speeds and 'g' factors of the carrier vehicles continue to increase and the performances required are more stringent. Technological research must therefore continue in order to provide materials and coatings to suit future requirements. Already material casting processes and hard coating techniques (in particular using diamonds) have made great steps in recent years.

6.6.2 Stabilization and scanning

Stabilization and scanning devices have a major effect on the resolution and precision capabilities of infrared systems. Gyroscopes and stabilization rate gyros have considerably improved and will continue to do so. We have already seen laser rate gyros and fibre optic rate gyros. A number of scanning devices have appeared including servo-controlled mirrors and galvanometric or piezoelectric devices. However, new technologies are being developed and in certain cases these may take over from present day techniques, for example acousto-optic and electro-optic processes, as well as certain liquid crystal or dynamic holography processes.

In spite of all these technological improvements, certain phenomena external to the system are still likely to interfere with it. Atmospheric fluctuations can modify the direction of the line of sight and the ability of a laser to focus on a target. New coherent optical adaptive techniques (COAT) are under development:

1. multiple servo-control mirrors;
2. deformable reflecting surface ('adaptive optics');
3. liquid crystals.

To these new optical technologies, suitable signal processing techniques must be added which can help with fine stabilization of the line-of-sight.

6.6.3 Optical systems

Military equipment suppliers are demanding high performance with low false alarm rates and optical design must be improved in the following ways:

1. materials and surface coatings which increase optical transmission;
2. combinations of materials enabling transmission of extended spectral band along the same optical path.

Such combinations will simplify multispectral systems, both in the field of passive infrared and active infrared as well as for combinations of the two. It is vital to achieve total control of aspherical surface production techniques in order to achieve satisfactory optical assemblies.

In addition to these traditional optical systems, we must now add the new possibilities provided by non-traditional optics and in particular non-linear optics. This field has continued to develop since the introduction of lasers with, in particular:

1. optical waveguides (integrated optics and fibre optics);
2. harmonic optical frequency generators;
3. non-linear and Raman effect amplifiers;
4. parametric tuneable oscillators;

all of which will produce new applications in the next decade.

6.6.4 Detectors

We have already said that IR detectors have represented a broad field of investigation for over two centuries and will remain so for some time. It would seem clear that in the 'near IR', the Vidicon-type tubes will give way to CCD matrices in the next ten years. In the medium IR band (3 to 5 μm), the new PtSi matrices (platinum doped silicide) offer further possible applications. For the first time infrared focal plane arrays of 500 × 500 pixels and 1000 × 1000 pixels are promised in the near future. The present state of material studies points to focal plane arrays with a high number of pixels in the medium term, sensitive in the band 8 to 12 μm.

As with optical materials the widening of the spectral sensitivity field of detectors should be studied in parallel. This would enable multispectral image processing at minimum cost.

6.6.5 Cooling devices

Considerable progress has been made in the last 15 years with a transition from the open circuit cooler (whose autonomy was limited) to the closed circuit miniature cooler enabling satisfactory reliability. There are two main aspects governing future developments:

1. the appearance of high number of pixels focal plane arrays giving a relatively large surface area to be cooled;
2. the creation of materials likely to present good detectivity at temperatures higher than that of liquid nitrogen which will make for technical simplifications.

6.6.6 Laser emitters

The multiplicity of studies and innovations in the field of laser emitters in the 1960s, was followed by a certain stabilization and a more careful orientation of research at the start of the 1970s. This led to the development of the main types of lasers described above.

Today, a new generation of lasers designed for military and civil applications is being developed in a number of laboratories both in USA, Europe, Asia, Australia and the Eastern countries. Without going into detail here about current research on high power lasers, which is usually classified in nature, we would nonetheless mention the broad trends.

One of these concerns the pumping of solid lasers by laser diodes, which gives considerable gains in both efficiency and lifetime. Another trend is the production of new crystals for laser emission in new infrared wavelengths which, combined with non-linear optics, open up the way to tuneable lasers.

6.6.7 Processing devices

By processing, in this context, we mean signal processing, and image processing.

Image processing can, moreover, be carried out for a number of reasons:

1. improvement of images (apparent resolution, contrast, etc.);
2. specific effects: rotation, zooming, etc.;

3. electronic stabilization;
4. target extraction;
5. tracking;
6. automatic target recognition (ATR), etc.

Infrared processing is often similar to the processing used in the radar field and which is also dealt with in the present volume. Improvements in the technology of electronic components and state-of-the-art processors are as important for the development of radars as they are for that of infrared systems.

Image processing is assuming increasing importance as resolution and therefore the number of pixels is increasing and processing times must be reduced in order to meet weapon system requirements.

6.6.8 Display

The receiver circuit of an infrared system frequently terminates in a display device which may be a simple black and white TV monitor, a colour monitor (with false colours image encoding) or the head-up display of an aircraft fire control system, etc. In all these cases, the display device must be properly matched to the image acquisition and processing system or it will affect system quality.

The increasing definition achieved by sensors calls for the use of display devices with improved resolution. New display technologies like liquid crystals will also need to be suited to the requirements of infrared systems.

7

Industrial, scientific and medical (ISM) applications of microwaves present and prospective

Bernard Epsztein, Yves Leroy, J. Vindevoghel and Eugene Constant

7.1 INTRODUCTION

The premier applications of microwaves are presented in different chapters of this book; they are terrestrial and satellite links, radio and TV transmission (Chapters 1–3), radars and remote sensing (Chapter 4), countermeasures (Chapter 5), infrared detection (Chapter 6) and radioastronomy (Chapter 8). Other applications also exist and present a growing interest, which are related to the so-called industrial, scientific and medical (ISM) applications. The corresponding processes can be easily described once several basic properties related to interactions between microwaves and different kinds of material have been brought to mind.

The properties of dielectric materials in microwaves are usually described by their relative complex permittivity (a frequency dependent parameter):

$$\varepsilon^*(f) = \varepsilon'(f) - j\varepsilon''(f). \tag{7.1}$$

Two other important parameters are derived from the complex permittivity, with respect to propagation effects: the refractive index $n^*(f)$ and the absorption coefficient $\alpha(f)$ (in amplitude). In the most simple case of a plane wave, we have:

$$\sqrt{\varepsilon^*(f)} = n - jk \tag{7.2}$$

$$\alpha(f) \simeq \frac{\pi\varepsilon''(f)f}{n(f)c} \tag{7.3}$$

where c is the velocity of light.

As a matter of fact, although electromagnetics is a very complicated science, its most important consequences with respect to the phenomena we are concerned with can be summarized by the following rules of thumb.

First, the amplitude of a plane, or TEM wave, propagating on a thickness l of a bulk material is affected by an attenuation factor exp $(-\alpha(f)l)$; in other words for any path

$$\delta(f) = \frac{1}{\alpha(f)} \tag{7.4}$$

the amplitude is reduced to a ratio $1/e = 0.37$. Consequently the penetration of waves is governed by δ, called the penetration depth or skin depth. However, for most real situations (non perpendicular incidence, multimode propagation), the penetration of waves is smaller than δ.

Second, an incident plane wave reaching a dioptre (interface) limiting two materials (subscripted 1 and 2) is reflected and transmitted with respect to Snell's law. For example for low loss materials, the reflexion and transmission coefficients are:

$$\rho'(f) \simeq \frac{n_2(f) - n_1(f)}{n_2(f) + n_1(f)} \tag{7.5}$$

$$t(f) \simeq \frac{2n_2(f)}{n_2(f) + n_1(f)} \tag{7.6}$$

As a matter of fact these relations apply to flat dioptres. The phenomenon is modified in the case of a rough surface.

These rules of thumb, a crude simplification of complicated phenomena, present the advantage to point out in a simple way that the losses of materials govern their penetration by waves, and their refractive index governs their transmission from one material into another one. For a more complete understanding of the phenomena, the reader can refer to the book "Principles of Optics" (M. Born, 1975). Anyway these basic properties govern most of the applications presented hereafter.

If we classify the common materials in terms of an increasing absorption, we have to mention first air which is quite transparent to microwaves; smoke, moisture, hydrocarbons and other cases do not really affect its transparency. Note however the existence of absorption lines in the microwave spectrum (such as 24 and 180 GHz for atmospheric water, 60 and 120 GHz for oxygen, etc.) (Ulaby, 1986).

Many solid materials are also generally transparent such as, ceramics, minerals, textiles, plastics (on the condition that they are dry) and ice. Water exhibits high losses in microwaves; consequently the absorption of matter is highly conditioned by its water content. Considering for example, the biological tissues, a difference of water content explains that muscle tissues are much more lossy, giving higher microwave absorption than bones or fat tissues.

Note also that the refractive index generally increases with the losses; consequently, in many cases, the greater the difference between the absorption of two materials, the smaller is the microwave power which can be transmitted from one material to the other, due to a high reflection coefficient at the interface.

Considering at last very lossy materials, metals are quite opaque to microwaves with a skin depth given by the expression:

$$\delta(f) = \frac{1}{\sqrt{\pi\mu\sigma f}} \tag{7.7}$$

where σ is the conductivity and μ is the magnetic permeability. They also have a very high reflection effect with respect to air (the reflection coefficient amplitude is equal to one).

Table 7.1 Properties of typical materials in microwaves [from (Ulaby 1986), (Burdette 1980)]

Materials	*Frequency* (GHz)	ε'	ε''	*Real part of refractive index* n	*Penetration depth for plane wave propagation* (Voltage δ)
Atmosphere	10			1	160 km
Heavy rain	10			~1	8 km
Atmosphere	94			~1	800 m
Teflon	2	2.2	0.0003	1.5	240 m
Dry snow (0 °C)	10	2	0.002	1.4	6.7 m
Ice (0 °C)	10	3.15	0.003	1.8	5.7 m
Silica (dry)	10	4.1	0.07	2.0	27 cm
Low water content biological tissues (bones, fat, etc.)	3	5.55	0.8	2.4	9.4 cm
High water content biological tissues (muscles, etc.) (20 °C)	3	46	12	6.8	1.8 cm
Earth material (loam 0.1 moisture)	12	5	1	2.2	1.80 cm
Earth material (loam 0.35 moisture)	12	18	3	4.26	1.12 cm
Water (20 °C)	3	78	14.6	8.9	1.9 cm
Gold	2				1.7 μm

In Table 7.1, we present the values of the complex permittivity, real part of refractive index and penetration depth at different frequencies for typical materials, in view of most of applications described hereafter.

As a matter of fact, the fields associated with a microwave signal produce alternating forces applied upon the electric charges and dipoles existing in the material and consequently produce a dissipation of power which is transformed into heat (Section 7.2.1). Moreover, a high microwave power applied to matter can also ionize matter and create plasmas (Section 7.2.2).

Another aspect of the interactions of microwaves with matter is the thermal noise emission. The spontaneous emission by matter of a random signal in a wide frequency range (radio frequencies, microwaves, infrared, etc.) described by Planck's law (black-body radiation) is temperature dependent. Its most visible consequence is sunlight (the sun's temperature is 15×10^6 K). Any material around us transmits such a signal even if not visible. Well known at infrared wavelengths (IR thermography (Chapter 6)) this effect also exists in microwaves (Evans and Mcleish 1977): the spectral density is independent of the frequency and proportional to the absolute temperature of matter (white noise). Note also a consequence of the second principle of thermodynamics: the absorption of a material is a necessary condition for a thermal noise emission (principle of detailed balancing). Moreover, such signals transmitted by any subvolume of a lossy material are subject to the consequences of skin effect (equations (7.3)(7.4) and (7.7)) and of reflections at the interfaces (equation (7.5)). The applications of microwaves which are presented in the following sections are easily obtained from these properties. They can be divided into these categories:

1. the high power applications, related to the heating of materials, the creation and heating of plasmas and the acceleration of electrons, and other charged particles;
2. the active sensors, related to the measurement of the attenuation, phase shift or delay of a microwave signal in interaction:
   (a) either with a reflective material, leading to applications related to radar type sensors and short range systems;
   (b) or with a moderately lossy (or semi-transparent) material, leading to investigations related to dielectric properties by transmission or reflection;
3. the passive sensors related to the measurement of the thermal noise emitted by matter (or received after reflection) leading mainly to applications devoted to contactless thermometry.

7.2 HIGH POWER APPLICATIONS

We describe applications for industry, medicine etc. based on the capability of microwaves to heat matter (section 7.2.1). For more complete information the reader can refer to Metaxas (1983). The applications devoted to the generation and heating of plasmas for thermonuclear fusion are also reported in section 7.2.2.

7.2.1 Microwave heating

Principle

We first explain the process of power deposition in a material to which a microwave signal is applied. Let us consider the following situation (Fig. 7.1): a sine voltage $v(t)$ (amplitude V_{max}, frequency f) is applied to an ideal plane capacitor (no fringing effects) (area S, thickness d). When the capacitor is in air, its admittance is $jC2\pi f$ with:

$$C = \frac{\varepsilon_0 S}{d} \tag{7.8}$$

ε_0 the dielectric constant of free space.

When the capacitor is filled with a material with a relative complex permittivity.

$$\varepsilon^* = \varepsilon' - j\varepsilon'' \tag{7.9}$$

the admittance becomes:

$$y = jC2\pi f\varepsilon' + C2\pi f\varepsilon''. \tag{7.10}$$

Consequently, a power P is dissipated in the material such as:

$$P = \frac{V_{max}^2}{2} C2\pi f\varepsilon''. \tag{7.11}$$

In the ideal situation, the electric field in the material being uniform (amplitude $E_{max} = V_{max}/d$), the expression of the power dissipated per unit volume is:

$$P/\text{Vol} = \varepsilon_0\varepsilon'' E_{max}^2 \pi f. \tag{7.12}$$

In this relation, we can notice that the quantity

$$\sigma = \varepsilon_0\varepsilon'' 2\pi f \tag{7.13}$$

is equivalent to the conductivity of the material.

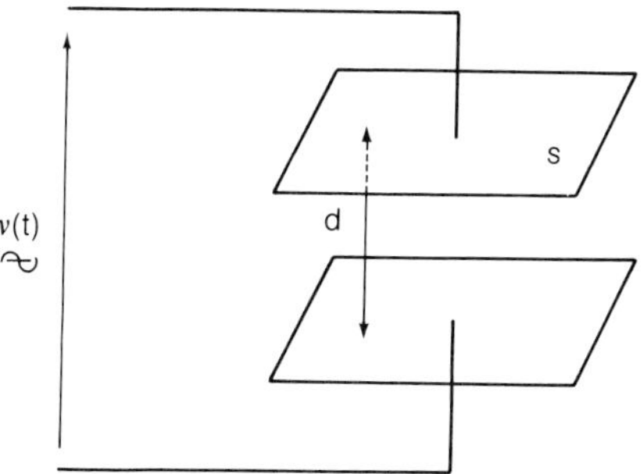

Fig. 7.1 Sine voltage applied to a plane capacitor and determination of the power deposition in a lossy material.

Anyway it can be shown, in a more general way, by application of Maxwell's equations, that the power deposition at any point of a lossy material may be described by the relation (7.12), that is depending on f, ε^*, and on the electric field E_{max} at the considered point. In a lot of practical cases, the physical model of power transfer can be conveniently described in terms of a travelling wave interacting with the material, such as described in the introduction of this section. Consequently, at first, reflection effects occur at the interface, afterwards the main heating process occurs in the skin depth of the material.

Note also a fundamental difference between a heating process by microwaves and by a classical way such as by infrared (for example in cooking): due to a greater penetration, microwaves generally heat quickly to a depth several centimetres, while heating by infrared is much more superficial.

We consider now the increase of the temperature ΔT as a function of time t in a material heated by microwaves. In the simplest case, when no transport phenomena of heat is occurring, then we can write ΔT as a function of the specific heat c such and of the specific mass ρ such as:

$$\Delta T = \frac{\varepsilon_0 \pi}{J} \frac{1}{\rho c} \varepsilon'' E_{max}^2 f t \tag{7.14}$$

with J the mechanical equivalence of a calorie.

As a matter of fact, an accurate knowledge of the temperature increase at any point of the material can only be obtained by the resolution of the heat transfer equation, a difficult problem requiring a good knowledge of the field distribution and of the heat transport parameters. A detailed description of the phenomena at the origin of dielectric absorption is out of the scope of this textbook. However the most common process arises from interactions of electric dipoles (such as polar molecules) with the a.c. field: the so-called dielectric losses result from phenomena which behave as mechanical frictions of the dipoles with the surrounding

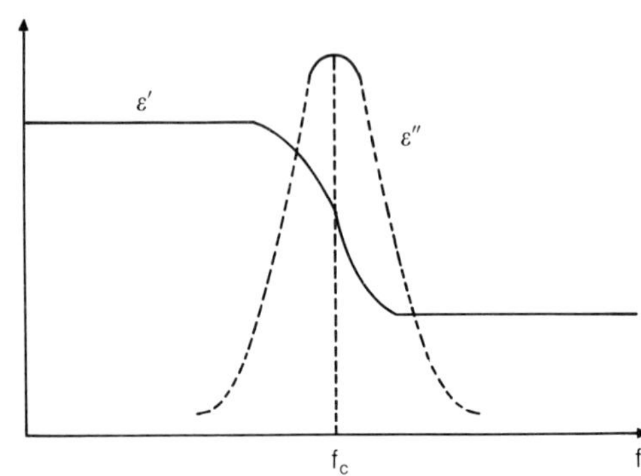

Fig. 7.2 Typical variation of the complex permittivity versus frequency (dipolar absorption process).

molecules. Other absorption effects result from charges accumulated in an heterogeneous material (Maxwell–Wagner effects) or from dipolar transition between two potential wells.

Anyway, the evolution of ε' with frequency is generally monotonic with a decrease near a frequency (characteristic frequency f_c), corresponding with a given phenomenon. Simultaneously, ε'' goes through a maximal value at f_c (Fig. 7.2).

Coupling microwaves with matter: applicators, the microwave oven

The applicator (or oven) is a necessary link between the microwave power generator and the material to be heated. It derives from previously described transmission devices such as waveguides (Volume 1), antennas (Volume 2) or resonant cavities (Volume 1). However these devices operate under different conditions which depend on the material to be heated.

An applicator must fulfil several conditions:

1. it must be fitted to the shape of the material to heat (bulk material, web, wire);
2. it must be matched to the generator in order to optimize the transmission of power, if necessary, a matching two port device is inserted between the generator and the applicator;
3. it must avoid any leakage of energy towards the outside.

The coupling devices can be classified as: travelling-wave applicators, monomode cavity applicators, multimode cavities, near field antennas.

Travelling-wave applicators These use standard waveguides in which the material (web or wire) is introduced. The research of a maximal coupling leads to position

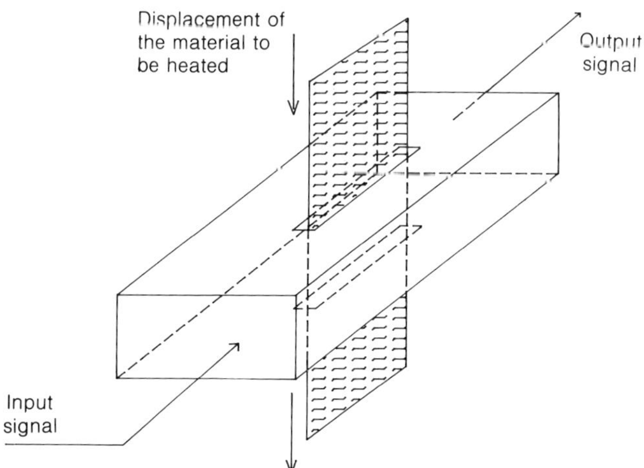

Fig. 7.3 An example of applicator for a web material (rectangular waveguide).

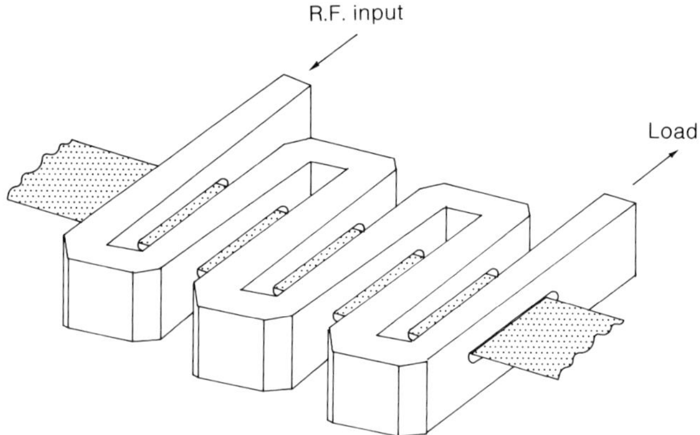

Fig. 7.4 An example of meander applicator for a web material.

the material where the field intensity is maximum. One can often assume the field distribution in the empty waveguide to be only slightly modified by the introduction of the material. Note also that the slits made in the waveguide, which allow the introduction of the material, must be designed such that they do not modify the fundamental propagation characteristics (the electric current lines inside the waveguide walls must not to be disturbed). These basic data lead to the design of applicators such as shown in Fig. 7.3. In an advanced version, a meander applicator ensures an increase of coupling between the wave and the material to be heated (Fig. 7.4). In a particular situation, a ridged waveguide is perferred to a rectangular waveguide in order to take advantage of a higher field intensity, whih produces a better coupling.

Monomode cavity applicators Starting from the same idea as for the travelling-wave applicators, a single mode cavity (rectangular or cylindrical) can also work as an applicator.

Let us recall that for a single mode of propagation one or more waves propagate in a transmission line:

1. a forward wave (such as in a travelling-wave applicator);
2. two waves (forward and reflected) which produce standing waves;
3. an infinity of waves (back and forth) in the case of cavity.

A cavity is defined by:

(a) a resonant frequency, depending on its dimensions and on the real part $\varepsilon'(f)$ of the permittivity of the material filling it (completely or partly);
(b) a quality factor Q defined such as:

$$Q = 2\pi \frac{\text{energy stored in the cavity}}{\text{energy dissipated by cycle}}. \tag{7.15}$$

Cavity applicators are generally rectangular or cylindrical waveguides. In the same way as for travelling-wave applicators, the material is positioned such that it receives a maximal amount of power. The power is generally transmitted from the generator to the cavity either through a coupling aperture of appropriate size, or by means of a monopole antenna or of a loop. A matching to the generator impedance depends on the cavity dimensions, on the losses of the material to heat and on the coupling device.

Multimode cavities (the microwave oven) The size of the previous applicators is governed by the wavelength of the microwave signal delivered by the generator: for example, travelling-wave applicators at 2450 MHz are standard S-band waveguides (10.92 cm × 5.46 cm). Consequently, such applicators are fitted and used only for bulk materials of small dimensions. For big samples, the dimensions of the oven may need to be greater than several wavelengths. In this situation, many propagation modes occur (Turner *et al.*1984). The well-known microwave oven for food applications is based on this principle. The coupling of the multimode cavity to the generator is ensured either by a horn or by waveguide with slits arranged such that the power is transmitted to the multimode cavity.

As a matter of fact, the power deposition distribution in the material to be heated is difficult to forecast because it depends both on its size and on its position in the oven. Consequently a non-uniform heating generally occurs in the material. Different solutions are applied. Ovens for cooking are equipped with a so-called mode stirrer (a metallic moving device which modifies the field distribution continuously) and with a rotating turntable. Industrial furnaces often use a conveyor belt, and are fed by several or many magnetrons.

Near field antennas Another situation consists of heating only a part of the lossy material by means of one or several applicators in contact with it. Such applicators are mainly used for biomedical engineering, in the treatment of cancers by heat (hyperthermia).

These applicators are horns, open-ended waveguides (Guy *et al.*, 1978) or openings made in the mass plane of a microstrip line (patch antennas) (Ledée *et al.*, 1985). The power deposition can be computed from the near-field effects: it depends both on the attenuation of the material and on diffraction effects produced by the aperture (Guy *et al.*, 1978, Robillard *et al.*, 1982, Mamouni *et al.*, 1988).

Advantages and applications of the microwave heating

Heating by means of microwave power is devoted to industrial applications, to cooking and also to medical applications.

In the first case, the microwave power presents the following advantages:

1. a quicker heating than with conventional processes (the penetration is much greater than in infrared; the heating process is more advantageous than with hot air);

2. cleaner systems;
3. a possibility of combination with a conventional process (microwave pre-heating).

The most current industrial applications are (Metaxas and Meredolth, 1983):

1. pasteurization of vegetables;
2. baking or doughnut frying;
3. thawing and heating food;
4. drying of paper, printing ink, food, fruit juice, textiles (Fig. 7.5);
5. thermal treatment of materials (Fig. 7.6), of pharmaceutical products;
6. retreatment of asphalt, engine oil, rubber;
7. vulcanization of rubber and elastomers;
8. heating of semiconductors.

Medical applications are mainly related to hyperthermia therapy for the treatment of cancer. In this process, the tumorous tissues must be brought to a temperature of 43 °C. A suitable cooling of the superficial tissues makes it possible to conveniently heat tissue volumes at a depth of up to several centimetres, while avoiding superficial burns. Figures 7.7 and 7.8 show hyperthermia systems developed at this time respectively by BSD Medical Corporation (USA) and Odam-Bruker (France).

Note also that the amount of power considered in these applications is roughly:

1. from several ten watts to several 100 kW in industrial applications (generally by addition of modular sources of about 1 kW);

Fig. 7.5 Microwave drier for textile bobbins (the power 12 kW is provided by 10 magnetrons) (by courtesy of Sairem Corporation France).

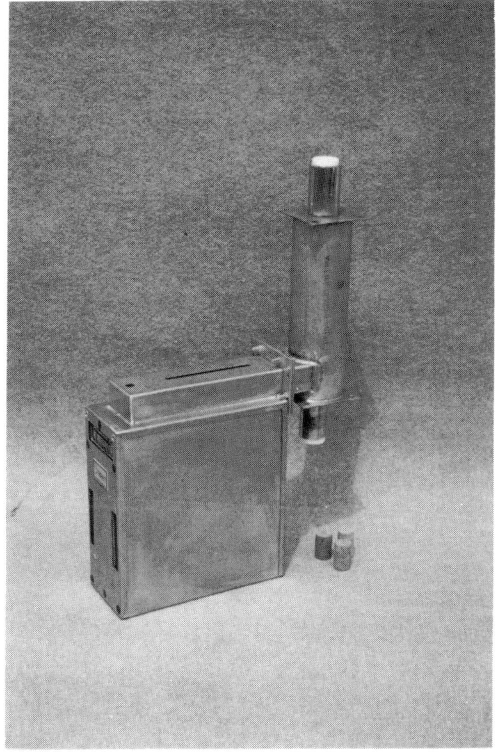

Fig. 7.6 Microwave source (1200 W) equipped with a monomode cylindrical cavity. This device takes its place in a system devoted to on-line reheating of tops (6000 tops per hour) (by courtesy of Sairem Corporation France).

2. about 800 W for the domestic microwave oven;
3. from several 10 W to several hundred watts for medical applications.

7.2.2 High energy scientific applications

Plasma for thermonuclear fusion

Among all possible sources of energy, only two offer the prospect of lasting at least as long as mankind, while abundant enough to satisfy all the needs: thermonuclear fusion energy and solar energy (which is also of thermonuclear origin). This explains why, since the 1950s, so much effort has been and still is devoted to thermonuclear fusion.

Basically, the mechanism is as follows. When two light nuclei collide, if their relative kinetic energy is sufficient for them to get over or even only to get through,

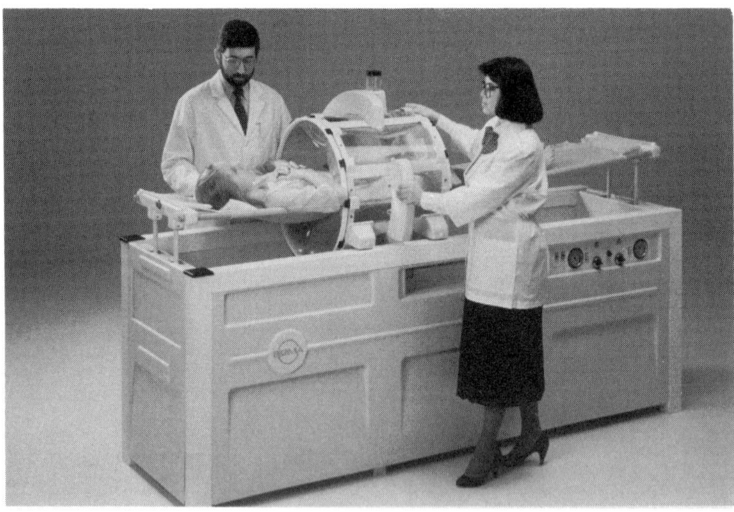

Fig. 7.7 Hyperthermia system BSD 2000 (by courtesy of BSD Medical Corporation USA).

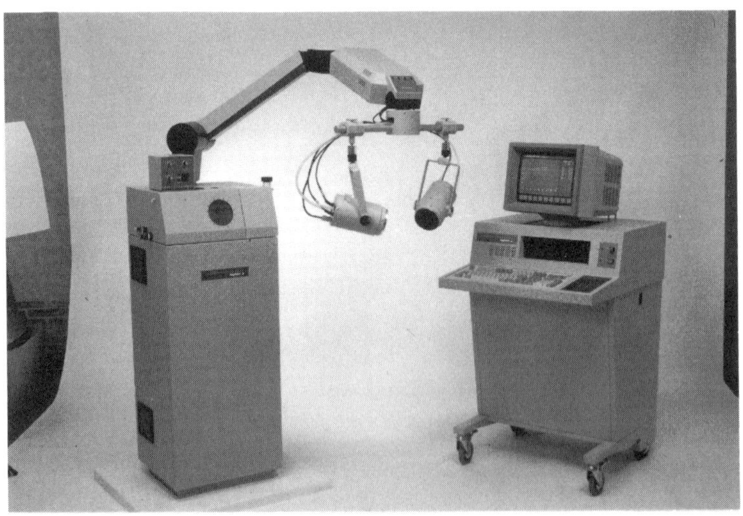

Fig. 7.8 Hyperthermia system (HYLCAR II) (by courtesy of Odam-Bruker, France).

by tunnel effect, the potential barrier which separates them, they will fuse and in the process release an amount of energy which can be quite large, as exemplified by the deuterium tritium reaction:

$$D + T \rightarrow {}^4He + n + 17.6\,MeV.$$

In order to make use of this property, one is led first to ionize the gases so as to be able to heat them up by external electromagnetic fields and at the same time keep the hot plasma thus obtained confined by a magnetic field, without any contact with the walls of the vacuum envelope which contains it. These ideas are behind the main approach to thermonuclear fusion, namely the magnetic confinement of plasmas, which is mostly implemented by Tokamaks (Fig. 7.9).

In this device, the plasma contained in a toroidal vacuum-tight envelop (Fig. 7.10) acts as the secondary winding of a pulse transformer. A large current flows in the azimuthal direction inducing a strong magnetic field surrounding the plasma which confines it very effectively. Some extra magnetic coils (not shown on the figure) are wound on the torus, creating an azimuthal magnetic field which perfects the confinement and stabilizes the plasma.

The circulating current heats up the plasma by the Joule effect. As the plasma temperature rises, however, its resistance decreases as $T_e^{-3/2}$, T_e being the electronic temperature, so that this mechanism becomes inoperative for $T_e > 1$ or a few keV (1 keV is equivalent to 11 600 K). In order to obtain a substantial fusion reaction, a temperature of at least 10 keV is necessary in the DT case; much higher temperatures are required for other mixtures.

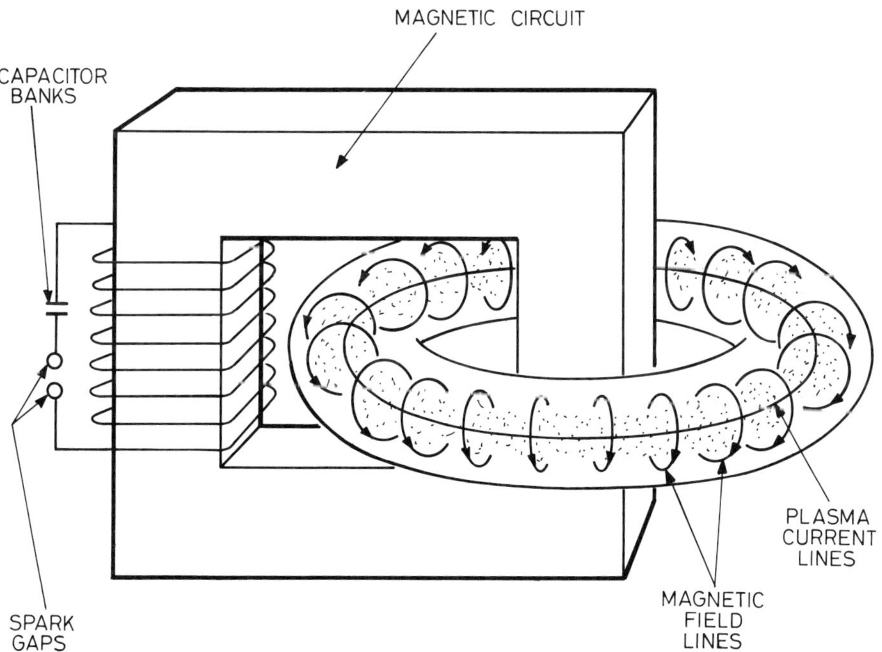

Fig. 7.9 Plasma confined by a magnetic field acts as the secondary winding of a pulse transformer.

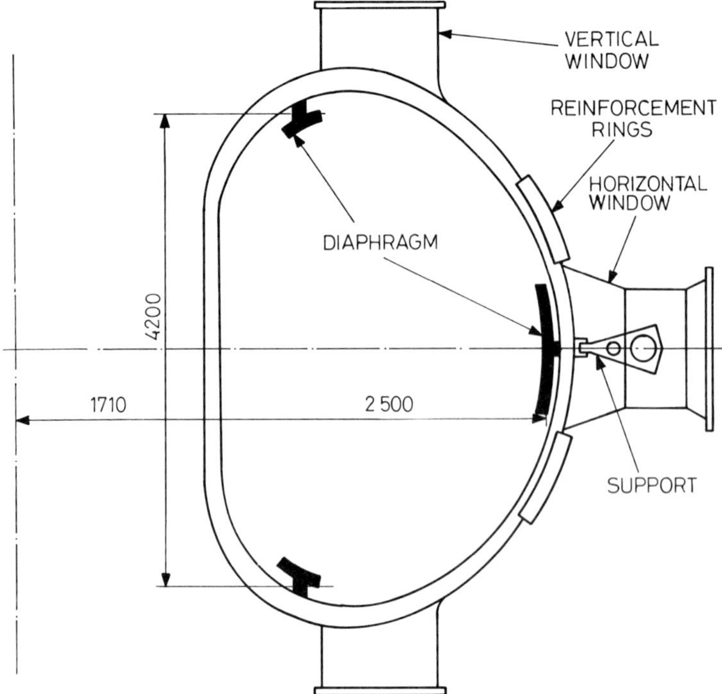

Fig. 7.10 Vacuum tight toroidal chamber of a Tokamak thermonuclear fusion machine.

This is where high frequency power makes its appearance for this application, as an extra means for heating the plasma. Three types of interaction, corresponding to three frequency domains are being used.

1. Ion cyclotron resonance heating in the 40 to 150 MHz range excites the cyclotron resonance of the ions in the local magnetic field. The RF power is generated by high power conventional tetrodes, each delivering from several hundred kW to 2.5 MW in CW operation. Recently (end of 1991), the JET (Joint European Torus) team, by using this method has obtained a thermonuclear reaction of several MW lasting 2 s, the plasma temperature rising to 30 keV.
2. Lower hybrid resonance heating in the 1 to 8 GHz range. This is a resonance which affects both electrons and ions, and which is approximately located at the geometric mean of the ion cyclotron and the electron cyclotron resonances. The sources are high power klystrons delivering several hundred kW up to 1 MW in long pulses (several seconds) and also high power gyrotrons (1 MW at 8 GHz).
3. Electron cyclotron resonance heating in the 60 to 150 GHz range. This type of heating has not yet been well explored for lack of high power sources in

this range until the advent of the gyrotron (industrially, at the beginning of the eighties). For long pulses ($t > 0.1$ s), the sources are still limited to a few hundred kW, mostly by window problems.

Another application of microwaves in Tokamaks is the following. It has been shown that it is possible to sustain a permanent direct current by means of an electromagnetic wave interacting with the electrons of the plasma. So, while the basic Tokamak is inherently a pulsed device, it should be possible to convert it into a permanent power generator. Gyrotrons seem most promising for this application, the interaction being all the more efficient the higher the frequency.

Particle accelerators

Early in the twentieth century, Rutherford had shown that it was possible to analyse the atomic structure of matter by bombarding it with subatomic particles and observing their scattering, as well as the fragments released during the bombardment. At the time, only natural radioactive sources were available and the need for more flexible sources of particles of higher energy and higher density was soon felt. In order to satisfy this need, particle accelerators began to appear in the late 1920s. The electrically charged particles were first accelerated by electrostatic fields (Van de Graaff accelerators, for example), but difficulties with high voltages led rapidly to the invention of machines making use of high frequency fields.

Here, the idea is to apply to the particle a short duration accelerating field of high intensity but limited voltage, a large number of times along the particle trajectory. In order to achieve very high energies (GeV or even TeV, that is, 10^9 to 10^{12} eV) which are necessary to explore the substructure of elementary particles (quarks, gluons, etc.), very long acceleration trajectories must be used. These long trajectories can be of two types: either a circular or re-entrant trajectory so that each particle passes a large number of times across a small number of accelerating RF gaps disposed along the path (the case of the cyclotron or the synchrotron, for example); or a straight line accelerating trajectory, which requires a large number of accelerating RF gaps (linear accelerator).

The linear accelerator The linear accelerator (linac) can best be considered as a big coupled-cavity travelling-wave tube (TWT) operating in reverse, that is, giving up microwave energy to the particle beam, thus increasing its kinetic energy.

In a good vacuum (10^{-8} to 10^{-11} torr), a particle beam is injected in a fashion similar to that of the TWT, in fact identical in the case of electrons, the only difference being that the linac gun has a much lower perveance (higher voltage, lower current) than that for a TWT. The beam goes through a first section which gathers the particles into bunches while accelerating them. Because the microwave energy injected into the section is very large (several MW of peak power), the fields in the structure are considerable (10^7 to 10^8 V/m) and bunch the beam very tightly: there is one bunch per wavelength in the structure, each bunch being about 1% of the wavelength long at its end. By a careful variation of the cell length as well as of the size of the coupling holes, the wave remains in step with the beam while its field remains constant along the structure in spite of the absorption of energy by the beam and ohmic losses.

To be more specific, let us consider the case of an electron linac. The basic accelerating structure (Fig. 7.11) consists of a disk-loaded waveguide. The fields of a travelling electromagnetic wave are shown in the figure. The coupling between cavities is accomplished by the electric field through the centre holes. Since the electron is a light particle, it very quickly reaches a velocity extremely close to the speed of light so that all sections except the buncher have a constant cell length and the wave a constant velocity equal to the speed of light. The narrow bunch rides the wave in a phase region where it is strongly accelerated and also phase focused, that is, where the last electrons of the bunch are somewhat more accelerated than those in front, so that the bunch gets tighter as it progresses. Figure 7.12 shows a simplified block diagram of a linac.

The energy of the linac increases linearly with its length. The largest linac in the world is the SLAC (Stanford Linear Accelerator Center), a two-mile long machine made of 960 sections 3.05 m long, fed by 245 klystrons delivering 60 MW each at a frequency of 2856 MHz and a repetition rate of 180 Hz with a peak beam current of 50 mA.

Because most of the microwave power gets consumed in ohmic losses for conventional linacs, the tendancy is now to replace the regular copper sections by sections made of a superconducting material such as niobium, so that the losses become negligible. Despite the complications arising from having to operate the sections in liquid helium, the savings in operating costs are considerable.

Linear accelerators also exist for the acceleration of ions, mostly as injectors used to feed synchrotrons. They operate at a much lower frequency (200 MHz), and are limited in energy (a few hundred MeV).

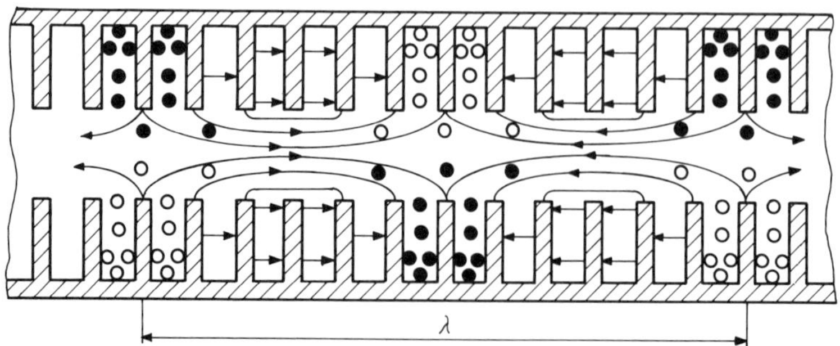

Fig. 7.11 Formation of a travelling wave in the iris-loaded resonator of an RF linear accelerator.

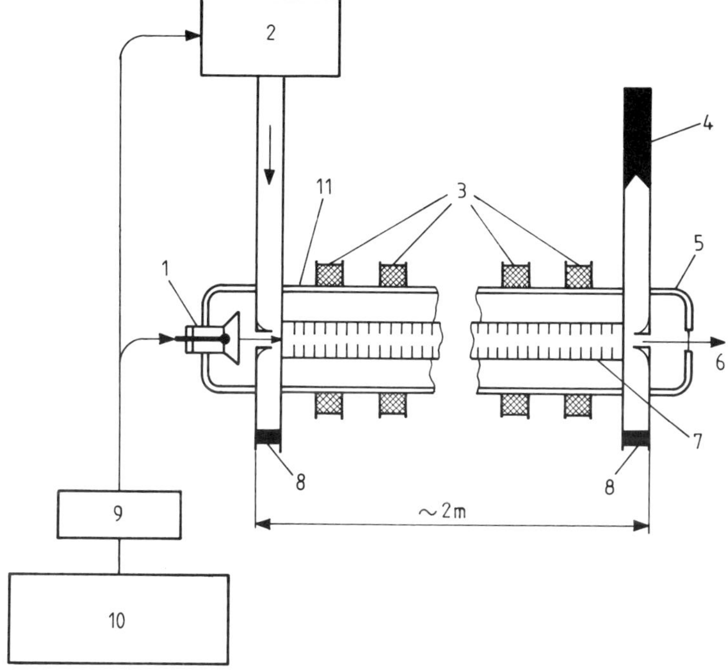

Fig. 7.12 Simplified block diagram of a travelling-wave electron linac: (1) electron gun; (2) 3 GHz klystron; (3) magnetic focusing coils; (4) load resistance; (5) vacuum tight chamber; (6) electron beam; (7) cavity resonators; (8) impedance matching (9) synchronization; (10) control system; (11) bunch resonator.

Circular and re-entrant accelerators

THE CYCLOTRON

In the cyclotron (Figure 7.13), the particles—usually positive ions—follow a spiral trajectory under the influence of a large transverse magnetic field associated with an accelerating RF field whose frequency is equal to the natural rotational frequency associated with the magnetic field, the cyclotron frequency. The expression for the cyclotron angular frequency ω_c is as follows:

$$\omega_c = qB/m$$

where q is the electric charge (C) m is the mass of the particle (kg) and B is the magnetic induction (T).

For most usual particles, since the magnetic induction cannot exceed 2 T because of saturation of the magnet, the frequency is 10 to 30 MHz. The RF power originating from conventional tetrodes is applied to two dees (hollow semi-circular accelerating electrodes) as shown in Fig. 7.14. The particle is accele-

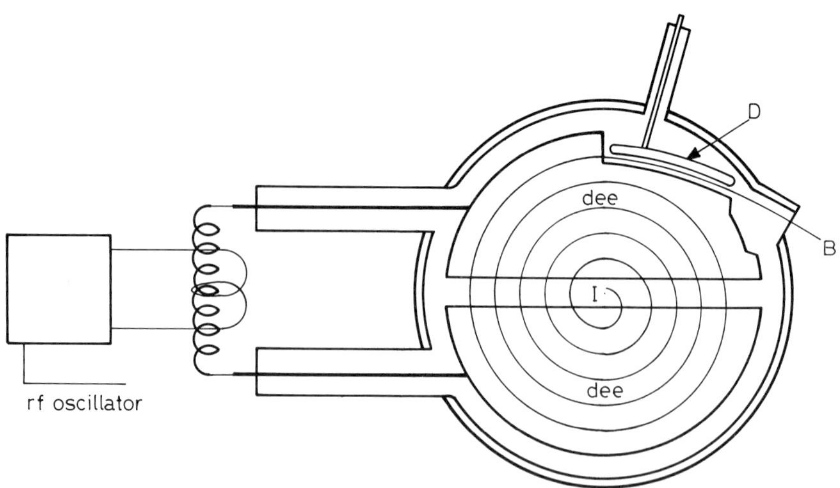

Fig. 7.13 The particles in a cyclotron follow a spiral path starting from the ion source I at the centre of the machine, and are accelerated by an externally applied RF field each time they cross the gap between the dees until they are extracted by the deflector plate D and exit the machine at B.

rated each time it passes the gap between the dees, the RF field reversing itself while the particle describes a half-circle in the (RF field-free) space inside each dee.

Due to relativistic effects (change of particle mass, thus variation of the cyclotron frequency), the maximum energy obtainable with a conventional cyclotron is limited, and this constraint is more severe the lighter the particle: 10 MeV for protons, 20 MeV for deuterons, 40 MeV for alpha particles (He^4 nuclei). Modification of the shape of the pole pieces allows higher limits to be reached for the isochronous cyclotron.

THE MICROTRON

This type of accelerator, devoted solely to electrons, is derived from the cyclotron with an important difference: its design takes relativistic effects into account, which are particularly important in the case of electrons (Fig. 7.15).

Due to the low mass of the electron, the frequency is now in the microwave range (3 GHz for example). The beam goes through the gap of a single cavity with an RF drive such that every time the particle crosses the gap, it is given an impulse of energy of exactly 511 keV or an integral multiple thereof. In this way, the relativistic mass of the electron is always an integral multiple of its rest mass, and the time taken by an electron to complete a full circle is also an integral multiple of the basic cyclotron period, so that every time it passes the gap it is in phase with the accelerating field.

Energies higher than 30 MeV have been obtained with some 30 orbits and pulsed currents of 30 mA. These accelerators are powered either by magnetrons

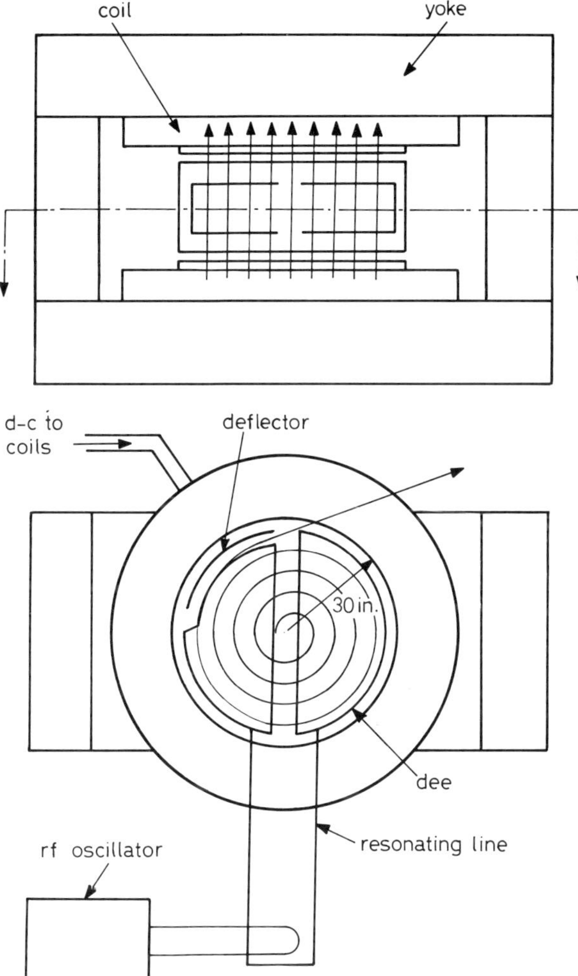

Fig. 7.14 Schematic views of a cyclotron, showing the magnetic circuit and pole pieces (top), and the RF circuit with particle trajectory (bottom).

or klystrons. More powerful versions have been devised where the cavity is replaced by a linear accelerator section, the magnet being divided into parts such as in the racetrack microtron (Fig. 7.16).

THE SYNCHROTRON

The sychrotron is another circular machine whose principle consists in accelerating particles while they move in a stable orbit, which means that the magnetic field must increase with time. When the particles are electrons, they very quickly reach the speed of light, so that the time necessary for a complete revolution

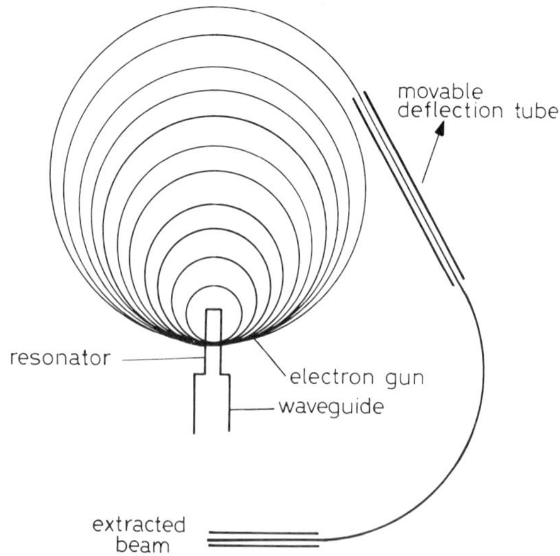

Fig. 7.15 Schematic diagram of a circular microtron accelerator.

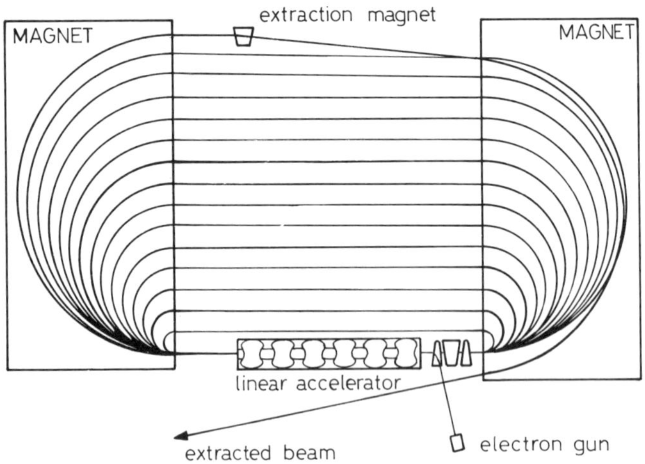

Fig. 7.16 Schematic diagram of a racetrack microtron accelerator.

stays constant and so does the RF frequency during the pulse. This is not true for a heavier particle such as a proton, in which case both the magnetic field and the frequency vary during the pulse.

The acceleration is achieved by single cavities or by sections of linac, either straight or bent, fed by tetrodes or klystrons, followed by bent drift tubes inserted between alternating gradient pole pieces which provide the guidance for the

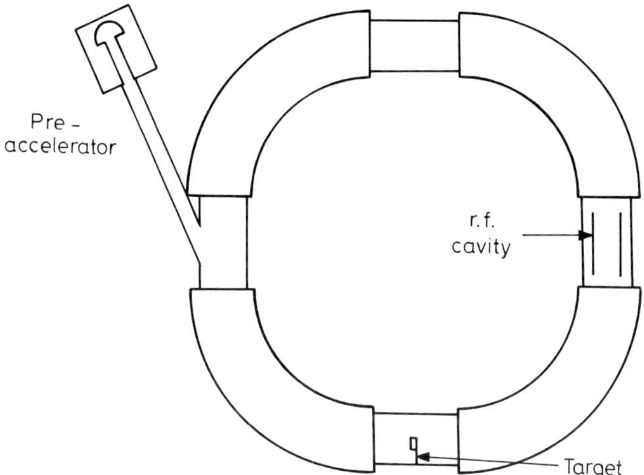

Fig. 7.17 Schematic diagram of a proton synchrotron.

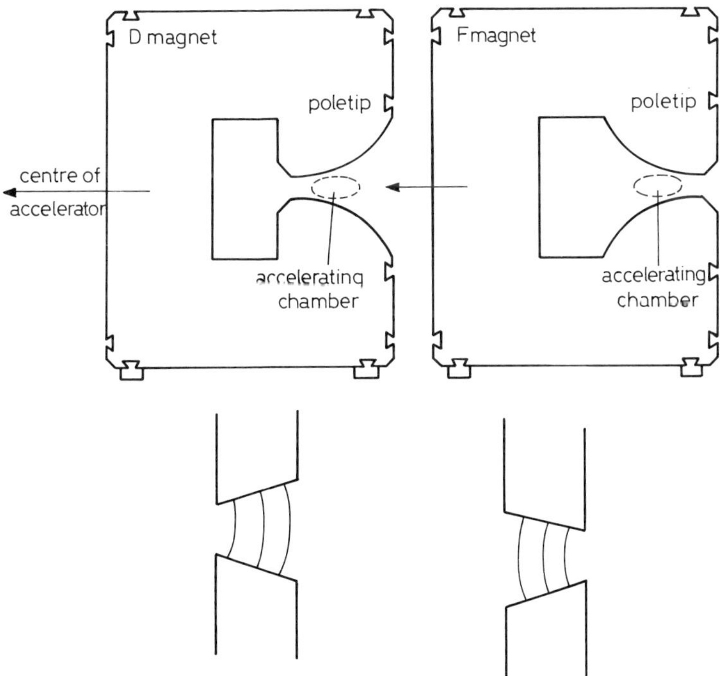

Fig. 7.18 Profiles of poletips of synchrotron magnets for strong focusing of beam: axial (vertical) focusing, D magnet (left); radial (horizontal) focusing, F magnet (right).

particles (Figs. 7.17 and 7.18). The injection of particles into the machine is achieved by a linac or a linac followed by a smaller synchrotron. By means of a fast varying magnetic field, the beam is extracted at the end of the pulse and sent onto a target. Here too, the trend is to use superconducting cavities and also superconducting magnets. At the present time, the largest proton synchrotron in the world is Fermilab at the 1 TeV level, using superconducting magnets.

Colliding beams When a very energetic particle strikes another particle at rest, most of the energy after the collision is devoted to the motion of the centre of gravity of the two particles. This becomes overwhelming when the incident particle is ultrarelativistic, for its mass has increased by the ratio of its energy to the rest energy of the particle. Thus for a 1 GeV electron, this ratio is about 2000. Therefore only a small part of the total energy available is being spent in collision processes.

If, however, two particles of identical mass and energy collide head on, the energy corresponding to the motion of the centre of gravity becomes nil, hence the idea of colliding beams. For example, electrons at 50 GeV striking positrons having the same energy (case of LEP—large electron–positron ring at CERN), cause the same reactions as a beam of almost 10 000 TeV hitting a stationary target.

The price to be paid is the small probability of collision, hence the small number of events per unit of time and the need for very large and very sophisticated detectors. Nevertheless, it is at present the only way to obtain such energetic reactions. Two approaches are being used: either the use of counter-rotating beams of particles and antiparticles in a synchrotron (case of LEP), or the acceleration of a beam of particles and a beam of antiparticles with the same linac, followed by separation of the beam of particles from the beam of antiparticles using a magnetic field, with subsequent guidance by bending magnets such as to bring the two beams into a head on collision (case of SLC at Stanford, using the SLAC).

Applications

SCIENTIFIC

It is fair to say that, without accelerators, our knowledge of the structure of matter would still be standing still where it was 50 years ago. Quantum electrodynamics, quantum chromodynamics, electroweak theory are all based on measurements made possible by accelerators. Except for the electron and the neutrino, all other leptons as well as quarks and gluons, that is all other fundamental particles are due to the accelerator. Furthermore, at lower energies, one can mention the production of beams of new particles (neutrino beams, meson factories), radiation chemistry where the application of very short pulses of high energy X-rays allows the study of transient chemical reactions via their fluorescence, the creation of transuranium elements (so far having very short lifetimes), the use as a source for synchrotron radiation, the most powerful source of X-rays known to date, etc.

MEDICAL

The accelerator is a powerful tool for the treatment of cancer tumours by bombarding them either directly with electrons, or more often, by producing X-rays of high energy. A large number of machines are installed in hospitals delivering radiation from 4 MeV up to 40 MeV. The accelerators are mostly linacs powered either by magnetrons or by klystrons, all in the 3 GHz range. Cyclotrons are used to produce radioisotopes useful in other types of radiotherapy, taking advantage of the specific affinity of certain tissues for certain elements of the periodic table.

INDUSTRIAL

A number of applications are being implemented, but at a slow pace, the investment cost being the limiting factor despite the fact that the total hourly cost is usually quite competitive. One can mention the following: food preservation, medical sterilization (syringes), destruction of pests (insects), vulcanization of rubber or silicones, cross linkage of polymers, X-ray radiography of thick parts, neutron radiography, activation analysis.

Most of these applications are performed with low energy electron beams (a few MeV), taking care not to exceed 10 MeV where induced radioactivity may appear. The accelerators are high intensity linacs; in the case of irradiation, the material is carried on a conveyor belt in a continuous process.

7.2.3 The problems of leakage: the personnel exposure standards

Undesirable microwave radiations

Different bandwidths are allocated for ISM applications. Consequently any frequency shift of the generators must be corrected (for example the bandwidth allocated around the ISM 2.45 GHz is ± 50 MHz). Harmonics also need to be avoided (for example the fifth harmonic from microwave ovens can interfere with a 12 GHz direct broadcasting satellite band in Europe) (Harada *et al.* 1987). The amount of power handled in the systems described here may be great enough that particular precautions must be taken in order to avoid the emission of spurious signals. For example, in microwave ovens, choked door seals make use of lossy materials and also of the impedance transformation of a quarter wavelength transmission line (Metaxas and Meredith, 1983). In on-line multimode applicators, the radiation by the openings is avoided by the positioning of absorbing loads.

Radiation monitors are systems which have been designed in order to verify that the spurious signals are sufficiently low. Moreover, standard safety levels have been defined in order to ensure the safety of citizens and operators. These points are reviewed hereafter.

Radiation monitors

These systems measure the field radiation potentially produced by the leakage of the microwave sources. At the distance of many wavelengths from the source, the propagation is TEM and the normalized plane power density (power by m^2) is directly determined by a measurement of the electric field E or magnetic field H:

$$P = E_{\mathrm{RMS}}^2/Z_0 = H_{\mathrm{RMS}}^2 Z_0 \tag{7.16}$$

with $Z_0 = 377\,\Omega$, the characteristic impedance of free space. Consequently, the power density can be directly determined by a measurement of E_{RMS} or H_{RMS}.

An electric field probe consists of three dipoles, associated with a lossy transmission line, which carries the signal to a square law detector (Fig. 7.19). The three dipoles are perpendicular, thus the total energy is provided from the addition of the three d.c. signals (Narda, 1988). However, in the region close to the source, situations may exist in which relation (7.16) no longer applies: a characterization of the power needs a separate determination of E_{RMS} and H_{RMS}.

Standard safety levels

Most countries have defined safety standards with respect to human exposure to radio frequency and microwave electromagnetic fields. The recommendations of the American National Standards Institute are explained in detail in the Radiofrequency protection guide ANSI C95.1 1982 (BEMS, 1982). In brief, the maximal power density as a function of the frequency is shown in Fig. 7.20.

This power density is limited to $10\,\mathrm{mW/cm^2}$ in microwaves (in another words, the specific absorption rate SAR or power/kg is lower than 0.4 W/kg). A smaller limit at lower frequencies comes from a possible resonance of the body working as a wire antenna: for example, the common human size corresponds to half a wavelength at about 100 MHz.

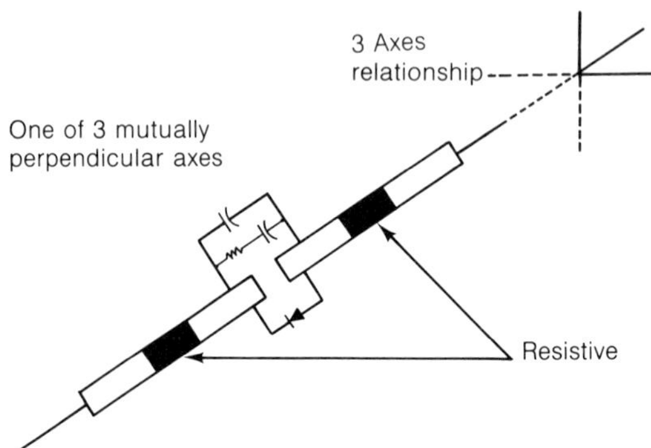

Fig. 7.19 A view of one of the three dipoles in a radiation monitor (by courtesy of Narda).

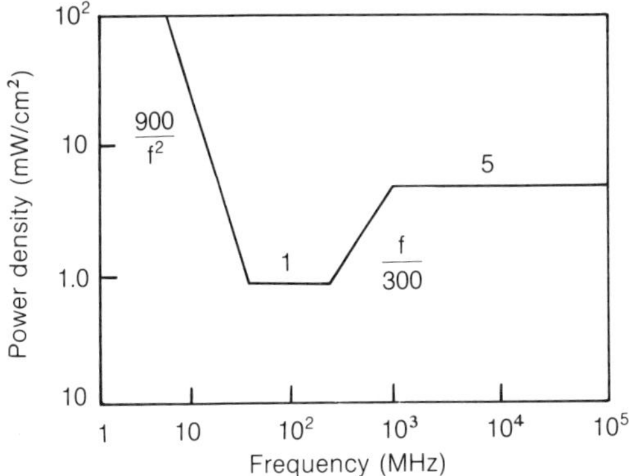

Fig. 7.20 Protection guide for whole body exposure (BEMS, 1982).

7.2.4 Conclusion

This section reviews the most salient high power applications of microwaves: other power applications will probably be developed in the future. Let us mention recent other new examples in quite different domains.

1. The projects SHARP (stationary high altitude relay platform) which consists of an aircraft powered by microwave radiation transmitted from ground (500 kW) and received by thin film antennas placed under its surface (Resnick and Stiglitz, 1988). The possible applications include surveillance, geological and agricultural surveys, atmospheric monitoring and coastal observations for search and rescue operations.
2. Another industrial application is devoted to ceramic sintering by means of a gyrotron (15 kW–28 GHz) (Schneiderman, 1989).
3. A microwave scalpel facilitates surgery on highly vascular organs by cauterizing blood vessels during cutting (Schneiderman 1989).

7.3 ACTIVE SENSORS AND SYSTEMS

The processes described in this section are qualified as active because they require a microwave source (unlike the systems described in the next section, called passive, which do not). Working at distances generally smaller than several tens of metres, they require a microwave power between several milliwatts and several watts. Some of them are radar type sensors, or short range transmission and identification systems; others are related to the interaction of microwaves with non-metallic matter and are devoted to non-destructive control and imaging.

7.3.1 Radar type sensors and miscellaneous

Principle and technology of radars have been reported in Chapter 4. Similar systems, but much less sophisticated are working at short distances. At the present

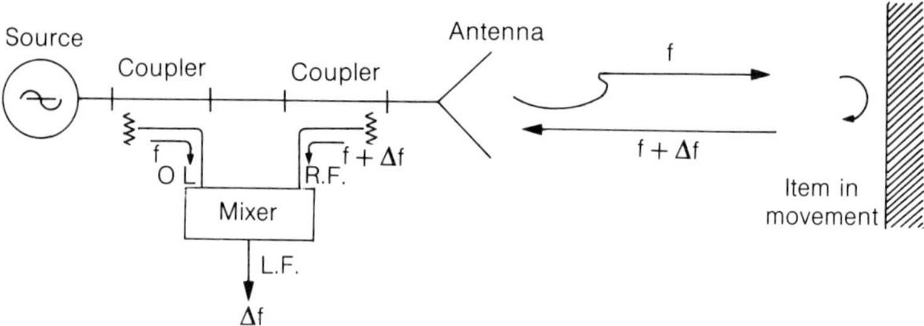

Fig. 7.21 Synoptic representation of a Doppler radar.

time, pulse radars are not used in these conditions which correspond to very short delay times and consequently need to work in a relatively wide bandwidth. The most current type of such sensors (Schilz and Schick, 1981) is the Doppler radar.

The corresponding basic system (Fig. 7.21) delivers an output signal of frequency Δf which is the difference between the transmitted frequency f and the received signal after reflection on the moving target. Then the velocity v of the target is deduced from the Doppler frequency by means of the expression:

$$\Delta f = 2\frac{v}{C} f \cos\alpha \tag{7.17}$$

where α is the angle of incidence.

Such a signal can be used on a qualitative point of view; a low frequency output signal indicates the presence of a moving item (volumetric detection, sensors for door-opening or traffic lights) and on a quantitative point of view; they measure the velocity of vehicles. Some modifications of this system lead to different possibilities.

In the frequency modulated continuous wave (FMCW) radar, such as applied for navigation (Chapter 4) a sawtooth modulated signal (frequency between f_1 and f_2, modulation period T) is transmitted. The measurement of the frequency:

$$\Delta f = 2\frac{f_2 - f_1}{Tc} l \tag{7.18}$$

of the output signal makes possible the determination of the distance l to the target.

Such a process is commercially available. It is used for the measurement of the level of a liquid in a container (industrial tank gauging). The precision announced by the firm Saab is ± 1 mm for a distance between 1 and 28 m.

In another application, the source being at a constant frequency f, the signal is transmitted towards a rough surface (Ishimaru, 1978). Then, a part of the beam is reflected following Snell's law, and moreover an important amount of signal is also scattered in all directions (Fig. 7.22). The part of the signal considered

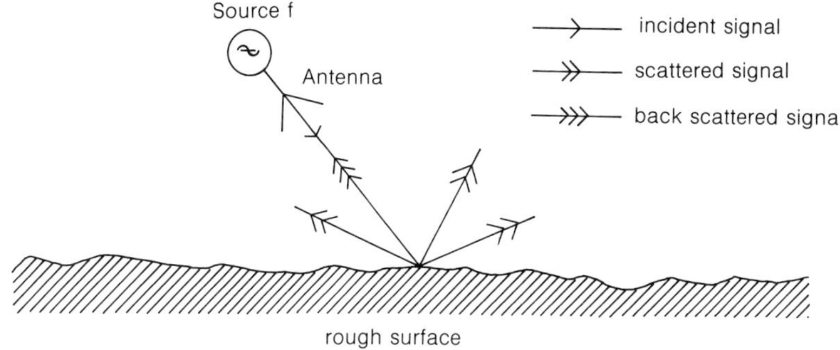

Fig. 7.22 Scattering of a signal by a rough surface.

in this application follows back the same path as the incident signal (back-scattered signal). When the system is moving with respect to the rough surface, the back-scattered signal exhibits a peak intensity at the frequency $f + \Delta f$, with Δf the frequency shift associated to the Doppler effect (Vindevoghel *et al.*, 1987). An appropriate data processing leads to the determination of Δf and consequently to the relative velocity of the mobile object on which the microwave sensor is fastened. An integration of the velocity with respect to time also provides a measurement of the distance.

A drawback of this method is that, unlike the classical Doppler method, the amplitude of the back-scattered signal is very weak (attenuation between -40 and -60 dB). Moreover, different causes of inaccuracy also result from:

1. the distribution of the transmitted signals in the antenna aperture;
2. the distribution of the back-scattering random surface;
3. the fact that the coherence time is limited to the time interval during which the same back-scattering material is in the area included in the radiation pattern of the antenna.

Such a process makes possible vehicle guidance and distance measurements even when a spinning of the wheels occurs. This process is being used presently by French Railways (ASTREE project) and by Dickey Jones Ltd (guidance of agricultural tractors). Dumoulin (1989) has obtained by this method an accuracy of 1/1000 over 1 km.

Note also the existence of short range transmission systems based on the transmission of digital messages between vehicles or between ground and vehicle, using a carrier in the microwave frequency range. Typical examples are devoted to the automation of subways (RATP, RER in France) and to toll motorways. Other similar kinds of sensors are being used for the identification of vehicles (French Railways) and the control of the operations in production line of cars (System Premid, developed by Philips Ltd).

7.3.2 Non-destructive control

Properties of non-metallic materials in microwaves are described by the complex permittivity:

$$\varepsilon^*(f) = \varepsilon'(f) - j\varepsilon''(f) \tag{7.1}$$

The knowledge of this parameter is of interest for different purposes such as:

1. to estimate its losses in order to forecast an object's ability to be heated by microwaves (Metaxas 1983) (section 7.2.1);
2. to measure its water content and its evolution during a process of drying;
3. for modelling of the power deposition in a complex structure (such as the human body);
4. for active microwave imaging (section 7.3.3) or radiometry (section 7.4);
5. for fundamental studies about matter.

An interest has been recently brought to lossy magnetic materials, characterized by a complex permeability

$$\mu^*(f) = \mu'(f) - j'\mu''(f). \tag{7.19}$$

This parameter can be measured by methods similar to dielectric measurements. Note that such materials are being used for shielding and construction of non-reflective areas (Stealth).

Measurement cells

The basic measurement cells of dielectric materials are made of sections of coaxial lines or waveguides (open-ended or short-circuited), which are filled up with the material under test. The permittivity of the material deduced from the impedance of the cell, requires at least two measurements (ε^* is a complex parameter) i.e. with different cell lengths or frequencies.

In situ measurements (or *in vivo* in the case of biomedical engineering) require different kinds of cells. Some of them consist in an open-ended waveguide (Ramachandraiah and Decreton, 1975) or coaxial line (Burdette, 1981) put flush on the material under test. Appropriate data processing has been elaborated for such sensors devoted mainly to non-destructive control and the measurement of the dielectric properties of living tissues.

Measurement systems

The measurement systems can be automatic network analysers such as described in Volume 2. However, thinking about a sensor, one generally refers to a simple system, working in limited conditions, mainly in a narrow frequency range.

As a matter of fact, such simple measurement systems exist; they are called six port networks (Evans, 1977) (Fig. 7.23). A basic measurement consists in the

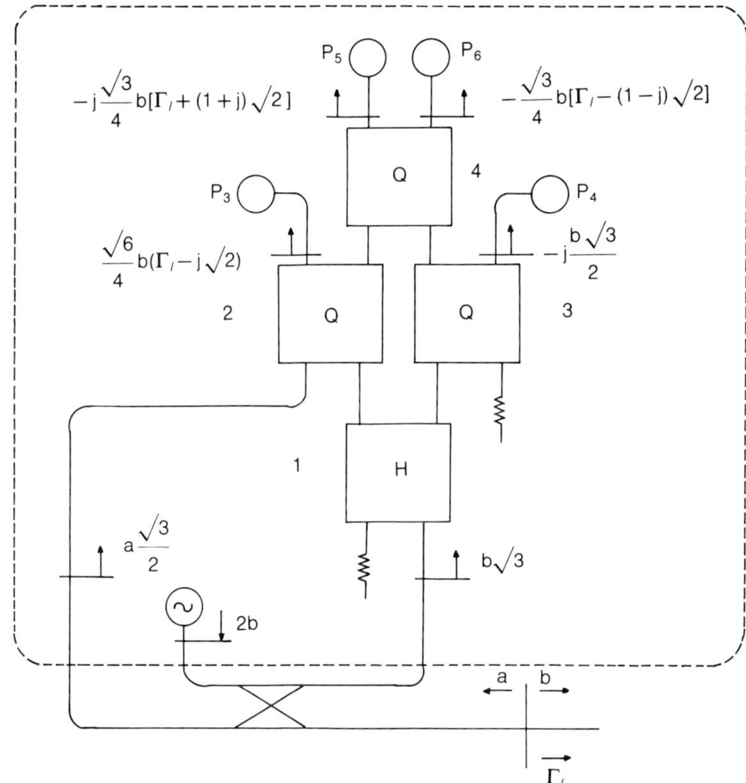

Fig. 7.23 Scheme of a six-port network (Q: 90° hybrid; H: 180° hybrid). The output signals measured by square-law detectors P_3, P_4, P_5, P_6 lead to the determination of the reflection coefficient Γ_l (Engen, 1977).

determination of the modulus $|\rho|$ and the phase Φ of the reflection coefficient Γ_l of the load. The solution derives from the four square law detectors, output signals which are proportional to the combination of quantities proportional to $|\rho|^2$, $|\rho| \cos \Phi$, $|\rho| \sin \Phi$. The combination of such two devices is also able to determine the scattering parameters of a quadripole, and consequently the parameters defining the transmission of a material.

Applications

Many authors have established tables of permittivity for different materials such as: food (Tran and Stuchly, 1987), cereal grains (Nelson, 1987); living tissues (Stuchly, 1980); other natural materials (ice, snow, brine, soils, vegetations)(Ulaby *et al.* 1986).

7.3.3 Microwave active imaging

A field of studies, which has developed in the last ten years, is related to the vision of objects in the microwave frequency range, in other words to the achievement of microwave radiography or tomography experiments, also called microwave active imaging. The term active introduces a differentiation with respect to passive or radiometric techniques (section 7.4). These studies in active imaging have mainly concerned medical applications.

The first significant result was obtained by the Walter Read Army Institute of Research (Larsen and Jacobi, 1979). According to this process, the images which are synthesized are two-dimensional arrays of microwave transmission coefficients (magnitude and phase): they depict the relative insertion loss and phase shift of a 3.9 GHz signal as it propagates through the structure under test: the examples considered are a phantom brain and an isolated canine kidney. The target and antenna are immersed in water, a material of high dielectric constant (Table 7.1), leading to a contraction of the wavelength of the interrogating radiations. The two advantages of this process are firstly a resolution (in

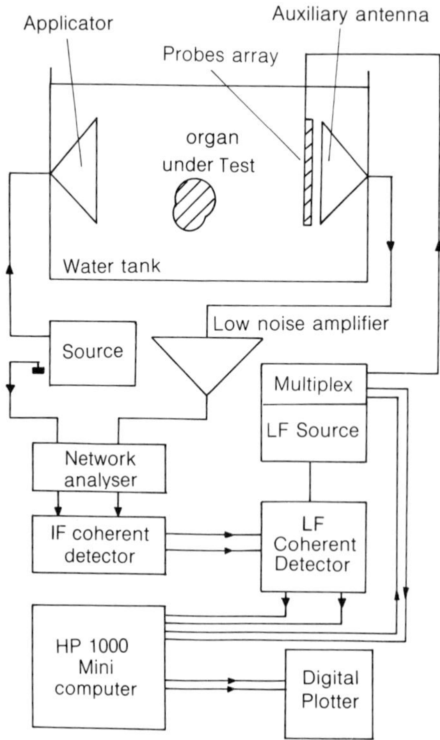

Fig. 7.24 Scheme of an active imaging system (Bolomey *et al.*, 1982): the field probing is done by means of the modulated scattered technique.

terms of separating two closely spaced objects) between 5 and 10 mm, and secondly the elimination of the need of an anechoic chamber (the free wave attenuation in water is 382 dB/m at 3 GHz). The images, which are represented with pseudo-colours, reveal details of the anatomy of the kidney, but the process is very long (4.5 hour) due to collection of 4096 pixel by means of an electromechanical scanner.

Another prototype (Bolomey *et al.*, 1982) introduces significant improvements. Firstly, the mechanical scanning is replaced by the modulated scattering technique: an array is constituted of probes loaded by non-linear elements. The signal resulting from the control of one element of the array is collected by an auxiliary antenna (Fig. 7.24); the phase and amplitude of the corresponding signal are measured and recorded. With such an electronic processing, an image of 1024 points is obtained after a time interval of only about 10 s.

Secondly, instead of assuming a straight line propagation, such as in the previous project, diffraction effects are taken into account: a tomographic process simulates a variable focal length; for each focal length the image of a thin organ slice is obtained in terms of the reconstructed equivalent currents.

Such systems, working at 3 and 2.45 GHz are devoted to medical applications (mainly the control of deep hyperthermia based on the variation of the permittivity of the living tissues as a function of temperature), and also to the test of antennas and the non-destructive testing of materials.

7.3.4 Conclusion

This presentation shows that if only a few sensors are working in the microwave frequency range, a lot of possibilities exist, which are still being studied at the present time. The development of microwave hybrid and monolithic integrated circuits (MMIC) (Volume 2) leading to a miniaturization, an easier feasibility of complex functions and a decrease of the cost is an element favourable to the development of such sensors and systems.

7.4 PASSIVE SENSORS AND SYSTEMS

7.4.1 Principles

Physical bases

The spectral radiation density transmitted by a lossy material (black-body) at an absolute temperature T is described by Planck's law:

$$B(f) = \frac{2hf^3}{C^2} \frac{1}{\exp\dfrac{hf}{kT} - 1}. \tag{7.19}$$

In microwaves, for temperatures greater than 100 K, this spectral density reduces to:

$$B(f) = \frac{2kTf^2}{C^2} \tag{7.20}$$

(Rayleigh Jean's law).

Further developments show that the noise power transmitted by a matched load at T is the same as received by an antenna surrounded by a lossy material at T:

$$P = kT\Delta f \tag{7.21}$$

with Δf the considered bandwidth (Evans and Leish, 1977). Consequently, in these ideal conditions of a matched impedance, we get a white noise (not depending on the frequency, but only on the bandwidth and temperature).

The scheme of Fig. 7.25(a) represents this situation. R_c is a matched detector; when the system is at a temperature T, the power transmitted by the lossy material surrounding the antenna balances the power transmitted by the matched load. This is a consequence of the second principle of thermodynamics (principle of detailed balancing). Moreover, for an increase ΔT of the temperature of the material, the power measured by the detector increases by an amount $k\Delta T\Delta f$.

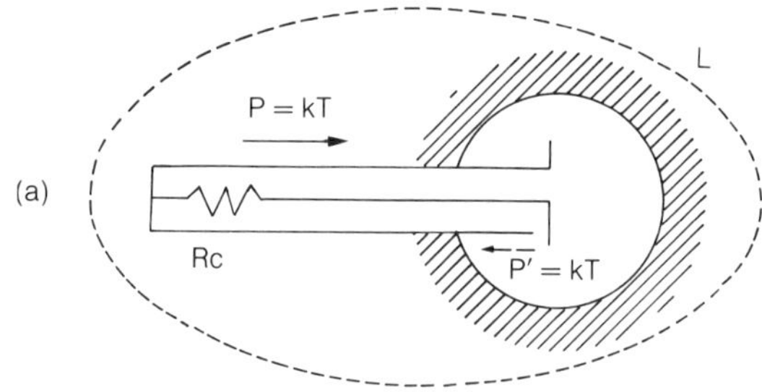

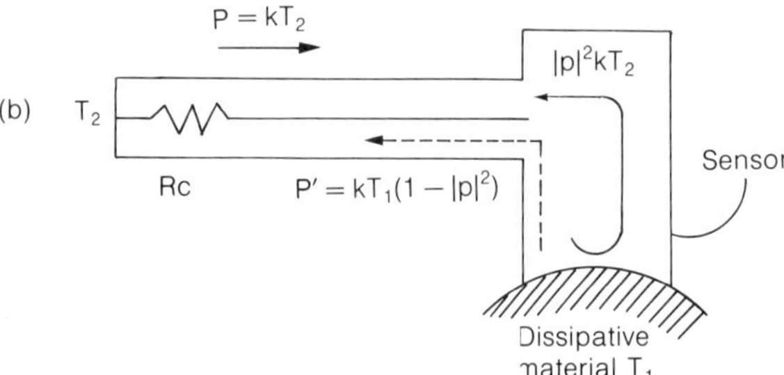

Fig. 7.25 Thermal noise power exchanged between a lossy material and a matched resistor; (a) case of a matched antenna, (b) case of a mismatched antenna (Leroy, 1987).

Consequently this process is able to define a non-contact thermometric process (Leroy, 1987).

However, as a matter of fact, the process is sometimes more complex. In Fig. 7.25(b) we represent a situation similar to Fig. 7.25(a) but in which the interface air-material presents a reflection coefficient ρ. Taking into account this reflection effect, and by application of the principle of detailed balancing, we get now a detected signal different from $kT_1\ \Delta f$ i.e.:

$$kT_1(1-|\rho|^2)\Delta f + kT_2|\rho|^2\Delta f \tag{7.22}$$

where T_1 is the temperature of the material, T_2 is the temperature of Rc; and $1-|\rho|^2$ is called the emissivity of the material.

In a similar way, in passive remote sensing an antenna pointed towards the ground receives a thermal noise signal which depends on the reflection coefficients of the ground, on the temperature of the ground, and on the temperature of the sky.

Computation of the radiometric signals

The expressions (7.21) and (7.22) define the radiometric signal for a material at uniform temperature. For a non-uniform temperature distribution, we have to consider the contribution of the different subvolumes ΔV_i of the lossy material coupled to the antenna. This contribution, proportional to the temperature T_i at this point, is also affected by a coupling parameter, or weighting function $C_i(f)$. The determination of this weighting function is also a consequence of the second principle of thermodynamics, and of the antenna reciprocity theorem. In this way, the weighting function is the same both in the forward path antenna to subvolume as in the inverse path. This weighting function is computed when considering the antenna to be active; in other words, $C_i(f)$ is proportional to the power deposited in ΔV_i by the antenna (Robillard *et al.*, 1982).

$$C_i(f) = A\sigma_i(f)E_i^2(f)\Delta V_i \tag{7.23}$$

where $E_i(f)$ is the field intensity in ΔV_i at f, $\sigma_i(f)$ is the conductivity of the material and A is a normalization parameter.

In these conditions, the contribution of the material to the radiometric signal is:

$$P(f) = \sum_i C_i T_i \Delta f \tag{7.24}$$

with

$$k(1-|\rho(f)|^2) = \sum_i C_i. \tag{7.25}$$

Conditions needed for a thermometric measurement

The thermal noise emission exists only in the case of a lossy material; this is a consequence of the principle of detailed balancing and of the reciprocity theorem.

Also, a high reflection coefficient at the interfance air–material limits the noise emission. So, in the case of metals, ρ is nearly equal to one and the emissivity is very weak. Furthermore, a thermometric process based on microwave radiometry can be carried out for materials at temperatures higher or lower than room temperature.

7.4.2 Radiometric receivers

In principle, the receiver is a square law detector with a very high sensitivity (for $\Delta f = 1$ GHz and $\Delta T = 1$ °C, the noise power variation is about 10^{-14} W). The temperature variation which can be detected is (Evans and Leish, 1977):

$$\Delta T_{\min} \approx \frac{T_1 + T_R}{\sqrt{\Delta f \Delta t}} \tag{7.26}$$

where T_1 is the temperature of the material, T_R is the noise temperature of the receiver, Δf is the bandwidth and Δt is the measurement time (or time constant of the integrator).

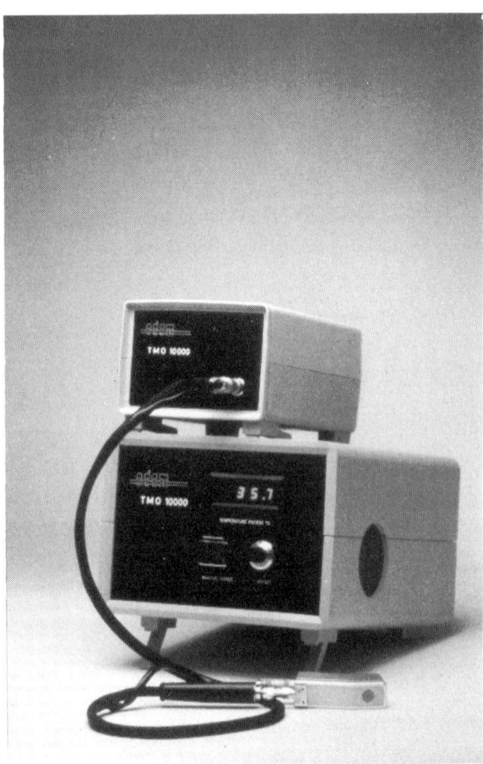

Fig. 7.26 Microwave radiometer at 10 GHz (by courtesy of Odam-Bruker).

The tangential sensitivity threshold (TST) of the square law detectors implies the use of an amplifier. In order to get a small value for ΔT_{min} the noise factor F of this amplifier must be optimal. Remember the expression of the noise factor:

$$F = \frac{T_R}{T_0} + 1 \tag{7.27}$$

where T_0 is the room temperature.

A typical radiometer has: a bandwidth $\sim$ 1 GHz, a gain $\sim$ 50 dB, and a noise factor $\sim$ several dB. In this way, for $\Delta t \sim 1$ s we get $\Delta T_{min} < 0.1$ °C.

As a matter of fact, the $1/f$ noise (low frequency gain fluctuation) has not been considered in the previous explanation. However, the $1/f$ noise can be limited by a switching process which continuously compares the noise signal to be measured and a reference noise signal (Dicke radiometer) (Dicke, 1946).

Appropriate improvements of the system have made it possible to measure the temperature T_1 of materials (uniform temperature) independently of the emissivity effects (zero method) (Mamouni *et al.*, 1976) (Fig. 7.26).

Note also the existence of correlation radiometers which carry out the correlation function of the noise signal received by two antennas. In this way, the classical radiometers measure a signal depending on the temperature in the bulk material under test; in a different way the correlation radiometers are sensitive to a temperature gradient in this material (Mamouni *et al.*, 1981).

7.4.3 Applications of radiometry

Undertaken mainly in the last ten years, these methods begin to interest research, medicine and industry.

Medical diagnosis and imaging

The relative transparency of living tissues to microwaves makes thermological investigations possible by means of microwave radiometry. However, the small temperature variations (at most several degrees) limits these observations to a depth of up to several centimetres.

Barret and P. C. Myers (Barret *et al.*, 1977) are the pioneers in this field. In the 1970s, they had already defined a method of breast cancer detection by a combination of microwave radiometry and infrared data. Since that time, other investigations have been carried out in this field (Land, 1983) (Mizushina *et al.*, 1986) (Bardatti, 1985).

A substantial improvement in such diagnostic investigations has been made possible by microwave radiometric imaging (Loery, 1987), a method based on the use of a multiprobe radiometer (Enel *et al.*, 1984). In this process, several antennas are sequentially connected to the radiometer. The multiprobe is for example, made of six rectangular waveguide apertures placed side by side. A given positioning of the multiprobe, associated to an appropriate processing of the radiometric data leads to a radiometric or thermal imaging. An evaluation of this process points out the possibility of detecting malignant non-palpable breast tumours, information which cannot be forecast by other classical methods of diagnosis (palpation, punction, mammography) (Giaux *et al.*, 1988).

Temperature control in hyperthermia

Microwave radiometry can also work in combination with heating by microwaves, such as in the treatment of cancer by heat (hyperthermia). This application requires that the radiometric receiver be made insensitive to signals transmitted by the microwave source devoted to the heating process.

Several methods leading to a satisfying decoupling between source and radiometers have been proposed (N'Guyen *et al.*, 1980). The hyperthermia system Hylcar developed by Odam-Bruker Corporation (Fig. 7.8) includes such a system of temperature control.

Thermometry of web materials and other applications

For web materials, waveguide apertures or patch antennas cannot work. A possible way consists of considering the same device as for heating webs by microwaves (section 7.2.1) (Fig. 7.3): as a consequence of reciprocity theorem, this device is suitable to perform radiometry. This interest of such a process consists

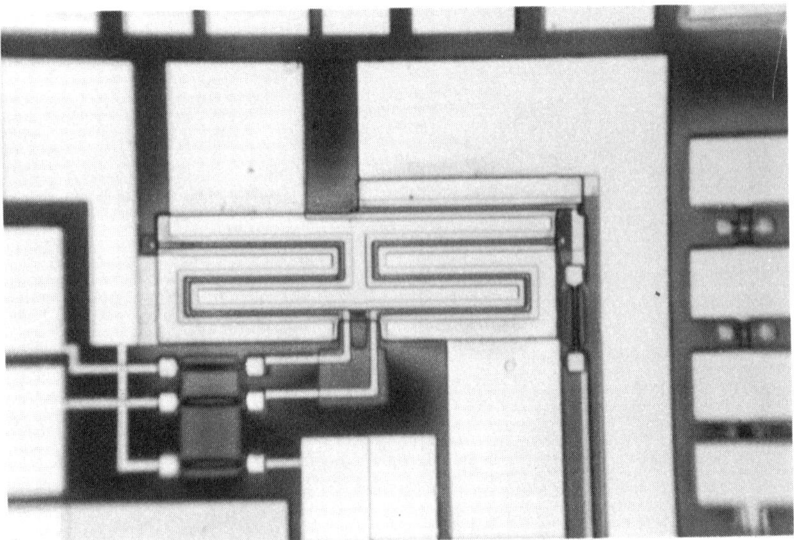

Fig. 7.27 Microwave GaAs monolithic integrated circuit used in the medical thermometer (Constant *et al.*, 1987).

of the definition of an on-line, non-contact thermometric sensor (Leroy *et al.*, 1986), a measurement needed in the quality treatments of textiles.

Other applications also concern the thermometry of bulk materials such as bituminous gravels: this problem is important because, considering its mechanical properties, asphalt used for roadworks must not be worked by road rollers when its temperature is lower than 80 °C; moreover no other thermometric sensor is able to work in this situation.

A quite different application is devoted to the road traffic management (Garceau *et al.*, 1987): when looking at the road, the radiometer measures approximately the ground temperature; when a vehicle enters the radiation pattern of the antenna, the radiometer measures the sky temperature (reflected by a metallic area).

7.4.4 Conclusion

The ISM applications of microwave radiometry are relatively recent; their interest for medical and industrial applications is growing. Future development of chip microwave radiometers using MMIC techniques probably will lead to many other new applications in such different areas as cooking control, the determination at home of body temperature, the temperature measurement of the tyres of cars, etc. As an example, we show in Fig. 7.27 an example of one such MMIC devoted to a microwave radiometer for medical applications (Constant *et al.*, 1987).

7.5 CONCLUSION

In conclusion, on one hand it seems that the high power applications, already existing for several decades, are still in progress; on the other hand the low power applications (active and passive systems and sensors) correspond, at this time, to a moderate development but their growing interest is due to the industrial conception of the MMIC, which decrease both the cost and size of the systems and make possible devices which were unthinkable several years ago.

REFERENCES

Bardatti, F., Montiardo, M. and Solimini, D. (1985) Inversion of microwave thermographic data by the singular function method *IEEE MTTS Digest*, 75–77.

Barret, A. H., Myers, P. C. and Sadowsky, N. L. (1977) Detection of breast cancer by microwave radiometry *Radioscience*, **12**, 167–171.

Bioelectromagnetics society newsletter (1982), **34**, 1–2.

Bolomey J. Ch., Izadnegahdar, A., Joffre, L., Pichot, C., Peronnet, G. and Solaimani, M. (1982) Microwave diffraction tomography for biomedical applications *IEEE MTT*, **30**, 11.

Born, M. and Wolf, E. (1975) *Principle of optics*, Pergamon Press, Oxford.

Burdette, E. C. (1981) In vivo probe measurement technique for determining dielectric permittivities at V.H.F. through microwave frequencies *IEEE MTT*, **28** (4), 414.

Constant, E., Leroy, Y., Van de Velde, J. C. (1987) Method and apparatus for measuring microwave noise US Patent 4.667.988, July 7 1987. Assignee CNRS Paris France.

Dicke, R. H. (1946) *Rev. of Scientific Instruments*, **17**, 268.

Dumoulin, G. (1989) Etude et réalisation d' une centrale cinémométrique hyperfréquence pour applications ferroviares Thèse Université-Lille, France.

Enel, L., Leroy, Y. Vandevelde, J. C. and Mamouni, A. (1984) Improved recognition of thermal structures by microwave radiometry *Electronics Letters*, 20.

Evans, G. and McLeish C. W. (1977) R. F. radiometer handbook Artech House, Washington.

Gargeau, R., Le Dinh, C. T., Loyolau, M. and Terreault G. (1987) Microwave vehicule sensor. *IEEE MTTS Symposium*, 200–204.

Giaux G., Delannoy, J., Delvalee, D., Leroy, Y., Bocquet, B., Mamouni, A., Van de Velde, J. C. (1988) IEEE MTTS Symposium, N. York, May.

Guy, A. N., Lehmann, J. F., Stonebridgee J. B., and Sorensen, C. C. (1978) Development of a 915 MHz direct contact applicator for therapeutic heating of tissues IEEE Trans MTT, **26 (8)**, 550–556.

Harada, A., Kitakaze, S. and Ogurot, T. (1987) Reduction of the 5th harmonic electromagnetic interference from magnetrons and microwave ovens *Jal of Microwave Power*, **22 (1)**, 3–11.

Ishimaru, A. (1978) *Wave propagation and scattering in random media Vol 1 (single scattering and transport theory)*, Academic Press, New York.

Land, D. V. (1983) Radiometer receivers for microwave thermography *Microwave journal*, **26 (5)**.

Larsen L. E., Jacobi, J. H. (1979) Microwave scattering parameter imagery of an isolated canine kidney *Medical Physics*, **6 (5)**.

Ledee, R., Chive, M. and Plancot, M. (1985) Microstrip microslot antennas for biomedical applications *EL. Letters*, **21 (7)**.

Leroy, Y., Van de Elde, J. C., Mamouni, A., Meyer, B. and Rochas, J. F. (1986) Contactless thermometry of a textile web by microwave radiometry 16th European Microwave Conference, Dublin, Microwave exhibitions and Publishers Ltd.

Leroy, Y. (1987) Radiométrie et thermographie microonde (T.M.O) Techniques de l'ingenieur (Paris) Mesures thermiques R 3030,1–8

Leroy, Y., Mamouni, A., Van de Velde, J. C., Bocquet, B. and Dujardin, B. (1987) Microwave radiometry for non invasive thermometry *Automedica*, **8 (4)**, 181–202.

Mamouni, A., Bliot, F., Leroy, Y. and Moschetto, Y. (1977) 7th European Microwave Conference Proceeding (Copenhagen) Sept. 1977 Microwave exhibitions and Publishers Ltd.

Mamouni, A., Van de Velde, J. C., Leroy, Y. (1981) New correlation radiometer for microwave thermography *El. Letters*, **17 (16)**.

Mamouni, A., Gelin, Ph., Leroy, Y. (1988) Modeling of radiometrics signals for medical applications European Microwave Conference Proceedings (Stockholm).

Metaxas, A. C. and Meredith, R. J. (1983) Industrial Microwave heating, Peter Peregrinus Ltd, London.

Mizushima, M., Hamamura, T., and Siguera, T. (1986) 3 band microwave radiometer system for non invasive measurement of the temperature at various depth *Proc. I.E.E.E. M.T.T. Digest*, 759–762.

N'Guyen, D. D., Chive, M., Leroy, Y. and Constant, E. (1980) Combination of local heating and radiometry by microwaves *I.E.E.E. T.I.M.*, **IM29 (2)**.

The Narda Microwave Corporation (1988) Microwave products and instruments Catalog 25, 329–390.

Nelson, N. O. (1987) Models for the dielectric constants of cereal grains and soybeans *Journal of Microwave Power*, **22 (1)**, 35.

Ramachandraiah, M. S. and Decreton, M. C. (1975) A resonant cavity approach for the determination of complex permittivity *I.E.E.E. Trans. I.M.*, **IM24**, 287–291.

Resnick, L. D. and Stiglitz, M. R. (1988) An airplane powered by microwave radiation *Microwave journal*, **3 (2)**66–71.

Robillard, M., Chive, M., Leroy, Y., Pichot, Ch. and Bolomey, J. Ch. (1982) Microwave thermography-Characteristics of waveguide applicators and signature of thermal structures *Journal of Microwave Power*, **17 (2)**, 97–105.

Rosenblatt, J. (1968) *Particle acceleration*, Methuen and Co., London.

Scharf, W. (1986) *Particle accelerators and their uses (Vol 2)*, Harwood Academic Publishers.

Schneiderman, R. (1989) Emerging Commercial applications *Microwaves and R.F.*, **28-3**, 35–44.

Schil, Z. and Schick, W. B. (1981) Microwave systems for industrial measurements *Ad. in Electro. and Elec. Physics*, **55**, 309–381.

Segre, E. (1988) *Nuclei and Particules*, W. A. Benjamin Inc., (Ch. 4).

Stuchly, M. A. (1980) Dielectric properties of biological substances tabulated *Journal of Microwave Power*, **15 (1)**, 181.

Tran., V. N. and Stuchly, S. S. (1987) Dielectric properties of beef, beef liver, chicken and salmon at frequencies from 100 to 2500 MHz *Journal of Microwave Power*, **22 (1)**, 29.

Turner, R., Voss, W., Tinga, W. and Baltes, H. (1984) On the counting of modes in rectangular cavities, *Journal of Microwave Power*, **19 (3)**, 199–208.

Ulaby, F. T., Moore, R. K., and Fung, A. K. (1986) *Microwave remote sensing active and passive Volume 3*, Artech House, Washington.

Vindevoghel, J., Baudet, J. and Deloof, P. (1987) Cinémomètre à effet Doppler; Colloque P.R.D.T.T.T.–Paris–Février.

8

Radioastronomy

Nguyen-Quang Rieu

8.1 INTRODUCTION

The universe has always been an object of curiosity for mankind. The phenomena in the sky were interpreted by our ancestors in ancient days as signs which governed their destiny. Astronomy took a giant step and became a science at the beginning of the 17th century, with the invention of the first astronomical instrument. Radioastronomy which consists of detecting radio signals from the universe was born only recently. The first radio emission of extraterrestrial origin came from the Milky Way and was discovered accidentally in 1932 by Karl Jansky while working at the Bell Telephone Laboratories. The radio signals from the sun, the closest star, were detected only a decade later. After these serendipitous discoveries, radioastronomy expanded rapidly after the Second World War, thanks to the development of more and more sophisticated antennas and receivers. Radio emission as weak as $10^{-12}\,\mu W$, coming from remote celestial objects can be detected by large modern radio telescopes which provide both high angular resolution and sensitivity.

Since the detection of solar radio emission, many more galactic radio sources associated with comets, planets, stars, gaseous nebulae, and pulsars, as well as other galaxies and quasars have been discovered. One of the most important discoveries in astronomy in the late 20th century, was the detection, also serendipitous, of the 3 K cosmic microwave background by two researchers of the Bell Telephone Laboratories, Penzias and Wilson, in 1965. This sea of black-body radiation is the residue of a much hotter universe which is believed to have been created by the giant explosion, the Big Bang, some 10–20 billion years ago. More recently, since 1970, the spectroscopic exploration in the millimetre range has revealed the existence of spectral lines emitted by many molecules, mostly organic, including the long carbon chain molecules (Table 8.1 in section 8.6). The discovery of intrastellar molecules is another milestone in modern astronomy and allows astrophysicists to investigate the dark component of the universe which has long escaped detection. In the following, we shall give an overview of the nature of radioastronomy along with some of its important contributions to the knowledge of the universe, using state-of-the-art technology.

8.2 RADIO TELESCOPES

The electromagnetic spectrum emitted by celestial objects extends from the ultra short wavelengths, the γ-, X- and ultraviolet (UV) rays, through the optical and infrared domain, to radio waves. However, most of the cosmic radiation in the electromagnetic spectrum is absorbed by the water vapour and carbon dioxide in the atmospheric layers or reflected back in space by the ionosphere, surrounding the earth. Only a few narrow 'windows' in the visible, infrared and radio wavelengths allow us to peer at the sky from the earth. The radio astronomical bands which are, in principle, protected from man-made emission, range from ~ 10 MHz (decametric waves) to ~ 300 GHz (millimetric waves).

8.2.1 Single dish

The cosmic signal is stochastic and obeys Gaussian statistics. It is recorded by a radio telescope which generally consists of a single antenna (single dish, as opposed to an interferometer, see section 8.2.2) equipped with a receiver. The amplified output is detected and then analysed by a spectrometer which is either an autocorrelator, a multi-channel filter bank or an acousto-optical device. The receiver is cooled in a cryogenic container in order to have the maximum performance. The receiver temperature can be as low as a few kelvin. Superconductor–insulator–superconductor (SIS) diodes are currently used to build low-noise receivers.

The accuracy of the surface of the antenna must be of the order of a few hundredths of the observed wavelength to achieve a good telescope efficiency. Millimetre radio telescopes have a surface accuracy of $\sim 30\,\mu$m. Fully steerable antennas are required to track radio sources during their diurnal rotation about the polar axis in order to detect their signals (Fig. 8.1). The integration time can be as long as 12 hours for weak sources.

The radio signal is characterized by its frequency and flux density. Some radio sources emit a broad continuum spectrum while others radiate narrow emission or absorption lines, depending on their physical nature (section 8.3). Since the cosmic signals are extremely weak, their flux density, S, is expressed in a very small unit, the jansky (1 Jy $= 10^{-26}$ W m^{-2} Hz^{-1}). The flux density is obtained by integrating the brightness, B, over the radio source

$$S = \iint B \, \mathrm{d}\Omega$$

$\mathrm{d}\Omega$ is an element of solid angle.

Another physical parameter used in radio astronomy to define the intensity is the antenna temperature, T_A. The antenna can be considered as a resistor. At a temperature T, the resistor delivers in a bandwidth, $\Delta\nu$, a power $P = kT\Delta\nu$ (k is Boltzmann's constant). The antenna temperature T_A (expressed in kelvin) is

Fig. 8.1 The radio telescope of IRAM (Institut de Radio Astronomie Millimétrique) installed in the Spanish Sierra Nevada (Pico Veleta) at an elevation of 2850 m is operated by the German Max-Planck Gesellschaft and the French Centre National de la Recherche Scientifique. It consists of an alt-azimuth mounted paraboloid reflector of 30 m diameter and is designed to work up to 350 GHz (0.85 mm). A temperature control system keeps the temperature of the back structure and the quadrupod constant to within 0.5° in order to avoid too large a variation of the beam shape and to get a good pointing accuracy in various weather conditions.

defined as equal to $T = P/(k\Delta\nu)$, which is precisely the temperature of the resistor which would produce the same power as that received by the antenna.

The antenna temperature from a direction of the sky of celestial coordinates, x_0, y_0, is given by the convolution of the true brightness distribution, $T_B(x, y)$, of the emitting region with the telescope beam pattern, $g(x, y)$, which is the response of the antenna to a point source (of negligible size with respect to the main beam of the antenna):

$$T_A(x_0, y_0) = \iint g(x_0 - x, y_0 - y).\, T_B(x, y)\,dx\,dy. \tag{8.1}$$

The gain is normalized so that

$$\iint g(x, y)\,dx\,dy = 1.$$

As an example, for a uniformly illuminated one-dimensional aperture of size D,

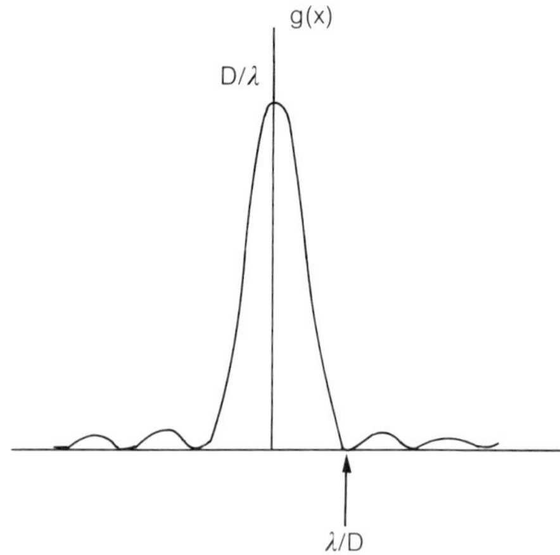

Fig. 8.2 The beam pattern of a finite one-dimensional antenna. It corresponds to the response to a point source.

the beam shape is

$$g(x) = \frac{D}{\lambda}\left[\frac{\sin(\pi Dx/\lambda)}{(\pi Dx/\lambda)}\right]^2. \tag{8.2}$$

This pattern consists of a main beam and a series of weak sidelobes (Fig. 8.2). The width to half-power of the main beam of such an antenna is 0.89 λ/D, where λ is the observed wavelength. The existence of sidelobes may be a disadvantage because they can pick up spurious signals from the neighbourhood of the observed direction. They could be attenuated by a tapered illumination of the antenna. The main beam would be broadened, however, thereby lowering the angular resolution. The noise fluctuations ΔT of a radioastronomical record depend on the temperature T_R and the bandwidth W of the receiver as well as the integration time τ spent in observing the source. The statistical theory shows that $\Delta T = T_R/\sqrt{W\tau}$. This expression represents the smallest antenna temperature detectable.

8.2.2 Interferometry and aperture synthesis

The goal of modern astronomy is to increase the angular resolution and sensitivity in order to detect more remote and therefore smaller (in angular size) objects. At 1 mm wavelength for example, the highest resolution achieved by the German–French 30 m antenna of the Institut de Radio Astronomie Millimétrique (IRAM) is about 10 arcseconds. The vast majority of galaxies and stellar objects within detection range of this radio telescope are smaller in angular size. As seen above, the larger telescope provides the better angular resolution. However, it is conceivable that the construction of very large single-dish instruments of hundreds of metres is technically difficult. Several telescopes of moderate size working in the interferometric mode can achieve much higher spatial resolutions, which only depend on the separation between the individual antennas. The details distinguished by the interferometer are as fine as those detected by a single dish which would have a diameter equal to the largest antenna separation (baseline). Interferometers of moderate baselines of hundreds of metres to tens of kilometres working at millimetre and centimetre wavelengths can give angular resolutions of a few arcseconds to a few hundredths of an arcsecond. Very long baseline interferometry (VLBI) involving an international network of radio telescopes located in different continents at distances of 10 000 km can perform angular resolutions of 10^{-4} arcsecond. Space VLBI projects will use both the ground-based radio telescopes and space antennas orbiting around the earth to achieve baselines up to 100 000 km and angular resolutions of the order of 10^{-5} arcsecond.

The basic principle

Let us consider two identical antennas working in the interferometric mode, by correlating their signals by the use of cables, waveguides or radio links (Fig. 8.3). The output voltages U result from the incoming signals E. Since the signal arriving at antenna 1 from a direction θ has a delay $\tau = (L \sin \theta)/c$ with respect to antenna 2,

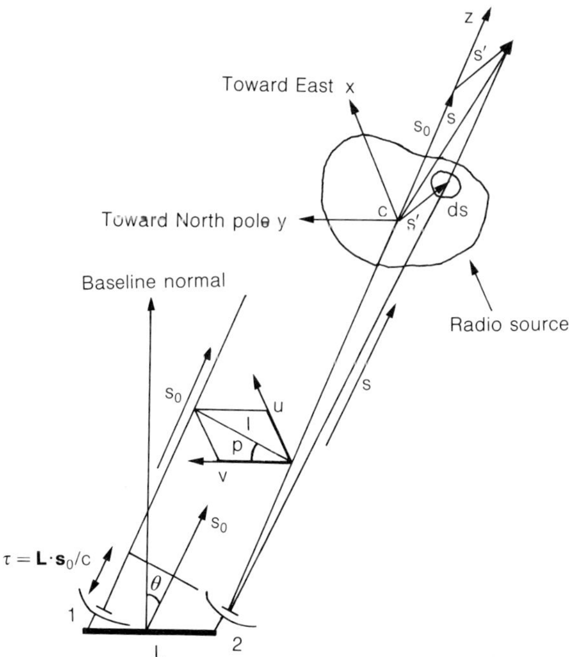

Fig. 8.3 Geometry of the interferometric system.

the output voltages are

$$U_1 \propto E\cos(\omega t) \qquad U_2 \propto E\cos[\omega(t-\tau)]$$

where L is the interferometer baseline, and ω is the frequency of the monochromatic signal.

The interferometer response to a point source results from the multiplication of the outputs

$$U_1 \times U_2 \propto \frac{E^2}{2}[\cos(2\omega t - \omega\tau) + \cos(\omega\tau)].$$

After filtering out the high-frequency term, the output is related to the source intensity I by

$$R(t) \propto I\cos[(2\pi L/\lambda)\sin\theta(t)].$$

It is more convenient to write the above expression in the complex form

$$R(t) \propto I\exp[(\mathrm{j}2\pi L/\lambda)\sin\theta(t)].$$

The phase term in the exponential contains the scalar product $\boldsymbol{L}\cdot\boldsymbol{s}_0$ which is the projection $\boldsymbol{l}$ of the baseline vector $\boldsymbol{L}$ on to the source direction of unit vector $\boldsymbol{s}_0$. We can write $R(t)$ in the vectorial form:

$$R(t) \propto I\exp[(\mathrm{j}2\pi/\lambda)\boldsymbol{L}\cdot\boldsymbol{s}_0]. \tag{8.3}$$

Owing to the Earth's rotation, θ, and hence the signal vary as a function of time giving rise to a fringe pattern. In fact, the fully steerable antennas follow the radio source in the sky to have the maximum intensity. The angular resolution is given by the fringe spacing which is the separation, λ/L, between two maxima or minima.

An extended radio source can be considered as an ensemble of point sources. The response of the interferometer is obtained by integrating the individual responses over the source in the sky and generalizing equation (8.3)

$$R(t) = \int B(\boldsymbol{s})\exp[(\mathrm{j}2\pi/\lambda)\boldsymbol{L}\cdot\boldsymbol{s})]\,\mathrm{d}\boldsymbol{s}.$$

If $\boldsymbol{s} = \boldsymbol{s}_0 + \boldsymbol{s}'$

$$R(t) = \int B(\boldsymbol{s}')\exp[(\mathrm{j}2\pi/\lambda)\boldsymbol{L}\cdot(\boldsymbol{s}_0(t) + \boldsymbol{s}')]\mathrm{d}\boldsymbol{s}' \tag{8.4}$$

B is the brightness of the extended source, $\boldsymbol{s}$ is the unit vector in the direction of an element $\mathrm{d}\boldsymbol{s}$ of the source, $\boldsymbol{s}_0$ is the unit vector in the direction z toward the field centre C of the observed field and $\boldsymbol{s}'$ is the vector determining the position of the element $\mathrm{d}\boldsymbol{s}$ relative to C (Fig. 3). Equation (A.4) can be written as:

$$R(t) = V\exp[(\mathrm{j}2\pi/\lambda)\boldsymbol{L}\cdot\boldsymbol{s}_0(t)] \tag{8.5}$$

where V is called the visibility function

$$V = \int B(s')\exp[(\mathrm{j}2\pi/\lambda)\boldsymbol{L}\cdot\boldsymbol{s}']\,\mathrm{d}\boldsymbol{s}'. \tag{8.6}$$

The argument of the exponential in equation (8.5) is similar to that in equation (8.3) and corresponds to the response of a point source at the phase centre C. The visibility function, V, represents the fringe pattern and its argument stands for the phase shift relative to the phase centre.

Let us define a Cartesian (u, v, w) coordinate system, with the u, v, axes perpendicular to the direction of the field centre taken as the w-axis. The x, y axes lie in the plane tangent to the celestial sphere at the field centre C and the z-axis is the same as the w-axis. The u and x-axes are oriented towards the east and the v and y-axes towards the north celestial pole (see Figs. 8.3 and 8.4). If the field of view is small, the (x, y) plane is almost the same as the portion of the sky in the neighbourhood of C. The distance from the field centre C is measured by x and y which are the components of $\boldsymbol{s}'$ in the u, v system. If the components of the projection $\boldsymbol{l}$ of the baseline $\boldsymbol{L}$ on to the plane of the sky are u, v, equation (8.6) can be written as:

$$V(u,v) = \int_{-\infty}^{\infty}\int_{-\infty}^{\infty} B(x,y)\exp[(\mathrm{j}2\pi/\lambda)(ux+vy)]\,\mathrm{d}x\,\mathrm{d}y \tag{8.7}$$

The visibility function is the two-dimensional Fourier transform of the source brightness distribution. Since B is a real function, its Fourier transform V is hermitian, that is $V(-u, -v) = V^*(u, v)$; V^* stands for the complex conjugate of V. Therefore it is sufficient to measure one half of the u, v plane; the other half is derived by symmetry with respect to the origin of the u, v coordinates. The brightness distribution is restored by an inverse Fourier transform of the visibility function

$$B(x,y) = \int_{-\infty}^{\infty}\int_{-\infty}^{\infty} V(u,v)\exp[(-\mathrm{j}2\pi/\lambda)(ux+vy)]\,\mathrm{d}u\,\mathrm{d}v \tag{8.8}$$

The spatial frequencies u, v are usually expressed in wavelengths.

The track in the spatial frequency plane

For an interferometer, the track in the spatial frequency u, v plane of the baseline vector as the Earth rotates depends on the coordinates of the source in the sky and the orientation of the baseline on the Earth. The geometry of the baseline–source system on the celestial sphere is characterized by the following parameters (Fig. 8.4).

The position C of the source in the celestial sphere is defined by its hour angle, h and its declination, δ. h is the angle between the meridian circle (passing from the north point N through the zenith Z to the south point S of the horizon) and

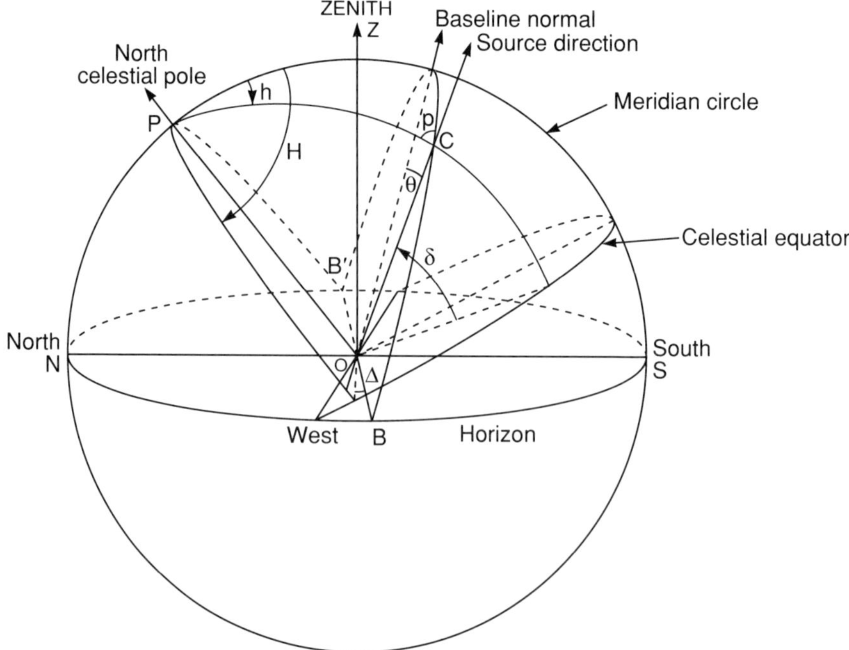

Fig. 8.4 Coordinates of the baseline vector and of the radio source in the celestial coordinate system.

the circle passing from the North celestial pole P through the radio source C. δ is the angle between the direction OC of the source and the celestial equator. The declination angle is counted positively from the celestial equator to the north pole. The interferometer baseline $\boldsymbol{L}$ which lies in the horizon plane intersects the celestial sphere at B and B′.

Likewise, H and Δ are the hour angle and the declination of the baseline, respectively. p is the angle between CP and CB, that is approximately the angle between the direction of the north pole and the projection of the baseline on to the plane of the sky at the source position C. θ is the angle between the source direction and the meridian circle. In the plane tangent to the celestial sphere at C, the position of the source, we can derive (Figs 8.3 and 8.4)

$$u = L \cos\theta \sin p \tag{8.9}$$

$$v = L \cos\theta \cos p. \tag{8.10}$$

The law of cosines in the spherical triangle CPB gives

$$\cos \mathrm{CB} = \cos \mathrm{PC} \cos \mathrm{PB} + \sin \mathrm{PC} \sin \mathrm{PB} \cos (H - h)$$

or in terms of θ, δ, Δ

$$\sin\theta = -\sin\delta \sin\Delta + \cos\delta \cos\Delta \cos(H - h) \tag{8.11}$$

and

$$\cos \mathrm{PB} = \cos \mathrm{CP} \cos \mathrm{CB} + \sin \mathrm{CP} \sin \mathrm{CB} \cos(\pi - p)$$

or

$$-\sin \Delta = \sin \delta \sin \theta - \cos \delta \cos \theta \cos p. \tag{8.12}$$

By using the law of sines in the same triangle, we derive

$$\sin(\pi - p) \sin \mathrm{CB} = \sin \mathrm{PB} \sin(H - h)$$

or

$$\sin p \cos \theta = \cos \Delta \sin(H - h). \tag{8.13}$$

Combining equation (8.9) with (8.13), and equation (8.10) with (8.11) and (8.12), we derive, respectively, u and v as a function of the hour angles and declinations of the radio source and of the baseline

$$u = L \cos \Delta \sin(H - h) \tag{8.14}$$

$$v = L \cos \Delta \sin \delta \cos(H - h) + L \sin \Delta \cos \delta. \tag{8.15}$$

Equations (8.14) and (8.15) show that as the source moves in the sky, the baseline vector describes in the u, v plane an ellipse centred at $u = 0$ and $v = L \sin \Delta \cos \delta$ and whose semi-major and semi-minor axes are respectively $L \cos \Delta$ and $L \cos \Delta \sin \delta$ (Fig. 8.5). For a source located at the pole, $\delta = 90°$, the u, v track becomes a circle. At the celestial equator, $\delta = 0$, the track is a straight line.

The two antennas can be moved during successive observing sessions to have several baselines oriented in different directions. The resolving power obtained with this aperture synthesis technique is equivalent to that obtained by a filled aperture much larger than the two elementary antennas. In fact, owing to the discrete spacings and hence a discrete coverage in the u, v plane, the map of the radio source is affected by spurious ripples. A satisfactory image restoration requires special treatments which clean the synthesized map. In practice, to speed up the mapping procedure, an array of N movable individual antennas is used to get $N(N-1)/2$ baselines at once. A T-shaped array oriented east–west and north–south or a Y-shaped array is currently used. Furthermore, the diurnal rotation of the earth also contributes to move the antennas and hence to fill the u, v plane. The largest and most famous array, the 'Very Large Array' (VLA) at Socorro in New Mexico, working at centimetre wavelengths consists of 27 movable antennas of 25 m diametre and is equivalent to 351 interferometers. The most remote antenna is located at 21 km from the array centre. Interferometers with more extended baselines are under construction. The very long baseline array (VLBA) has antennas placed at sites across the USA.

Very long baseline interferometry (VLBI) networks use a system of antennas widely separated by thousands of kilometres. The synchronization between antennas is performed by local atomic clocks. The data from simultaneous obser-

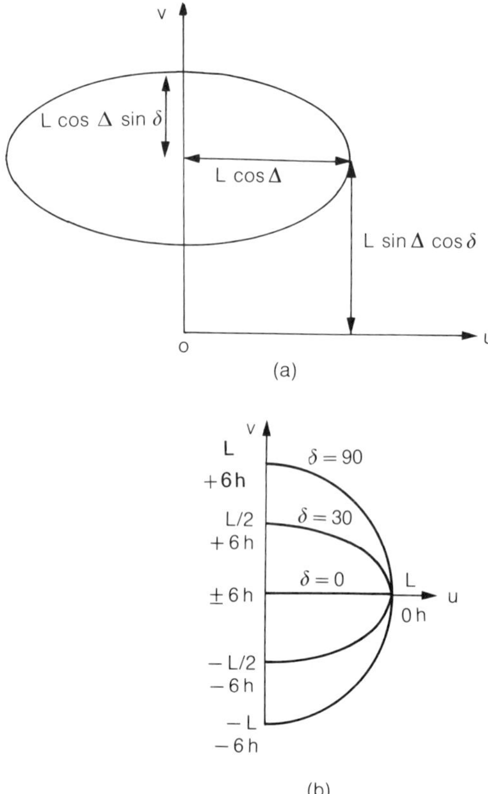

Fig. 8.5 The tracks in the spatial frequency (u, v) plane of a two-element interferometer during the earth's rotation. Each curve represents the locus of the projection of the baseline vector on to a plane perpendicular to the line of sight (see also Fig. 8.3); (a) skew baseline: the u, v track is an ellipse, the coverage of the other half u, v plane is obtained by symmetry with respect to the origin, o; (b) east–west baseline ($\Delta = 0°$; $H = 6h = 90°$) for radio sources at declinations $\delta = 0°$, $30°$, $90°$, and for hour angles from -6 h to $+6$ h. For hour angles from $+6$ h to -6 h, flip the curves around the v axis.

vations are recorded separately at each site and then cross-correlated with each other. Spatial resolutions of tenths of a milliarcsecond are currently achieved.

8.3 COSMIC RADIO EMISSION

Cosmic radio emission is essentially of two kinds: the continuum emission which occurs over a very broad frequency band, and the line emission radiated in a narrow frequency band. The analysis of radio lines requires high spectral resolution performed by a multi-channel spectrometer.

8.3.1 Continuum emission

Black-body radiation

Black-body radiation is a well-known process in which an object, heated at a temperature, T, emits electromagnetic waves. For instance, the planets or the interstellar dust grains heated by the sun or stars emit in the infrared and microwave bands. It is also the cause of the isotropic and ubiquitous 3 K cosmic black-body radiation which can be detected in the millimetre wavelengths. The intensity of the radiation can be derived from Planck's formula:

$$I = (2h\nu^3/c^2)\frac{1}{\exp(h\nu/kT) - 1} \tag{8.16}$$

ν is the observing frequency, c is the velocity of light, h and k are Planck's and Boltzmann's constants, respectively.

Thermal emission from ionized gas

The interstellar gas heated and ionized by ultraviolet stellar photons emit thermal radiation similar to the noise produced by a heated resistor. Thermal radiation from ionized gas is produced by the interaction between charged particles. The brightness temperature, T_b, defined by equation (8.16), depends on the electronic temperature, T_e, (temperature of electrons) which reflects the random motion of the particles. The volume of ionized gas is not completely transparent to the emitted radiation. Let us consider an elementary slab of thickness ds, inside an ionized gas cloud. The brightness temperature of this element is:

$$\mathrm{d}T_b = T_e \kappa \mathrm{d}s$$

where κ is the absorption coefficient per unit length, and κ ds is the opacity, dτ. Inside the cloud, at a distance s from the elementary slab along the line of sight, the total opacity is given by:

$$\tau = \int_0^s \kappa \mathrm{d}s.$$

At the distance s, the brightness temperature of the slab is attenuated by a factor of $\exp(-\tau)$, and becomes:

$$\mathrm{d}T_b = T_e \exp(-\tau) d\tau.$$

By integrating along the line of sight, and assuming that the electronic temperature is uniform in the cloud, we get for the output brightness:

$$T_b = T_e[1 - \exp(-\tau)] \tag{8.17}$$

Equation (8.17) shows that the brightness temperature of a radio source emitting thermal emission cannot exceed the electronic temperature. For an optically thick source ($\tau \gg 1$) the brightness and electronic temperatures are equal. The optically thin limit for T_b is $\tau\ T_e$. Thermal emission is detected in gaseous nebulae in which one or several hot stars are embedded and where the electronic temperature is usually smaller than approximately 10^4 K.

Synchrotron radiation

High brightness temperatures, approximately 10^6 K, arise from non-thermal processes involving the magnetic field. When an electron moves in a magnetic field at a relativistic velocity (close to the light velocity), it emits a radiation similar to that observed in a synchrotron. The gyration frequency of a relativistic electron spiralling along a line of magnetic force is:

$$\nu_H = \frac{eH_\perp}{2\pi mc} \cdot \frac{mc^2}{E} \tag{8.18}$$

e, m, and E are the charge, the mass and the energy of the electron. The ratio mc^2/E is the relativistic term $(1 - v^2/c^2)^{1/2}$ including the electron velocity v. $H_\perp$ is the component of the magnetic field perpendicular to the electron velocity.

The emission is confined to a narrow cone of aperture $\psi = mc^2/E$, whose axis coincides with the velocity vector. For an electron of 1 Gev (giga electronvolt), the cone angle ψ is only 1.8 arc minutes. An observer receives pulses of short duration. A Fourier analysis of this series of pulses gives rise to a quasi-continuous spectrum consisting of a large number of high overtones. The maximum density of emission is concentrated at $\nu_{max} \approx 16 H_\perp E^2$; ν is expressed in MHz, H in μG, and E in Gev. The emission frequency of a 1 Gev electron moving in a 10 μG galactic magnetic field is 160 MHz.

Galaxies and quasars radiate powerful synchrotron emission. The intensity depends on the distribution of the electron energy and of the configuration of the magnetic field in the source. The energy distribution of the relativistic electrons is usually assumed to follow a power law, $N(E) \propto E^{-\gamma}$. Two simple cases, an isotropic or a homogeneous magnetic field, are generally considered. The synchrotron emission can be highly polarized in the latter case.

8.3.2 Line emission

Atomic and molecular emission

An atom or a molecule emits or absorbs radiation at a series of well-determined frequencies. Each chemical species has its own signature in the spectrum. Spectroscopy therefore allows the determination of the chemical composition of the interstellar gas.

The energy of an atom or molecule is quantized in a series of discrete levels. The energy jumps from one level to another correspond to the energetic rearrangement depending upon the electron orbits or rotation and vibration of the molecule.

A photon of frequency ν is emitted or absorbed when the atomic or molecular energy changes from one quantum state to another. The frequency of these transitions is proportional to the energy difference ΔE between the two states, according to the relation $\Delta E = h\nu$. A photon is absorbed when the energy increases and is emitted when the energy decreases. The emission can be spontaneous or induced by an external source of energy. In principle, atomic and molecular emission is monochromatic. However, the atoms and molecules in an interstellar cloud move randomly and give rise to a broadened line centred at the rest frequency (corresponding to a particle at rest). The motion of the whole cloud with respect to the surroundings also shifts the line centre from the rest frequency. The line shift in frequency, $\Delta\nu$, is related to the cloud radial velocity, v, (velocity along the line of sight) through the Doppler formula, $v = -c(\Delta\nu/\nu)$, where ν is the observing frequency. By convention, v is positive for a source which recedes from us, and the line becomes redshifted ($\Delta\nu < 0$) with respect to the rest frequency. When the source is approaching, v is negative and the line is blueshifted ($\Delta\nu > 0$). The redshift of remote quasars and galaxies which recede with large radial velocities is $z = [(c+v)/(c-v)]^{1/2} - 1$. For closer objects with modest recession velocities, $v \ll c$, the redshift becomes $z \sim v/c$. The spectrum that is the line profile of an astronomical object represents the intensity against the radial velocity (Fig. 8.6).

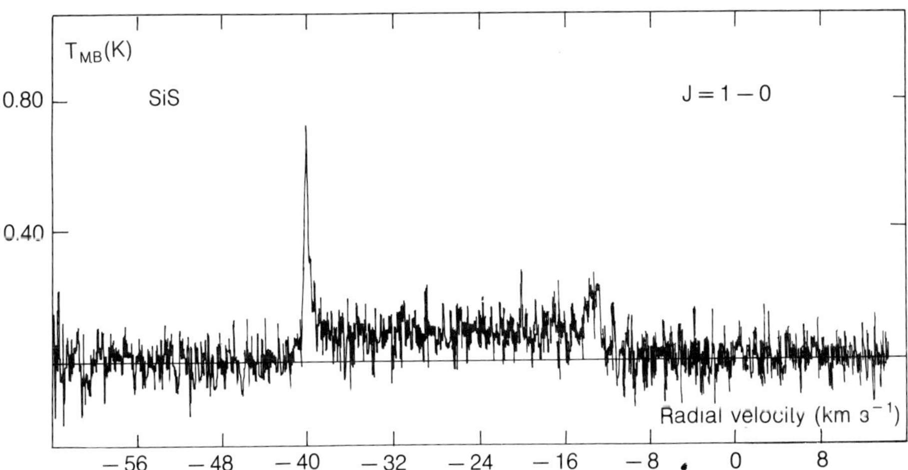

Fig. 8.6 The spectrum of the carbon star, IRC + 10216, corresponding to the transition at 18 GHz of the molecule SiS, observed by Nguyen-Q-Rieu, V. Bujarrabal, H. Olofsson, L. E. B. Johansson, and B. E. Turner (1985), using the 140 foot radio telescope of the NRAO (Green Bank, West Virginia). The y-axis represents the brightness temperature and the x-axis represents the line-of-sight (radial) velocity which is proportional to the frequency (Doppler effect). The expansion of the envelope produces a broad line profile, from ≈ -40 to ≈ -12 km/s. The narrow spike at about -40 km/s is a maser emission line arising from the amplification of the background stellar core by the circumstellar maser cloud.

When an electron is close to an ion, they can be bound together and recombine. The electrons cascade down through a series of intermediate lower levels and emit the so-called recombination lines. While atomic and molecular lines are detected in predominantly neutral regions, recombination lines are present in ionized gas.

Molecular line observations

Molecules are relatively abundant in the interstellar medium. The line intensity depends on the energy distribution of the ensemble of molecules present in the cloud. The probability of finding a molecule in the level n of energy E_n is given by the Boltzmann law: $g_n \exp(-E_n/kT)$. The 'statistical weight', g_n, denotes the number of sublevels. For a transition of frequency ν between the upper state u and the lower state l, the ratio of the populations of the upper and lower states is governed by the Boltzmann law

$$\frac{n_\mathrm{u}}{n_\mathrm{l}} = \frac{(N_\mathrm{u}/g_\mathrm{u})}{(N_\mathrm{l}/g_\mathrm{l})} = \exp(-h\nu/kT_\mathrm{ex}) \tag{8.19}$$

where N and n are the populations of a level and a sublevel, respectively.

The excitation temperature, T_ex, which is defined by equation (8.19), depends on the excitation conditions of the gas cloud. Collisions with other particles, namely atomic and molecular hydrogen, and bombardment by infrared photons from stars and warm dust grains are the main excitation processes. The line intensity radiated by a cloud is given by a relation similar to equation (8.17)

$$I(\nu) = I_\mathrm{ex}\{1 - \exp[-\tau(\nu)]\} \tag{8.20}$$

I_ex is the intrinsic cloud intensity which is self-absorbed inside the cloud of opacity τ. The intrinsic intensity I_ex is related to T_ex by the Planck formula (equation (8.16)). The opacity $\tau = \int \kappa \, \mathrm{d}s$ (section 8.3.1) depends on the populations n_u and n_l (section 8.6.4). The observation of the line spectrum leads to the determination of T_ex and τ, and hence n_u and n_l. These parameters allow us to explore the physical conditions of the interstellar cloud whose kinematics determines the shape of the line profile.

8.4 CONTINUUM RADIO SOURCES

8.4.1 The Galaxy

The radio emission of our Galaxy, the Milky Way, is made up of thermal and synchrotron radiation from the numerous complexes of discrete radio sources and the large scale interstellar medium.

The solar system

The sun, like more distant stars, emits thermal and synchrotron radiation. Different layers of the solar atmosphere can be investigated, depending on the observing frequency. The critical frequency depends on the electron density which varies with the altitude in the solar atmosphere. The dense inner region at 6000 K, the chromosphere, can be reached at millimetre wavelengths, whereas the upper layer at 10^6 K, the corona, can be observed in the metric waveband. The emission intensity follows an 11-year cycle. During the period of activity, solar 'bursts' a million times stronger than the intensity of the 'quiet sun' can occur.

The thermal emission from planets depends on their temperature (section 8.3.1). In some circumstances, thermal emission from the earth can increase the noise level of radio telescopes. Jupiter and the earth have a magnetic field in which relativistic electrons are trapped. Both planets are surrounded by a radiation belt emitting synchrotron radiation.

Protostars

Ionized gas clouds appear in the visible as luminous gaseous nebulae. They are called HII regions (as opposed to HI regions of neutral gas) and consist mostly of electrons and ions resulting from the ionization of atomic hydrogen and carbon by the radiation of embedded hot stars. They are sometimes bounded by an ionization front which represents the zone of interaction between the ionizing photons and the ambient neutral interstellar medium. Thermal emission of very compact HII regions has been detected by interferometers. These radio sources are probably associated with the embryos of stars, the protostars. In this environment, the density of material is high enough to trigger gravitational collapse, leading to the star formation.

Supernova remnants

Stars in the ultimate phase of their evolution eject material into the interstellar medium through a 'stellar wind'. Evolved massive stars can undergo a spectacular explosive mass loss, the supernova explosion. The interaction with the ambient gas produces shock waves compressing gas and magnetic fields in which relativistic particles are trapped. Synchrotron radiation detected in these supernova remnants is generally distributed in a hollow shell The rate of supernova explosion in our galaxy is about one every 50 years. The most famous supernova explosion occurred in the Milky Way in 1054 and its remnant christened the 'Crab Nebula' is nowadays a popular astronomical object. Bright filaments of material were ejected into interstellar space at a velocity of thousands of kilometres per second while the highly compressed core became a pulsar. The Crab Nebula is also a strong synchrotron source. The first naked-eye supernova explosion since the event detected by Kepler in 1604 was discovered in 1987. It happened in the closest galaxy, the Large Magellanic Cloud, located at 1.6×10^5 light years from the earth. A velocity outflow as high as 25 000 km/s has been detected.

Stars are formed by the gravitational collapse of dense interstellar clouds. The interstellar medium is chemically enriched by supernova explosions since the original material is processed in the stellar interior through a network of thermonuclear reactions.

Pulsars and Einstein's general relativity

Pulsars are known to be formed in the catastrophic supernova explosions. According to the classical scenario, an implosion occurs due to the gravitational collapse and creates a core of ultradense matter consisting of neutrons (neutron star), while the debris flows into interstellar space. The rotating neutron star has a strong magnetic dipole field, approximately 10^8 to 10^{12} G, whose axis is tilted with respect to the rotation axis. Relativistic charged particles accelerated in such a field radiate an anisotropic synchrotron radiation confined in a cone whose axis coincides with the magnetic axis. Since the star is spinning, usually with a period of about 0.2 to 2 s, a pulsed emission is produced, resulting in a 'lighthouse' effect. Most pulsars are a few million years old, and correspond to a class of young objects on the astronomical time scale. During their evolution both the magnetic field and the spin rate decrease.

Recently, a new class of pulsars with extremely short rotation periods, of the order of a few milliseconds, has been discovered. The fastest millisecond pulsar has a period of 1.558 ms (Fig. 8.7). Its associated neutron star with a typical mass similar to that of our sun but a size of only about 10 km rotates with an extraordinary rate of 642 turns per second. These objects are believed to be older (about 10^8 years) than pulsars with moderate spin frequencies but their magnetic field is weaker (approximately 10^9 G). If the pulsar belongs to a binary system, the associated neutron star can capture material from the atmosphere of the companion. The mass accretion increases the angular momentum and the slowly rotating neutron star can be accelerated to become a millisecond pulsar.

The basic property of millisecond pulsars is their high rotation frequency stability. Using a network of the world's best atomic clocks as time reference, the pulsar timing leads to a period stability of a few parts in 10^{14}. Millisecond pulsars can be considered as excellent natural time standards, which are at least as accurate as the best man-made clocks so far available. The first derivative of the period which gives the slow-down rotation rate is of the order of 10^{-19} s/s. In fact, the recorded arrival times of radio pulses are affected by several factors in their propagation to the Earth. The partially ionized intervening interstellar medium produces a delay in the arrival times, depending on the observing frequency. The motion of the Earth in the solar system also affects the arrival time of the radio signal. Cosmic gravitational waves emitted in the vicinity of massive objects like black holes or by the vibrating 'cosmic strings' can sweep over the pulsar and thereby modify the pulse arrival times.

According to the prediction of Einstein's theory of general relativity, gravitational waves travel with the speed of light and are produced in regions where the gravitational field is strong and where the motion of the bodies is relativistic. The environment of massive objects like black holes or binary neutron stars

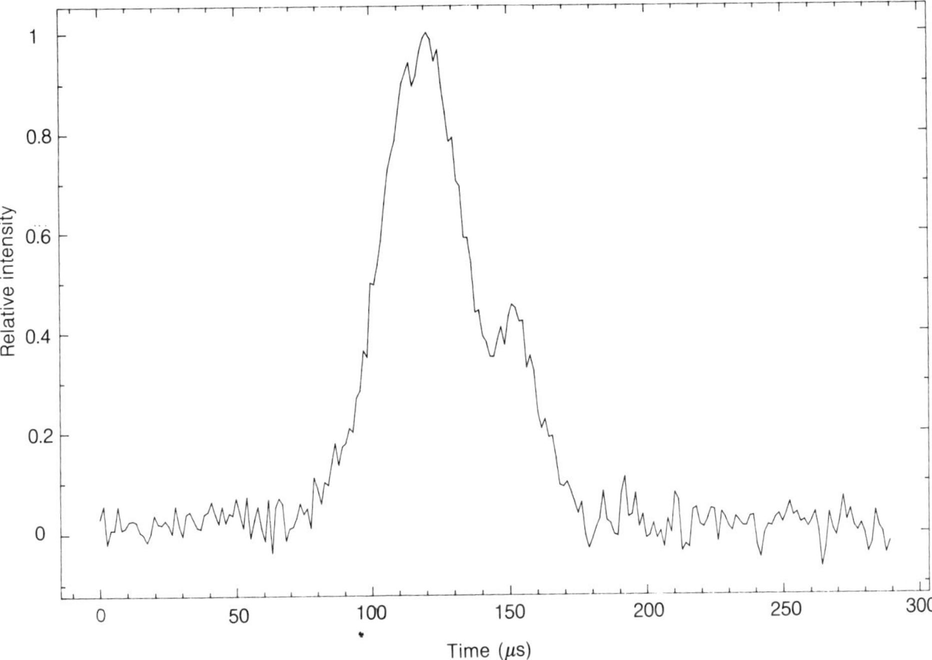

Figure 8.7 Emission from the millisecond pulsar 1937 + 21 in the constellation of Cygnus observed at 1410 MHz with radio telescope at Nançay (France) by F. Biraud, G. Bourgois, I. Cognard, J. F. Lestrade, D. Aubry, B. Darchy, and J. P. Drouhin. The intensity (in arbitrary units) of the pulse is plotted against time (in μs). The individual pulses observed during one hour have been added together to increase the signal-to-noise ratio. The width of the pulse is $\sim 33\,\mu$s and the period of the pulsar is 1.5578 ms. This is the fastest millisecond pulsar ever detected.

which accrete matter is a source of gravitational waves. Vibrating cosmic strings which are believed to be the inhomogeneities in the structure of the primordial Universe, and which are considered as the fundamental constituents for the formation of galaxies, can also generate gravitational waves. Cosmic gravitational radiation is, however, so weak that it has not been detected so far by any sophisticated devices built in the laboratory. The discovery of the class of pulsars with ultrashort period offers a new and promising cosmic laboratory to achieve this ultimate aim. In principle, when a gravitational wave reaches the pulsar or the Earth, a change can be detected in the pulse arrival time. In order to extract this information, systematic uncertainties due to other factors should be reduced. More specifically, the accuracy of the time standards which play a major role in this experiment appears to be improvable. An international effort is under way to combine the best atomic clocks in order to create a better time reference. So far, the pulsar timing experiments give an upper limit for the gravitational energy density equivalent to a mass density of approximately 10^{-35} g/cm^3. This value is of the same order of magnitude as that predicted for the stochastic background of gravitational waves due to a network of cosmic strings. It is however only an extremely small fraction of the critical density (about 10^{-29} g/cm^{-3}) required for closure of the universe, that is for gravitational attraction to balance the expansion.

8.4.2 Extragalactic radio sources

Structure

External galaxies with active nuclei and quasars (quasi-stellar radio sources) are powerful synchrotron radio sources. They are the most remote objects that can be detected to probe the deep universe at its early stage. In this respect, they play an important role in cosmology.

Interferometric and VLBI experiments have shown that the radio sources associated with galaxies and quasars consist typically of three components:

1. a central compact source;
2. two diffuse 'lobes' distributed on either side of the nucleus;
3. two jets joining the nucleus to the lobes.

Figure 8.8 shows the radio image of a radio galaxy, 3C 111, obtained at 1612 MHz by supersynthesis with the very large array. The synthesized beam is 4 arcseconds. The distance of the radio source estimated from its redshift $z = 0.0485$ (sections 8.3.2 and 8.5) is approximately 200 Mpc (megaparsecs). The central core coincides with the position of the very faint optical counterpart (not visible in Fig. 8.8). A highly collimated jet connecting the core and the north-east component is clearly detected, but no obvious south-west counterjet is apparent. A pseudo-elliptical halo surrounding the whole complex is also seen. The unresolved core which coincides with the nucleus of the optical galaxy is the central energy engine. Relativistic particles are injected into the magnetized lobes from the core via the highly supersonic jets which impinge on the intergalactic medium at a velocity of thousands of km/s. The apparent absence of the south-west counterjet may be due to the fact that the local magnetic field is not capable of confining the plasma (ionized matter). The larger southern lobe can consist of more than one individual lobe resulting from different events, suggesting that the energy injection from the core any be quasi-continuous.

This radio structure is typical of both radio galaxies and quasars. Radio galaxies correspond generally to giant optical galaxies with bright nuclei. The optical counterparts of quasars are faint remote starlike objects. The mechanism of energy production of radio galaxies and quasars remains unclear. The energy output of these objects can be hundreds of billion times that of the sun. Yet this extraordinarily large reservoir of energy is confined in a very small volume with a diameter of about 10^{-5} of the overall size of our Galaxy. The origin of such a

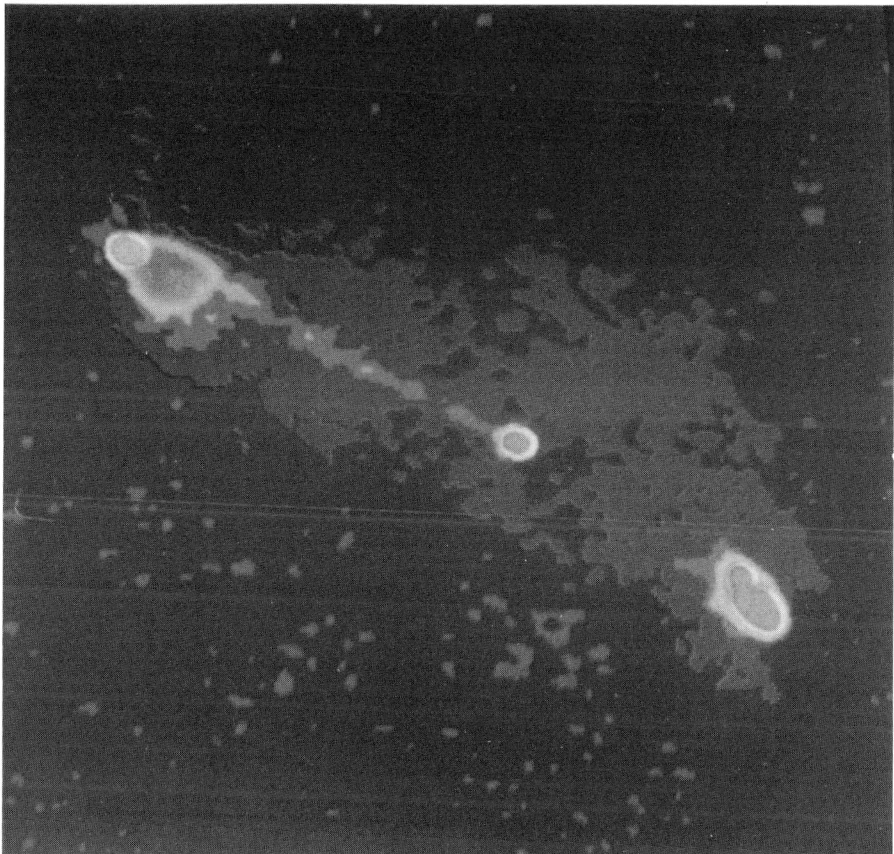

Fig. 8.8 (Shown in colour as the frontispiece) A radio photograph (in false colours) of the radio galaxy 3C 111 obtained at 1612 MHz ($\lambda = 18$ cm) by Nguyen Quang Rieu and Anders Winnberg with the Very Large Array (VLA) of the National Radio Astronomy Observatory (NRAO) at Socorro, New Mexico. The radio source emits a strong synchrotron radiation. The intensity of the central 'hot spots' and the two radio lobes (in red) is the highest. An inhomogeneous jet (in green) is moderately strong. A faint halo in dark green surrounding the whole complex is also detected. Up is north and left is east. The central hot spot which coincides with an optical galaxy injects relativistic particles into the intergalactic medium. The electrons are confined in a collimated jet before being trapped in the highly compressed intergalactic magnetic field giving rise to the two radio lobes. The south-west counterjet may have existed in the past and is now too weak to the detectable.

tremendous source of energy may be explained by a high density of massive stars which can trigger an important rate of supernova explosions. A more plausible possibility is that the radio source contains a huge black hole of billions of solar masses which accretes neighbouring stars and matter, thereby producing intense X-ray emission as energy supply.

The discovery of a double quasar has raised the interesting question about its origin. The twin radio sources are, in fact, the images of a same parent quasar. Einstein's theory of general relativity predicts that light rays from a remote quasar are bent by the gravitational field of an intervening massive object such as a galaxy which is located by chance in front of the observed quasar. The deflection angle α varies with the radial distance r of the incident ray and the mass M of the gravitational lens as $\alpha \propto GM/r$; G is the gravitational constant. The gravitational lensing results in a system of multiple images. Although a coincidental superposition along the line of sight between quasars and galaxies is rare, almost ten multiple quasar systems due to gravitational lensing have been detected.

Superluminal radio sources

The intensity of extragalactic radio sources and quasars varies over a time scale of the order of years, due to quasi-periodic outbursts. Long-term monitorings of bright remote radio sources by means of high spatial resolution interferometry have revealed that their angular position in the sky can change. If the distance of the object is known by spectroscopic measurements (sections 8.3.2 and 8.5), the motion velocity can be calculated. It turns out that, in many instances, the observed velocity exceeds the speed of light! It is well-known that physical particles with mass cannot propagate faster than light. This paradox of super-light motion is actually an 'optical' illusion effect when the motion of the body is ultrarelativistic, with velocities very close to but not exceeding the speed of light.

Let us consider an explosion occurring at a point S in the sky. The debris of the explosion emitting synchrotron radiation is distributed on a surface surrounding the explosion centre S. It is assumed to be ejected isotropically with a uniform velocity v. The observer measures an apparent velocity $v_\perp$ projected on to the plane of the sky. It can be shown that

$$v_\perp = v/[1-(v^2/c^2)]^{1/2}. \tag{8.21}$$

For a relativistic ejection velocity, $v = 0.995c$, equation (8.21) gives an apparent expansion velocity $v_\perp$ ten times larger than the velocity of light.

Long-term radio interferometry measurements with transcontinental baselines made over periods of years showed that the jet component moves away from the core with apparent velocities up to 45 times the speed of light. This value corresponds to an impressive actual relativistic expansion velocity of $0.99975c$, very close to the speed of light.

8.5 THE 21 CM HYDROGEN LINE

The radio line emitted by neutral hydrogen atoms in the interstellar medium occurs at 21 cm wavelength (approximately 1420 MHz). It is a transition between the two closely spaced energy sublevels in the ground (lowest) energy state. The energy difference between the two states arises from the different orientations of the magnetic moments of the electron and the proton. The emission of a quantum of 21 cm radiation corresponds to the change from one orientation (parallel) to the other (antiparallel). However, for each hydrogen atom, a spontaneous change is very rare and occurs every 11 million years. Collisions with other particles, especially with electrons, can trigger an induced emission and enhance considerably the emission rate. These collisional events are still rare in the interstellar medium and happen every 400 years. Yet, the 21 cm hydrogen line emission, called HI emission, is the strongest thermal line emission in the Milky Way as well as in the external galaxies. This is due to the fact that hydrogen is the most abundant constituent of the interstellar gases.

The 21 cm line is used, in particular, to trace the large-scale structure of galaxies. The spiral structure of our Galaxy was revealed by the 21 cm observations. Galaxies rotate about their minor axis. In the plane of the galaxy, the gas motion about the galactic centre is assumed to be circular. From the observations, one can determine the rotation curve which shows the rotational velocity about the centre as a function of the galactic radius. For our galaxy, the circular velocity increases from about 200 km/s in the inner region to about 250 km/s in the vicinity of the sun which is at approximately 10 kpc (kiloparsec) from the galactic centre and then decreases outwards. This strong differential rotation indicates that the galaxy does not rotate like a solid body.

The distance of galaxies can also be determined by the observation of the 21 cm line. The line emitted by a galaxy receding away from the observer undergoes a redshift that is related to the recession velocity through the Doppler formula (section 8.3.2). According to the Hubble law, the radial velocity increases with the distance d of the galaxy, $v = Hd$. The Hubble constant H is 75 ± 25 km/s/Mpc. Galaxies in the Virgo cluster at about 13 Mpc from the sun are receding with velocities of approximately 1000 km/s. The Hubble's finding is the consequence of an expanding universe. Deep surveys of more and more remote galaxies can contribute to an estimate of the density of the early universe.

8.6 INTERSTELLAR MOLECULES

8.6.1 The discovery

The discovery of molecules in space makes it possible to investigate the cold and dark universe so far inaccessible. This is because, in dark interstellar clouds, most of the hydrogen is in the molecular form. As a result, it does not emit the 21 cm atomic hydrogen line. Molecular lines serve as diagnostic probes to investigate the physical conditions of the cold and dense interstellar matter where gas and dust are intimately mingled. The dust component can be detected through its far infrared continuum radiation by telescopes installed on the ground or on board of airplanes or spacecrafts.

The average temperature and density of the interstellar medium are about a few tens of kelvin, and a few tens of particles per cubic centimetre, respectively.

They are very low compared to the standard values on the Earth where the temperature is usually above 273 K and the density is a few 10^{19} atoms and molecules per cm^3. The vacua performed in terrestrial laboratories still correspond to a few millions of molecules per cm^3. The cold and dilute interstellar environment is apparently not very favourable to the synthesis of molecules. However, interstellar space is far from homogeneous and clouds as dense as 10^6 particles per cm^3 can exist. Furthermore, the ultraviolet radiation field from stars favour some chemical reactions leading to the formation of complex molecules. This environment of cold gas and dust can in fact facilitate the formation of giant molecular clouds and yound stars.

The first interstellar molecule, the diatomic CH radical, was detected in 1937 in the visible, at 4300 Å. Molecular transitions also occur in the radio range. After many unsuccessful searches, another diatomic species, the OH radical, was detected in the interstellar medium at 18 cm, in 1963. Ammonia (NH_3) and water vapour (H_2O) were found a few years later. The radio transitions arise between low rotational energy levels which are easy to excite by collisions with molecular hydrogen or by infrared photons from dust grains. Most molecular rotational transitions occur in the millimetre wave band. The number of detected interstellar molecules has increased steadily since 1970, with the advent of sensitive millimetre-wave radio telescopes. Nearly 90 molecules and many of their isotopes have been discovered to date (1991), in our Galaxy and some in other galaxies as well (see Table 8.1). The heaviest molecule so far detected in the interstellar medium, $HC_{11}N$ (cyano-deca-penta-yne) with the structural formula H—C≡C—C≡C—C≡C—C≡C—C≡C—C≡N, belongs to the family of long carbon chain molecules HC_xN, the cyanopolyynes. Some species like the hydrocarbons, C_3H and C_3H_2, are ring molecules. Their molecular geometry consists of three carbon atoms located at the vertices of a triangle where hydrogen atoms are attached. A number of interstellar emission features observed in the near and mid infrared have recently been attributed to large organic ring molecules called polycyclic aromatic hydrocarbons (PAHs) containing about 50 atoms or more. These large molecular complexes are believed to be interstellar dust grains in which hydrogen atoms are tied up in the graphitic planes at the periphery.

The search for interstellar molecules can be performed by observing molecular transitions whose frequencies are known by laboratory measurements or by theoretical quantum mechanical calculations. A second method consists of observing systematically a frequency band to detect any lines which will be identified subsequently in the laboratory. The last procedure has the advantage of discovering unsuspected 'exotic' species. A number of lines still remain unidentified.

Molecules are concentrated in the central region of our Galaxy and of other galaxies. The giant cold dark clouds, the environment of protostars and the circumstellar envelopes of evolved stars are favourable targets for molecular searches. Comets are reservoirs of molecules which are produced by evaporation in the vicinity of the sun. Water vapour is the main constituent of the cometary

Table 8.1 Molecules detected in the Milky Way

Molecule	Name
2 atoms	
H_2	molecular hydrogen
C_2	molecular carbon
CH^+	methylidyne ion
CH	methylidyne radical*
OH	hydroxyl radical
CO	carbon monoxide
CN	cyanogen radical
CS	carbon monosulphide
NO	nitric oxide
NS	nitrogen sulphide
NH	nitrogen hydride
SO	sulphur monoxide
SiO	silicon monoxide
SiS	silicon monosulphide
SiC	silicon carbide
SiN	silicon nitride
PN	phosphorus nitride
PC	phosphorus carbide
HCl	hydrogen chloride
NaCl	sodium chloride
KCl	potassium chloride
AlCl	aluminium chloride.
3 atoms	
C_3	tricarbon
H_2O	water
CCH	ethynyl radical
HCN	hydrogen cyanide
HNC	hydrogen isocyanide (isomer[†] of HCN)
HCO	formyl radical
HCO^+	formyl ion
HOC^+	isoformyl ion
N_2H^+	diazenylium
H_2S	hydrogen sulphide
HNO	nitroxyl
OCS	carbonyl sulphide
SO_2	sulphur dioxide
HCS^+	thioformyl ion
SiC_2	silicon dicarbide radical
C_2O	dicarbon monoxide
C_2S	dicarbon sulphide
4 atoms	
NH_3	ammonia
C_2H_2	acetylene
H_2CO	formaldehyde
HNCO	isocyanic acid
$HOCO^+$	protonated carbon dioxide
H_2CS	thioformaldehyde
C_3N	cyanoethynyl radical
HNCS	isothiocyanic acid
C_3H	propynylidyne
C_3O	tricarbon monoxide
C_3S	tricarbon sulphide
$HCNH^+$	protonated hydrogen cyanide
H_3O^+	hydroxonium ion
5 atoms	
C_5	pentacarbon
CH_4	methane
CH_2NH	methanimine
H_2CCO	ketene
NH_2CN	cyanamide
HCOOH	formic acid
C_4H	butadiynyl radical
HC_3N	cyanoacetylene
SiH_4	silane
C_3H_2	cyclopropynylidene
CH_2CN	cyanomethyl radical
SiC_4	silicon tetracarhide
6 atoms	
CH_3OH	methanol
CH_3CN	methyl cyanide
NH_2CHO	formamide
CH_3SH	methyl mercaptan
C_2H_4	ethylene
C_4H_2	diacetylene
C_5H	pentynylidyne radical
HC_2CHO	propynal
7 atoms	
CH_3NH_2	methylamine
CH_3CCH	methyl acetylene
CH_3CHO	acetaldehyde

Table 8.1 *contd*

CH_2CHCN	vinyl cyanide	*10 atoms*	
HC_5N	cyanobutadiyne	CH_3COCH_3	acetone
C_6H	hexatriynyl radical		
8 atoms		*11 atoms*	
$HCOOCH_3$	methyl formate	HC_9N	cyanooctatetrayne
CH_3C_3N	methylcyanoacetylene		
9 atoms		*13 atoms*	
CH_3CH_2OH	ethanol	$HC_{11}N$	cyanodecapentayne
CH_3OCH_3	dimethyl ether		
CH_3C_4H	methyldiacetylene		
CH_3CH_2CN	ethyl cyanide		
HC_7N	cyanohexatriyne		

*A radical is a species which reacts very actively with other molecules and is extremely unstable in the physical conditions of terrestrial laboratories.

†The isomers consist of the same atoms but their molecular structure and chemical properties are different.

gas. During its close approach to the sun and the earth in March 1986, Halley's comet produced water vapour at a rate of about 15 tons per second.

Figure 8.9 displays the radio images of the 'Egg Nebula' obtained in the NH_3 and HC_7N lines at 23.7 GHz by supersynthesis with the Very Large Array. The superimposed optical image shows two bright lobes produced by the scattering of the light of a star located at the centre of an edge-on toroid of dust which is invisible. The NH_3 emission is confined in the toroid while the HC_7N emission is distributed in an extended halo. It is expected that NH_3 molecules are formed in regions where the density of gas and dust is high. Important outflow of ammonia gas proceeds through the central hole of the expanding and rotating toroid, on both sides of the toroid plane. Such a bipolar flow is a phenomenon commonly detected in the vicinity of young and evolved stars. The halo distribution of HC_7N is however unclear. It has been argued that HC_7N is produced by fragmentation of carbon grains ejected at supersonic velocity from the central star.

8.6.2 Astrochemistry

The synthesis of interstellar molecules is widely believed to proceed in the gas phase chemistry. Reactions of ions with neutral species seem to prevail in the chemical synthesis networks. Ions are produced from neutral atoms and molecules by cosmic rays and ultraviolet starlight. For example, the reactions leading to

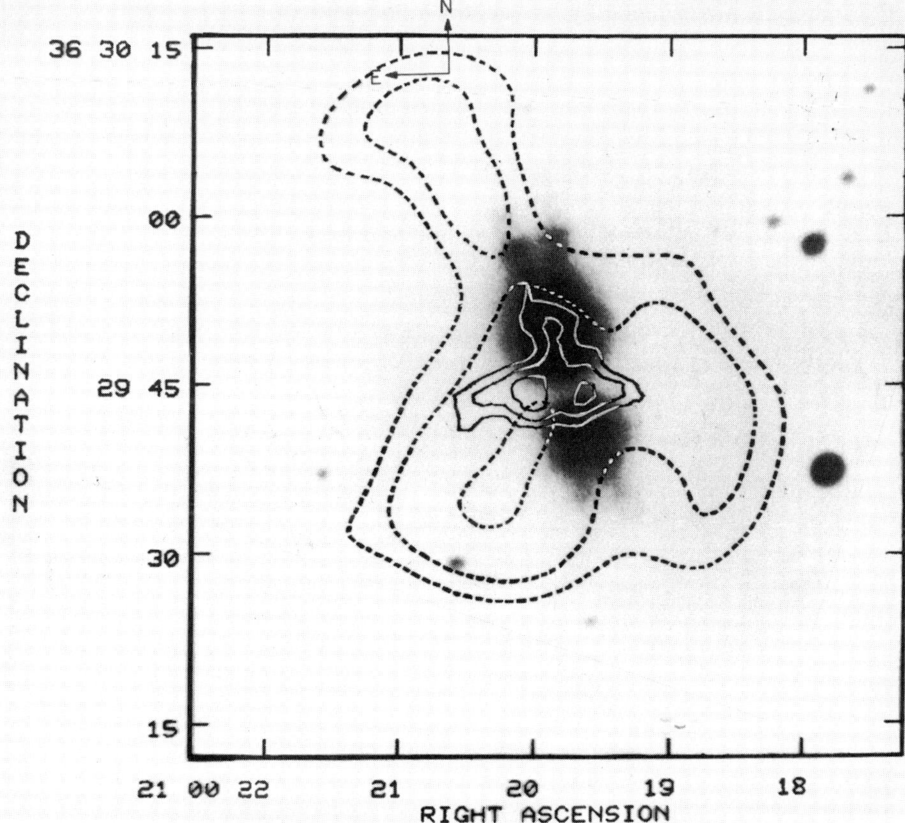

Fig. 8.9 The contour maps (curves of equal intensity) of the emission of NH_3 (ammonia, full curves) and HC_7N (cyanohexatriyne, dashed curves) observed by Nguyen-Q Rieu, A. Winnberg, and V. Bujaarabal (1986) at 23.7 GHz with the Very Large Array (VLA) of the national Radio Astronomy Observatory (NRAO) in the direction of the 'Egg Nebula' are superimposed on an optical photograph (negative image, in black). The maps are drawn in the celestial orthogonal coordinates (declination and right ascension axes). Ammonia emission is confined in the (invisible) dense, dusty, flat toroid whose equatorial plane is perpendicular to the plane of the sky. The ammonia toroid is expanding at a velocity of 15 km/s. An isotropic outflow of ammonia proceeds through the hole at the toroid centre where a star is concealed. The ejection of matter occurs on both sides of the plane of the toroid at a velocity of 35 km/s. Cyanohexatriyne is distributed in a halo and is probably created from fragmentation of dust.

carbon-bearing molecules such as CH and C_2H can start with

$$C^+ + H_2 \rightarrow CH_2{}^+ + h\nu$$

and subsequently CH and CH_2, which are formed via further ion–molecule reactions and recombination with electrons, react in turn with C^+ to produce

C_2 and C_2H. More generally, a reaction of the type

$$AX^+ + BY \rightarrow AB^+ + XY$$

leads to a more complex ion, AB^+, which can be used to synthesize heavier molecules. The cyanopolyyne, HC_3N, is thought to be created by the reaction between $C_3H_3{}^+$ and nitrogen to give $H_2C_3N^+$ followed by a dissociative recombination with an electron. The chemical study of more complex species still suffers from the lack of rate coefficients for major reactions.

Dust grains also play an important role in interstellar chemistry. They shield the molecular clouds from photodestruction by ultraviolet radiation. Furthermore, molecular hydrogen is likely to be formed on surfaces of grains, contrary to most molecules which are created in the gas phase.

8.6.3 Cosmic maser amplification

The most spectacular characteristic of interstellar clouds is their ability to amplify radiation of a specific frequency which depends on the molecular content and the physical properties of the cloud. Water vapour clouds in some circumstellar envelopes can amplify as much as 10^{14} times the celestial input signals at 22 GHz. The input radiation field can originate from an external source or from spontaneous emission generated inside the cloud itself. These cosmic masers (microwave amplification by stimulated emission of radiation) which operate like the man-made maser amplifiers are excited by a 'pump'. The maser concept is as follows.

In the 'local thermodynamic equilibrium' (LTE) condition, the population distribution of molecules in a gas cloud is governed by the Boltzmann law (see equation 8.19). According to this law, the population of any upper energy state n_u is smaller than that of any lower state n_1. Most molecular clouds are almost in this situation. The principle of the maser action is to upset the local thermodynamic equilibrium state by pumping the population from lower to higher levels. This population inversion can be produced in the gas cloud by collisions with molecular hydrogen (collisional process) or by infrared photons radiated by stars and dust (radiative process). If the pumping is efficient enough, n_u becomes greater than n_1. Equation (8.19) indicates that, in this case, the excitation temperature, T_{ex}, is negative. The absorption coefficient, κ, and the opacity, $\int \kappa \, dx$, of the maser line which are both proportional to $n_1 - n_u$ (section 8.6.4) are also negative. A negative absorption of radiation corresponds therefore to maser emission. Radio astronomical observations combined with a theoretical treatment involving radiative transfer allow the detection of maser lines through the determination of the excitation temperature and opacity and hence the level population distribution.

The pumping involves, in fact, more than two energy levels. In a simplified three-level maser (Fig. 8.10), molecules are pumped from level 1 to level 3 and then cascade from there to populate level 2 via spontaneous emission. The over-population in level 2 with respect to level 1 results in a maser emission $2 \rightarrow 1$,

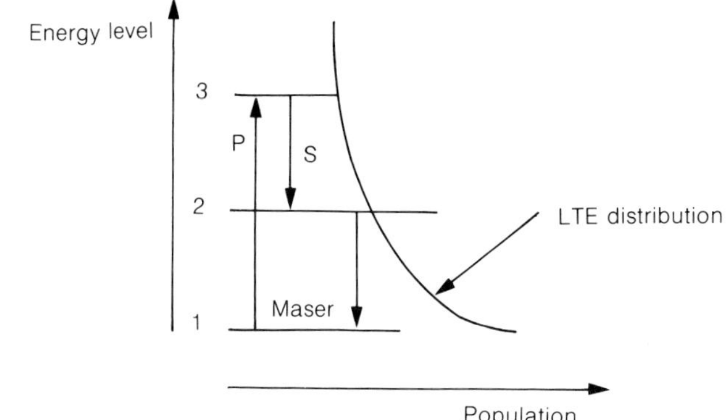

Fig. 8.10 Schematic pumping scheme of a three-level maser. The level population is initially distributed according to the Boltzmann law (LTE distribution). Collision and radiation operate as a pump P to excite the molecules from level 1 to level 3. They subsequently cascade down through spontaneous emission to overpopulate level 2, thereby creating a maser transition between levels 2 and 1.

which is triggered by spontaneous emission or by an external radiation source of frequency corresponding to the $2 \rightarrow 1$ transition.

Cosmic maser emission is often detected in regions of star formation as well as in the envelopes of old red giant and supergiant stars. The physical conditions of these environments, such as high temperatures and densities of gas and dust are quite appropriate for molecular masers to be pumped. Powerful maser emissions from hydroxyl (OH) at 1.665 GHz, water vapour (H_2O) at 22 GHz, methanol (CH_3OH) at 25 GHz, and silicon monoxide (SiO) at 43 GHz and 86 GHz are commonly detected. Hydrogen cyanide (HCN) and silicon monosulphide (SiS) can also exhibit weak maser emission. An example of a maser emission in the envelope of a carbon star, IRC + 10216, is shown in Fig. 8.6. The narrow spike in the SiS spectrum (18 GHz) is a maser line arising from the circumstellar cloud which amplifies the radiation of the central star concealed in the thick circumstellar envelope.

Cosmic masers are easily detectable because of their path lengths, at least a few astronomical units ($\approx 1.5 \times 10^{11}$ m), a very large size indeed, as compared with the dimension of man-made masers. A narrow line, a high variability, and a high degree of polarization are the main features of a cosmic maser emission. Long-term monitorings of H_2O masers in star-forming regions have shown that there exist luminous outbursts of brightness temperatures reaching 10^{15} K. The polarization measurements suggest embedded magnetic fields of about 10 mG, a value several orders of magnitude stronger than the average interstellar field (approximately a few microgauss).

8.6.4 The concept of a two-level maser

In the interstellar conditions, the molecules can be excited to very high energy levels. The populations of the many levels involved in the pumping scheme are governed by the absorption and the emission of photons and by collisions. As a result, molecules are excited towards high levels and fall down to lower levels through a series of cascades. It is possible to describe a simplified maser theory by representing all details of the population transfer in a two-level maser. In steady state, the rate of populations leaving the upper level u downward (de-excitation) is balanced by the rate of exit from the lower level 1 upward (excitation). The de-excitation is due to spontaneous and stimulated emission and downward collisions. The excitation is caused by photon absorption and upward collisions. For simplicity, we shall, in the following, ignore the statistical weights g_u and g_l (see equation 8.19) and write the statistical equilibrium equation which controls the flux of population transfer between the two energy states as follows:

$$n_u\{A_{ul} + B_{ul}\Omega I/(4\pi) + C_{ul} + P_{ul}\} = n_l\{B_{lu}\Omega I/(4\pi) + C_{lu} + P_{lu}\} \qquad (8.22)$$

where A_{ul}, B_{ul} are the Einstein coefficients for spontaneous and stimulated emissions, B_{lu} is the Einstein absorption coefficient; C_{ul} and C_{lu} are the downward and upward collision rates; the pump rates P_{ul} and P_{lu} represent the population transfer between levels u and l via other higher levels which are omitted here; I is the maser (stimulated) intensity and Ω the solid angle of the maser emission.

We shall further neglect spontaneous emission which is usually weak, and assume that B_{lu} and B_{ul} are equal; their indices will be omitted. Collisions are also ignored. We define $P = P_{ul} + P_{lu}$ which is the total net pump rate, and $\Delta P = P_{lu} - P_{ul}$. From equation (8.22) we then derive the rate of fractional population inversion:

$$\frac{n_u - n_l}{n_u + n^l} = \frac{\Delta P}{P + B\Omega I/(2\pi)} \qquad (8.23)$$

Population inversion requires that ΔP be positive. The maser is unsaturated when the maser output I is too weak so that the stimulated emission rate $B\Omega I/(2\pi)$, is dominated by the pump rate P. In the saturated regime, $B\Omega I/(2\pi) \gg P$, the population inversion decreases inversely proportional to the stimulated emission rate (see equation (8.23)). Then the pump is the most efficient, since the population of the upper level is transferred as fast as possible to the lower level by emitting maser photons. We define the saturation intensity (stimulated emission rate equal to the pump rate) as:

$$I_s = 2\pi P/(B\Omega). \qquad (8.24)$$

The unsaturated population inversion rate Δn_0 can be derived from equation (8.23) by neglecting the stimulated emission:

$$\Delta n_0 = (\Delta P/P)(n_u + n_l). \qquad (8.25)$$

Combining equations (8.23), (8.24) and (8.25) we express the population inversion rate as:

$$\Delta n = n_u - n_l = \Delta n_0 \frac{1}{1+(I/I_s)}. \tag{8.26}$$

We now investigate the dependence of the maser intensity, I, on the path length, x. It can be obtained from the one-dimensional radiative transfer equation:

$$\frac{dI}{dx} = -\kappa I. \tag{8.27}$$

The absorption coefficient κ is a function of the population inversion rate Δn:

$$\kappa = -\frac{Bh\nu}{4\pi\Delta\nu}\Delta n \tag{8.28}$$

$h\nu$ is the level energy separation of the maser transition and $\Delta\nu$ is the line width. The solution of equation (8.27) in the unsaturated regime is:

$$I = I_0 \exp(-\kappa_0 x) \tag{8.29}$$

κ_0 which is derived from equation (8.28) denotes the unsaturated absorption coefficient. Note that κ_0 is negative when population inversion occurs ($\Delta n_0 > 0$). In the unsaturated region, the maser intensity therefore increases exponentially with the maser gain ($-\kappa_0 x$). As discussed previously, the population inversion cannot be maintained at a constant level indefinitely. Consequently, the exponential growth should stop and the maser begins to saturate. Equations (8.26) and (8.28) allow us to define κ as:

$$\kappa = \frac{\kappa_0}{1+(I/I_s)}. \tag{8.30}$$

The maser intensity in the saturated case is derived from equations (8.27) and (8.30)

$$\frac{dI}{dx} = -\frac{\kappa_0}{1+(I/I_s)} I \approx -\kappa_0 I_s. \tag{8.31}$$

By integrating equation (8.31):

$$\int_{I_s}^{I} dI = -\kappa_0 I_s \int_{x_s}^{x} dx$$

one gets:

$$I = I_s\{1 - \kappa_0(x - x_s)\}. \tag{8.32}$$

The maser is unsaturated up to a path length of x_s and then turns to a saturated regime. The maser intensity in the saturated case continues to grow but only linearly with x (Fig. 8.11). Interstellar masers can contain an unsaturated core surrounded by a saturated halo.

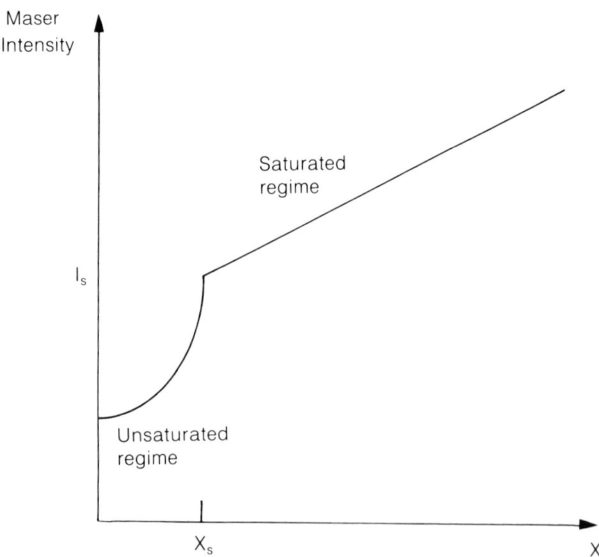

Fig. 8.11 Maser amplification along the path length x. The maser intensity grows exponentially with x in the unsaturated regime and then linearly when the maser becomes saturated.

8.7 CONCLUSION AND PROSPECTS

This brief survey has shown that the objective of radio astronomy is not only confined to the investigation of the universe, but it concerns a great variety of other branches of science.

The instruments used in the detection of weak signals from remote radio sources require a high standard in sensitivity and stability. In this respect, radio astronomy contributes to stimulate the development of large millimetre antennas and low-noise receivers capable of operating at high frequencies. Large single dishes such as the French–German 30 m radio telescope of IRAM (Institut de Radio Astronomie millimétrique) at Pico Veleta in Spain, and the Japanese 45 m radio telescope at Nobeyama in Japan are actively working at 1 and 3 mm wavelengths. Millimetre-wave interferometers installed at Nobeyama, Hat Creek and Owens Valley (California) as well as the IRAM interferometer at Plateau de Bure (France) are currently operating to investigate the small-scale structure of our Galaxy and of other extragalactic systems. The study of fascinating objects such as pulsars and quasars which are believed to be associated with neutron stars and black holes is related to the physics of superdense matter and to Einstein's theory of general relativity.

The observations of molecules in the interstellar medium where the physical conditions are extreme allow the investigation of the chemical species which are otherwise unstable in the laboratory conditions. The discovery of complex molecules in the Milky Way offers a new field of research not only in astrophysics and astrochemistry, but also in astrobiology. The detection of molecules in comets, like Halley's comet, whose atmosphere may be the frozen remnant of the primordial solar system, can reveal valuable infomation on the origin of the sun and the earth. It is however premature to consider that the detection of complex interstellar organic molecules serves as a clue to the existence of any form of extraterrestrial life. The simplest amino- acid, the glycine NH_2CH_2COOH, which is a basic component of most proteins, has not yet been detected in space.

Molecules can serve as probes of the terrestrial atmosphere and are therefore of great importance to ecology and meteorology. The complex chemistry governing the density of the ozone (O_3) layer which shields life on earth from the lethal ultraviolet solar radiation can be understood by observing not only ozone but also many other stratospheric molecules in the millimetre waveband. Millimetre observations aboard aircrafts and satellites are used in meterology to explore the large-scale vertical distribution of temperature and water vapour in the upper atmosphere.

The transcontinental VLBI experiments on remote quasars as radio 'beacons' can measure the distances between different terrestrial sites within a centimetre. These data are of great importance for geophysicists in deriving the relative motion of tectonic plates, and thereby helping to predict earthquakes. Long-term measurements of the baselines of transcontinental interferometers across the North Atlantic have shown that the continents are drifting at a rate of 1.5 cm per year.

Space VLBI projects consisting of orbiting radio telescopes used in conjunction with ground-based antennas provide baselines of about ten times the earth diameter. The corresponding 10 microarcseconds resolutions would allow us to distinguish details as fine as the light of a candle at the distance of the moon. Such high resolutions will be needed especially to investigate the still-unresolved core of quasars and understand the powerful reservoir of energy it contains. The study of interstellar masers can also be undertaken to improve our knowledge on the processes of star formation.

APPENDIX 8.A UNITS AND CONSTANTS IN ASTRONOMY

Astronomical unit (AU) = $1.495\,979 \times 10^{11}$ m
light year (l.y.) = $9.460\,53 \times 10^{15}$ m
parsec (pc) = $3.085\,678 \times 10^{16}$ m

(distance at which the angular radius of the terrestrial orbit is 1 arcsecond)

solar mass ($M_\odot$) = 1.989×10^{30} kg
solar radius ($R_\odot$) = 6.9599×10^{8} m
mass of the Earth ($M_\oplus$) = 5.976×10^{24} kg
equatorial radius of the Earth ($R_\oplus$) = 6378.164 km
velocity of light (c) = $2.997\,924\,562 \times 10^{8}$ m/s

gravitational constant (G) $= 6.672 \times 10^{-11}\,N \cdot m^2 \cdot kg^{-2}$
Jansky (Jy) $= 10^{-26}\,W \cdot m^{-2} \cdot Hz^{-1}$
Hubble constant (H) $= 75 \pm 25$ km/s/Mpc

REFERENCES

Backer, D. C., and Hellings, R. W. (1986) Pulsar timing and General Relativity *Annual Review of Astronomy and Astrophysics*, **24**, 537.

Christiansen, W. N. and Högbom, J. A. (1985) Radio telescopes 2nd Edn, Cambridge University Press.

Fomalont, E. B. and Wright M. C. H. (1974) Interferometer and Aperture Synthesis Galactic and Extragalactic Radio Astronomy, (eds G. L. Verschuur and K. I. Kellermann); Springer-Verlag p. 256.

Ginzburg, V. L. and Syrovatskii, S. I. (1969) Developments in the Theory of Synchrotron radiation *Annual Review of Astronomy and Astrophysics*, **7**, 375.

Napier, P. J., Thompson, A. R. and Ekers, R. D. (1983) The Very Large Array *Procedings of I.E.E.E.* **71 (11)**, 1295.

Nguyen-Quang-Rieu (1986) Circumstellar Radio Molecular lines *Monograph Series on Non-Thermal Phenomena in Stellar Atmospheres, The M-Type Stars*, ed. NASA-CNRS (NASA SP-492), p. 209.

PROBLEMS

1. Radio astronomical measurements are limited by the noise fluctuations of the receiving system. The system temperature T_{sys} includes the noise of the receiver and those of the background radiation and of the atmosphere.

   Calculate the number of photons required for a celestial source to be detected at 115 GHz, using a radio telescope tracking the source during 1 hour. The system temperature T_{sys} is 300 K, and the bandwidth W of the receiver is 1 MHz. (section 8.2.1)

2. An astronomical site is characterized by its latitude ϕ, defined as the angle between the plane of the celestial equator and the zenith axis (see Fig. 8.4). A two-element interferometer of baseline L is oriented along the BB′ axis in the horizon plane.

   a) Give the expressions of the spatial frequencies u, v as a function of L, ϕ, and the hour angle, h, and the declination δ of the source.

   b) Draw the track in the u, v, plane for a twelve hour observation ($h = \pm 6$ hours) on a source at $\delta = 30°$ and for a latitude of 40° (section 8.2.2).

3. In a molecular transition of frequency ν, the population distribution between the upper level u and lower level l is governed by spontaneous emission, stimulated emission and collisions. Assume that the molecular cloud is in local thermodynamic equilibrium state (no maser emission).

   a) Write the statistical equilibrium equation for a transition between the upper level u and the lower level l, in the case of an isotropic stimulated emission. Take account of spontaneous and stimulated emissions and collisions but ignore the pump (section 8.6.4).

b) Assume that the statistical weights g_u and g_l are equal. The gas kinetic temperature T_k is derived from the collision rates by $C_{lu}/C_{ul} = \exp[-h\nu/(kT_k)]$ and the radiation temperature T_R is obtained from the intensity of the stimulated emission by the Planck formula. The relation between the stimulated and spontaneous emission rates is $B_{ul} = B_{lu} = A_{ul}c^2/(2h\nu^3)$. Calculate the population ratio, n_u/n_l, as a function of T_R, T_k, A_{ul}, and C_{ul}.

c) In the centimetre wavebands, $h\nu/k$ is negligibly small compared to the temperatures (Rayleigh–Jeans approximation). In this approximation, show that the excitation, kinetic, and radiation temperatures are related by the following expression

$$T_{ex} \approx T_k(T_R + T_c)/(T_k + T_c)$$

where $T_c = (h\nu/k)C_{ul}/A_{ul}$. What will become the excitation temperature T_{ex}?
i) when collision dominates;
ii) when radiation dominates.

Index

21 cm hydrogen line 532–3

Abbe's Law 386–8, 395
Aberrations, optical systems 392
Accelerators, circular and re-entrant 487–92
Acceptance solid angle, detectors 411
Access, microwave links 1–2
Accuracy, laser systems 438–44
Acousto-optical processes, spectrum analysers 338
Active antennas, radar 327–9
Active array antennas 328
Active imaging, microwave 500–1
Active infrared systems 380–1, 383, 384–5
  military applications 453–62
Active laser systems 428–9
Active sensors 495–501
ACTS 189
Actuators, low thrust, satellites 172
Aerial reconnaissance 448
Airborne early warning radar 322–3
Aircraft, radar systems on board 325–6
Air defence radar 293–4
Air strike, active support 373
Air-to-ground attack, laser application 463–5
Airy's diffraction spot 388–9
American National Standards Institute 494
Amplifiers
  assemblies 251
  broad band 334
  high power 115–17, 130, 146, 185–7, 191, 198, 200, 217–22
  low noise 17–19, 20, 130, 143, 145, 191, 198, 200, 212–17, 287–8
  microwave links 16–19
Amplitude correctors, analogue
  microwave links 40–1
Amplitude measurement, direction-finding 338–9
Amplitude modulation
  infrared signals 421
  residual 220
Amplitude modulation to phase modulation conversion 61–5, 220, 221–2
Amplitude-phase shift keying 162
Amplitude shift keying 162
Analogue microwave links 2, 10–11
  comparability with digital 94–7
  modulation 12–16
  operating aid facilities 26–8
  performances 28–40
  quality improvement 40–5
  signals transmitted 11–12
  transmission quality 45–9
Analogue modulation 151–3, 158–62
Analogue-to-digital modulation 313–15
Angular measurement
  laser systems 440–1, 442–4
  radar 269–70, 328–9
    perturbation 346–7
Angular resolution, diffraction
  limited instruments 389
ANIK 137
Antenna feed 10, 181, 201–5
Antenna gain 128, 139–41, 198, 199
Antenna noise temperature 142, 143–5, 198, 199, 280–1
Antenna polarization, *see* Polarization
Antenna radiation pattern 141–2
Antennas
  accesses 16
  adaptive 353–4
  branching of transmissions and receptions 25–6
  directivity 4
  high gain 113–15, 350
  master television 133

Antennas (*cont.*)
  near-field 479
  phased-array 273
  phase measurement 339
  pointing error 208
  radars 280–1, 286–7, 327–9
    ECCM 357–62
  radio relay links 7–10
  radio telescopes, *see* Single dish radio telescopes
  reflector 200
Antenna side lobes, *see* Side lobe
Antenna systems
  earth stations 198–211
  satellites 133, 139–42, 176–81
Antenna temperature, radio telescopes 512–14
Anti-aircraft defence, countermeasures 374
Antipersonnel radars 326
Antiradar paint 296
Aperture synthesis, radio telescopes 514–20
Apogee motor 172
Applicators, microwave heating systems 477–9
ARABSAT 136–7, 174
Arrays, antennas 519–20
Artificial sources, emissivity 405–7
Astigmatism 392
ASTRA 137
Astrochemistry 534–6
Atmospheric attenuation 30–3, 109–12, 144, 150–1, 350
  outages 106–9
Atmospheric scattering, laser light 455
Atomic clocks 526–8
Atomic emissions, cosmic 522–4
Attenuation, sensitivity time control, radar 288
Attitude stabilization 168–70
Audio broadcasting, digital 224
AURORA 137
AUSSAT 137
Autoadaptive equalizers 100–1
Automatic frequency control, radars 284
Automatic gain control 17
  radars 291–2, 347
  radio repeaters 16
Automatic transmitted power control 105–6
Auxiliary signals, transmissions 80–1
Availability objectives
  analogue microwave links 28
  digital microwave links 87
Azimuth and elevation (AZEL) antenna mount 206, 208

Backoff, amplifiers 147
Ballistic missiles, radar detection 293
Bandpass filters 23–5
Barrage jammers 344
Baseband
  accesses 2
  digital signal 52–3
  filtering 75
  interference effect 37
  switching matrices 190
Basic equipment noise 30, 45–6
Beam separation, frequency reuse 148–9
Beam waveguide (BWG) antenna feed 200, 201
Beamwidth, antennas 358–9
Binary phase shift keying (BPSK) 162–3
Bit error rate ratio 50, 51, 85–7, 88, 106, 153, 188, 221–2
Bit rates 50, 51
Bit stream processing 190
Black bodies 397–8, 405
  radiation 521
Body-fixed stabilization 168
Boltzmann Law 536
Bragg cells 338
BRAZILSAT 137
Broadcasting satellite service (BSS) 127, 133, 176, 224
Bulk acoustic wave devices 307
Butterworth functions 25–6

C3, countermeasures 374
Cable television, networks 133
Cameras
  infrared 384, 422–5
  thermal 384, 448, 459
Cancers, heat treatment 479, 480, 506
Carbon arc, spectrometric source 407
Carrier frequencies, utilization, radio links 4–5, 7
Carrier recovery, digital microwave links 67–73
Carrier-to-noise ratio (C/N) 139, 151
Cascaded networks 143
Cassegrain antenna 177, 200, 206
Cavity applicators, microwave heaters 478–9

CCD matrices 467–8
CCIR (Comité International des Radiocommunications) 5, 28, 38, 85, 94, 176
CFAR chains, radar 291, 367–9
Channel switching, digital microwave links 81
Chemistry, gas phase 534–6
Chirp system, pulse compression 304–7
Chromatic aberrations 392
Circuit noise 30
Circular polarization 176, 181, 204
Circular waveguides
  antenna polarization 181
  earth station antennas 202
  microwave links 10
Circulators, bandpass filters 24–5
Clarke, Arthur C. 135
Closed loop carrier recovery 70
Clutter
  effect of pulse compression 311–12
  radar 272, 287, 290–1, 315, 320
CO_2 lasers 385, 442, 458–61
Coast earth stations 136
Coded radars 270–1, 276, 282
Coherence bands 111–12
Coherent laser systems 430, 431–3, 436–40
Coherent oscillators, radar 285
Colliding beams, accelerators 492
Colling devices, infrared 468
Coma 392
Comets 532, 534
COMINT 332, 371–2, 374
Command line of sight, weapons guidance 450
Communication, military 331–2
Communication equipment, satellites 132–3
Communication intelligence 332, 371–2, 374
Communication links, satellites 137–9
Communication satellite systems 125–6
Companding 160–2
Constant false alarm receivers, radar 291, 367–9
CONTEL-ASC 137
Continuous wave
  laser systems 441–2, 460–1
  radar 268, 284–5, 442
Continuum emission, cosmic radio emission 520–2
Continuum radio sources 524–30
Control, command and communications, countermeasures 374
Conversion loss, mixers 19–20
Corrugated horn 201–2
Cosmic masers, amplification 536–7
Cosmic radio emission 511, 520–4
Counter-mobility 373–4
Coverage
  satellite communications 127–30
  surveillance radar 286–7
Crab Nebula 525
Cross-field amplifiers 287
Cross-modulation, power amplifiers 246–9
Cross-modulation correction 245
Crystal video detectors 334–5
CS-2 137
Cyclotrons 487–8, 493

Data transmission 2
  satellies 132–3, 224
Decoys, radar jamming 342, 343, 346
Deformations, laser beams 455
Delay, control, radar antennas 283–4
Demand assigned multiple access 131
Demodulation
  analogue microwave signals 14–16
  digital microwave links 79–80
  threshold 153
Demultiplexers
  input 172–4, 183–5
  polarization 9
Descramblers, digital microwave links 82–5
Despun mode, radar beams 324
Despun satellite antenna 176
Detectivity, flux detectors 411
Detector responsivity 410
Detectors, infrared 467–8
DFS-Kopernicus 137
Dichroic surfaces 182
Dicke fix 367
Dielectric absorption, microwaves 476–7
Dielectric materials, microwaves 471–4
Difference diagrams 208, 210
Differential gain 248
Differential phase corrections 245, 248–9
Diffraction 388–9
Diffraction limited resolution 389

Digital audio broadcasting 224
Digital beam forming, radar 328
Digital memories, jamming systems 347, 349
Digital microwave links 2
 comparability with analogue 94–7
 functions 80–5
 improving quality 98–106
 modulation 61–74
 optimizing cost and performance 75–80
 performances 85–98
 signals, characteristics 49–61
Digital modulation 153, 162–3
Digital processing, radar 313–15, 327
Digital signal, baseband 52–3
Digital transmission, satellites 132–3
Diode balanced microwave mixers 78–9
Dirac pulses 55–6, 58, 278
Direct detection, laser systems 430, 431, 434–6, 440–1
Direct detection receivers 334–5
Direction-finding, electronic techniques 338–40
Direct modulation, digital signals 122
Direct-to-home distribution, television 224
Dispersive filters 305, 307–9
Display devices, infrared systems 469
Diversity techniques, countering fading 101–6
Domestic systems, satellite links 137
Doppler filtering 287, 318
 radar 271–2, 277, 283, 284, 315–23, 366–7, 496
Doppler frequency, offset, jamming 345–6
Doppler measurement, laser systems 441, 442, 459–61
Doppler shift, chirp systems 307
Double frequency change, television 229–31
Down-converters 19, 182, 191, 198, 215, 222–4
Down links 126, 138, 155–6, 174, 188, 190
Dual-polarization frequency reuse 149–50, 210
Duplexers, antennas 284
Dynamic switching matrices 188

Early warning radars 332
Earth coverage 126, 127–8, 177–8
Earth segments 126, 131
Earth stations 126, 131, 146, 150, 191–8
 antenna systems 198–211
Earth triaxiality 170
Echoes
 Doppler frequency offset 345–6
 time offset 344–5
Echo suppressors 297, 346
Egg Nebula 534, 535
Einstein, Albert 384, 526, 530
EKRAN 137
Electromagnetic detection 267–8
Electron cyclotron resonance heating 484–5
Electronically scanned transmitters 349
Electronic confrontation 331–3
 system design 370–5
Electronic counter-countermeasures 333, 347
 radars 355–69
 radio frequency links 350–5
Electronic countermeasures 332, 342–49, 373
Electronic intelligence 332, 371–2, 374
Electronic scanning 272–3, 283, 324–5
Electronic support measures 332, 333–4, 374–5
 direction-finding techniques 338–40
 reception techniques 334–8, 354–5
Electronic warfare 331–3, 350
 support systems 372–3
ELINT 332, 371–2, 374
Elliptical polarization 204
Elliptical waveguides, semi-flexible, microwave links 10
Emissivity
 artificial sources 405–7
 materials 398–407
 natural backgrounds 401–5
Encoding, spectrum spreading 352–3, 355
Engineering order wires 26–7
Environmental aspects, infrared equipment 445–6
Equipment signatures 89–92
Equivalent isotropically radiated power 130
Error bursts 87
Error correction codes, protection against jamming 351
Error correction, *see* Forward error correction

Error-free seconds, data communications 87
Error rates, digital signals 81
Escort jammers 343
EUTELSAT 136–7
Excitation temperatures 524

Fading
  deep and fast 117
  outages 107–9
  reduction 100–6
False echoes, jammers 344–6
FASCOD 455
Federal Communications Commission 5, 34–5, 94
Feeders, microwave links 10
Feed, *see* Antenna feed
Field curvature 392
Field-effect transistors 183
  high power amplifiers 219
  low noise amplifiers 213–15
  RF amplifiers 17
Field of view, optical instruments 386
Figure of merit at reception, *see* Gain-to-noise ratio
Filters
  digital microwave links 75–7
  dispersive 305, 307–9
  Doppler 318
  matched, radar 305–6
  RF 22–5
  surface-acoustic-wave 242
  transfer functions 58–9
  transverse 308–9
Fire control, radar systems 325
Fixed satellite service (FSS) 127
Flat fading outages 107
FLIRs 384, 422–5, 448, 459
Flux, emission 391–5
Flux detectors 410, 415–16
  design 410–12
Forward error correction 98, 100, 163–7, 190
Forward looking infrared 384, 422–5, 448, 459
Frequency
  cyclotrons 487
  infrared systems 416
Frequency allocations
  analogue microwave links 33–5
  digital microwave links 92–4
  satellite communications 126–7
Frequency changes
  double, radar ECCM 364–5
  television 229–31
Frequency conversion 222
Frequency division multiple access (FDMA) 130, 146, 156, 157, 190
Frequency division multiplex (FDM) 2
Frequency domain equalizers 100
Frequency jump 353, 354–5, 362–3
Frequency modulation 158–9
  analogue microwave links 12–16, 122
  distortion 27–8
  infrared signals 421
  radars 271, 284–5, 287, 310, 347, 442
Frequency reuse 93–4, 148–50, 176, 201
  dual polarization 210
Frequency sensitive surfaces 181–3
Frequency shift keying 162
Frequency shift modulation 74
Frequency spectrum, broadcasting channels 257–66

Gain control
  radar ECCM 365
  television transmission 233–7
Gain-to-noise ratio 146, 198, 215–17
Galaxies
  distance 531
  radio sources 528–30
  structure 531
  synchrotron radiation 522
GALAXY 137
Gallium arsenide 213–14, 287, 328
Gas, interstellar, thermal emission 521–2
Gases, effects on propagation attenuation 33
Gas phase chemistry 534–6
Gaussian noise 273–4
General relativity, theory of 526, 530
Geometrical aberrations 392
Geostationary orbits, satellites 1, 126, 127, 131, 135, 170
Global beam antennas 177
Global coverage, satellites 127–8
Global positioning by satellite 225
Globalstar 226
Globars 407
GORIZON 137
Gregory antennas 177, 200
Ground communication equipment 126
Group delay equalizers, analogue microwave links 41–3

G-STAR 137
Gunn oscillators 21–2
Gyroscopes 466
Gyrotrons 485

Heat treatment, cancers 479, 480
Height-finders, radars 269
Herschel, Sir William 382
Hertz, Heinrich 267
Heterodyne detection, laser systems 430, 431–3, 436–40, 459–61
Heterodyne repeaters 7, 47–8
High bit error ratio value 87
High definition television 224
High electron mobility transistors 183, 214
Higher order multiplexing 50–1
High gain antennas 113–15
High path loss, satellites 130
High power amplifiers 115–17, 130, 146, 185–7, 191, 198, 200, 217–22
High sensitivity receivers 117
High speed baseband switching matrices 190
HII regions 525
Hitless switching 81–2
  combiners 103
Horns microwave 181, 200, 208, 210
Hulfsmeyer 267
Hydrogen line 530–1
Hydrometeors, effects on propagation attenuation 33, 38
Hylcar 506
Hyperthermia 479, 480, 506
Hypothetical reference circuits 28, 29, 85–7

Ideal receivers, radar 274, 276–8
Identification, infrared systems 447–8
Identification of friend and foe (IFF) 324
Image detectors 410, 415–16
Image of extended objects, optical instruments 389–92
Image frequency 20–1
Image of point source 388–9
Image processing, infrared 468–9
Imaging, radiometric, medical applications 505–6
IMUX 172–4, 183–6
Incidental phase correction 245
Incoherent laser systems 430, 431, 434–6, 440–1
Indirect modulation, digital signals 122
Industrial applications
  accelerators 493
  infrared 452
  microwave heating 480
  radiometry 506–7
Infrared 377
  classification of systems 380–2
  detectors 408–15
  developments and trends 466–9
  historical background 382–5
  instrument design 415–22
  optical quantities and relationships 386–92
  performance 42–8
  photometry and radiometry 392–5
  sources of radiation 395–407
  spectral bands 378–9
  see also Active infrared systems; Lasers; Passive infrared systems; Semi-active infrared systems
INMARSAT 135, 136, 226
In-phase combiners 103–4
Input demultiplexers 172–4, 183–6
Input stage linearity, television transmission 231–3
INSAT 137, 174–5
Instrumentation radars 326
Instantaneous frequency measurement 335
Integrated services digital network (ISDN) 133
INTELSAT 131, 132, 135–6, 137, 148–9, 159, 162, 174, 175, 181, 188, 205–6, 218
Intensity, point sources 393
Interference
  analogue microwave links 35–8
  digital microwave links 92
Interferometry
  networks, phase measurement 339
  radio telescopes 514–20, 528
Intermediate frequency
  accesses 1
  amplifiers 16, 17, 238–9, 288–9
  filtering 75–6
  minimum dispersion combiners 104
  modulation 77–8, 222
    television transmission 240–2
  non-linearity corrections of vision channel 245
  radars 282, 347, 364–5
  variable amplification 233–5
Intermodulation 220–1
  noise 30, 41–4, 47–8, 156

non-linear amplifiers 146–8
power amplifiers 246–9
International Frequency Registration Board 5
International Radio Consultative Committee, *see* CCIR
International systems, satellite links 135–6
International Telecommunication Union (ITU) 126–7, 133
Intersatellite links 190–1
INTERSPUTNIK 135, 136
Interstellar molecules 531–40
Intersymbol distortion, digital microwave links 55–8
Intruder detection, infrared 452
Ion cyclotron resonance heating 484
Ionized gas, thermal emission 521–2
IR domes 466
IRIDIUM 226
Irradiance 393, 394
ISDN, *see* Integrated services digital network
ITALSAT 189

Jammers 373–4
countermeasures 357
expendable 349
protection techniques 350–5
radar 281–2, 287, 328, 332, 342–50
Jansky, Karl 511
Jitter, bit error origins 88
Junctions, microwave devices 85

Kastler, Alfred 384
Kepler, Johann 525
Kirchhoff's Law 400
Klystrons 115, 219, 221, 287, 489, 490, 493

Lagrange's constant 386–8
Lambert's Law 394–5, 397
Lamps, infrared sources 406–7
Land mobile satellite service 225
LANDSAT 226
Large Magellanic Cloud 525
Lasers 338, 384–5
detection techniques 428–33
emitters 468
performance 433–44
rangefinding 455–61
solid-state, rangefinders 456–8
see also Infrared
Least jammed frequency, radar ECCM 365
Left-hand circular polarization 176, 204
Lidars 431, 461–2
Linear accelerators 485–7, 493
Linear distortion, analogue microwave links 38–40
Linearization, analogue microwave signals 14–16
Linearizers 98, 99, 147–8, 187
signal combination 117–19
Linear magnification 386
Line emissions, cosmic radio emissions 522–4
Line-of-sight links 1, 2–3, 7, 16, 17
Line-scan 452
Link budget 139, 151–8
Link quality 158–67
Lin-log structure, IF amplifiers 288–9
Localization
infrared systems 446–7
satellites 225
Local oscillators 21–2, 212, 222, 224, 317
Local thermodynamic equilibrium, molecular distribution 536
Location measurement, for radar transmitters 341–2
Logarithmic chains, radar signals 289–90
Logic circuits 58
Log–PLD chains, radars 290–1
Long pulses, radar transmitters 287
Long term error free seconds 87
Low altitude detection, radar 294
Low bit error ratio value 85, 87, 225
Low earth orbit satellites 226
Lower hybrid resonance heating 484
Low-frequency corrections 244–5
Low level amplifiers 16–19
Low noise amplifiers 17–19, 20, 130, 143, 145, 191, 198, 200, 212–17, 287–8
Low noise converters 215, 222
Low side lobe radiation level 198, 199
Luminance 389–91, 393
black bodies 397–8
conservation 395
natural backgrounds 401–5
useful sources 416–18
Lunisolar attraction 170

Magnetic fields, synchrotron radiation 522

Magnetrons 488, 493
  radars 270–1, 287
Magnifying power 386
Maiman, Theodore 385
Manufacturing defects, optical instruments 392
Marconi, Guglielmo 267
Maritime mobile satellite service 224–5
Masers
  cosmic, amplification 536–7
  two-level 538–40
Masking, jamming 343
Master TV antennas 133
Matrices
  dynamic switching 188
  high speed baseband switching 190
Measurement cells 498
Measurement systems, sensors 498–9
Mechanical design, earth stations 205–7
Medical applications
  accelerators 493
  infrared 452
  microwave heating 479, 480
  radiometry 505–6
METEOSAT 226
Microstations, *see* Very small antenna terminals (VSATs)
Microtrons 488–9
Microwave active imaging 500–1
Microwave frequency, modulation 78–9
Microwave heating 475–81
Microwave horns 181, 200, 208, 210
Microwave intersatellite links 190–1
Microwave links
  analogue, *see* Analogue microwave links
  digital, *see* Digital microwave links
  principles 2–10
  subunits 16–26
Microwave ovens 479, 493
Microwave radiation 493–4
Microwaves
  dielectric materials 471–4
  particle accelerators 485–93
  plasma for nuclear fusion 481–5
Military applications
  active infrared systems 453–62
  passive infrared systems 445–53
  semi-active infrared 462–5
Millisecond pulsars 526–8
Minimum resolvable temperature
  difference, infrared detectors 424–5, 426
Mirrors, passive, microwave links 9–10
Mixers
  microwave frequency 19–21
  television transmission 238
Mobile satellite service 127, 224–6
Mobile target indication devices 283, 285
Modified reflectors 200
Modulation
  digital microwave links 61–74, 77–80
  television transmitters 240–6
Modulation transfer function
  infrared detectors 424, 425
  lens quality 391
Molecular emissions, cosmic 522–4
Molecules, interstellar 531–40
MOLNYAs 125, 137
Monitoring, signal quality, analogue microwave links 26
Monomode cavity applicators 478–9
Monopulse angular measurements, radar 270
Monopulse tracking systems 208–10
Moon, emissivity 403, 405
MOSKVA 137
Moving target indication, radar 271–2, 358, 366–7
Multimode cavity applicators, microwave heating 479
Multiple access, satellite systems 130–1
Multiple path propagation 31–3, 88–9
Multiple spot beam antennas 178–9
Multiplexers
  output 174, 183–5
  polarization 9
Multiprobe radiometers 505–6

Narrow spectrum jamming 344
NASA 189
National satellite systems 137
Natural sources, light 40–5
Near-field antennas 479
Nernst glower 406
Networks, synchronized 51–2
Night vision devices 383
Nodal stations, interference 37
Noise
  direct detection systems 434, 436
  gaussian 273–4

Noise contributions, evaluation 47–8
Noise equipment temperature perceptible 424, 425
Noise equivalent power 410–11
  laser systems 431, 433
Noise equivalent temperature difference 424, 425
Noise factor, antinomic couple, optimization 231–3
Noise figure 146, 233–5
Noise jammers 343–4, 346
Noise temperature 142, 143–5, 215–17
Non-destructive control, microwaves 498–9
Non-linear correction, vision channels 243–5
Non-linear effects 99
  high power amplifiers 220–2
Non-linearity correctors 249–51
Non-linearity distortion, synchronization signal 39–40
Normal beam antennas 176–8
Notch in-phase combiners 104
Nuclear fusion, plasmas 481–5
Nyquist formula 142, 163

Oblique imagery, aerial reconnaisance 448
Omnidirectional infrared search 446
Omnitracs 225
On-board processing 188
On-board regeneration 188
On-board switching 188
Open loop carrier recovery 68–70
Optical fibres 224
Optical instruments
  characteristics 386
  performance 389–92
Optical intersatellite links 191
Optical materials, choice 408, 409
Optical power 386
Optical pumping 384
Optical systems, infrared 467
Optical windows, infrared 466
Optics
  microwave links 9
  quantities and relationship 386–92
Optronics 332
ORBITA 137
Orbit control, *see* Station keeping
Orbit/spectrum utilization 126
Orthomode junction 202, 204
Outages, propagation 106–9
Output multiplexers (OMUX) 174, 183–5
Over-the-horizon microwave links, *see* Troposcatter links
Over-the-horizon radar 292–4
Ozone layer 541

Paint, antiradar 296
PALAPA 137
Panoramic plane indicators 324
Panoramic surveillance radars 269–70
Paraboidal reflector 200
Parametric amplifiers 212–13, 287
Parasitic noise, radar 280–2
Parasitic radiation, radars 272, 273, 320
Parity bit, measuring error rates 81
Particle accelerators 485–93
Passive infrared systems 382, 383–4
  instrument design 415–22
  military applications 445–53
  performance 422–8
Passive mirrors, microwave links 9–10
Passive sensors 501–7
Path loss, high, satellites 130
Payload, satellites 172–5, 187–91
Peltier effect 452
Perigee motor 172
Periodic permanent magnets 218, 219
Permittivity 498–9
  dielectric materials 471–4
Phase ambiguity suppression 73
Phase coded signals, radar 312–13
Phased-array antennas, radar 273, 283
Phase measurement, direction finding 339–40
Phase modulation
  analogue microwave links 122
  coded radars 282, 284–5, 287
  infrared signals 421–2
Phase shift keying 73–4, 163, 190
Photodetectors, laser systems 430
Photometric parameters, optical systems 388
Photometry 392–5
Photonic detectors 410, 415
Planck's Law 397–8
Plane interferometric networks, phase measurement 339
Planets, radiation 525
Plasmas, nuclear fusion 481–5
Plesiochronous multiplexing and demultiplexing 50–1

Point source, image 388–9
Point-to-point IR transmission systems 380, 465–6
Polarization
  antennas 88, 176, 181
  elliptical 204
  microwave links 4–5
Polarization discrimination, frequency reuse 149–50
Polarization sensitive surfaces 181–3
Polarizers 181, 204
  see also Frequency reuse
Pollution, infrared measurement 452
Power, radiation, by sources 392–3
Power amplifiers, distortions 246–9
Power amplifiers, *see* High power amplifiers
Power deposition, microwaves 475–6
Power flux density 141–2
Power supplies, satellites 171
PPI infrared search 445–6
Preamplification, television transmission 237–40
Pre-assigned multiple access 131
Precision approach radars 326
Pre-emphasis 161
Primary multiplex 49–50
Primary sources
  antennas, *see* Antenna feed
  light 400
Propagation effects, countering 44–5
Propagation, *see* Atmospheric attenuation
Propulsion, satellites 172
Protection switching, analogue microwave links 26
Protostars 525, 532
Pseudo-random sequence generators 82–5
Pulsars 526–8
Pulse code modulation 132, 270–1
Pulse compression
  effect on clutter 311–12
  radar 270–1, 276, 305–13, 322, 327, 363–4, 367
Pulse laser systems 441
Pulse length discriminators, radars 290–1
Pulse radars 268–9, 276, 284
  Doppler 315–23
Pulse repetition frequencies, radars 287
Pulse-width modulation 17
  infrared signals 422

Quadratic-phase shift keying 162–3
Quadrature amplitude modulation, digital microwave links 20, 65–7, 70–2, 99
Quality
  analogue microwave links 40–5
  digital signals 81, 85–7
  transmission 28
Quantum detectors 410, 415
Quasars
  radio sources 528–30
  synchrotron radiation 522

Radar
  applications 323–6
  basic principles 273–80
  block diagram 282–5
  cross-section 294–5
  electronic counter-countermeasures 355–69
  history 267–73
  location of transmitters 341–2
  multifunction, analysis, problem 296–304
  multifunction and multimode 326–7
  pulse 268–9
  pulse compression 270–1, 276, 305–13
  range in the presence of jamming 356–7
  receiver circuits, perturbation 347
  sensors 495–7
  wavelengths 292–4
Radiation
  microwave 493–4
  monitoring 494
  power, point sources 392–3
  sources 395–407
Radiation beams, satellite antennas 176
Radiation diagrams, *see* Antenna diagrams
Radiation level, low side lobe 198, 199
Radioastronomy 511
  prospects 540–1
Radioelectrical design, earth station antennas 198–9
Radioelectricity, military applications, *see* Electronic warfare
Radioelectric superiority 374
Radio emission, cosmic 520–4
Radio-frequency
  accesses 1

amplifiers 17–19, 287–8
channels
arrangement 5
branching 23
characteristics, satellites 176
direct amplification 7
filters 22–5, 237–9
links, ECCM 350–5
satellite communications 126–7
Radiography, microwave 500
Radioisotopes 493
Radiometry 392–5, 501–3
applications 505–7
infrared 452
receivers 504–5
signals 503
thermometric measurement 503–4
Radio-navigation 332
Radio regulations 5
Radio relay links, *see* Microwave links
Radio signals 512
Radio sources
continuum 524–30
extragalactic 528–30
superluminal 530
Radio telescopes 512–20
RADUGA 137
Range
laser systems 434–8
passive infrared systems 426–8
Rangefinding, lasers 455–61
Range gate pull off 346
Receiver noise 142–3
Receiver-only stations 131
Receivers
high sensitivity 117
radar 287–92
ECCM 364–9
radiometric 504–5
wideband 183
Reconnaissance, infrared systems 447–8
Rectangular waveguides
earth station antennas 202
microwave links 10
Reference circuits, hypothetical 28, 29
Reference diagrams, antennas 208
Reflectance, diffuse surfaces 395
Reflector antennas 200
Refractive index, dielectric materials 472, 474
Regeneration, on-board 188
Regenerators, digital microwave links 59–61
Regional systems, satellite links 136–7
Relativity, theory of general 526, 530
Remodulating repeaters 6
Repeaters, microwave links 6–7
Residual amplitude modulation 220
Residual carrier transmission 67–8, 315, 317
Resonance heating 484–5
Reticles, infrared systems 418, 421–2
Retransmission, television 229–31
Right-hand circular polarization 176, 204
Ring modulators, double balanced 77–8
Road traffic management, radiometry 507
Ruby lasers 385, 456
Rutherford, Ernest 485

Safety standards, microwaves 494
SATCOM 137
Satellite multi-service system 137
Satellite news gathering 133
Satellites
antenna systems 133, 139–42, 176–81
characteristics 127–32
communication links 137–9
communication systems 125–6
construction 167–72
data transmission 132–3
existing systems 135–7
geostationary orbits 1
historical overview 133–5
non-geostationary 225, 226
payload 172–5, 187–91
prospects 224–6
radio-frequencies 126–7
service quality objectives 151–8
telephony 132
television 133
Saturation, laser systems 444
SBS 137
Scanning, infrared systems 466–7
Schawlow, A.L. 385
Schottky effect 18
Scientific applications
accelerators 492–3
infrared 453
microwaves, thermonuclear fusion 481–5
Scramblers, digital microwave links 82–5
Secondary radar 324

Secondary sources, light 400
Sector infrared search 445
Security, infrared applications 452
Selective fading 88, 89
  outages 107–9
Self-guidance, weapons 450–1
Self-protection jammers 343
Self-protection range, radars 343
Self-synchronizing scramblers 84–5
Semi-active infrared systems 381
  military applications 462–5
Semi-active laser systems 428–9
Semiconductor amplifiers 287–8
Semi-flexible elliptical waveguides, microwave links 10
Sensitivity time control, radars 288
Sensors
  active 495–501
  passive 501–7
Shaped beam antennas 176, 179–81
Shaped reflectors 200
Ship earth stations 136, 191, 225
Ships, surveillance Doppler radar 320–3
Short term error free seconds 87
Shot noise 431, 432, 433
Side-lobe blanking, radar 292, 361–2
Side-lobe cancellation, radar 292, 359–62
Side-lobe reduction, chirp systems 309–10
Signal quality, monitoring, microwave links 26
Signal-to-noise ratio 231
  analogue microwave links 45–9
  direct detection lasers 436
  heterodyne lasers 440
  infrared systems 415, 416
  radars 274, 275, 329
Sine law, optical systems 388
Single channel per carrier 138, 147, 160, 224
Single dish radio telescopes 512–14
Single frequency change, television 229–31
Six-port networks 498–9
Sky backgrounds, emissivity 401–2
Solar arrays 171
Solar cells 171
Solar energy 481
Solar radiation pressure 170
Solar system, radiation 524–5
Solid angle, detectors 411
Solid-state attenuation devices 288
Solid-state lasers, rangefinders 456–8
Solid-state power amplifiers 147, 185, 219
Sound RF modulation 240–2
SPACENET 137
Space segment 126
Spatial filtering 415–16, 418–22
Spatial frequency
  optical instruments 389
  tracks, interferometry 517–20
Spatial isolation frequency reuse 148–9
Spatial response, detectors 411
Specific detectivity 411
Spectral bands, infrared 378–9, 416–18
Spectral emissivity 398
Spectral filtering 416–18
Spectral improvement 92–4
Spectral luminance 398
Spectral response, detectors 411
Spectral transmittance 388
Spectral units 393–4
Spectrum analysers 335, 338
Spectrum spreading, encoding 352–3
Spherical aberration 392
Spill-over sidelobes, radar 272
Spin stabilization 168
SPOT 226
Spot beam antennas 128, 177
  multiple 178–9
Spot noise, jamming 343–4
Stabilization
  infrared systems 466
  satellites 168–70
Stand forward jammers 343
Stand-off jammers 343
Stanford Linear Accelerator Center 486
Stars, formation 537
Station keeping, satellites 125–6, 208
STATIONSAR 136
Stealth planes, radar detection 296
Stefan–Boltzmann's Law 398
Stellar wind 525
Step track system 210
Strategic intelligence 371–2
STW 137
Sun
  emissivity 403
  radio emissions 511, 524–5
Superconductor–insulator–superconductor (SIS) diodes 512

Superheterodyne detection receivers 335, 338
Supernovas, remnants 525–6
Suppression of enemy air defence 374
Surface-acoustic-wave
  devices 442
  filters 242, 307–8
  receivers, 335–6
Surface backgrounds, emissivity 401
Surveillance radars 286, 323–5
Switching, *see* On-board switching, processing, regeneration
Symbols, digital signals 52–3
Synchronization signal, non-linearity distortion 39–40
Synchronized networks 51–2
Synchronous scramblers 82, 84
Synchrotron radiation 522, 524, 525
Synchrotrons 489–93
Synthetic aperture radars 325

Tactical intelligence 372
Targets, fluctuation, radar 295–6
Target search and acquisition, infrared systems 445–8, 462–5
TELECOM 1 137, 174
Telecommunications, military, ECCM 350–5
Telemetry, tracking and command 171–2
Telephony 132, 224
  digital, public service 137
  linear distortion 38
  modulating signals 14
  multi-channel 2
  noise 30
  signals, analogue links 11
  transmission quality 45–8
TELESTAR 137
Television
  channels, grouping 10–11
  colour, transmission 2
  high definition 224
  linear distortion 38–9
  modulating signals 14
  noise 30
  satellites 133, 224
  signals, analogue links 11–12
  translators 229–40
  transmission quality 48–9
  transmitter modulation stages 240–6
  transmitter output design
    optimization 246–57
Television receive only station (TVRO) 131, 133, 137, 191, 200, 215, 222
TELE-X 137
Temperature, bit error origins 88
Temperature control, hyperthermia, radiometry 506
Temperature differences, passive infrared detection 424–6
Terrain avoidance 325
Thermal cameras 384, 448, 459
Thermal control, satellites 170–1
Thermal detectors 141, 410, 414
Thermal emission
  ionized gas 521–2
  solar system 524–5
Thermal noise 30, 45, 48–9, 53–5, 88, 153–6, 431, 474
Thermodynamic equilibrium 536
Thermoelectric cooling 452
Thermography, infrared 452
Thermometric measurement 503–4
Thermonuclear fusion, *see* Nuclear fusion
Three-axes stabilization 168
Threshold, demodulation 160
Threshold extension demodulator 160
Time of arrival measurement, direction finding 339
Time division multiple access 130, 148, 190
Time division multiplex 2, 49–52, 188, 190
Time-domain equalizers 100
Time offset, echoes, jamming 344–5
Time standards, pulsars 526–8
Tokamaks 483, 485
Tomography, microwave 500, 501
Total field of view, optical instruments 386
Total luminance 398
Townes, C.H. 385
Tracking, infrared systems 446–7
Tracking mode couplers 201, 202
Tracking radars 269, 270, 325, 369–71
Tracking systems, earth station antennas 201, 207–11
Transfer functions, filters 58–9
Transfer orbit 168, 172
Transistors
  power amplifiers 115
  RF amplifiers 17–18
Translators, television 229–40
Transmission quality 122
  analogue microwave links 45–9

Tranmitters
  electronically scanned, jamming 349
  radar 287
    ECCM 362–4
Transmitting–receiving modules, radars 287
Transponders 218
  regenerative 188
  satellites 126, 130, 138, 146, 147, 156, 157, 176
Transverse filters 308–9
Travelling-wave applicators 477–8
Travelling-wave tubes 17
  amplifiers 115, 147, 185
  high power amplifiers 218–19, 221
  linear accelerators 485–7
  radar 287
Triodes 287
Troposcatter links 1, 109–22
Tropospheric radio-frequencies 350
Tsiolkovsky, Konstantin 135
Tungsten lamps 407
Tunnel diode amplifiers 287

Up-converters 19, 191, 198, 222–4
Up links 126, 137–8, 153–4, 172, 188, 190

Van de Graaf accelerators 485
Varactors 21, 212–13
Velocity gate pull off 346
Vertical imagery, aerial reconnaissance 448
Very long baseline arrays 519
Very long baseline interferometry 515, 519, 528, 541
Very small aperture terminals (VSATs) 133, 191, 206, 215
Vignetting, optical systems 388
Vision channels, non-linearity correction 243–5
Vision correctors, IF 242
Vision IF modulation 240–2
VISTA 136
Voice activation 160
Voltage sensitivity 410

Warning detectors 372–3
Watson-Watt, Sir Robert 267, 268
Waveguides, microwave links 10
Wavelengths, new, radar 330
Weapon guidance
  infrared application 450–1
  laser application 461–5
Weapon systems, self-protection 372–3
Weather, effects on propagation attenuation 110–11
Weighting
  intrinsic, chirp systems 310
  telephony 14
  television 14, 161
WESTAR 137
Wideband converters, output 246
Wideband receivers 183
Wien's displacement law 398

Xenon arc lamps 407
X-rays 493

YAG lasers 385, 456